Reliability Engineering Handbook

QUALITY AND RELIABILITY

A Series Edited by

EDWARD G. SCHILLING
Founding Editor
Center for Quality and Applied Statistics
Rochester Institute of Technology
Rochester, New York

1. Designing for Minimal Maintenance Expense: The Practical Application of Reliability and Maintainability, *Marvin A. Moss*
2. Quality Control for Profit: Second Edition, Revised and Expanded, *Ronald H. Lester, Norbert L. Enrick, and Harry E. Mottley, Jr.*
3. QCPAC: Statistical Quality Control on the IBM PC, *Steven M. Zimmerman and Leo M. Conrad*
4. Quality by Experimental Design, *Thomas B. Barker*
5. Applications of Quality Control in the Service Industry, *A. C. Rosander*
6. Integrated Product Testing and Evaluating: A Systems Approach to Improve Reliability and Quality, Revised Edition, *Harold L. Gilmore and Herbert C. Schwartz*
7. Quality Management Handbook, *edited by Loren Walsh, Ralph Wurster, and Raymond J. Kimber*
8. Statistical Process Control: A Guide for Implementation, *Roger W. Berger and Thomas Hart*
9. Quality Circles: Selected Readings, *edited by Roger W. Berger and David L. Shores*
10. Quality and Productivity for Bankers and Financial Managers, *William J. Latzko*
11. Poor-Quality Cost, *H. James Harrington*
12. Human Resources Management, *edited by Jill P. Kern, John J. Riley, and Louis N. Jones*
13. The Good and the Bad News About Quality, *Edward M. Schrock and Henry L. Lefevre*
14. Engineering Design for Producibility and Reliability, *John W. Priest*
15. Statistical Process Control in Automated Manufacturing, *J. Bert Keats and Norma Faris Hubele*
16. Automated Inspection and Quality Assurance, *Stanley L. Robinson and Richard K. Miller*
17. Defect Prevention: Use of Simple Statistical Tools, *Victor E. Kane*
18. Defect Prevention: Use of Simple Statistical Tools, Solutions Manual, *Victor E. Kane*
19. Purchasing and Quality, *Max McRobb*
20. Specification Writing and Management, *Max McRobb*
21. Quality Function Deployment: A Practitioner's Approach, *James L. Bossert*
22. The Quality Promise, *Lester Jay Wollschlaeger*
23. Statistical Process Control in Manufacturing, *edited by J. Bert Keats and Douglas C. Montgomery*
24. Total Manufacturing Assurance, *Douglas C. Brauer and John Cesarone*
25. Deming's 14 Points Applied to Services, *A. C. Rosander*
26. Evaluation and Control of Measurements, *John Mandel*
27. Achieving Excellence in Business: A Practical Guide to the Total Quality Transformation Process, *Kenneth E. Ebel*
28. Statistical Methods for the Process Industries, *William H. McNeese and Robert A. Klein*
29. Quality Engineering Handbook, *edited by Thomas Pyzdek and Roger W. Berger*
30. Managing for World-Class Quality: A Primer for Executives and Managers, *Edwin S. Shecter*
31. A Leader's Journey to Quality, *Dana M. Cound*
32. ISO 9000: Preparing for Registration, *James L. Lamprecht*
33. Statistical Problem Solving, *Wendell E. Carr*
34. Quality Control for Profit: Gaining the Competitive Edge. Third Edition, Revised and Expanded, *Ronald H. Lester, Norbert L. Enrick, and Harry E. Mottley, Jr.*
35. Probability and Its Applications for Engineers, *David H. Evans*
36. An Introduction to Quality Control for the Apparel Industry, *Pradip V. Mehta*
37. Total Engineering Quality Management, *Ronald J. Cottman*
38. Ensuring Software Reliability, *Ann Marie Neufelder*
39. Guidelines for Laboratory Quality Auditing, *Donald C. Singer and Ronald P. Upton*
40. Implementing the ISO 9000 Series, *James L. Lamprecht*
41. Reliability Improvement with Design of Experiments, *Lloyd W. Condra*
42. The Next Phase of Total Quality Management: TQM II and the Focus on Profitability, *Robert E. Stein*
43. Quality by Experimental Design: Second Edition, Revised and Expanded, *Thomas B. Barker*

44. Quality Planning, Control, and Improvement in Research and Development, *edited by George W. Roberts*
45. Understanding ISO 9000 and Implementing the Basics to Quality, *D. H. Stamatis*
46. Applying TQM to Product Design and Development, *Marvin A. Moss*
47. Statistical Applications in Process Control, *edited by J. Bert Keats and Douglas C. Montgomery*
48. How to Achieve ISO 9000 Registration Economically and Efficiently, *Gurmeet Naroola and Robert Mac Connell*
49. QS-9000 Implementation and Registration, *Gurmeet Naroola*
50. The Theory of Constraints: Applications in Quality and Manufacturing: Second Edition, Revised and Expanded, *Robert E. Stein*
51. Guide to Preparing the Corporate Quality Manual, *Bernard Froman*
52. TQM Engineering Handbook, *D. H. Stamatis*
53. Quality Management Handbook: Second Edition, Revised and Expanded, *edited by Raymond J. Kimber, Robert W. Grenier, and John Jourdan Heldt*
54. Multivariate Quality Control: Theory and Applications, *Camil Fuchs and Ron S. Kenett*
55. Reliability Engineering and Risk Analysis: A Practical Guide, *Mohammad Modarres, Mark Kaminskiy, and Vasiliy Krivtsov*
56. Reliability Engineering Handbook, *Bryan Dodson and Dennis Nolan*
57. Quality Engineering Handbook, *Thomas Pyzdek*
58. Product Development and Design for Manufacturing: A Collaborative Approach to Producibility and Reliability: Second Edition, Revised and Expanded, *John W. Priest and José M. Sánchez*
59. Reliability Improvement with Design of Experiments: Second Edition, Revised and Expanded, *Lloyd W. Condra*
60. Quality Engineering Handbook: Second Edition, Revised and Expanded, *Thomas Pyzdek, edited by Paul Keller*

ADDITIONAL VOLUMES IN PREPARATION

Reliability Engineering Handbook

BRYAN DODSON
DENNIS NOLAN

CRC Taylor & Francis
Taylor & Francis Group
Boca Raton London New York

CRC is an imprint of the Taylor & Francis Group,
an informa business

QA Publishing, LLC
7650 East Broadway, Suite 208
Tucson, AZ 85710

Published 2002
Printed in the United States of America

ISBN 0-8247-0364-2

To Mirian, Lois, Matthew and Furnie.
B.D.

To my loving daughters and granddaughter, Christina, Andrea, and Karissa.
D.N.

Contents

Preface **ix**
1. Introduction **1**
1.1 Life cycle costs 1
1.2 Reliability data sources 3
1.3 Data Collection, Analysis, and Reporting 5
1.4 Customer expectations 17
1.5 Exercises 26
1.6 References 28
2. Distributions **33**
2.1 Distribution characteristics 33
2.2 Normal and lognormal distributions 40
2.3 Exponential, Poisson, binomial, hypergeometric, and geometric distributions 49
2.4 The Weibull distribution 59
2.5 The gamma and beta distributions 64
2.6 Chi-square, *F*, and *t*-distributions 72
2.7 Bayes' probability analysis 77
2.8 Propagation of errors 81
2.9 Exercises 90
2.10 References 93
3. Parameter Estimation **95**
3.1 Censored data 96
3.2 Maximum likelihood estimation 97
3.3 Hazard plotting 97
3.4 Probability plotting 98
3.5 Moment estimation 98
3.6 Exponential distribution 99
3.7 Normal distribution 108
3.8 Lognormal distribution 118
3.9 Weibull distribution 118
3.10 Goodness of fit 125
3.11 Exercises 128
3.12 References 129
4. Statistical Inference **131**
4.1 Confidence intervals for the mean and variance 131

4.2 Confidence Intervals for proportions 133
4.3 Tolerance Intervals 136
4.4 Hypothesis testing 137
4.5 Exercises 142
4.6 References 143
5. Reliability Modeling **145**
5.1 Models with standard data 145
5.2 Mathematical models 147
5.3 Fault trees 162
5.4 Monte Carlo Simulation 172
5.5 Markov analysis 187
5.6 Reliability allocation 191
5.7 Software reliability 196
5.8 Reliability growth 201
5.9 Sneak analysis 207
5.10 Exercises 211
5.11 References 219
6. Reliability Testing **221**
6.1 Accelerated testing 222
6.2 Sequential testing 232
6.3 Exercises 249
6.4 References 251
7. Maintainability and Availability **253**
7.1 Quantifying maintainability and availability 256
7.2 Preventive maintenance 263
7.3 Designing for maintainability 269
7.4 System effectiveness 270
7.5 Exercises 272
7.6 References 273
8. Failure Mode, Effects and Criticality Analysis **275**
8.1 Criticality analysis 275
8.2 Environmental requirements and influence 285
8.3 Differences between part function, part failure mechanism, failure effect, failure detection, and failure mode 285
8.4 Exercises 290
8.5 References 293
9. Part Selection and Derating **295**
9.1 Evaluation of parts for reliability 295
9.2 Reliability testing for parts 297
9.3 Vendor evaluation of part reliability 301
9.4 Derating principles 301
9.5 Summary 304
9.6 Exercises 304
9.7 References 307
10. Reliability Design and Management Control **309**
10.1 Reliability program management principles 309
10.2 Design review teams 315
10.3 Follow-up and corrective-action 315

10.4 Reliability needs 318
10.5 Managing the program 320
10.6 General considerations 321
10.7 Summary 322
10.8 Exercises 323
10.9 References 323

11. Product Safety 325

11.1 Contingency analysis 325
11.2 Malfunction and multiple failure analysis 332
11.3 Exercises 336
11.4 References 339

12. Human Factors In Reliability 341

12.1 Human factors in design and design principles 342
12.2 Man-machine function allocation 356
12.3 Human factors in production 361
12.4 Human factors in field testing 366
12.5 Training and education 367
12.6 Exercises 401
12.7 References 403

13. Statistical Process Control 407

13.1 Sources of variation 408
13.2 Types of control charts 409
13.3 Advanced control charts 430
13.4 Implementation 437
13.5 Tampering effects and diagnostics 448
13.6 Process capability studies and indices 450
13.7 Exercises 455
13.8 References 456

14. Experimentation 461

14.1 Experimental design 461
14.2 Analyzing experimental data 471
14.3 Regression 485
14.4 Surface response methodology 500
14.5 Exercises 509
14.6 References 516

Appendix A: Area Under the Standard Normal Curve 519

Appendix B: The Gamma Function 523

Appendix C: Rank Tables 525

Appendix D: Beta Table 541

Appendix E: The *F* Distribution 545

Appendix F: Critical Values of the *t*-Distribution 551

Appendix G: Poisson Probability Sums 553

Appendix H: Critical Values of the Chi-Square Distribution 557

Appendix I: Tolerance Interval Factors 561

Appendix J: Useful Integrals 565

Appendix K: Government Documents 567

Appendix L: Financial Formulas 569

Appendix M: **For Further Information...** **571**
Appendix N: **Glossary** **573**

Preface

The primary purpose of this text is to present the entire body of knowledge for reliability engineering. The material is presented with the practitioner in mind; thus, proofs and derivations have been omitted, and numerous examples have been included. In addition, most examples have been worked in Microsoft Excel and are included in the accompanying disk. For those desiring proofs and derivations, references are provided at the end of each chapter. It is the authors' goal that individuals with a basic knowledge of statistics will have a comprehensive knowledge of the field of reliability engineering after studying this text.

Practitioners will find this text valuable as a comprehensive reference book, but this book is also ideal for college courses. It is recommended that this text be used as a two-semester sequence in reliability for college courses or as an 80-hour seminar for practicing engineers. Students should have a familiarity with basic probability and statistics before attempting this material.

The text consists of six basic parts. Chapter 1 provides an introduction, Chapters 2-4 cover fundamental statistical principles, Chapters 5–7 cover reliability mathematics, Chapters 8–12 describe the non-quantitative aspects of reliability, and Chapters 13 and 14 describe statistical process control and designed experiments.

The authors are grateful to many individuals for helping prepare this book. Most notable are the reviewers. Several reviewers were anonymous, but you know who you are, and we thank you for your comments. A special thanks goes to John Miller, James McLinn, Tom Pyzdek, and Peter Worden whose many comments greatly improved the manuscript. We are also grateful to Shahid Faruqui who prepared the Sneak Circuit section.

♦♦♦

CHAPTER

1

Introduction

The need for reliability engineering became apparent during World War II. The availability of Naval electronic equipment was approximately 70%; the availability of Army equipment was less than 40%; over 60% of airborne equipment shipped to the far east arrived damaged; the mean time to fail for bomber electronics was less than 20 hours; the cost of repair and maintenance was over ten times the original purchase price, and so on.

A portion of the problems was due to the increased complexity of the equipment. The fields of communications and transportation experienced rapid increases in complexity, and manufacturers capitalized on advances in electronics and control systems. The Department of Defense (DOD) recognized the problem, realizing that equipment would continue to become more complex. In 1950, the DOD created the Advisory Group on the Reliability of Electronic Equipment (AGREE). The DOD has continued to be a leader in reliability engineering. Much of reliability engineering theory has been developed directly by the DOD, or through research sponsored by the DOD. This is evidenced by the large number of DOD documents on the subject of reliability. A listing of some of these documents is given in Appendix K.

1.1 Life cycle costs

As a result of the problems encountered in World War II, the concept of **life cycle cost** was developed.

The original purchase price of equipment is only a fraction of the total cost of the equipment over its useful life. The total cost of operating a piece of equipment over its lifetime also includes the following post purchase costs:

- start-up
- planned maintenance
- repairs
- downtime
- spare parts
- fuel or energy

Life cycle costs are displayed graphically in Figure 1.1.

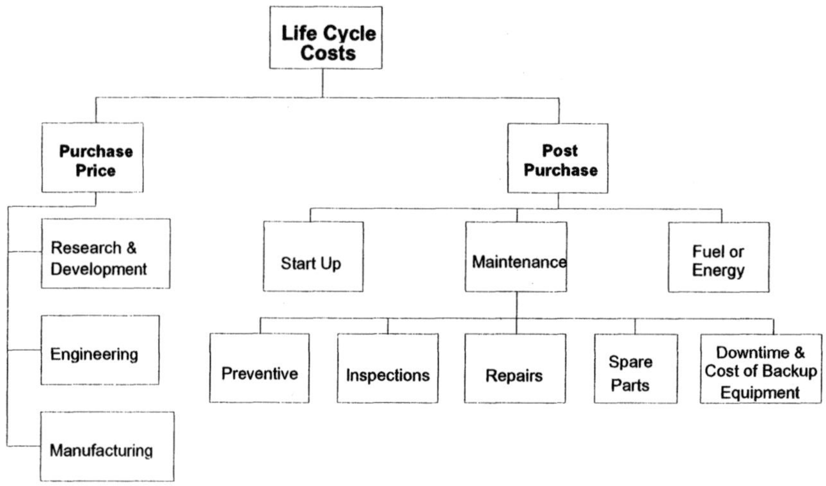

Figure 1.1. Life cycle costs.

By expending more on research, engineering, and manufacturing, the purchase price of an item is increased, but the post purchase expenses are decreased because of increased reliability, maintainability and availability. There exists an optimum level of reliability; this is shown in Figure 1.2.

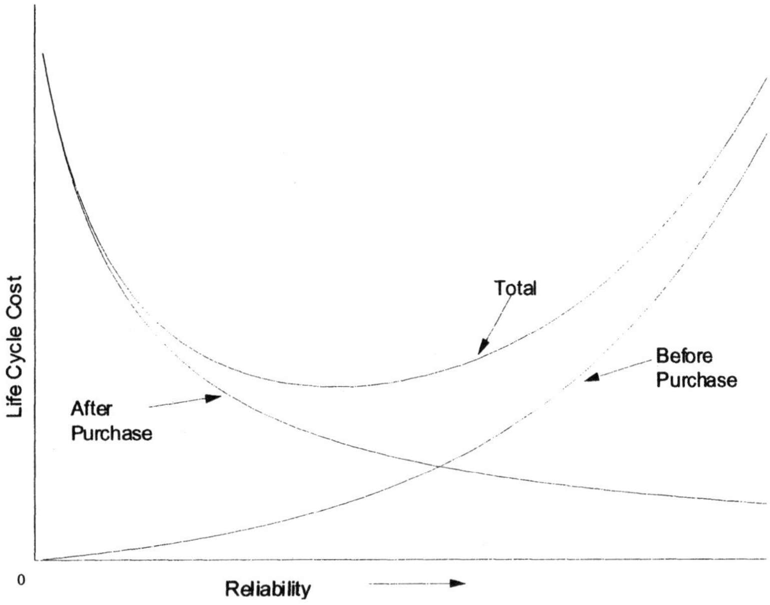

Figure 1.2. The optimum level of reliability.

The optimum level of reliability may not be the same for both the seller and the buyer. The life cycle costs discussed above are from the buyer's point of view. The post purchase costs of the seller include warranty costs and goodwill costs, such as loss of market share.

1.2 Reliability data sources

When attempting to determine the reliability of a system being designed, it may be infeasible to test all components considered for the design. Component data must be supplied from other sources. A good source of reliability data is often the component manufacturer. In addition to the manufacturer, there are several databases available to the public. Beware that when using reliability data from outside sources environmental conditions can cause great error. Correction factors are typically used to compensate for environmental conditions. Some databases include correction factors, but some do not.

Databases available to the public contain three types of data:

1. Raw Data—times to failure, operating hours and/or number of failures are provided. The user must compute failure statistics.
2. Primary, Consolidated Data—failure rates and failure probabilities derived from raw data are provided. Point estimates for mean life with confidence intervals are usually provided.
3. Expert Opinion Data—failure rates and failure probabilities estimated by experts.

Public databases contain a great deal of information, but care must be taken to ensure accuracy. Most databases provide sources of data, limitations, and the methods of statistical analysis used (if any). Be careful to note this information. Some common shortcomings of public databases are:

1. Independent and dependent failures may not be differentiated
2. Primary and secondary failures may not be differentiated
3. Failures due to human error may not be noted
4. Different organizations have different levels of reporting accuracy

Some public databases are

- **FARADA** (Failure Rate Data)—component failure probabilities provided by NASA and the Department of Defense. Available from ***Fleet Missile Systems Analysis and Evaluation Group***, Department of Defense, Corona, CA.
- **GIDEP** (Government and Industry Data Exchange Program)—a computerized database with information from a wide variety of components used in the defense industry. The database is operated by GIDEP Operations Center, Department of Defense, Corona, CA, 91720-5000, (714) 736-4677.
- **IEEE Data**—failure probabilities of electrical components based on expert opinion. Available from the ***Institute of Electrical and Electronics Engineers***, 345 East 47 Street, New York, NY, 10017-2394, (212) 705-7900.
- **LER Data** (Licensee Event Reports)—failure summaries of components from nuclear power plants. Available from the ***Nuclear Regulatory Commission***, Washington, DC
- **NERC** (National Electrical Reliability Council)—component failure data from power plants. Available from the ***National Electrical Reliability Council***, New York, NY.
- **NPRD** (Non-electric Parts Reliability Data)—failure data of non-electric components used in military systems. Available from ***Rome Air Development Center***, Griffins Air Force Base, NY, 13441-5700, (315) 330-4151.
- **NPRDS** (Nuclear Plant Reliability Data System)—a computerized database of failure data from nuclear power plants. The database is operated by ***Southwest Research Institute***, San Antonio, TX.
- **PVP** (Pressure Vessel and Piping)—failure data concerning pressure vessels and piping. This is a technical paper by Bush (1985) and updated periodically.

Another useful source of failure data is Military Handbook - 217. This handbook has been enhanced by the *Reliability Analysis Center*; the enhanced version is titled "Consolidated Reliability Assessment Model (CRAM)". The purpose of these handbooks is to establish and maintain consistent and uniform methods for estimating the inherent reliability of military electronic equipment systems. It provides a common basis for reliability predictions during acquisitions programs and establishes a common basis for comparing and evaluating reliability predictions of related or competitive designs.

1.3 Data Collection, Analysis, and Reporting

Data should be collected and analyzed prior to making decisions that affect the quality of a product, process or operation. Data provides the foundation for forming and testing hypotheses. The outcome of decisions derived from collected data is positively correlated to the quality of data on which the decision is made. Therefore, it is essential to develop a system for retaining and retrieving accurate information in order to maintain or improve product reliability.

Before starting a good data collecting program the objectives for collecting data should be understood. The basic objectives are to:

1. Identify the problem
2. Report the problem
3. Verify the problem
4. Analyze the problem
5. Correct the problem

According to Ireson (1988) the benefits of any good data collecting, analysis, and reporting system are:

1. Reduction of research or redundant test time
2. Assurance that proven parts are specified
3. Assistance for quality assurance and purchasing in obtaining parts
4. Assurance that products are capable of meeting specifications
5. Reduction in time and expense of testing
6. Improvement of capability to estimate reliability

Other important considerations are the accuracy needed, methods to use and who will do the collecting. Before formulating questions and attempting to answer these questions it is important to distinguish between the basic types of data; continuous and discrete. Anything that can be measured; weight, height, length, time, etc., are all examples of **continuous** data. Anything that can be counted; number of accidents, number of failures, or defects are examples of **discrete** data.

The purpose of a failure reporting, analysis, and corrective-action system (FRACAS) is to determine the root cause of the problem, document the problem, and record the corrective-action taken. Considerations for establishing such an information gathering program are outlined in this chapter. Suggested reading for further study is, "*A Reliability Guide to Failure Reporting, Analysis, and Corrective-action Systems.*"

1.3.1 Recording, reporting, and processing

Before a quality analysis can be made a closed loop system that collects and records failures that occur at specified levels must be in place. A "closed loop" system is a feedback loop that determines what failed, how it failed, why it failed, and tracks corrective-action. A program should be developed first to initiate the recording, reporting, and processing of all failures into the design, test and manufacturing processes. The first thing to consider in the recording process is the recording form. It should be clear, simple, and easy to understand by the user. As a minimum the form should include the following information:

- Equipment identification
- Failed part identification
- Type or category of failure
- Time of failure
- Who reported the failure
- Possible causes
- Action taken
- Who initiated action if any

Training should be provided for the more complex forms as necessary. Always provide valid reasons for collecting data. If the data collectors understand why they are taking data they will ultimately understand the need for accuracy when filling in the forms.

A Failure Review Board (FRB) should be established to monitor failure trends, significant failures and corrective-actions. The mission of the FRB is to review functional failure-data from appropriate inspections and testing including qualification, reliability, and acceptance test failures. All failure occurrence information should be made available to the FRB, including a description of conditions at the time of failure, symptoms of failure, failure isolation procedures, and known or suspected causes of the failure. Reports should include data on successes as well as failures. Details of any action taken must also be reported to enable the review board to make exceptions for items that may have been wrongfully replaced during test or service. Open items should then be followed up by the FRB until failure mechanisms have been satisfactorily identified and corrective-action initiated. FRB members should include appropriate representatives from design, reliability, safety, maintainability, manufacturing, and quality assurance. The leader of this group is usually a reliability engineer.

It is important that the term "failure" be defined. For example, ask what a failure means to the organization or process? Failure is defined as, "The event, or inoperable state, in which any item or part of an item does not, or would not, perform as previously specified."

Failures should also be defined and sorted into categories; relevant, irrelevant, chargeable, non-chargeable, and alternate failures (refer to Sections 2.1.2 and 2.1.3). All failures occurring during reliability qualification and production reliability acceptance tests should be classified as relevant or irrelevant. Relevant failures should be further classified as chargeable or non-chargeable. Rules for classification of failures should be in agreement with approved failure definitions or other criteria. Flow diagrams or other modeling techniques depicting failed hardware and data flow should be utilized where possible. Figure 1.3 is a flow diagram for failure categories.

Costs should be included in the recording process, where management approval is needed. Dollars and cents are sometimes the most important factors in management decisions. For example, the costs of redesigning or replacing a certain item could be compared to the cost of not replacing the item. The cost of not replacing the item, thus allowing a failure, should include warranty, liability, customer satisfaction, etc. These costs are usually only qualified estimates since they are difficult or impossible to quantify precisely. In any case include as many cost comparisons as possible for the decision making process.

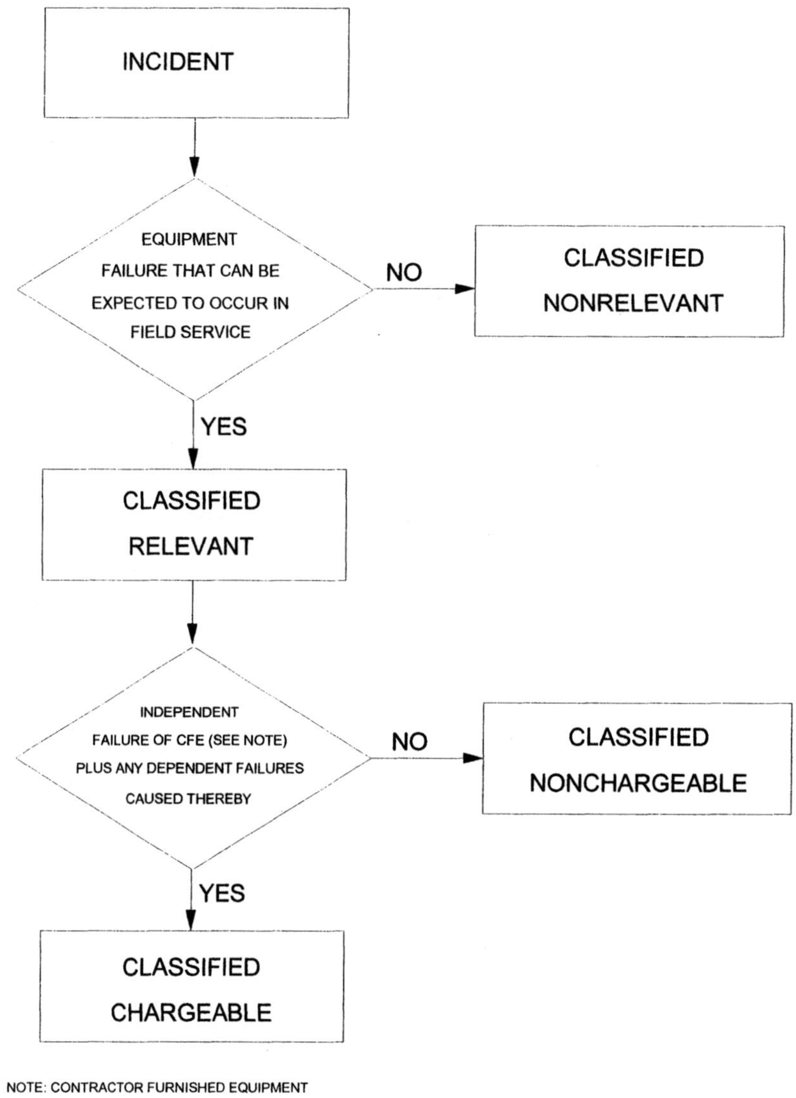

Figure 1.3. Flow diagram of failure categories.

1.3.1.1 Failure terms and definitions

Catastrophic failure—A failure that can cause item loss.

Critical failure—A failure, or combination of failures, that prevents an item from performing a specified mission.

Dependent failure—Failure caused by the failure of an associated item(s).

Independent failure—Failure that occurs without being caused by the failure of any other item.

Equipment design failure—Any failure that can be traced directly to the design of the equipment; that is, the design of the equipment caused the part in question to degrade or fail, resulting in an equipment failure.

Equipment manufacturing failure—A failure caused by poor workmanship or inadequate manufacturing process control during equipment construction, testing, or repair prior to the start of testing. For example, the failure due to cold solder joints.

Intermittent failure—Failure for a limited period of time, followed by the item's recovery or its ability to perform within limits without any remedial action.

Mechanism failure—The physical, chemical, electrical, thermal or other process that results in failure.

Multiple failures—The simultaneous occurrence of two or more independent failures. When two or more failed parts are found during troubleshooting and failures cannot be shown to be dependent, multiple failures are presumed to have occurred.

Pattern failures—The occurrence of two or more failures of the same part in identical or equivalent applications when the failures are caused by the same basic failure mechanism and the failures occur at a rate that is inconsistent with the part's predicted failure rate.

Part design failure—Failures that can be traced directly to inadequate design.

Part manufacturing failure—These failures are the result of poor workmanship or inadequate manufacturing process control during part assembly, inadequate inspection, or improper testing.

Primary failure—This type of failure is the result of a component deficiency. These failures result while operating within design limits.

Secondary failure—This type of failure produces operating conditions outside of design limits thereby causing primary failures. The primary failure would not have occurred if the secondary failure had not caused stress outside the design limits.

Software error failure—A failure caused by an error in the computer program.

Unverified failure—The opposite of a verified failure. Lack of failure verification, by itself, is not sufficient rationale to conclude the absence of a failure.

Verified failure—A failure that is determined either by repeating the failure mode on the reported item or by physical or electrical evidence of failure.

1.3.1.2 Failure categories

Relevant failures—Relevant failures are specified in 1 through 4:

1. Intermittent failures
2. Unverified failures (failures that cannot be duplicated, which are still under investigation, or for which no cause could be determined)
3. Verified failures not otherwise excluded under approved failure categories
4. Pattern failures

Irrelevant failures—Irrelevant failures are specified in 1 through 7:

1. Installation damage
2. Accident or mishandling
3. Failures of the test facility or test-peculiar instrumentation
4. Equipment failures caused by an externally applied overstress condition in excess of the approved test requirements
5. Normal operating adjustments (non-failures) specified in the approved equipment operating instructions
6. Secondary failures within the equipment, which are directly caused by irrelevant or relevant primary failures (the secondary failures must be proved to be dependent on the primary failure)
7. Failures caused by human errors

Chargeable failures—A relevant, independent failure of equipment under test and any dependent failures caused thereby which are classified as one failure and used to determine contractual compliance with acceptance and rejection criteria. Chargeable failures are specified in 1 through 4:

1. Intermittent failures
2. Unverified or verified failures
3. Independent failures
    a. Equipment design
    b. Equipment manufacturing
    c. Part design
    d. Part manufacturing

   e. Software errors identified, corrected, and verified during the pretest, the system verification, and the test, should not be chargeable as equipment failures
   f. Contractor-furnished equipment (CFE)
4. Relevant failures

Non-chargeable failures—Non-chargeable failures are specified in 1–3:

1. Irrelevant failures
2. Failures induced by operating, maintenance, or repair procedures
3. Failures of items having a specified life expectancy and operated beyond the specified replacement time of the item

Alternate failure categories—The alternate failure categories specified in 1 through 3 can be used to categorize both hardware and software failures for systems.

1. ***Critical failure*** is a failure which prevents the system from performing its mission.
2. ***Major failures*** are specified in a through f:
   a. Any failure that reduces performance of on-line hardware or software so that the system mission capability is degraded but not eliminated
   b. Any failure that prevents built-in test equipment (BITE) from detecting major failures in on-line hardware or software
   c. Any failure which causes BITE to continuously indicate a major failure in on-line hardware or software when such a failure does not exist
   d. Any failure that prevents BITE from isolating and localizing on-line hardware or software major failures to the lowest group
   e. Any failure that causes BITE to continuously isolate and localize a major failure in on-line hardware or software when such failure does not exist
   f. Any failure that causes BITE false alarm rate to exceed the specified level

3. ***Minor failures*** are as specified in a and b:
   a. Any failure that impairs on-line hardware or software performance, but permits the system to fully perform its objective
   b. Any failure that impairs BITE performance (creates false alarms) but permits BITE, when recycled, to isolate and localize to the required level, except when false alarm rate exceeds specified levels

1.3.1.3 Determining what information to collect

Deciding what data to collect will depend on the phase of the project—the conceptual, design, production, or maintenance phase. In any case, data should include failures due to equipment failure and human error. The conceptual phase will require the use of data from similar products. The design phase will require research or actual test data for the specific product. The production phase requires the use of a more historical type data derived sometimes from the design stages. The maintenance phase requires the use of actual failure data that may have been acquired with various failure analysis techniques. In short, all failures must be included from development to acceptance.

Five basic steps are outlined below that will help determine what data to collect:

1. Find out what happened. Be as specific as possible. At what level in the overall system, product or process was the event discovered?
2. Method of detection. Internally? Externally?
3. Find out when the event happened. During testing? During production run?
4. Find out if there is a similar event in historical records. If the answer is "yes," it could save time by eliminating some data collection.
5. Find out if there have been any recent changes. Check vendor materials, test conditions, etc.

1.3.1.4 How will data be collected and reported

Data may be collected by either a manual or automatic means. Most test results or observations are recorded manually on forms customized to collect specific information then input into a computer database. An example of a typical manual data collecting form is shown in Figure 1.4. Data is sometimes taken automatically through the use of electronic devices that send information directly to a database.

1. Report No.:	2. Report Date:	3. Time of Failure:	4. Equip Type:	5. Equip S/N:

6. Failed Part Description:	7. Failed Part ID:	8. Failed Part S/N:	9. Failed Part Mod No.:

ITEM USAGE:	10. Hrs:	11. Mins:	12. Secs:	13. Cycle Time.	14. Cal. Time:	15. Miles:

16. Failure Occurred During:	17. Action Taken:	18. Replacement:	19. Critical Cause:	20. Date of Failure:
❑ 1. Inspection	❑ 1. Adjustment	❑ 1. Complete	❑ 1. Priority 1	________
❑ 2. Production	❑ 2. Replaced	❑ 2. Partial	Name: ________	
❑ 3. Maintenance	❑ 3. Repaired	❑ 3. Tested		21. Time of Failure:
❑ 4. Shipping	❑ 4. Retested	❑ 4. Not Tested	❑ 2. Priority 2	
❑ 5. Field Use	❑ 5. Other	❑ 5. None	Name: ________	________

REASON FOR REPORT (check one) ➡

22. Removal or Maintenance Action Required as a Result of:
❑ 1. Suspected failure or malfunction ❑ 2. Due to improper maintenance ❑ 3. Damaged accidentally

23. Symptom(s):

❑ A. Excessive vibration
❑ B. High fuel consumption
❑ C. High oil consumption
❑ D. Inoperative
❑ E. Interference
❑ F. Leakage
❑ G. Noisy
❑ H. Out of balance
❑ I. Overheating
❑ J. Out of limits
❑ K. Unstable op

24. Condition(s):

		❑ 120. Shorted
❑ 110. Arced	❑ 009. Frayed	❑ 130. Stripped
❑ 990. Bent	❑ 008. Loose	❑ 140. Worn
❑ 880. Cracked	❑ 005. Open	❑ 150. Corroded
❑ 440. Galled	❑ 555. Plugged	❑ 160. Dented
❑ 505. Chafed	❑ 444. Ruptured	❑ 170. Chipped
❑ 606. Peeled	❑ 333. Split	❑ 101. Burned
❑ 707. Eroded	❑ 222. Sheared	❑ 202. Other

25. Cause of Trouble:

❑ A. Design deficiency	❑ E. Faulty packaging	❑ I. Faulty replacement	❑ M. Wrong part used
❑ B. Faulty maintenance	❑ F. Foreign object	❑ J. Operator adjustment	❑ N. Unknown contaminates
❑ C. Faulty manufacturing	❑ G. Fluid contamination	❑ K. Undetermined	
❑ D. Faulty overhaul	❑ H. Faulty installation	❑ L. Weather conditions	❑ O. Other (amplify)

26. Disposition or Corrective-action (select appropriate code(s) from list below and check boxes at left to indicate action taken in respect to failed item):

REASON FOR: NO ACTION	ACTION
❑ Hold for 90 days	❑ Adjusted
❑ Lack of facilities	❑ Removed and replaced
❑ Lack of repair parts	❑ Repaired and reinstalled
❑ Lack of personnel	❑ Retest and hold
❑ Other (explain below):	❑ Other (explain below):
________	________
________	________

27. Maintainability Information:

	HOURS	MINUTES
Time to locate trouble	____	____
Time to repair/replace	____	____
Total time non-operable	____	____

28. Component/Subcomponent Replaced With:
Description: ________
Part no.: ________
Serial no.: ________

29. Remarks (Furnish additional information concerning failure or corrective-action not covered above—do not repeat information checked above.):

Location:	Department:	Signature:	Date:

Figure 1.4. Failure report form.

The automatic data information gathering technique is usually desirable where continuous monitoring is necessary.

There are no standards for how to record or store data. When data is input into a computer, manually or automatically, both retrieval and use become obviously enhanced. There are many software packages on the market that can be readily tailored to fit specific needs.

1.3.1.5 Who will collect the information

Deciding who will collect the information will depend on who will use the data, accuracy needed, time and cost constraints. Keep in mind that the person who collects data is not necessarily the one who will use it.

1.3.1.6 What level of accuracy is needed

Accuracy will depend on the product and its intended use. For example, a cook may only need to take time and temperature data at ± minutes and degrees, while a race car designer may want time and temperature in ± tenths of seconds and degrees. For another example, if someone is asked their age 10 days before their 40th birthday, they may reply 39 or 40. Which is more accurate? Which is accurate enough? It could be important enough to require an answer like: 39 years, 355 days, 12 hours, and 15 minutes. Of course, asking a person's age usually will not require that much detail but when asking how long a verified equipment problem has persisted, details do become important.

The program outlined below will help assure that accurate and complete data is collected which meets the objectives for data collecting—identifying, reporting, verifying, analyzing, and correcting problems.

1. **Identify and control failed items.**
   A tag should be affixed to the failed item immediately upon the detection of any failure or suspected failure. The failure tag should provide space for the failure report serial number and for other pertinent entries from the item failure record. All failed parts should be marked conspicuously and controlled to ensure disposal in accordance with all laws, rules or regulations. Failed parts should not be handled in any manner that may obliterate facts which might be pertinent to the analysis and stored pending disposition of the failure analysis agent.
2. **Reporting of problems or failures.**
   A failure report should be initiated at the occurrence of each problem or failure of hardware, software or equipment. The report should contain the information

required to permit determination of the origin and correction of failures. The following information should be included in the report:

a. Descriptions of failure symptoms, conditions surrounding the failure, failed hardware identification, and operating time (or cycles) at time of failure
b. Information on each independent and dependent failure and the extent of confirmation of the failure symptoms, the identification of failure modes, and a description of all repair action taken to return the item to operational readiness
c. Information describing the results of the investigation, the analysis of all part failures, an analysis of the item design, and the corrective-action taken to prevent failure recurrence (if no corrective-action is taken, the rationale for this decision should be recorded)

3. **Verify failures.**
Reported failures should be verified as actual failures or an acceptable explanation provided for lack of failure verification. Failure verification is determined either by repeating the failure mode on the reported item or by physical or electrical evidence of failure (leakage residue, damaged hardware, etc.) Lack of failure verification, by itself, is not sufficient rationale to conclude the absence of a failure.

4. **Investigation and analysis of problems or failures.**
An investigation and analysis of each reported failure should be performed. Investigation and analysis should be conducted to the level of hardware or software necessary to identify causes, mechanisms, and potential effects of the failure. Any applicable method (test, microscopic analysis, applications study, dissection, x-ray analysis, spectrographic analysis, etc.) of investigation and analysis that may be needed to determine failure cause shall be used. When the removed item is not defective or the cause of failure is external to the item, the analysis should be extended to include the circuit, higher hardware assembly, test procedures, and subsystem if necessary. Investigation and analysis of supplier failures should be limited to verifying that the supplier failure was not the result of the supplier's hardware, software, or procedures. This determination should be documented for notification of the procuring activities.

5. **Corrective-action and follow-up.**
When the cause of a failure has been determined, a corrective-action shall be developed to eliminate or reduce the recurrence of the failure. The procuring activity should review the corrective-actions at scheduled status reviews prior to

implementation. In all cases, the failure analysis and the resulting corrective-actions should be documented. The effectiveness of the corrective-action should be demonstrated by restarting the test at the beginning of the cycle in which the original failure occurred.

1.3.1.7 Who will use the information

Deciding who will use the data is probably of less concern than what data to use. Usually everyone involved in the project will use some portion of the information. Assuring that collected information is available to all as needed, and making it easily accessible is the key.

1.3.2 Corrective-action and follow-up

The purpose of a corrective-action plan is to eliminate or reduce failures. Prior to implementation, the corrective-action plan should be submitted to management for approval. The system for such action should include provisions to assure that effective corrective-actions are taken on a timely basis by a follow-up audit that reviews and reports all open failure delinquencies to management. The failure cause and classification for each failure must be clearly stated. It is not necessary to include the cost of failure in the corrective-action status report if it was included in the initial report to management for corrective-action approval. This also describes the duties of the FRB.

The purpose for corrective-action status or follow-up should be obvious—to determine the effectiveness of the corrective-action taken. There are four questions that must be answered during the follow-up phase:

1. Was the failure or problem completely eliminated?
2. Was the failure or problem partially eliminated?
3. Did the corrective-action cause an insignificant change or no change?
4. Did the corrective-action cause a worse condition?

In essence, effective follow-up closes the loop in the "closed loop" system. Each failure must be monitored by continued reporting and analysis until effective results are obtained.

1.3.3 Summary

Data should be collected, reported, and investigated on all failures. It is important that all failures be categorized as relevant or irrelevant while at the same time resisting the temptation to classify a failure as irrelevant due to cost or time constraints.

As changes are made to correct a failure, follow-up is essential to assure the effectiveness of the corrective-action. For example, the corrective-action taken may cause over-stressing of another item that would cause a secondary failure, etc. The effectiveness of any corrective-action should never be assumed without 100% confidence that the decision or change will prevent a like failure. Always keep in mind that the outcome of decisions derived from collected data is positively correlated to the quality of data on which the decision is made.

1.4 Customer expectations

A product's market determines the reliability design process. Consider the following products.

- Artificial hip – a very high reliability for the remaining life of the patient is desired.
- Computing devices - because of the time required for many computing devices to become obsolete, serviceability is of little importance, and the design life is often less than 2 years.
- Aircraft – with a high purchase cost and high availability requirement, rapid and affordable serviceability is required. A very long service time is also desired. Many military and commercial aircraft have more than 40 years in service. There must be a very low rate of catastrophic failures because of the potential for the loss of human life. Redundancy is required for many systems.
- Inexpensive children's toys – it is expected that the child will lose interest in the toy after a short period of time. Cost considerations are much more important than reliability.

It does little good to hit the bulls eye if you are aiming at the wrong target. This is exactly what happened in the United States in the post world war II era. In reliability the target is represented by the requirement. For several decades those requirements were determined by engineering, purchasing, marketing, manufacturing, quality or some other internal department. This internal focus created a disconnect between the firm and its customers and competitors. The result was a serious balance of trade problem as foreign competitors "invaded" the American markets. The invasion was successful in large part because the competition did a better job of meeting the customer's requirements.

There are a number of ways to arrive at the targets a firm will try to hit. Historically, one of the best is to look at the purpose, or mission, of the firm. What are the basic human values the firm seeks to provide? What is their *long-term* focus? The firm's leadership will develop strategic objectives that are innovative and ambitious. Planning to hit these targets will provide the firm's employees with a challenge that will inspire them to make their best efforts. Meeting the objective will result in a new product or service that will meet the customer's requirements better than any competitor. This will provide an improved standard of living for the customer and will stimulate improvement among the firm's competitors. Firms that take this approach to goal setting are called market leaders.

Another source of requirements is the customer. Although customers seldom spark true innovation (for example, they are usually unaware of state-of-the art developments), their input is extremely valuable. Obtaining valid customer input is a science itself. Market research firms use scientific methods such as critical incident analysis, focus groups, content analysis and surveys to identify the "voice of the customer. Noritaki Kano developed the following model of the relationship between customer satisfaction and quality.

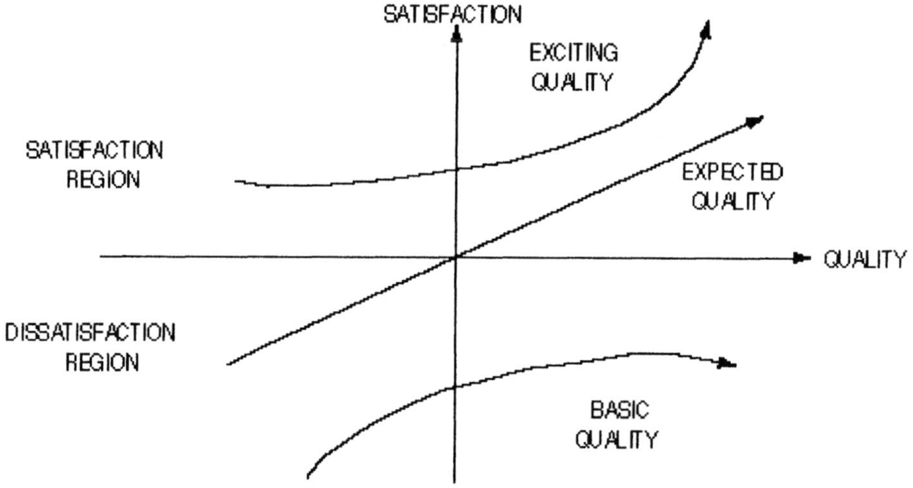

Figure 1.5. Kano model.

The Kano model shows that there is a basic level of quality that customers assume the product will have. For example, all automobiles have windows and tires. If asked, customers do not even mention the basic quality items, they take them for granted.

However, if this quality level isn't met the customer will be dissatisfied, note that the entire "Basic Quality" curve lies in the lower half of the chart, which represents dissatisfaction. However, providing basic quality isn't enough to create a satisfied customer.

The expected quality line represents those expectations which customers explicitly consider. For example, the length of time spent waiting in line at a checkout counter. The model shows that customers will be dissatisfied if their quality expectations are not met; satisfaction increases as more expectations are met.

The exciting quality curve lies entirely in the satisfaction region. This is the effect of innovation. Exciting quality represents *unexpected* quality items. The customer receives more than they expected. For example, Cadillac pioneered a system where the headlights stay on long enough for the owner to walk safely to the door.

Competitive pressure will constantly raise customer expectations. Today's exciting quality is tomorrow's basic quality. Firms that seek to lead the market must innovate constantly. Conversely, firms that seek to offer standard quality must constantly research customer expectations to determine the currently accepted quality levels. It is not enough to track competitors since expectations are influenced by outside factors as well. For example, the quality revolution in manufacturing has raised expectations for service quality as well.

Once information about customer expectations has been obtained techniques such as quality function deployment (QFD) can be used to link the voice of the customer directly to internal processes.

Benchmarking is another popular method for developing requirements. Benchmarking involves research into the best practices *at the process level.* Benchmarking goes far beyond a determination of the "industry standard," it breaks the firm's activities down to process operations and looks for the best in class for a particular operation. For example, Xerox corporation studied the retailer LL Bean to help them improve their parts distribution process. One pitfall of benchmarking is that it is based on copying others, rather than developing new and improved approaches. Since the process being copied is there for all to see, a firm will find that benchmarking cannot give them a sustained competitive advantage. Although helpful, benchmarking should never be the *primary* strategy for improvement.

Competitive analysis is an approach to goal setting used by many firms. This approach is essentially benchmarking confined to one's own industry. Although common, competitive analysis virtually guarantees second-rate quality because the firm will always be following their competition. If the entire industry employs the approach it will lead to stagnation for the entire industry, setting them up for eventual replacement by outside innovators. An example of this is the replacement of the mechanical watch industry by less expensive, more accurate and more reliable electronic timepieces.

Many firms set their goals according to "industry standards." This is a modified version of the competitive analysis approach. Whereas the competitive analysis set the requirement as the best in the industry group, industry standards are based on the average or even the minimum acceptable performance of an industrial group. As a long-term strategy this approach will inevitably lead to failure as outside innovators or industry innovators, run away from the standard, leaving those who merely meet it far behind.

Another method of setting long-term quality goals is not to set them at all. This approach, which will be referred to as the ad hoc approach, involves identifying problems and focusing on solving the problems. This amounts to no long-term planning at all. It is nothing more than crisis management.

1.4.1 Limits to planning

Consider the following proposed experiment: take a population that is homogeneous and divide it into two parts. One part of the population will operate a centrally planned economy, the other population will not. Compare the results after several decades.

Farfetched? Not at all! This experiment has been done, in fact it has been done more than once. Here are two typical results.

Table 1.1. Planned versus unplanned economies (1988 data).

Performance Measure	Hong Kong	People's Republic of China	West Germany	East Germany
GDP per capita	\$9,613	\$301	\$19,743	\$5,256
People per telephone	2.2	149.8	1.6	4.3
People per TV set	4.2	100.7	2.4	5.8
People per automobile	29.8	1,093.3	2.2	4.8

These results make thing clear: it is possible to do too much *centralized* planning. The word "centralized" is emphasized here because, in fact, more planning actually takes place in the "unplanned" economies! F.A. Hayek, a Nobel Prize-winning economist, points out that in the free market nations planning is done by individuals and firms rather than by central planning authorities. We are each engaged in economic planning when we discuss business at a cocktail party. In aggregate, the amount of planning in so-called "unplanned" economies is vastly greater than in planned economies.

The lesson here is that the scope of centralized quality planning should be restricted. The quality plan should allow plenty of room for individual maneuvering to meet unanticipated customer demands and future events. As much planning as possible should be done by those closest to the customer and the market. It is a serious mistake to think that staff personnel can prepare plans that cover all contingencies. In the realm of economics, Hayek calls such thinking a "fatal conceit."

1.4.2 Top down Planning

Top-down planning, as the name implies, begins with the objective or goal and works backwards to the activities needed to accomplish the goal. Goal setting was discussed earlier in this chapter.

Top-down planning involves developing a strategy for meeting the quality goals. The plan will describe the business reason for pursuing the quality objective. The strategic quality plan is part of, and is totally integrated with, the overall strategic business plan. Quality is but one aspect of the strategy of the firm. Quality objectives are subsumed under other goals relating to creating maximum value for customers and other stakeholders in the firm.

1. Tom Peters (1987) lists five characteristics of a good strategic planning process:
2. It gets everybody involved.
3. It is not constrained by corporate assumptions.
4. It is perpetually fresh.
5. It is not left to planners.
6. It requires lots of noodling time and vigorous debate.

Peters believes that there is no such thing as a good strategic plan. The chaotic business environment precludes the ability to see very far into the future. However, he describes desirable properties for strategic planning documents: they are succinct, they emphasize development of strategic skills, it is burned the day before it goes to the printers--it is a living document, not an icon.

In short, the benefit of strategic planning is in the *process*, not in the plan itself. A good strategic planning process is a means of breaking down inter-departmental barriers. It helps senior management communicate with one another and focuses their attention on the external factors that will determine the future success of the firm. It also provides a roadmap, albeit an inevitably out-of-date one, for the front line personnel.

A key to developing a strategic quality plan is the concept of total quality. The total quality concept originated with Feigenbaum (1956) and has been evolving ever since.

1.4.3 Customer and supplier roles in planning

Tactical quality planning involves developing an approach to implementing the strategic quality plan. One of the most promising developments in this area has been policy deployment. Sheridan (1993) describes policy deployment as the development of a measurement-based system as a means of planning for continuous quality improvement throughout all levels of an organization. Originally developed by the Japanese, American companies are also beginning to adopt policy deployment because it clearly defines the long-range direction of company development, as opposed to short-term.

Quality function deployment (QFD) is a customer-driven process for planning products and services. It starts with the voice of the customer, which becomes the basis for setting requirements. QFD matrices, sometimes called "The House of Quality," are graphical displays of the result of the planning process. QFD matrices vary a great deal and may show such things as competitive targets and process priorities. The matrices are created by inter-departmental teams, thus overcoming some of the barriers which exist in functionally organized systems.

QFD is a system for design of a product or service based on customer demands, a system that moves methodically from customer requirements to specifications for the product or service. QFD involves the entire company in the design and control activity. Finally, QFD provides documentation for the decision-making process. The QFD approach involves four distinct phases (King, 1987):

Organization phase. Management selects the product or service to be improved, the appropriate interdepartmental team, and defines the focus of the QFD study.

Descriptive phase. The team defines the product or service from several different directions such as customer demands, functions, parts, reliability, cost, and so on.

Breakthrough phase. The team selects areas for improvement and finds ways to make them better through new technology, new concepts, better reliability, cost reduction, etc., and monitors the bottleneck process.

Implementation phase. The team defines the new product and how it will be manufactured.

QFD is implemented through the development of a series of matrices. In its simplest form QFD involves a matrix that presents customer requirements as rows and product or service features as columns. The cell, where the row and column intersect, show the correlation between the individual customer requirement and the product or service requirement. This matrix is sometimes called the "requirement matrix." When the requirements matrix is enhanced by showing the correlation of the columns with one another, the result is called the "house of quality." Figure 1.6 shows one commonly used house of quality layout.

The house of quality relates, in a simple graphical format, customer requirements, product characteristics, and competitive analysis. It is crucial that this matrix be developed carefully since it becomes the basis of the entire QFD process. By using the QFD approach the customer's demands are "deployed" to the final process and product requirements.

One rendition of QFD, called the Macabe approach, proceeds by developing a series of four related matrices (King 1987): product planning matrix, part deployment matrix, process planning matrix, and production planning matrix. Each matrix is related to the previous matrix as shown in Figure 1.7.

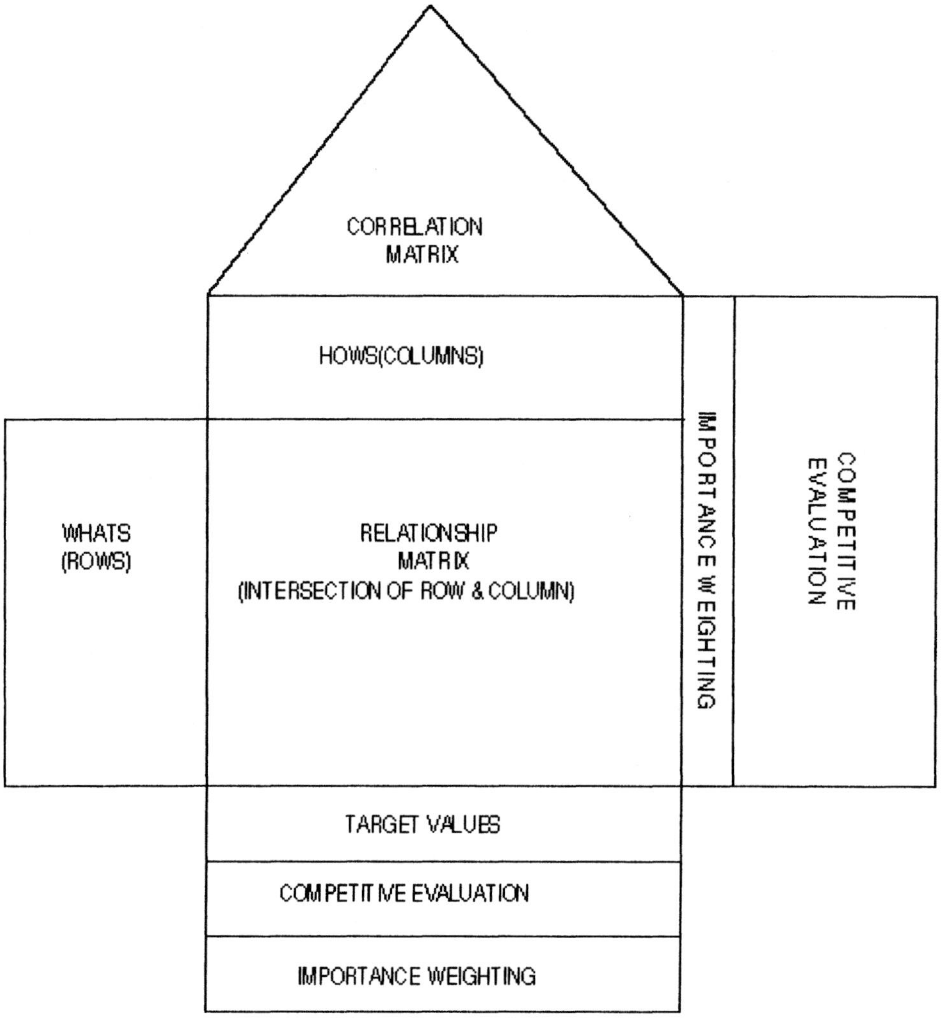

Figure 1.6. The house of quality.

Another approach to QFD is based on work done by Yoji Akao. Akao (1990, 7-8) presents the following 11 step plan for developing the quality plan and quality design using QFD.

1. First, survey both the expressed and latent quality demands of consumers in your target marketplace. Then decide what kinds of "things" to make.
2. Study the other important characteristics of your target market and make a demanded quality function deployment chart that reflects both the demands and characteristics of that market.
3. Conduct an analysis of competing products on the market, which we call a competitive analysis. Develop a quality plan and determine the selling features (sales points).

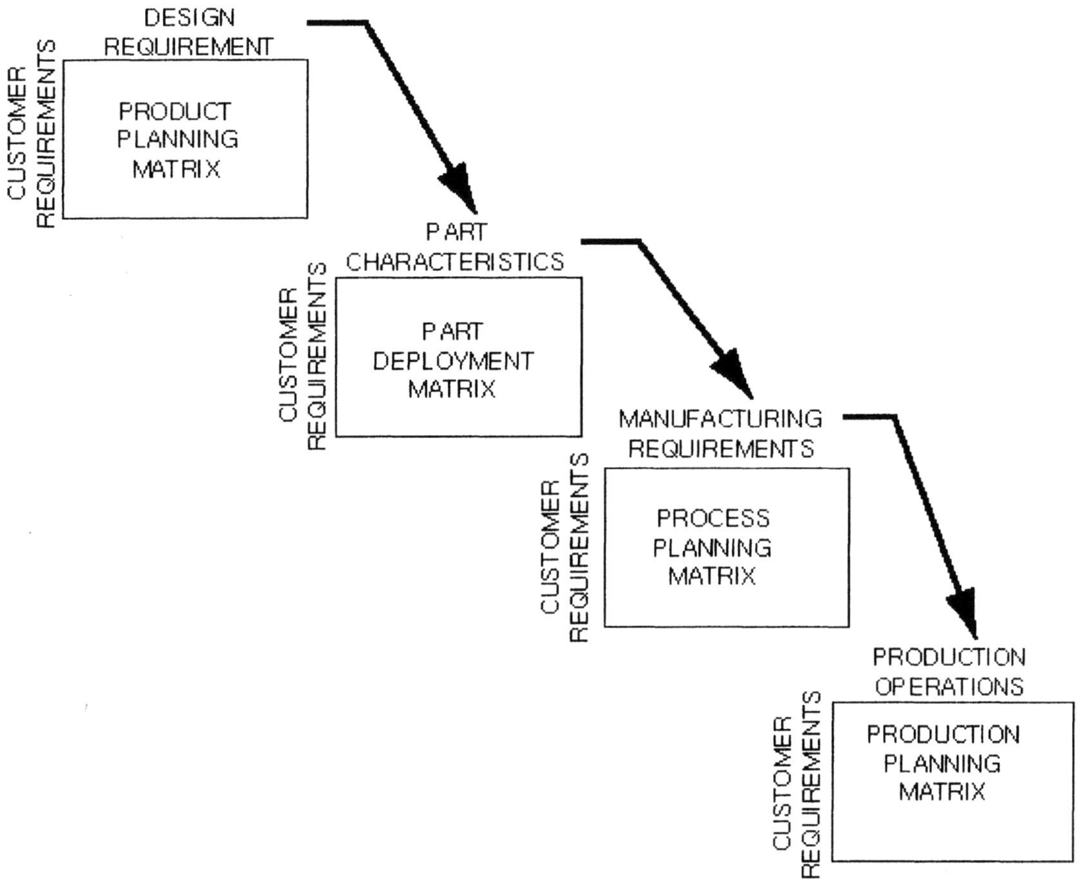

Figure 1.7. QFD methodology: Macabe approach.

4. Determine the degree of importance of each demanded quality.
5. List the quality elements and make a quality elements deployment chart.
6. Make a quality chart by combining the demanded quality deployment chart and the quality elements deployment chart.
7. Conduct an analysis of competing products to see how other companies perform in relation to each of these quality elements.
8. Analyze customer complaints.
9. Determine the most important quality elements as indicated by customer quality demands and complaints.
10. Determine the specific design quality by studying the quality characteristics and converting them into quality elements.
11. Determine the quality assurance method and the test methods.

1.5 Exercises

1. Explain how reliability affects the cost of owning and operating an item over its useful life.
2. Explain the relationship between quality control and reliability.
3. Item A costs $15,000 and has expected maintenance costs of $4,000 per year. Item B costs $22,000 and has expected maintenance costs of $1,500 per year. Using a cost of money (interest rate) of 10%, which item has the lowest life cycle cost assuming a useful life of 8 years with no residual value?
4. Item A costs $25,000. To obtain the required availability on item A, two mechanics at a cost of $21,000 per year each, supplies at a cost of $8,000 and $7,000 worth of spare parts must be maintained. Item B costs $55,000. To obtain the required availability on item B, one mechanic at a cost of $29,000 per year, supplies at a cost of $16,000 and $4,000 worth of spare parts must be maintained. Which item has the lowest life cycle cost assuming a useful life of 12 years with no residual value, and a 10% cost of money (interest rate)?
5. Which automobile has the lowest life cycle cost assuming a 10% cost of money (interest rate)?

	TYPE 1	TYPE 2
Purchase price	$21,000	$23,000
Maintenance cost, year 1	$0	$0
Maintenance cost, year 2	$0	$0
Maintenance cost, year 3	$0	$0
Maintenance cost, year 4	$500	$0
Maintenance cost, year 5	$1,000	$0
Maintenance cost, year 6	$2,000	$500
Trade in value	$11,000	$11,500

6. Before a failure is reported it must first be:
   a. Identified
   b. Verified
   c. Defined
   d. Recorded
7. The ultimate purpose behind failure data collecting is to:
   a. Identify the failure
   b. Verify that a failure has occurred
   c. Analyze the failure
   d. Prevent recurrence of the failure

8. The basic element of a failure reporting, analysis, and corrective-action system that is most likely to prevent the recurrence of a failure is the:
   a. Corrective-action system
   b. Analysis system
   c. Reporting system
   d. Follow-up system
9. A minor failure may be defined as:
   a. Any failure that causes built-in test equipment to continuously isolate a major failure in on-line hardware when such failure does not exist
   b. Any failure that causes built-in test equipment false alarm rates to exceed the specified levels
   c. Any failure that impairs built-in test equipment performance (creates false alarms), but permits equipment, when recycled, to isolate and localize to the required level, except when false alarm rate exceeds specified levels
   d. Any failure that causes built-in test equipment to continuously indicate a major failure in on-line hardware or software when such a failure does not exist
10. Irrelevant failures are:
    a. Intermittent failures
    b. Unverified failures that cannot be duplicated
    c. Verified failures not otherwise excluded under approved failure categories
    d. Failures caused by human errors
11. Describe the mission of a failure review board.
12. Failure costs should be included on the:
    a. Corrective-action status report
    b. Initial recording form
    c. The FRB's initial reporting form
    d. The failure analysis report
13. Successes should be reported as well as failures.
    a. True
    b. False
14. The effectiveness of corrective-action is best demonstrated by:
    a. Restarting the test at the beginning of the cycle in which the original failure occurred
    b. Follow-up to see if costs have been reduced
    c. Follow-up to see if customers are better satisfied.
    d. Follow-up to see if the problem was completely eliminated

e. a and d above

15. Describe the best way to verify a failure.
16. State the five objectives for data collecting?
17. A chargeable failure may be:
    a. A intermittent failure
    b. An independent failure
    c. Relevant failure
    d. Unverified failure
    e. All of the above

1.6 References

Akao, Y. Editor (1990). Quality Function Deployment: Integrating Customer Requirements into Product Design, Cambridge, MA: Productivity Press.

ASQ Reliability Division. 1977. ***A Reliability Guide to Failure Reporting, Analysis, and Corrective-action Systems.*** Milwaukee, WI: ASQ–Quality Press.

Day, R.G. (1993) Quality Function Deployment: Linking a Company with Its Customers. Milwaukee, WI: ASQ Quality Press.

Dhillon, B. S. 1985. ***Quality Control, Reliability, and Engineering Design.*** New York: Marcel Dekker.

Dodson, B. L., and M. D. Mulcahy. 1992. ***Certified Reliability Engineer Examination Study Guide.*** Tucson, AZ: Quality Publishing.

Dovich, R. A. 1990. ***Reliability Statistics.*** Milwaukee, WI: ASQ–Quality Press.

Edosomwan, J.A. (1993). Customer and Market-Driven Quality Management. Milwaukee, WI: ASQ Quality Press.

Edwards, C.D. (1968). "The Meaning of Quality," Quality Progress, October.

Hayes, Bob E. (1992). Measuring Customer Satisfaction: Development and Use of Questionnaires. Milwaukee, WI: ASQ Quality Press.

Holland, John H. (1996). Hidden Order: How Adaptation Builds Complexity Reading, MA: Perseus Books, p.56.

Hurley, H. and Loew, C. (1996). "A Quality Change for New Product Development," Quality Observer, January, pp. 10–13.

Hutton, D.W. (1994). The Change Agent's Handbook: A Survival Guide for Quality Improvement Champions. Milwaukee, WI: ASQ Quality Press.

Imai, M. (1986). Kaizen, New York: Random House.

Ireson, G. W., and C. F. Coombs. 1988. ***Handbook of Reliability Engineering and Management.*** New York: McGraw-Hill.

Iverson, Ken. (1998). Plain Talk: Lessons from a Business Maverick New York: John Wiley & Sons, p. 149.

Johnson, R.S. (1993a). TQM: Leadership for the Quality Transformation. Milwaukee, WI: ASQ Quality Press.

Johnson, R.S. (1993b). TQM: Management Processes for Quality Operations. Milwaukee, WI: ASQ Quality Press.

Joiner, B.L. (1994). Fourth Generation Management: The New Business Consciousness. New York: McGraw-Hill, Inc.

Joiner, B.L. (1996). "Quality, Innovation, and Spontaneous Democracy," Quality Progress, March, pp. 51–53.

Kano, N. (1993). "A Perspective on Quality Activities in American Firms." California Management Review, 35(3), Spring, pp. 12–31.

Kapur, K. C., and L. R. Lamberson. 1977. ***Reliability in Engineering Design.*** New York: John Wiley & Sons.

Kececioglu, D. 1991. ***Reliability Engineering Handbook, Volume 1.*** Englewood Cliffs, NJ: Prentice Hall.

King, B. (1987). Better Designs in Half the Time: Implementing QFD in America. Methuen, MA: Goal/QPC.

Kirkpatrick, J. (1992–1993). "Pragmatism and the Harvard Case Method of Business Education." The Intellectual Activist, 6(6) pp. 9–18, and 7(1) pp. 11–15.

Kotler, P. (1991). Marketing Management: Analysis, Planning, Implementation, and Control, 7th edition. Englewood Cliffs, NJ: Prentice-Hall.

Lewis, C.I. (1929), Mind and the World Order, New York: Scribners.

Lewis, E. E. 1987. ***Introduction To Reliability Engineering.*** New York: John Wiley & Sons.

Main, Jeremy (1994). Quality Wars: The Triumphs and Defeats of American Business, The Free Press, 173.

Mizuno, S., Editor. (1988). Management for Quality Improvement: The 7 New QC Tools, Cambridge, MA: Productivity Press.

Pyzdek, T. (1996b). "The Pursuit of Happiness: Use Market-based Management to Improve Performance and Go Beyond TQM." Quality Progress, May, pp. 92–96

Pyzdek, Thomas. (1999). The End of Management.

Rand, A. (1964). The Virtue of Selfishness, New York: Penguin Books.

Reichheld, F.F. (1996), The Loyalty Effect: The Hidden Force Behind Growth, Profits, and Lasting Value. New York: McGraw-Hill.

Roberts, L. (1994). Process Reengineering: The Key to Achieving Breakthrough Success. Milwaukee, WI: ASQ Quality Press.

Rose, K.H. (1995). "A Performance Measurement Model." Quality Progress, February, pp. 63–66.

Saaty, T.L. (1988). Decision Making for Leaders: The Analytic Hierarch Process for Decisions in a Complex World. Pittsburgh, PA: RWS Publications.

Senge, P.M. (1990), The Fifth Discipline–The Art and Practice Of The Learning Organization, New York: Doubleday.

Sheridan, B.M. (1993). Policy Deployment: The TQM Approach to Long-Range Planning. Milwaukee, WI: ASQ Quality Press.

Sheridan, John W. (1990), "America's Best Plants," Industry Week, vol. 239, p. 27 (19).

Shooman, M. L. 1990. ***Probabilistic Reliability: An Engineering Approach.*** Malabar, FL: Robert E. Krieger.

Stewart, T.A. (1995). "After All You've Done for Your Customers, Why are They Still NOT HAPPY?" Fortune, December 11, 1995.

US Department of Defense. ***MIL-STD-721: Definition of Terms for Reliability and Maintainability.*** Philadelphia: Naval Publications and Forms Center.

US Department of Defense. ***MIL-STD-785: Reliability Program for Systems and Equipment Development and Production.*** Philadelphia: Naval Publications and Forms Center.

Vaziri, H.K. (1992). "Using Competitive Benchmarking to Set Goals," Quality Progress, November.

♦♦♦

CHAPTER

2

Distributions

A basic step in many reliability studies is choosing a probability density function. This first part of this chapter defines several popular density functions along with their reliability and hazard functions. The relationships among the probability density function, the reliability function and the hazard function are shown, as well as methods for determining the mean and variance of distributions.

The second part of this chapter covers the techniques of parameter estimation, Bayesian analysis, transformation of random variables, hypothesis testing and confidence intervals.

2.1 Distribution characteristics

Probability density functions (also known as density functions) are grouped into two categories, discrete and continuous. The values a **discrete** density function can take on are limited to integers. For example, in five flips of a coin, the number of heads is limited to 1, 2, 3, 4, or 5; the value 3.7 is nonsense. **Continuous** density functions can take on all values in a specific range. For example, the time to fail may be 1 hour, 1.01 hours, 1.001 hours, 1.0001 hours, etc.

To qualify as a density function two criteria must be met:

1) $f(x) \geq 0$

for all values of x, and

2) $\int_{-\infty}^{\infty} f(x)dx = 1$

for continuous distributions, or for discrete distributions,

$$\sum_{n} f(x_n) = 1$$

where the sum is taken over all possible values of *n*.

The cumulative distribution function (CDF) is the area under the probability density function to the left of the specified value, and represents the probability of x being less than a specific value, $P(X < x)$. For continuous distributions, the cumulative distribution function is defined as

$$F(x) = \int_{-\infty}^{x} f(\tau)d\tau \tag{2.1}$$

The probability of x falling between two specified values, a and b, is the area under the probability density function from a to b, which is

$$P(a < X < b) = \int_{a}^{b} f(x)dx,$$

which is also equal to $F(b) - F(a)$. This is represented graphically in Figure 2.1.

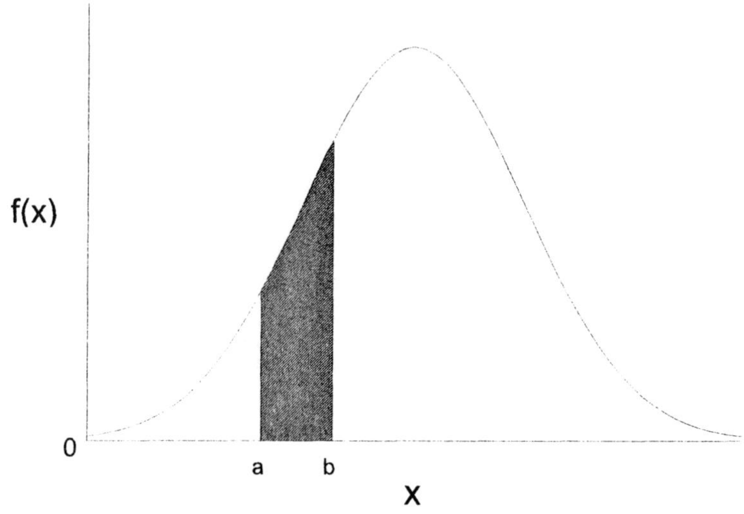

Figure 2.1. Probability of *x* falling between two specific values.

Note that for continuous distributions the probability of x being exactly equal to any specific value, for example a, is zero. As seen in Figure 2.1, as b moves closer to a, the area (which is equal to the probability) decreases. When b is equal to a, the area is zero.

For discrete distributions the probability of x being exactly equal to a specific value must be greater than zero for some values. For example, in four flips of a coin, where x represents the number of heads obtained, the probability of obtaining exactly 0 heads is 0.0625, the probability of obtaining exactly 1 head is 0.25, etc. This is the probability density function, and is shown in Figure 2.2.

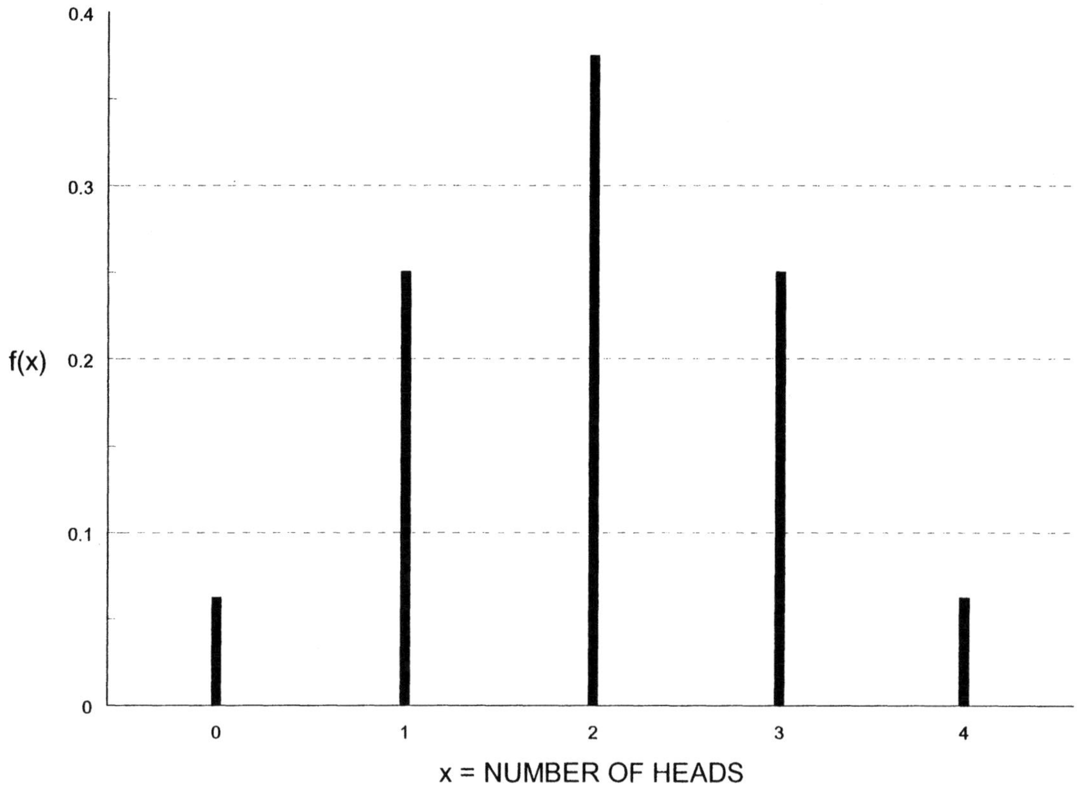

Figure 2.2. A discrete probability density function.

The cumulative distribution function for discrete distributions is

$$F(x) = \sum_{\tau \leq x} f(\tau) \tag{2.2}$$

The cumulative distribution function for the coin-flipping example is given in Figure 2.3.

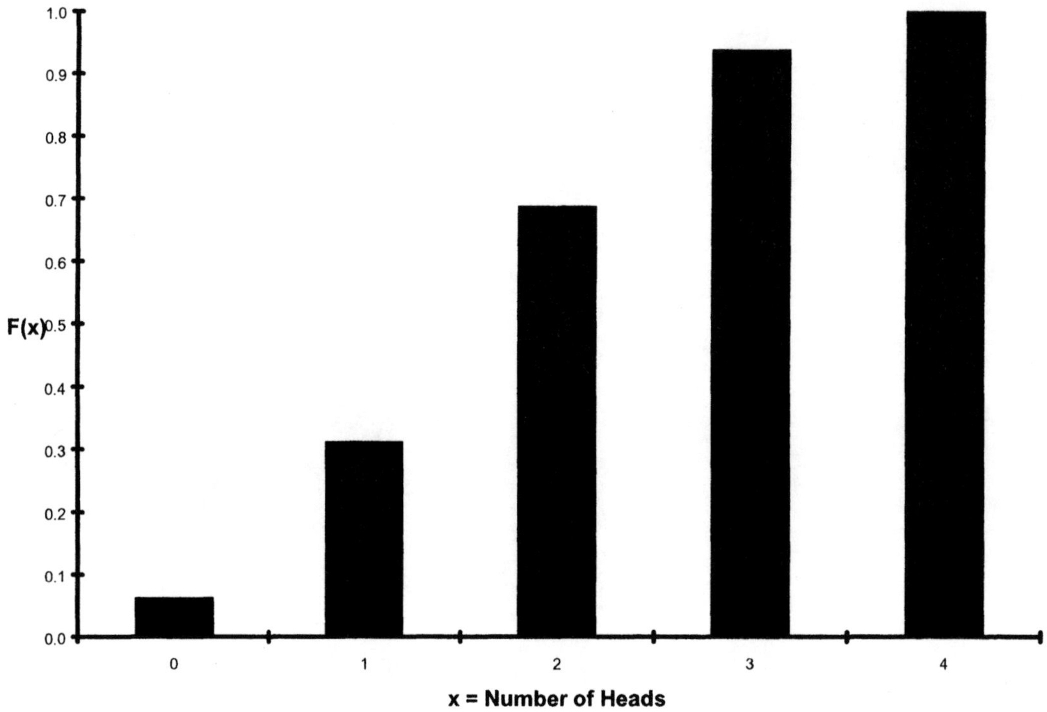

Figure 2.3. A discrete cumulative distribution function.

The reliability function is the complement of the cumulative distribution function, $F(x)$, and is obtained from the probability density function using the relationship

$$R(x) = 1 - F(x) = 1 - \int_{-\infty}^{x} f(\tau)d\tau = \int_{x}^{\infty} f(\tau)d\tau \tag{2.3}$$

The relationship between a reliability function and the cumulative distribution function is shown in Figure 2.4.

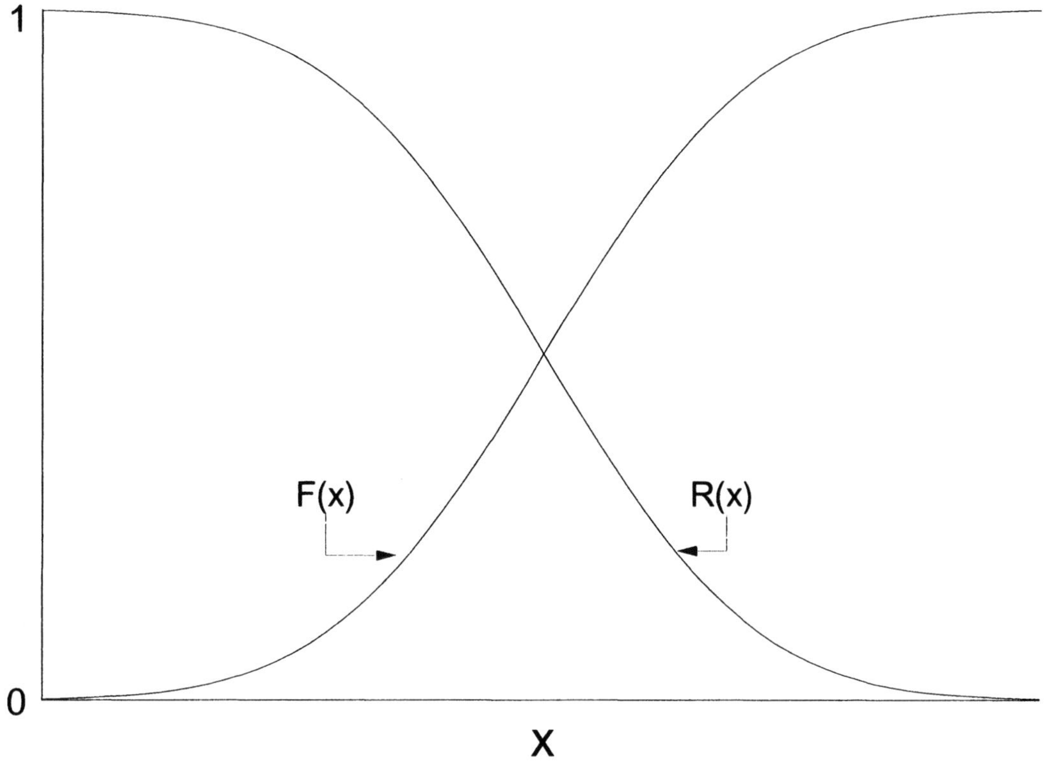

Figure 2.4. The relationship between the reliability function and the cumulative distribution function.

The hazard function is the instantaneous failure rate, and is a measure of proneness to fail. The hazard function is defined as

$$h(x) = \frac{f(x)}{R(x)} \tag{2.4}$$

The reliability function and the probability density function can be obtained from the hazard function using the expressions

$$R(t) = e^{-\int_{-\infty}^{t} h(\tau)d\tau} \tag{2.5}$$

and

$$f(x) = h(x)e^{-\int_{-\infty}^{x} h(\tau)d\tau} \tag{2.6}$$

The expected value, of a distribution can be determined from the expressions

$$E(x) = \int_{-\infty}^{\infty} xf(x)dx = \mu \tag{2.7}$$

or

$$E(x) = \int_0^{\infty} R(x)dx = \mu \tag{2.8}$$

The variance of a distribution can be found from the expression

$$V(x) = E(x^2) - \mu^2 \tag{2.9}$$

where μ is the distribution mean, and

$$E(x^2) = \int_{-\infty}^{\infty} x^2 f(x)dx \tag{2.10}$$

Example 2.1

Given the density function

$f(x) = ax, \ 0 \leq x \leq 10$

a. determine the value of a that makes $f(x)$ a valid density function,

b. determine the mean and variance of the distribution, and

c. derive expressions for the reliability and hazard functions.

Solution

For the density function to be valid, the area under the curve must equal one. Thus,

$$\int_0^{10} (ax)dx = 1 \Rightarrow \frac{a}{2}(x^2)\Big|_0^{10} = 1 \Rightarrow \frac{a}{2}(10^2 - 0^2) = 1 \Rightarrow a = \frac{1}{50}$$

From Equation 2.7, the mean of this distribution is

$$E(x) = \int_0^{10} x\frac{x}{50}dx = \int_0^{10} \frac{x^2}{50}dx = \frac{1}{150}(10^3) = 6.67$$

From Equation 2.9, the variance of this distribution is

$$V(x) = \int_0^{10} \frac{1}{50}x^3dx - (6.67)^2 = \frac{1}{20}(10^4) - 44.5 = 5.55$$

The reliability function is

$$R(x)=\int_x^{10}\frac{1}{50}\tau d\tau=1-\frac{x^2}{100}\ ,\ 0\le x\le 10$$

The hazard function is

$$h(x)=\frac{f(x)}{R(x)}=\frac{\frac{1}{50}x}{1-\frac{x^2}{100}}=\frac{2x}{100-x^2}\ ,\ 0\le x\le 10$$

Example 2.2

Given the hazard function

$$h(x)=c\ ,\ x\ge 0$$

Determine the reliability function and the probability density function.

Solution

The reliability function can be obtained from Equation 2.5.

$$R(x)=e^{-\int_0^x c d\tau}=e^{-cx}\ ,\ x\ge 0$$

The probability density function is

$$f(x)=h(x)R(x)=ce^{-cx}\ ,\ x\ge 0$$

This is the well known exponential probability density function, and will be discussed later.

Distribution characteristics are often obtained from moment-generating functions. If x is a continuous random variable, the nth moment about the origin is

$$E\left(x^n\right)=\int_{-\infty}^{\infty}x^n f(x)dx \tag{2.11}$$

If x is a discrete random variable, the nth moment about the origin is

$$E\left(x^n\right)=\sum_x x^n f(x) \tag{2.12}$$

The first moment about the **origin** is the distribution mean; recall that this is equivalent to Equation 2.7. The second moment about the **mean** is the variance; recall that this is equivalent to Equation 2.9. The third moment about the **mean** is the skewness of the distribution. If a single peaked distribution has a long tail to the right, it is said to be skewed right. The fourth moment about the **mean** is a measure of kurtosis, which is a measure of the peakedness of a distribution.

2.2 Normal and lognormal distributions

The normal distribution is a continuous distribution, and is also known as the Gaussian distribution. It is one of the easiest probability density functions to work with. Two useful properties of the normal distribution are additivity of variances and the tendency of sums of random variables, regardless of individual density functions, to be normal when added together (this is known as the central limit theorem or the theory of large numbers). These two properties make the normal distribution very useful when modeling a buildup of tolerances.

The lognormal distribution is also continuous, and is one of the most widely used life distributions. It is highly skewed and lends itself to situations when effects are multiplicative. The lognormal distribution has been used to model the cycles to failure for metals, the life of transistors, the life of bearings, as well as many other phenomenon.

2.2.1 The normal distribution

The normal probability density function is defined as

$$f(x)=\frac{1}{\sigma\sqrt{2\pi}}\exp\left[-\frac{1}{2}\left(\frac{x-\mu}{\sigma}\right)^2\right],\ -\infty<x<\infty \tag{2.13}$$

where: μ = the distribution mean (population mean), which is a measure of central tendency or location, and

σ = the distribution standard deviation (population standard deviation), which is a measure of dispersion.

The distribution mean is estimated from the sample mean

$$\hat{\mu}=\bar{x}=\frac{\sum_{i=1}^{n}x_i}{n} \tag{2.14}$$

where n is the sample size.

The distribution standard deviation is estimated from the sample standard deviation

$$\hat{\sigma}=s=\sqrt{\frac{n\sum_{i=1}^{n}x_i^2-\left(\sum_{i=1}^{n}x_i\right)^2}{n(n-1)}} \tag{2.15}$$

Another commonly used expression for the sample standard deviation is

$$\hat{\sigma} = s = \sqrt{\frac{\sum_{i=1}^{n}(x_i - \bar{x})^2}{(n-1)}} \tag{2.16}$$

Equation 2.15 is more accurate for numerical calculations than Equation 2.16 because of the small values obtained when subtracting individual values from the sample mean.

The normal distribution is not skewed, and the mean, median and mode are all equal. Figure 2.5 graphically represents the normal probability density function.

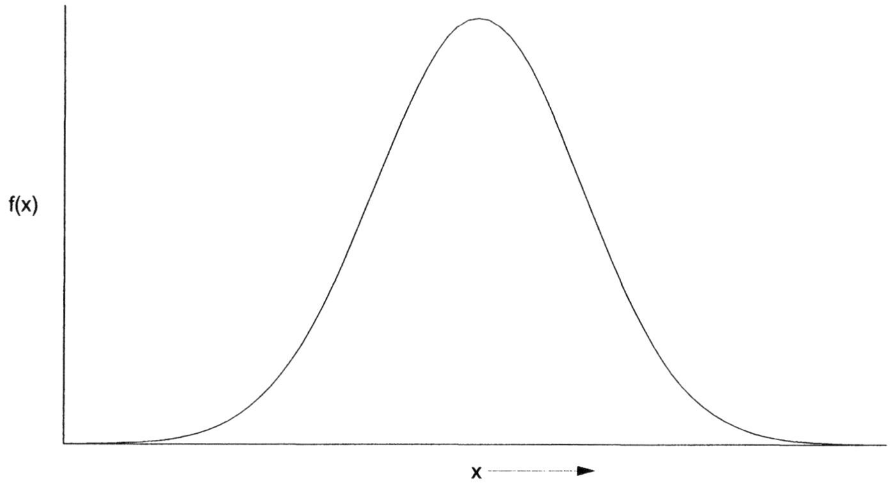

Figure 2.5. The normal probability density function.

The normal probability density function cannot be integrated implicitly. Thus, the reliability and cumulative distribution functions are often expressed as a function of the standard normal distribution. Transformation to the standard normal is achieved with the expression

$$z = \frac{x - \mu}{\sigma} \tag{2.17}$$

The standard normal random deviate, z, has a mean of 0 and a standard deviation of 1.0. The standard normal cumulative distribution function, $\Phi(z)$, is tabled in Appendix A and in virtually all statistical texts. However, these tables are becoming unnecessary, as electronic spreadsheets, such as Lotus™ 123 and Microsoft™ Excel have built-in statistical functions.

The normal reliability function is

$$R(x) = \int_x^{\infty} \frac{1}{\sigma\sqrt{2\pi}} \exp\left[-\frac{1}{2}\left(\frac{\tau-\mu}{\sigma}\right)^2\right] d\tau = 1 - \Phi(z), \; -\infty < x < \infty \tag{2.18}$$

where: $\Phi(z)$ = the standard normal cumulative distribution function (tabulated in Appendix A).

The standard normal reliability function along with the standard normal cumulative distribution function are shown in Figure 2.6.

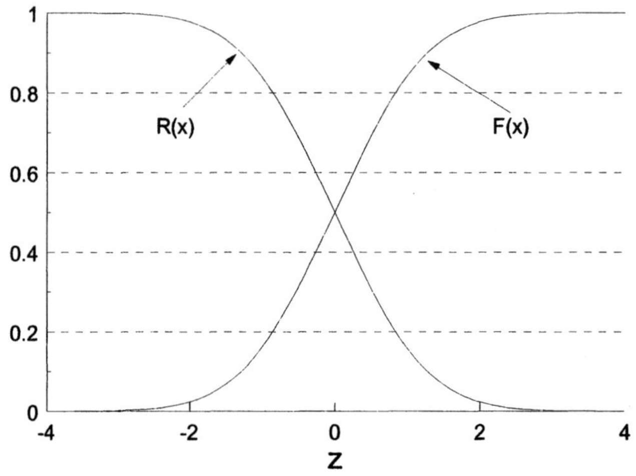

Figure 2.6. The standard normal reliability and cumulative distribution functions.

Example 2.3

A manufacturer produces metal beams with a mean strength of 40 units and a variance of 36. If the strength follows a normal distribution, what is the probability of a particular beam having a strength (a) greater than 52 units, and (b) less than 43 units?

Solution

Part a—the solution is the area under the normal probability density function to the right of $\boldsymbol{x} = 52$.

$$R(52) = \int_{52}^{\infty} \frac{1}{\sqrt{36}\sqrt{2\pi}} \exp\left[-\frac{1}{2}\left(\frac{x-40}{\sqrt{36}}\right)^2\right] dx$$

Transforming to standard normal gives

$$z = \frac{x-\mu}{\sigma} = \frac{52-40}{\sqrt{36}} = 2$$

From Appendix A, the area under the standard normal curve from $z = -\infty$ to $z = 2$ is $\Phi(2) = 0.9772$. If using Lotus 123, this value is found with the function @Normal(2,0,1,0). The first 0 indicates the mean of the distribution is zero, and the 1 indicates the standard deviation of the distribution is one, and the second 0 signifies the cumulative distribution is desired. In Microsoft™ Excel, this function is =NORMDIST(2,0,1,true). Like 123, the 0 and 1 are the mean and the standard deviation of the distribution. The true indicates the cumulative distribution function is desired. If false is entered, the probability density function is returned. If using Lotus 123 or Excel, transformation to standard normal is not required; simply enter the actual mean and standard deviation in the function.

R(52) = 1–0.9772 = 0.0228.

Part b—the solution is the area under the normal probability density function to the left of $\boldsymbol{x} = 43$.

$$P(\boldsymbol{x}<43) = 1 - R(43) = \int_{43}^{\infty} \frac{1}{\sqrt{36}\sqrt{2\pi}} \exp\left[-\frac{1}{2}\left(\frac{x-40}{\sqrt{36}}\right)^2\right] dx$$

Transforming to standard normal gives

$$z = \frac{43-40}{\sqrt{36}} = 0.5$$

From Appendix A, the area under the standard normal curve from $z = -\infty$ to $z = 0.5$ is Φ (0.5) = 0.6915. In Excel the solution is =NORMDIST(43,40,36^0.5,TRUE)

The normal hazard function is monotonically increasing, and is defined as

$$h(x) = \frac{f(x)}{R(x)} = \frac{\phi(z)}{\sigma[1-\Phi(z)]}, \quad -\infty < x < \infty \tag{2.19}$$

where: $\phi(z)$ is the standard normal probability density function.

The standard deviation appears in the denominator of Equation 2.19 because $\phi(z)$ represents the standard normal probability density function and has a standard deviation of 1. Figure 2.7 shows the standard normal hazard function.

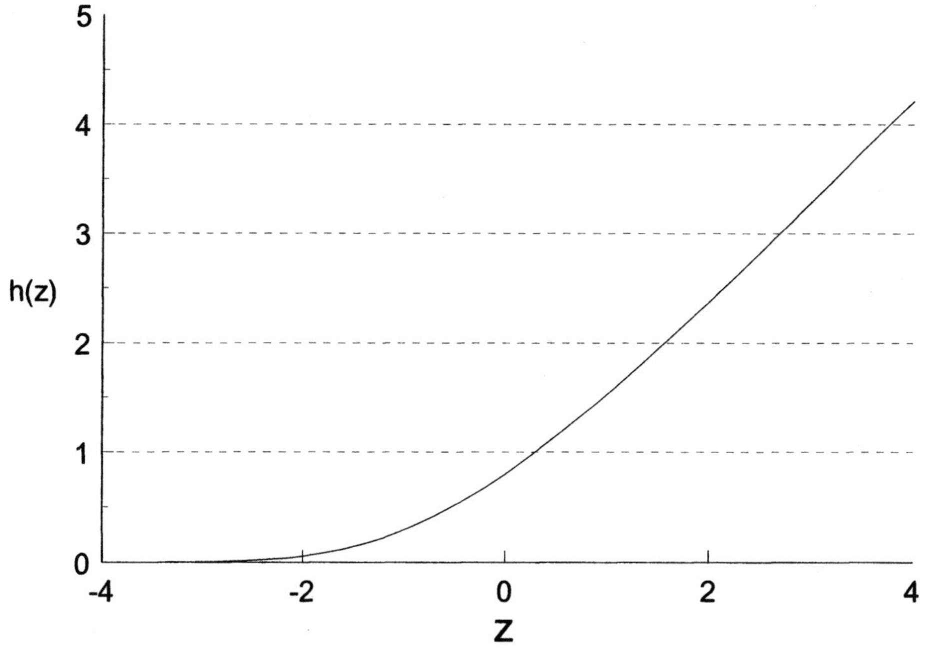

Figure 2.7. The standard normal hazard function.

Example 2.4

Given failure times of 28, 29, 31, 34, and 37 hours, what is the estimated instantaneous failure rate at 27 hours?

Solution

The sample mean is

$$\hat{\mu} = \frac{\sum_{i=1}^{n} x_i}{n} = \frac{28+29+31+34+37}{5} = 31.8$$

The sample standard deviation is

$$\hat{\sigma} = \sqrt{\frac{n\sum_{i=1}^{n} x_i^2 - \left(\sum_{i=1}^{n} x_i\right)^2}{n(n-1)}} = \sqrt{\frac{5(5111)-(159)^2}{5(5-1)}} = 3.701$$

The instantaneous failure rate is given by the hazard function, and is estimated by

$$h(27) = \frac{\phi(z)}{\hat{\sigma}\left[1-\Phi(z)\right]} = \frac{\phi\left(\frac{27-31.8}{3.701}\right)}{3.701\left[1-\Phi\left(\frac{27-31.8}{3.701}\right)\right]} = 0.0515$$

2.2.2 The lognormal distribution

The normal and lognormal distributions are closely related. If x is a lognormal random variable, then the variable $y=\ln(x)$ is a normal random variable. The lognormal probability density function is

$$f(x) = \frac{1}{\sigma x\sqrt{2\pi}} \exp\left[-\frac{1}{2}\left(\frac{\ln x - \mu}{\sigma}\right)^2\right], \; x>0 \tag{2.20}$$

The parameters of the lognormal distribution are the location parameter, μ, and the scale parameter σ. The scale parameter is sometimes referred to as the shape parameter. The location parameter can be estimated from the expression

$$\hat{\mu} = \frac{\sum_{i=1}^{n} \ln x_i}{n} \tag{2.21}$$

where n is the sample size.

The scale parameter is estimated from the expression

$$\hat{\sigma} = \sqrt{\frac{n\sum_{i=1}^{n} \ln x_i^2 - \left(\sum_{i=1}^{n} \ln x_i\right)^2}{n(n-1)}} \tag{2.22}$$

The term T_{50} represents the median of the lognormal distribution, and $\ln(T_{50})$ is interchangeable with μ. The lognormal probability density function is shown in Figure 2.8.

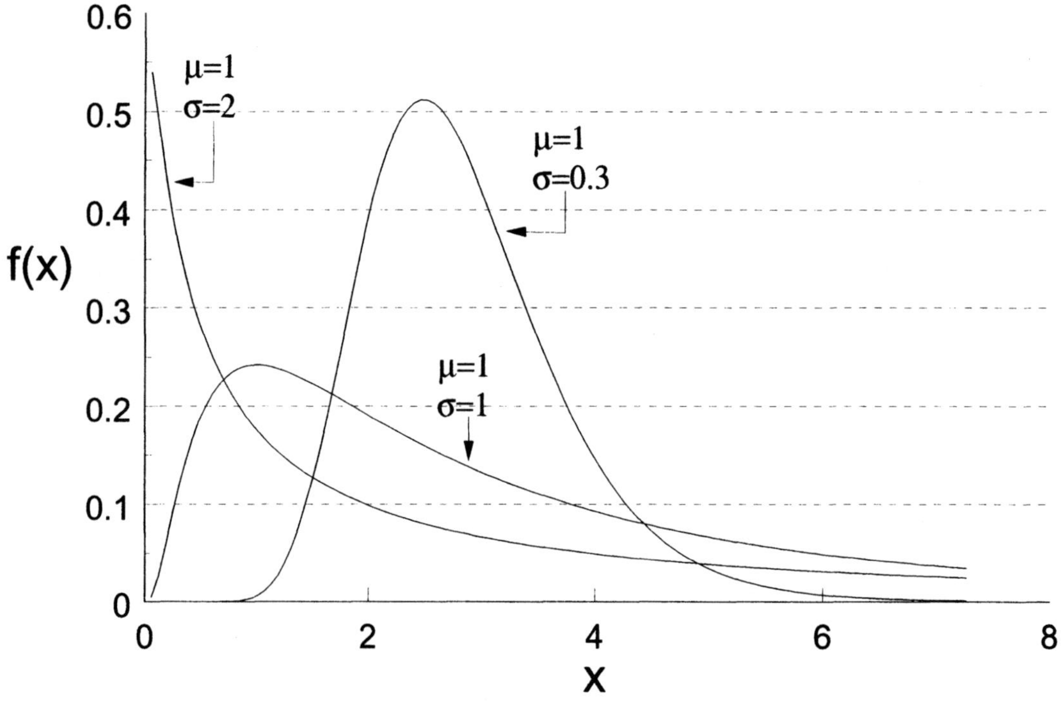

Figure 2.8. Lognormal probability density functions.

The mean of the lognormal distribution is

$$E(x) = \exp\left(\mu + \frac{\sigma^2}{2}\right) = T_{50}e^{\sigma^2/2} \tag{2.23}$$

The variance of the lognormal distribution is

$$V(x) = \left(e^{2\mu+\sigma^2}\right)\left(e^{\sigma^2} - 1\right) = T_{50}e^{\sigma^2}\left(e^{\sigma^2} - 1\right) \tag{2.24}$$

The lognormal reliability function is

$$R(x) = 1 - \Phi\left(\frac{\ln x - \mu}{\sigma}\right), \; x > 0 \tag{2.25}$$

The lognormal reliability function is shown in Figure 2.9.

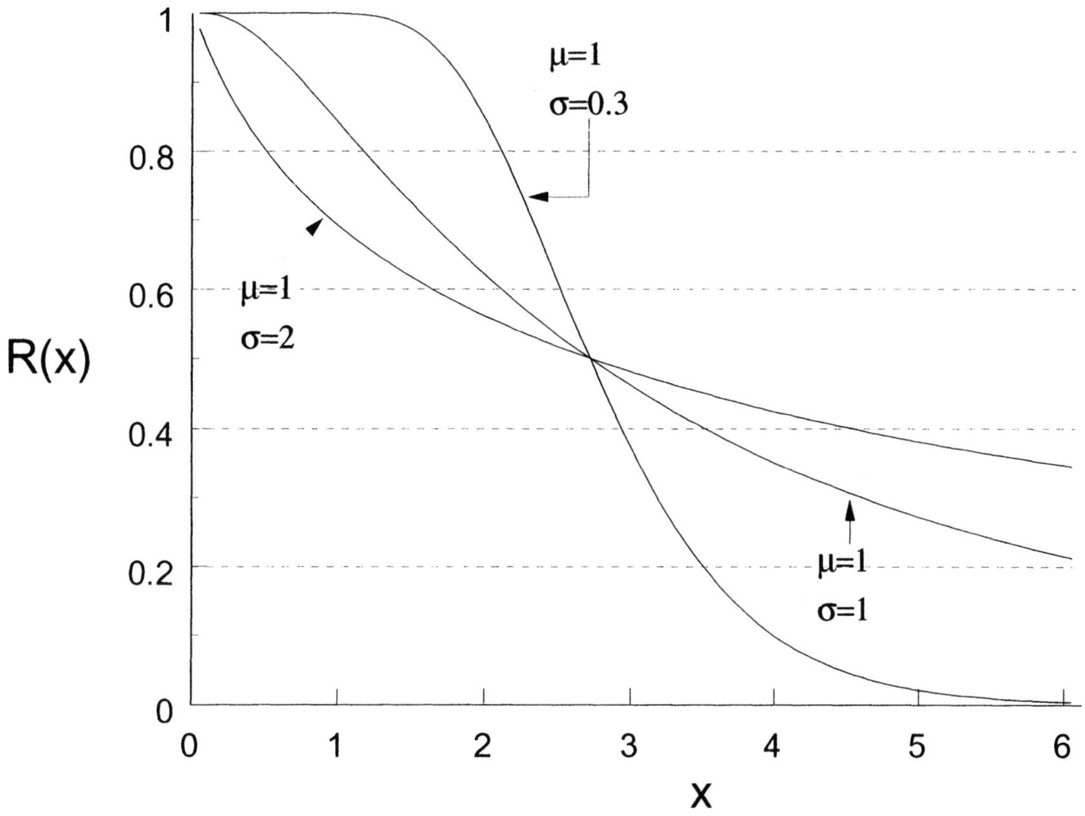

Figure 2.9. Lognormal reliability functions.

The lognormal hazard function is

$$h(x) = \frac{f(x)}{R(x)} = \frac{\phi\left(\frac{\ln x - \mu}{\sigma}\right)}{x\sigma\left[1 - \Phi\left(\frac{\ln x - \mu}{\sigma}\right)\right]}, \; x > 0 \tag{2.26}$$

The lognormal hazard function, regardless of the parameter values, is initially increasing, then decreases and approaches zero as infinity is approached. For large values of σ (σ > 1.5) the hazard function increases so quickly, that it may appear on a graph that it is initially decreasing, but this is not the case. The lognormal hazard function is shown in Figure 2.10.

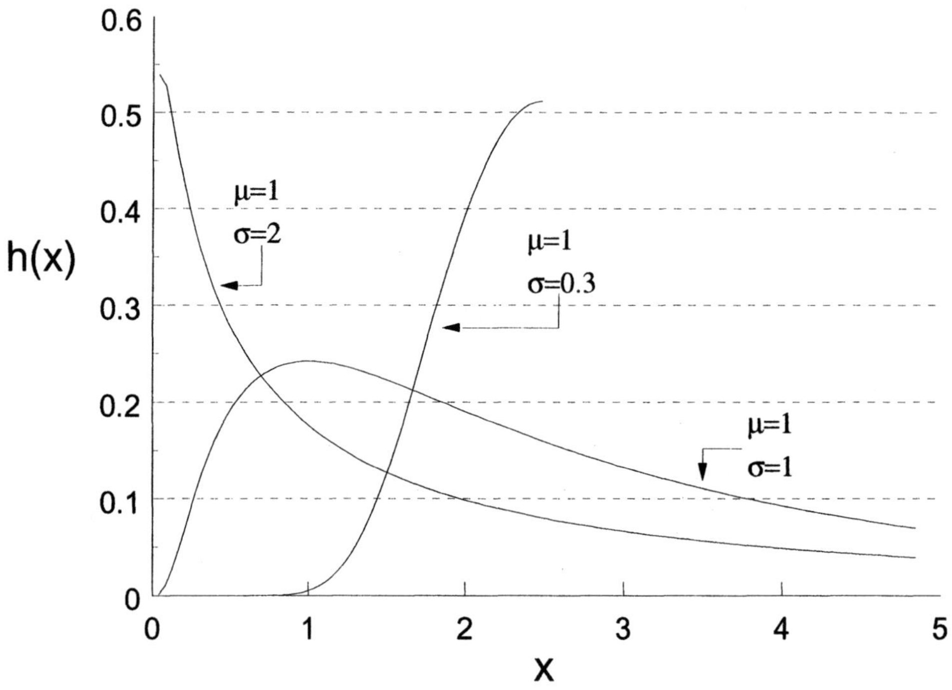

Figure 2.10. Lognormal hazard functions.

Example 2.5

The failure time of a component is lognormal distributed with parameters μ = 4 units and σ = 0.9. When should the component be replaced if the minimum required reliability for the component is 0.85? What is the value of the hazard function at this time?

Solution

Substituting into the lognormal reliability function gives

$$R(x) = 0.85 = 1 - \Phi\left(\frac{\ln x - 4}{0.9}\right)$$

Simplifying,

$$0.15 = \Phi\left(\frac{\ln x - 4}{0.9}\right)$$

From Appendix A, and linearly interpolating (or from an electronic spreadsheet),

$$\Phi(-1.036) = 0.15$$

The time the component must be replaced to ensure reliability of 0.85 is

$$-1.036 = \frac{\ln x - 4}{0.9}$$

$x = \exp[-1.036(0.9)+4] = 21.5$ units.

The component must be replaced after operating for 21.5 units.

The value of the lognormal hazard function at $x = 21.5$ is

$$h(21.5) = \frac{\phi\left(\dfrac{\ln(21.5)-4}{0.9}\right)}{(21.5)(0.9)\left[1-\Phi\left(\dfrac{\ln(21.5)-4}{0.9}\right)\right]} = 0.014 \text{ failures per unit time}.$$

2.3 Exponential, Poisson, binomial, hypergeometric, and geometric distributions

The exponential distribution is a continuous distribution that is related to the Poisson distribution which is discrete. If the number of failures per unit time is Poisson distributed, then the mean time between failures is exponentially distributed. The exponential distribution has a constant failure rate, and is used to model what is termed the "useful life" of many items. That is, the period after burn-in and before wear-out; the bottom portion of the bathtub curve, when the failure rate is constant. The Poisson distribution is used to describe rates, such as defects per computer chip, defects per automobile, failures per hour, etc.

The binomial, hypergeometric and geometric distributions are discrete distributions used to model situations that have only two possible outcomes. The binomial distribution is a Bernoulli process, and thus, has the following properties:

1. the experiment has n trials,
2. each experiment has two possible outcomes—success or failure,
3. the probability of success or failure is constant for all trials, and
4. each trial is independent.

The hypergeometric distribution is the same as the binomial distribution except the probability of success or failure is not constant for all trials. For example in a lot of 100 items with 10% being defective, the probability of selecting a defective item on the first trial is 10%, but the probability of selecting a defective item on the second trial is 10/99 if a defective item was not selected on the first trial and 9/99 if a defective item was selected on the first trial. To simplify, the binomial distribution assumes replacement, the hypergeometric does not. Another viewpoint is that the binomial distribution applies to sampling from an infinitely large population (replacement has no effect if the sample size is infinite), while the hypergeometric assumes a finite population size.

The geometric distribution is a special case of the negative-binomial distribution. It is used to model the number of trials required to achieve the first success. The negative-binomial is the distribution of the number of trials to achieve a given number of successes.

2.3.1 The exponential distribution

The exponential distribution is a one-parameter, continuous distribution. It is commonly expressed in terms of its mean, θ, and the inverse of its mean, λ. The exponential probability density function is

$$f(x)=\frac{1}{\theta}e^{-\frac{x}{\theta}}=\lambda e^{-\lambda x}\ ,\ x\geq 0 \tag{2.27}$$

where: θ = the distribution mean, and

$\lambda = 1/\theta$ = the failure rate.

The exponential probability density function is shown in Figure 2.11.

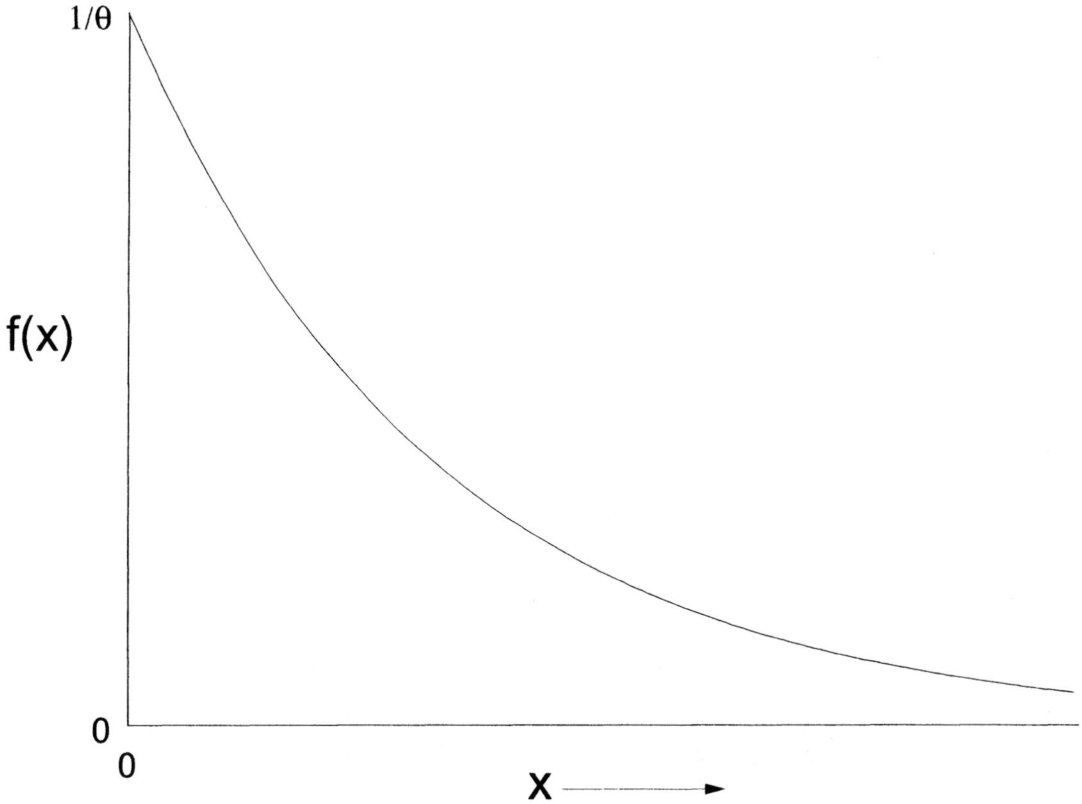

Figure 2.11. The exponential probability density function.

The exponential reliability function is

$$R(x)=e^{-\frac{x}{\theta}}=e^{-\lambda x}\ ,\ x\geq 0 \tag{2.28}$$

This is shown in Figure 2.12.

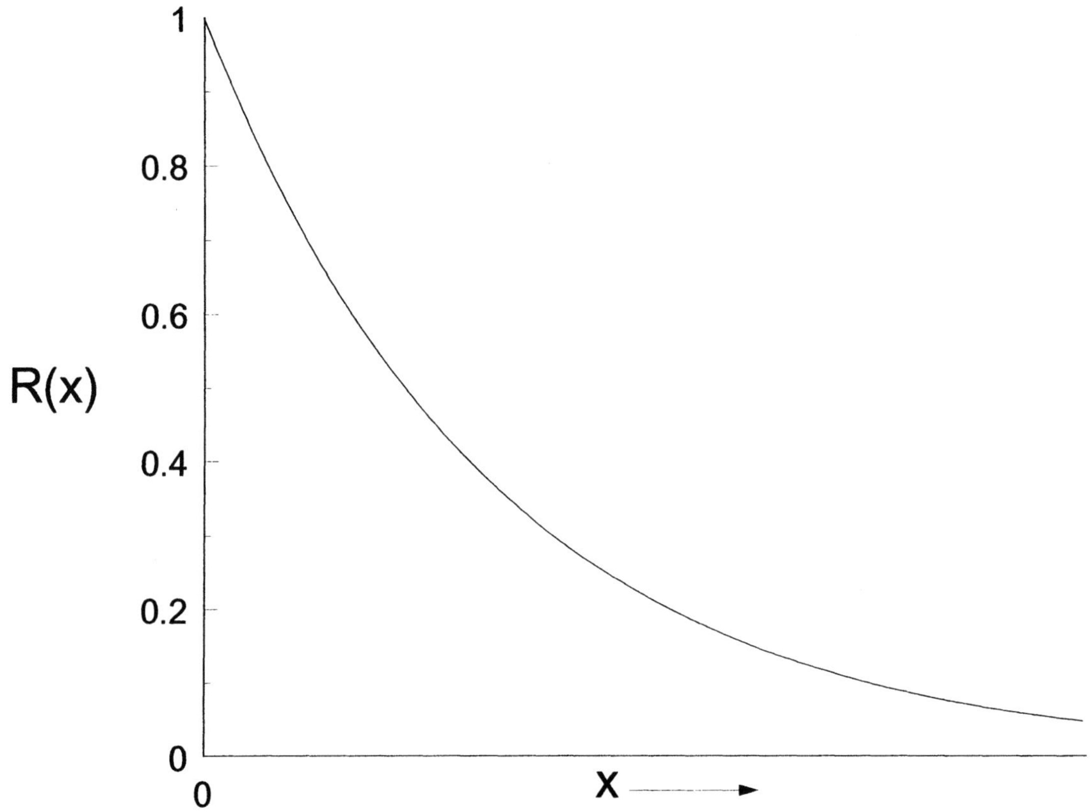

Figure 2.12. The exponential reliability function.

The exponential hazard function is

$$h(x)=\frac{f(x)}{R(x)}=\frac{1}{\theta}=\lambda \tag{2.29}$$

The exponential hazard function is constant, as shown in Figure 2.13. A constant failure rate is unique to the exponential distribution, and is responsible for the "lack of memory" ""property. "Lack of memory" means the probability of failure in a specific time interval is the same regardless of the starting point of that time interval.

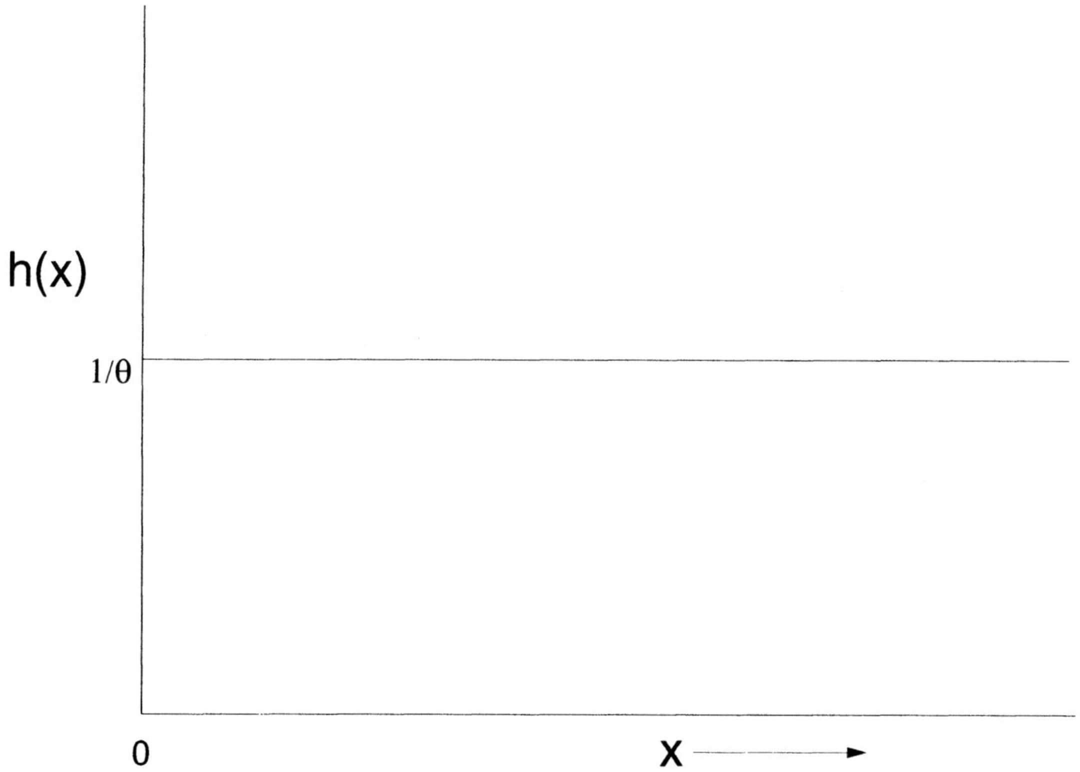

Figure 2.13. The exponential hazard function.

For example, if an item follows the exponential distribution, the probability of failure in the interval $x = 0$ to $x = 20$ is the same as the probability of failure in the interval $x = 100$ to $x = 120$.

The variance of the exponential distribution is

$$V(x) = \theta^2 = \frac{1}{\lambda^2} \tag{2.30}$$

Example 2.6

What is the probability of an item surviving until $t = 100$ units if the item is exponentially distributed with a mean time between failure of 80 units? Given that the item survived to 200 units, what is the probability of survival until $t = 300$ units? What is the value of the hazard function at 200 units, 300 units?

Solution

The probability of survival until $t = 100$ units is

$$R(100) = e^{-\left(\frac{100}{80}\right)} = 0.2865$$

The probability of survival until $t = 300$ units given survival until $t = 200$ units is[1]

$$R(300,200) = \frac{R(300)}{R(200)} = \frac{e^{-300/80}}{e^{-200/80}} = 0.2865$$

Note that this is equal to the probability of failure in the interval from t=0 to t=100. The value of the hazard function is equal to the failure rate and is constant

$$h(t) = \frac{1}{80} = 0.0125$$

2.3.2 The Poisson distribution

The Poisson distribution is used to model occurrence rates, such as defects per item or failures per hour. A Poisson experiment has the following properties:

1. the number of occurrences is independent from period to period—the Poisson process has no memory; and,
2. the probability of an occurrence in a time interval is proportional to the length of the interval.

The Poisson probability density function is

$$P(x,\lambda t) = \frac{e^{-\lambda t}(\lambda t)^x}{x!}, \; x = 0, 1, 2, 3, \ldots, n \tag{2.31}$$

where: $P(x,\lambda t)$ = the probability of exactly x occurrences in the interval λt, and
λ = the average occurrence rate per unit time.

The mean and variance of the Poisson distribution are equal and are

$$E(x) = \lambda t \tag{2.32}$$

$$V(x) = \lambda t \tag{2.33}$$

[1] Details on Bayesian analysis are covered in a later section.

Example 2.7

A computer chip manufacturer averages 7 defects per chip. What is the probability of a chip containing - (a) no defects, (b) 1 defect, (c) less than 3 defects?

Solution

Here, t, the number of computer chips is 1.0.

$\lambda t = 7$, thus, the probability of no defects on a chip is

$$P(0,7) = \frac{e^{-7}(7)^0}{0!} = 0.00091$$

The probability of a chip containing exactly one defect is

$$P(1,7) = \frac{e^{-7}(7)^1}{1!} = 0.00638$$

The probability of less than 3 defects is equal to the sum of exactly 0 defects, exactly 1 defect and exactly 2 defects. The probability of exactly 2 defects is

$$P(2,7) = \frac{e^{-7}(7)^2}{2!} = 0.02234$$

The probability of less than 3 defects is

$$P(x<3,7) = 0.00091 + 0.00638 + 0.02234 = 0.02963$$

Rather than calculating the probability of exactly 0 defects, the probability of exactly 1 defect and the probability of exactly 2 defects, the cumulative Poisson tables in Appendix G could have been utilized. Microsoft Excel contains a function for the Poisson distribution. The accompanying CD contains this example in Excel.

If the sample size is large and the probability of occurrence is near 0, the Poisson distribution may be used to approximate the binomial distribution. This will be discussed further in the next section.

2.3.3 The binomial distribution

The binomial probability density function is

$$P(x;n,p) = \binom{n}{x} p^x (1-p)^{n-x}, x = 0,1,2,3,\ldots,n \tag{2.34}$$

where: $P(x;n,p)$ = the probability of exactly x successes in n independent trials,

n = the sample size or number of trials, and

p = the probability of success on a single trial.

The mean of the binomial distribution is

$$E(x) = np \tag{2.35}$$

The variance of the binomial distribution is

$$V(x) = np(1-p) \tag{2.36}$$

Example 2.8

The probability for successfully launching a certain rocket is 89%. What is the probability of exactly two successful launches in the next three attempts?

Solution

The probability for a successful launch is p = 0.89, and the sample size is three attempts, n = 3. The probability of exactly two successful attempts is

$$P(2;3,0.89) = \binom{3}{2} 0.89^2 (1-0.89)^{(3-2)} = 0.2614$$

Note that the term $\binom{3}{2}$ is equal to $\left(\frac{3!}{2!(3-2)!}\right)$.

Example 2.9

To destroy a military target, at least three missiles must hit the target. The probability of each missile hitting the target is 90%. How many missiles must be launched to ensure, with 99% reliability, that the target will be destroyed?

Solution

The probability of a single missile hitting the target is p = 0.9. If four missiles are launched, n = 4, and the probability of exactly three missiles hitting the target is

$$P(3;4,0.9) = \binom{4}{3} 0.9^3 (1-0.9)^{(4-3)} = 0.2916$$

The probability of exactly four missiles hitting the target is

$$P(4;4,0.9) = \binom{4}{4} 0.9^4 (1-0.9)^{(4-4)} = 0.6561$$

The probability of at least three missiles hitting the target if four are launched is 0.9477 (0.2916 + 0.6561). Thus, launching four missiles does not meet the required reliability, 99%. If five missiles are launched, the probability of exactly three missiles hitting the target is

$$P(3;5,0.9)=\binom{5}{3}0.9^3(1-0.9)^{(5-3)}=0.0729$$

The probability of exactly four missiles hitting the target is

$$P(4;5,0.9)=\binom{5}{4}0.9^4(1-0.9)^{(5-4)}=0.3281$$

The probability of exactly five missiles hitting the target is

$$P(5;5,0.9)=\binom{5}{5}0.9^5(1-0.9)^{(5-5)}=0.5905$$

The probability of at least three missiles hitting the target if five are launched is 0.9915 (0.0729 + 0.3281 + 0.5905). Launching five missiles meets the objective of 99% reliability.

If p is small and n is large, the Poisson distribution can be used to approximate the binomial distribution by replacing λt with np. Note that if p is large the Poisson approximation can also be used by substituting $1-p$ for p and subtracting the resulting probability from 1.0. This approximation is no longer necessary. Electronic spreadsheets include Poisson distribution, and there is no need to approximate a distribution when exact values can be easily computed. See the example above solved using Microsoft Excel on the accompanying CD.

Example 2.10

A transistor manufacturer has a failure rate of 3 per 10,000. What is the probability of finding exactly one defect in a random sample of 200 transistors?

Solution

This is a binomial experiment with $n = 200$ and $p = 0.0003$. Using a Poisson approximation, $\lambda t = (200)(0.0003) = 0.06$. The probability of finding exactly one defect in the random sample is

$$P(1,0.06)=\frac{e^{-0.06}(0.06)^1}{1!}=0.0565$$

Note that it is not necessary for the value of λt to be an integer for the approximation to be valid.

2.3.4 The hypergeometric distribution

The hypergeometric distribution differs from the binomial distribution in that the random sample of n items is selected from a finite population of N items. With the hypergeometric distribution, there is no replacement. If N is large in respect to n (N>10n), the binomial distribution is a good approximation to the hypergeometric distribution. The hypergeometric probability density function is

$$P(x;N,n,k)=\frac{\binom{k}{x}\binom{N-k}{n-x}}{\binom{N}{n}}, \ x = 0, 1, 2, 3, \ldots, n. \tag{2.37}$$

where: $P(x;N,n,k)$ = the probability of exactly x successes in the sample size n, and k = the number of successes in the population, N.

The mean of the hypergeometric distribution is

$$E(x)=\frac{nk}{N} \tag{2.38}$$

The variance of the hypergeometric distribution is

$$V(x)=\left(\frac{nk}{N}\right)\left(1-\frac{k}{N}\right)\left(\frac{N-n}{N-1}\right) \tag{2.39}$$

Example 2.11

A manufacturer of televisions produced a lot of 20 sets in which 5 of the sets were shipped with incorrect documentation. If a distributor receives 10 televisions from this lot, what is the probability of the distributor receiving exactly 3 televisions with incorrect documentation?

Solution

The population size is N = 20, the sample size is n = 10, and the number of successes (a television shipped without documentation is defined as a success in this case) is k = 5. The probability of the distributor receiving exactly 3 television sets without documentation is

$$P(3;20,10,5)=\frac{\binom{5}{3}\binom{20-5}{10-3}}{\binom{20}{10}}=0.3483$$

2.3.5 The geometric distribution

The geometric distribution is used to model the number of trials required to obtain the first success. For the binomial and hypergeometric distributions the number of trials is fixed, and the random variable is the number of successes. For the geometric distribution the number of successes is fixed at one, and the random variable is the number of trials needed to obtain the success. The geometric probability density function is

$$P(x;p) = p(1-p)^{x-1}, \ x = 1, 2, 3, \ldots, n \tag{2.40}$$

where: $P(x;p)$ = the probability of obtaining the first success on trial number x, and

p = the probability of success on a single trial—all trials are independent.

The mean of the geometric distribution is

$$E(x) = \frac{1}{p} \tag{2.41}$$

The variance of the geometric distribution is

$$V(x) = \frac{1-p}{p^2} \tag{2.42}$$

The probability that the number of trials to achieve the first success is greater than a specified number, n is

$$p(x>n) = (1-p)^n \tag{2.43}$$

If n is large and p is small,

$$p(x>n) \approx e^{-np} \tag{2.44}$$

Example 2.12

A complex electronic system has self diagnostics and does a self check before initializing. On average, the system performs 1.8 self checks before initializing. If each self check is independent, what is the probability that the system will initialize on the third attempt?

Solution

The probability of success per each attempt is

p=1/1.8 = 0.5556

The probability of the first success on the third attempt is

$$P(3;0.5556) = 0.5556(1-0.5556)^{(3-1)} = 0.1097.$$

2.4 The Weibull distribution

The Weibull distribution is a continuous distribution that was publicized by Waloddi Weibull in 1951. Although initially met with skepticism, it has become widely used, especially in the reliability field. The Weibull distribution's popularity resulted from its ability to be used with small sample sizes and its flexibility. In addition to being the most useful density function for reliability calculations, analysis of the Weibull distribution provides the information needed for troubleshooting, classifying failure types, scheduling preventive maintenance and scheduling inspections. The Weibull probability density function is

$$f(x)=\frac{\beta x^{\beta-1}}{\theta^{\beta}}\exp\left[-\left(\frac{x}{\theta}\right)^{\beta}\right],\ x\geq 0 \tag{2.45}$$

where: β = the shape parameter, and
θ = the scale parameter.

Beta and θ are continuous. The acceptable ranges for these variables are $0 < \beta < \infty$, and $0 < \theta < \infty$.

Typically, β is between 0.5 and 8. As β increases, the mean of the Weibull distribution approaches θ and the variance approaches zero. Theta is also known as the characteristic life because 63.2% of the population fails by the characteristic life point regardless of the value of β. The estimation of β and θ is not straightforward; special techniques such as probability plotting, hazard plotting, or maximum likelihood estimation are required.

2.4.1 Effects of the shape parameter

By altering the shape parameter, β, the Weibull probability density function takes a variety of shapes. This is demonstrated in Figure 2.14.

Note that several of the probability density functions displayed in Figure 2.14 look familiar. The Weibull distribution can be used in a wide variety of situations and dependent on the value of β, is equal to or can approximate several other distributions. For example, if:

β = 1, the Weibull distribution is identical to the exponential distribution

β = 2, the Weibull distribution is identical to the Rayleigh distribution

β = 2.5, the Weibull distribution approximates the lognormal distribution

β = 3.6, the Weibull distribution approximates the normal distribution

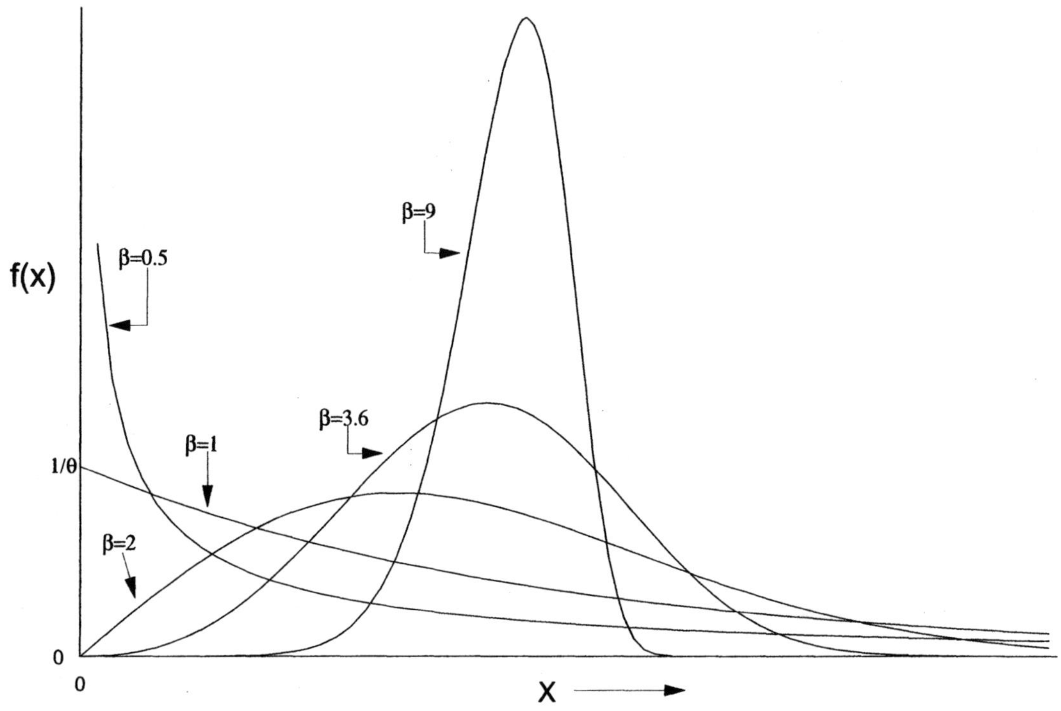

Figure 2.14. Weibull probability density functions.

Because of this flexibility, there are few observed failure rates that cannot be reasonably modeled by the Weibull distribution. Some specific uses have been:

- The breaking strength of components or the stress required to fatigue metals
- The time to fail for electronic components
- The time to fail for items that wear out, such as automobile tires
- Systems that fail when the weakest component in the system fails

2.4.2 Mean and variance of the Weibull distribution

The mean of the Weibull distribution is

$$E(x) = \theta\Gamma\left(1+\frac{1}{\beta}\right) \tag{2.46}$$

where: $\Gamma(x)$ is the Gamma function and is tabulated in Appendix B.

The variance of the Weibull distribution is

$$V(x) = \theta^2\left[\Gamma\left(1+\frac{2}{\beta}\right)-\Gamma^2\left(1+\frac{1}{\beta}\right)\right] \tag{2.47}$$

If $\beta < 1$, the mean of the Weibull distribution is greater than θ. If $\beta = 1$, the mean of the Weibull distribution is equal to θ. In the case, $\beta > 1$, the mean of the Weibull distribution is less than θ and approaches θ as x increases. The variance of the Weibull distribution decreases and approaches zero as β increases to infinity.

Example 2.13

Determine the mean and variance for a system having a Weibull time to fail distribution with β = 2 and θ = 100.

Solution

From Appendix B

$$\Gamma\left(1+\frac{1}{2}\right)=\Gamma(1.5)=0.88623$$

The mean of this distribution is

$$E(x)=(100)\Gamma\left(1+\frac{1}{2}\right)=100(0.88623)=88.6$$

The variance of this distribution is

$$V(x)=(100)^2\left[\Gamma\left(1+\frac{2}{2}\right)-\Gamma^2\left(1+\frac{1}{2}\right)\right]=100^2\left[\Gamma(2)-\Gamma^2(1.5)\right]=2146$$

2.4.3 The Weibull reliability and hazard functions

The Weibull reliability function is

$$R(x)=\exp\left[-\left(\frac{x}{\theta}\right)^{\beta}\right], x>0 \tag{2.48}$$

Figure 2.15 illustrates the effect of β on the reliability function.

For the case of β < 1, the reliability sharply decreases initially and then flattens out. This is the result of infant mortality failures. For the case of β = 1, the reliability gradually decreases; the result of a constant failure rate. When β > 1, the reliability slowly decreases initially and then decreases as the characteristic life is approached.

The Weibull hazard function is

$$h(x)=\frac{\beta x^{(\beta-1)}}{\theta^{\beta}}, x>0 \tag{2.49}$$

The effects of β on the hazard function are demonstrated in Figure 2.16.

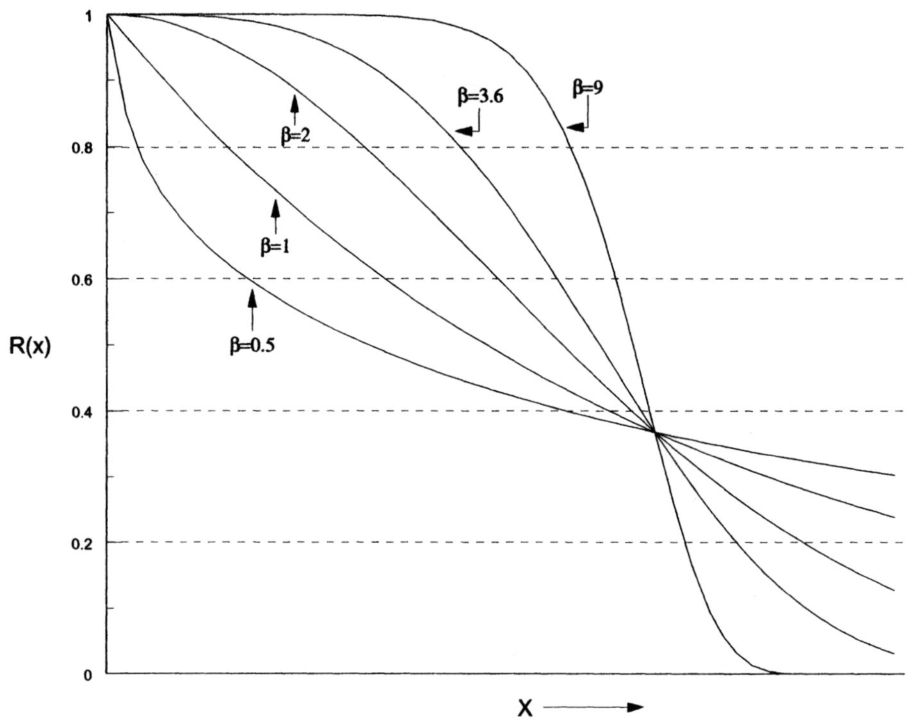

Figure 2.15. Weibull reliability functions.

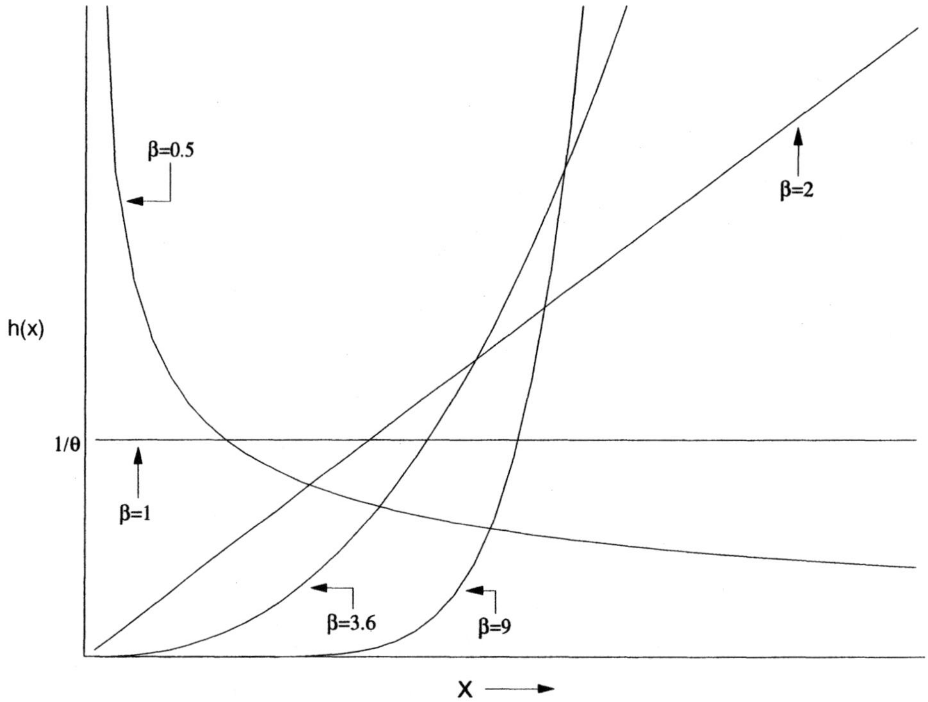

Figure 2.16. Weibull hazard functions.

A shape factor of less than one results in a decreasing failure rate. If $\beta = 1$ the failure rate is constant at a value of $1/\theta$. If $\beta > 1$ the failure rate is increasing; the higher the value of β, the faster the failure rate is increasing.

Example 2.14

What is the probability of survival until time = 40 for a system with a Weibull time to fail distribution having parameters $\beta = 1.8$ and $\theta = 115$? What is the instantaneous failure rate at this time?

Solution

The reliability at time = 40 is

$$R(40) = \exp\left[-\left(\frac{40}{115}\right)^{1.8}\right] = 0.8612$$

The value of the hazard function at time = 40 is

$$h(40) = \frac{(1.8)\left(40^{(1.8-1)}\right)}{115^{1.8}} = 0.0067$$

2.4.4 Relationship to the extreme value distribution

Consider a system consisting of n individual components in series. If one of the components fails, the system fails. In this situation the time to fail for the system is equal to the smallest of the n failure times. This situation may be modeled using extreme value distributions.

The type III asymptotic distribution for the minima is simply the Weibull distribution. The type I asymptotic distribution for maxima and minima, also known as the extreme value distribution (EVD), is closely related to the Weibull distribution.

The EVD reliability function is

$$R(x) = \exp\left[-\exp\left(\frac{x-\delta}{\alpha}\right)\right], \ -\infty < x < \infty \tag{2.50}$$

where: δ = the location parameter, and

α = the scale parameter.

If data that follows a Weibull distribution is transformed by taking the natural logarithm, the result is data that follows the EVD. If $x = \ln(t)$, where t follows a Weibull distribution, x follows the EVD. The parameters of the Weibull distribution and the EVD have the following relationships.

$$\delta = \ln(\theta) \tag{2.51}$$

$$\alpha = \frac{1}{\beta} \tag{2.52}$$

2.5 The gamma and beta distributions

The gamma distribution is a continuous distribution that describes random variables that are bounded at one end. The gamma distribution is often used to model the lifetime of systems. If an event takes place after *n* exponentially distributed events take place sequentially, the resulting random variable follows a gamma distribution. Mathematically, if

$$y_n = x_1 + x_2 + ... + x_n$$

and x_1, x_2, …, x_n are exponentially distributed with identical failure rates, then y_n follows the gamma distribution. Some examples are:

- The time to fail for a system consisting of *n* independent components, with *n-1* components being standby components, where system failure occurs when all *n* components fail
- The time between maintenance actions for a system that requires maintenance after a fixed number of uses

The beta distribution is a continuous distribution that describes the ratio $y_1/(y_1 + y_2)$, where y_1 and y_2 are independent variables that follow the gamma distribution. It is bounded on both ends, and is used to describe random variables constrained to an interval. Because of this, the beta distribution is useful for determining tolerance intervals and for modeling the completion time of tasks in PERT (project evaluation and review technique) simulations.

2.5.1 The gamma distribution

The gamma probability density function is

$$f(x) = \frac{\lambda^\eta}{\Gamma(\eta)} x^{(\eta-1)} e^{-\lambda x}, \quad x \geq 0, \eta \geq 0, \lambda \geq 0 \tag{2.53}$$

where: η = the shape parameter,
λ = the scale parameter, and
$\Gamma(x)$ = the gamma function and is tabled in Appendix B.

The mean of the gamma distribution is

$$E(x) = \frac{\eta}{\lambda} \tag{2.54}$$

The variance of the gamma distribution is

$$V(x) = \frac{\eta}{\lambda^2} \tag{2.55}$$

The parameters of the gamma distribution can be estimated from the expressions

$$\hat{\lambda} = \frac{\bar{x}}{s^2} \tag{2.56}$$

$$\hat{\eta} = \hat{\lambda}\bar{x} \tag{2.57}$$

where: s^2 = the sample variance and

$\bar{x}$ = the sample mean.

The above equations are based on the method of matching moments. Maximum likelihood estimates are more accurate, but are more complicated.

As seen in Figure 2.17, the gamma probability density function is versatile. When $\eta = 1$, the gamma distribution reduces to the exponential distribution. When η is a positive integer, the gamma distribution is an Erlang distribution, a distribution used in queuing theory. If $\eta = 2$, then the gamma distribution becomes the chi-square distribution with $\nu = 2\eta$ degrees of freedom. Also, the gamma distribution becomes normal as η becomes large.

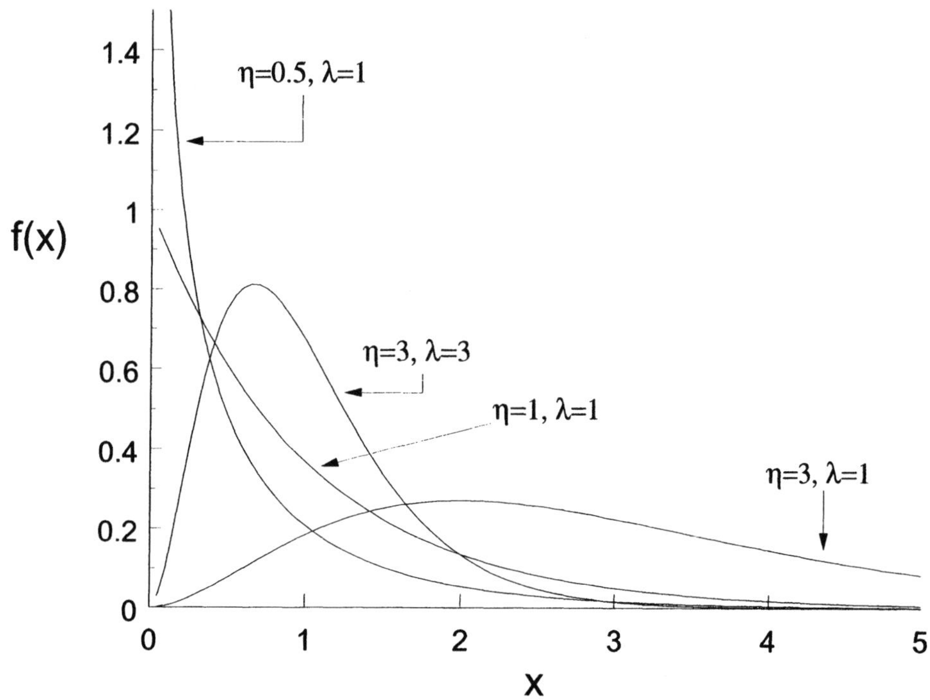

Figure 2.17. Gamma probability density functions.

The gamma reliability function does not exist in closed form unless η is restricted to integer values. For this special case, the gamma reliability function is

$$R(x)=\sum_{k=0}^{\eta-1}\frac{(\lambda x)^k e^{-\lambda x}}{k!}\quad, x\geq 0 \tag{2.58}$$

The gamma reliability function is shown in Figure 2.18 for $\lambda = 1$.

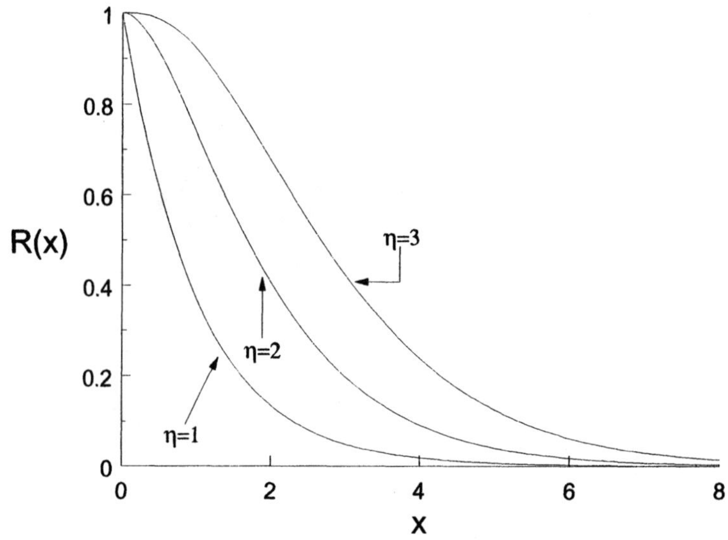

Figure 2.18. Gamma reliability functions.

The gamma hazard function does not exist in closed form unless η is restricted to integer values. For this special case, the gamma hazard function is

$$h(x)=\frac{\dfrac{\lambda^\eta}{\Gamma(\eta)}x^{(\eta-1)}e^{-\lambda x}}{\displaystyle\sum_{k=0}^{\eta-1}\frac{(\lambda x)^k e^{-\lambda x}}{k!}}\quad, x\geq 0 \tag{2.59}$$

The gamma hazard function is shown in Figure 2.19 for $\lambda = 1$.

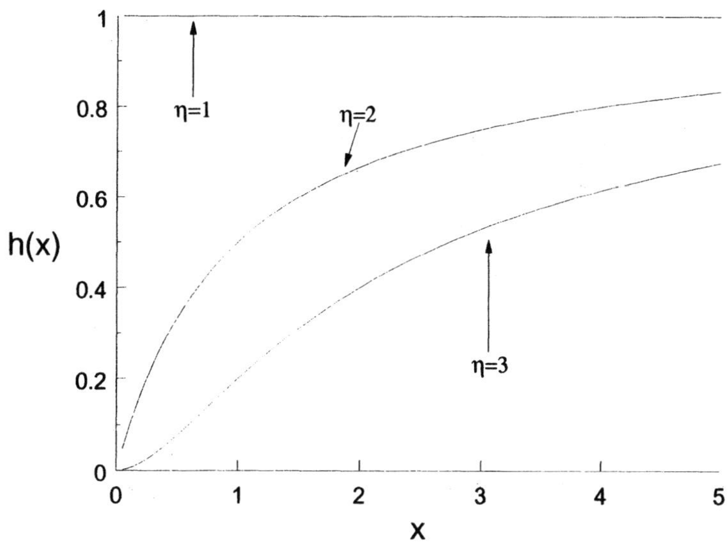

Figure 2.19. Gamma hazard functions.

Example 2.15

The time to fail for a system is gamma distributed with a shape parameter of 2 and a scale parameter of 0.03. What is the probability of survival until $x = 50$ units? What is the value of the hazard function at this time?

Solution

Since η is an integer, the reliability at $x = 50$ units is

$$R(50) = \sum_{k=0}^{2-1} \frac{[0.03(50)]^k \, e^{-0.03(50)}}{k!} = 0.5578$$

The value of the hazard function at $x = 50$ units is

$$h(50) = \frac{\dfrac{(0.03)^2}{\Gamma(2)} 50^{(2-1)} e^{-(0.03)50}}{0.5578} = 0.1800$$

Example 2.16

A system fails when there are no resistors available. The system is designed with 3 resistors, 2 being standby redundant (the system also has perfect switching). If the resistors have a constant failure rate of 0.0001 hours, what is the probability of system survival to a time of 12,800 hours?

Solution

The resistors have a constant failure rate, thus, they are exponentially distributed. Since the resistors are standby redundant, the time to fail for the system is the sum of the individual failure times for the 3 resistors. The time to fail for this system is described by a gamma distribution with $\lambda = 0.0001$ and $\eta = 3$. The reliability at $t =$ 12,800 hours is

$$R(12,800) = \sum_{k=0}^{3-1} \frac{[0.0001(12,800)]^k e^{-0.0001(12,800)}}{k!} = 0.862$$

2.5.2 The beta distribution

The standard beta probability density function is

$$f(x) = \frac{1}{B(p,q)} x^{p-1}(1-x)^{q-1}, 0 \le x \le 1, p > 0, q > 0 \tag{2.60}$$

where: p and q are shape parameters, and,

$B(p,q)$ is a beta function, and is defined as

$$B(p,q) = \int_0^1 x^{p-1}(1-x)^{q-1} dx = \frac{\Gamma(p)\Gamma(q)}{\Gamma(p+q)} \tag{2.61}$$

The beta probability density function is shown in Figure 2.20.

As seen from Figure 2.20, two special cases of the beta distribution are the uniform distribution and the triangular distribution. The beta distribution is quasi-symmetrical; $f(x;p,q)$ is identical to $[1-f(x;q,p)]$. This is shown in Figure 2.21.

The beta distribution is not limited to the range 0 to 1. If x follows the standard beta distribution, then $y = a + bx$ follows the beta distribution with identical parameters as x, but the range is (a,b).

The mean of the beta distribution is

$$E(x) = \frac{p}{(p+q)} \tag{2.62}$$

The variance of the beta distribution is

$$V(x) = \frac{pq}{(p+q)^2(p+q+1)} \tag{2.63}$$

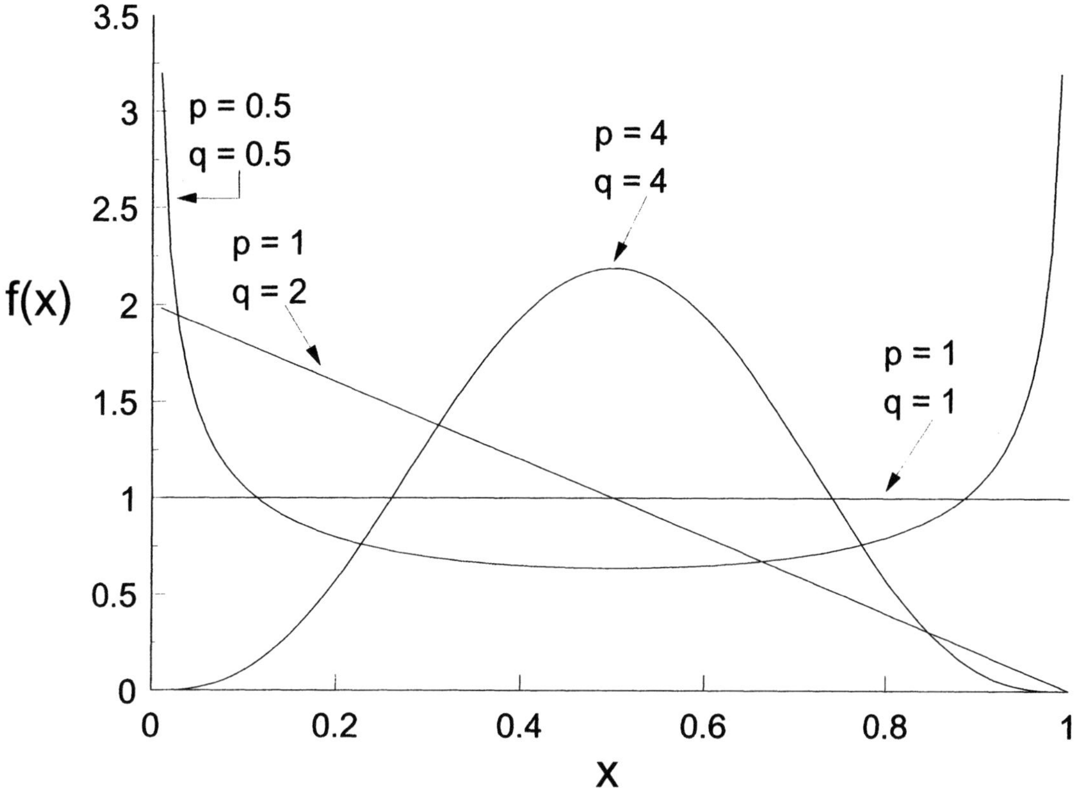

Figure 2.20. Beta probability density functions.

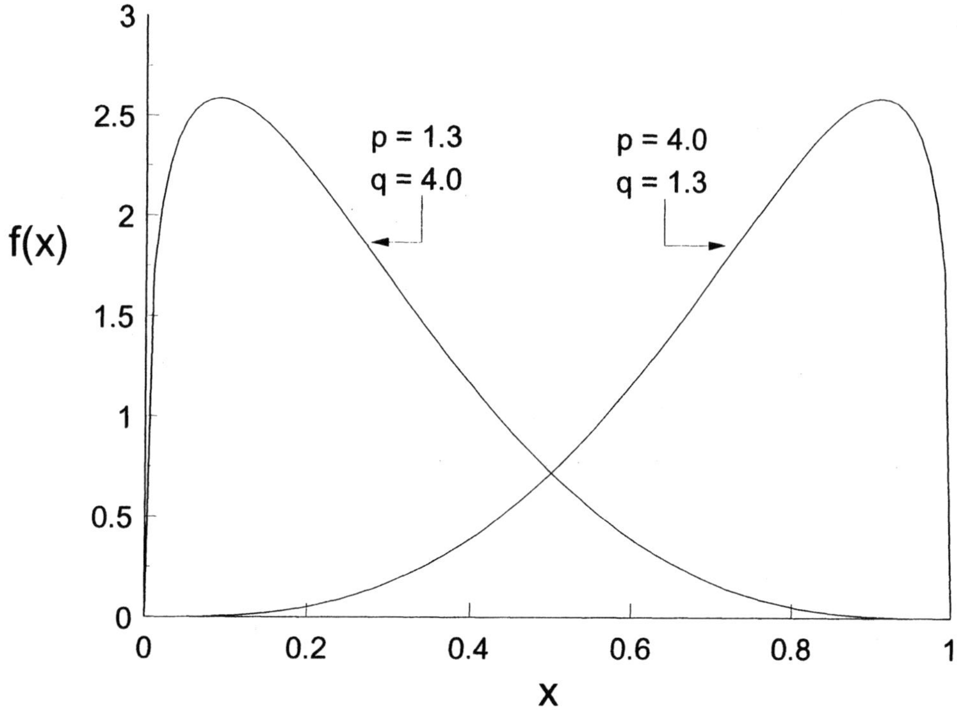

Figure 2.21. Quasi-symmetry of the beta distribution.

Like the gamma distribution, maximum likelihood estimates are the most accurate method of parameter estimation, but are difficult to work with. The moment estimators given below match the accuracy of maximum likelihood estimators if the sample size is large.

$$\hat{p} = \frac{\bar{x}\hat{q}}{1-\bar{x}} \tag{2.64}$$

$$\hat{q} = \frac{(1-\bar{x})}{s^2}\left[\bar{x}(1-\bar{x}) - s^2\right] \tag{2.65}$$

where: s^2 = the sample variance and

$\bar{x}$ = the sample mean.

The beta reliability function is

$$R(x) = 1 - \frac{\Gamma(p+q)}{\Gamma(p)\Gamma(q)}\int_0^x \tau^{p-1}(1-\tau)^{q-1}\,d\tau = 1 - I_x(p,q)\ ,\ 0 \le x \le 1 \tag{2.66}$$

where $I_x(p,q)$ is the incomplete beta function which is related to the binomial distribution. The incomplete beta function can be read from tables, but it is much easier to use functions contained in electronic spreadsheets (@betai if using Lotus™ 123).

$$I_x(p,q) = I_x(p,n-p+1) = \sum_{i=p}^{n}\binom{n}{i}x^i(1-x)^{n-i} \tag{2.67}$$

The beta reliability function is shown in Figure 2.22. For the special case when both shape parameters are equal to one, the reliability function is a straight line with slope = 1.

The beta hazard function is

$$h(x) = \frac{\dfrac{1}{B(p,q)}x^{p-1}(1-x)^{q-1}}{1-I_x(p,q)}\ ,\ 0 \le x \le 1 \tag{2.68}$$

The beta hazard function is shown in Figure 2.23. In general, the hazard function increases drastically as x approaches 1.

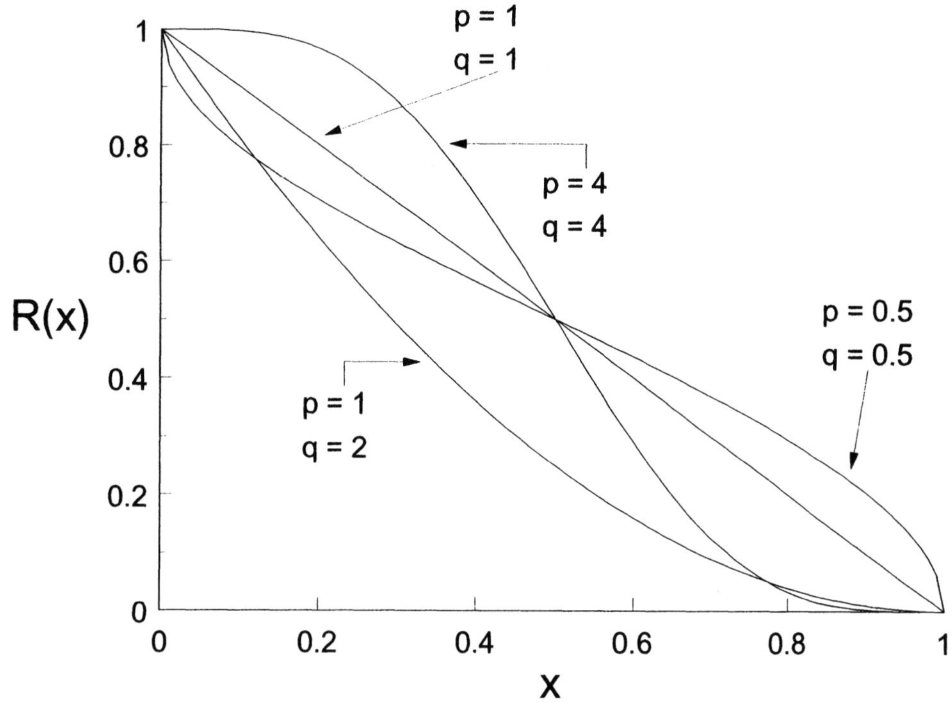

Figure 2.22. Beta reliability functions.

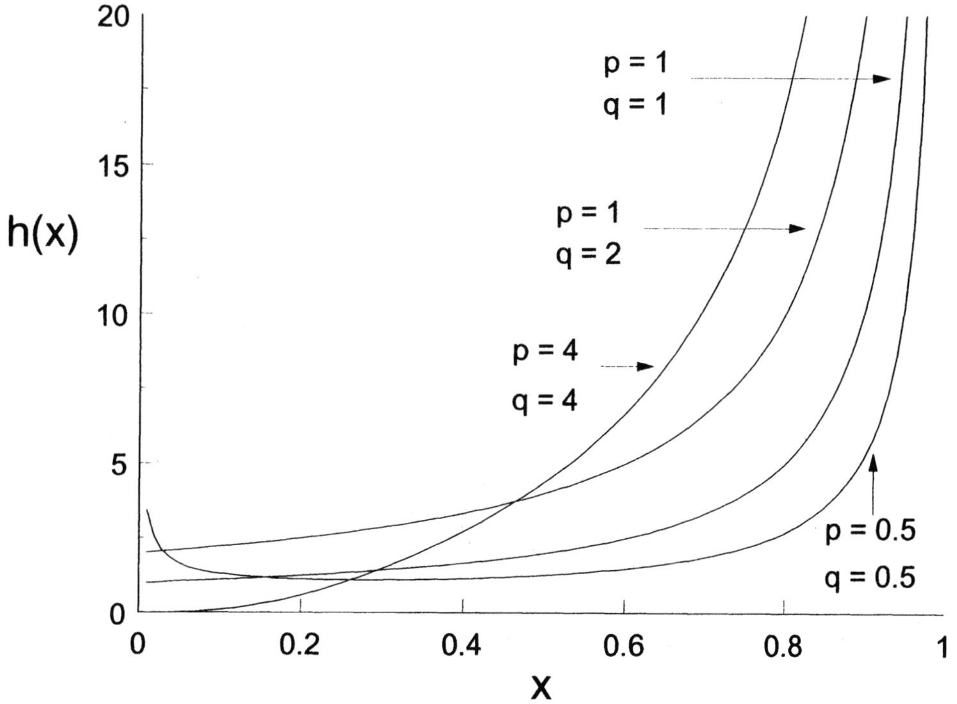

Figure 2.23. Beta hazard functions.

Example 2.17

The strength of a radio signal has been shown to follow a beta distribution over the range 10 units to 50 units with parameters p = 1.8 and q =2.5. What is the probability that at any given time the strength of the signal is greater than 38 units?

Solution

Transforming to a standardized beta distribution gives

$$y = \frac{x-10}{40}$$

For x = 38,

$$y = \frac{38-10}{40} = 0.7$$

The reliability at x = 38 is

$$R(0.7) = 1 - I_{0.7}(1.8, 2.5) = 1 - 0.8838 = 0.1162$$

2.6 Chi-square, *F*, and *t*-distributions

The chi-square, *F*, and *t*-distributions are primarily used as sampling distributions, rather than to model lifetimes. Among other things, these distributions are used to:

- Count the number of failures in an interval
- Determine goodness of fit
- Determine confidence limits
- Draw inferences concerning population means
- Draw inferences concerning population variances

2.6.1 The chi-square distribution

If $y_1, y_2, \ldots, y_v$ are independent, standard normally distributed variables, then

$$\chi^2 = y^2_1 + y^2_2 + \ldots + y^2_v$$

is chi-square distributed with v degrees of freedom (the number of independent variables whose squares are being added). The chi-square distribution is a special case of the gamma distribution with λ = 2 and η = 2/v. Since the number of failures in a given interval is an integer by definition, the chi-square distribution may be used to model this phenomenon. This leads to applications involving goodness of fit and the determination of confidence limits.

The chi-square probability density function is

$$f(x) = \frac{1}{2^{\nu/2}\Gamma(\nu/2)} x^{(\nu/2-1)} e^{-x/2} \; , x > 0 \tag{2.69}$$

The shape of the chi-square distribution is shown in Figure 2.24.

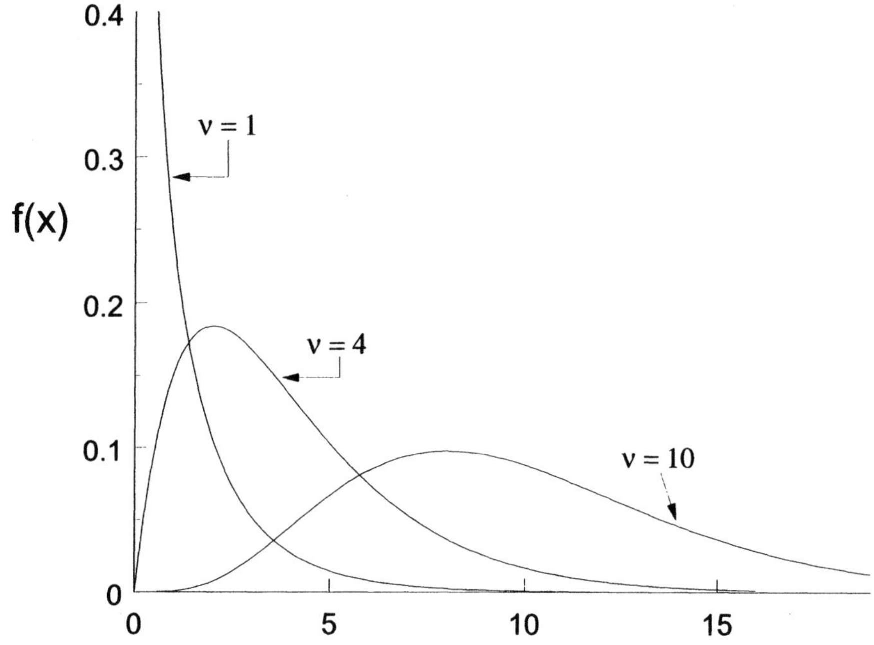

Figure 2.24. Chi-square probability density functions.

The mean of the chi-square distribution is

$$E(x) = \nu \tag{2.70}$$

The variance of the chi-square distribution is

$$V(x) = 2\nu \tag{2.71}$$

The chi-square distribution has several useful properties:

- The sum of two or more independent chi-square variables is also a chi-square variable with degrees of freedom equal to the sum of the degrees of freedom for the individual variables.
- The ratio of two independent chi-square variables, with degrees of freedom, ν_1 and ν_2 multiplied by (ν_2/ν_1) is F distributed.
- If two independent variables are chi-square distributed, the ratio of either of the variables to the sum of both variables is beta distributed.
- As ν becomes large, the chi-square distribution approaches normal with mean ν and variance 2ν.

2.6.2 The *F* Distribution

The F distribution is used to make inferences about variances and to construct confidence limits. Given two independent chi-square variables, χ_1^2 and χ_2^2, with degrees of freedom ν_1 and ν_2, the random variable

$$F = \frac{\chi_1^2 / \nu_1}{\chi_2^2 / \nu_2} \tag{2.72}$$

has an F distribution with degrees of freedom ν_1 and ν_2. The probability density function for this variable is

$$f(F) = \left(\frac{\Gamma\left[(\nu_1+\nu_2)/2\right](\nu_1/\nu_2)^{\nu_1/2}}{\Gamma(\nu_1/2)\Gamma(\nu_2/2)} \right) \left(\frac{F^{\nu_1/2-1}}{(1+\nu_1 F/\nu_2)^{(\nu_1+\nu_2)/2}} \right), F > 0 \tag{2.73}$$

The F probability density function is shown in Figure 2.25.

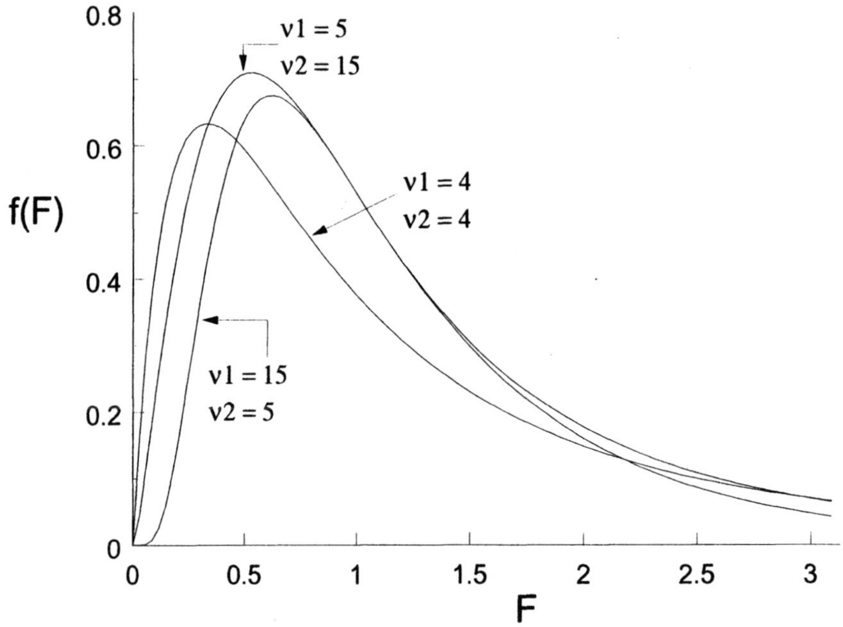

Figure 2.25. F probability density functions.

As seen from Figure 2.25, the shape of the F distribution is dependent on the order of degrees of freedom. The degrees of freedom of the chi-square variable in the numerator of F is stated first, followed by the degrees of freedom of the chi-square variable in the denominator. If $F_\alpha(\nu_1, \nu_2)$ represents the area under the F probability density function, with degrees of freedom ν_1, ν_2, to the right of α, then

$$F_{1-\alpha}(\nu_1,\nu_2) = \frac{1}{F_\alpha(\nu_2,\nu_1)} \tag{2.74}$$

The F distribution is tabled in Appendix E.

The mean of the F distribution is

$$E(F) = \frac{\nu_2}{\nu_2 - 2}, \nu_2 > 2 \tag{2.75}$$

Note that the mean of the F distribution is independent of ν_1. The variance of the F distribution is

$$V(F) = \frac{2\nu_2^2(\nu_1 + \nu_2 - 2)}{\nu_1(\nu_2 - 2)^2(\nu_2 - 4)}, \nu_2 > 4 \tag{2.76}$$

If s_1^2 and s_2^2 are the variances of two samples taken from independent, normally distributed populations with variances, σ_1^2 and σ_2^2, and sample sizes, n_1 and n_2, then the statistic

$$F = \frac{s_1^2 / \sigma_1^2}{s_2^2 / \sigma_2^2} \tag{2.77}$$

is F distributed with $\nu_1 = n_1 - 1$ and $\nu_2 = n_2 - 1$. Thus, the F distribution is also known as the variance ratio distribution.

Example 2.18

Given a random variable with an F distribution and degrees of freedom of $\nu_1 = 8$ and $\nu_2 = 15$, what is the value of F that gives a 5% chance that a random sample will yield a value greater than F? What is the value of F that gives a 95% chance that a random sample will yield a value greater than F?

Solution

For a 5% probability of selecting a value greater than F, this is equivalent to $F_{0.05}(8,15)$. From Appendix E, $F_{0.05}(8,15)$. = 2.64. For a 95% probability of selecting a value greater than F, this is equivalent to $F_{0.05}(8,15)$.

$$F_{0.95}(8,15) = \frac{1}{F_{0.05}(15,8)} = \frac{1}{3.22} = 0.311$$

2.6.3 The *t*-distribution

The t-distribution is used to draw inferences concerning means and to construct confidence intervals for means when the variance is unknown, or too few samples are available to utilize the normal distribution. When dealing with large sample sizes, $n > 30$, the sample variance, s^2, is a reasonable estimate of the variance, σ^2. When sample sizes are small, $n < 30$, the error introduced from using s^2 as an estimate of σ^2 causes the statistic

$$t = \frac{\bar{x} - \mu}{s / \sqrt{n}} \tag{2.78}$$

to deviate from the standard normal distribution*. The *t*-distribution is the distribution for the random variable

$$t = \frac{z}{\sqrt{\frac{\chi^2}{\nu}}} \tag{2.79}$$

where: z is a standard normal random variable, and

χ^2 is a chi-square random variable with γ degrees of freedom.

The *t*-distribution is a special case of the F distribution with $1,\gamma$ degrees of freedom. The t probability density function is

$$f(t) = \frac{\Gamma\left[(\nu+1)/2\right]}{\Gamma(\nu/2)\sqrt{\Pi \nu}}\left(1+\frac{t^2}{\nu}\right)^{-(\nu+1)/2}, -\infty < t < \infty \tag{2.80}$$

The t probability density function is shown in Figure 2.26.

As seen in Figure 2.26, the variance of the *t*-distribution decreases as ν increases. Once ν is greater than 30, there is almost no change in the variance as ν increases, and the *t*-distribution is practically identical to the standard normal distribution.

The mean of the *t*-distribution is

$$E(t) = 0 \tag{2.81}$$

The variance of the *t*-distribution is

$$V(t) = \frac{\nu}{\nu - 2}, \nu > 2 \tag{2.82}$$

* The standard normal deviate is $z = \frac{x - \mu}{\sigma}$. $\bar{x}$ is used to approximate μ, and s is used to approximate σ. The central limit theorem states that the standard deviation of an average is inversely proportional to the square root of the number sampled. Thus, the standard normal deviate for a sample of size n is $z = \frac{\bar{x} - \mu}{\sigma / \sqrt{n}}$. For more information see Section 2.2.

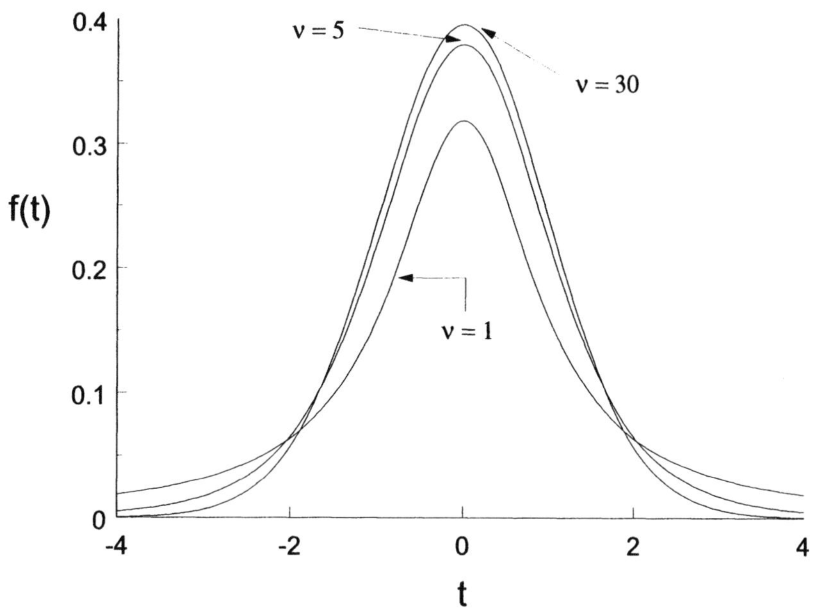

Figure 2.26. *t*-probability density functions.

Example 2.19

A cloth manufacturer guarantees the average strength of a specific material to be 45 units. Each month, 15 specimens are tested, and the computed *t*-statistic must be within $-t_{0.05}$ and $t_{0.05}$ to satisfy the manufacturer's customers. Recently, the strength of the 15 specimens averaged 40 units with a standard deviation of 8, were the specifications met?

Solution

The degrees of freedom is $\nu = n-1 = 14$. From Appendix F, $t_{0.05} = 1.761$. Thus, $-t_{0.05} = -1.761$. If the average strength of the material is 45 units then,

$$t = \frac{40-45}{8/\sqrt{15}} = -2.42$$

This is not within the range -1.761 to 1.761, the manufacturer's process is not producing to the specified strength.

2.7 Bayes' probability analysis

The probability of event B_r occurring, given event A has occurred, can be computed using Bayes' theorem

$$P(B_r \backslash A) = \frac{P(B_r)P(A \backslash B_r)}{\sum_{i=1}^{k} P(B_i)P(A \backslash B_i)}, r = 1, 2, ..., k \tag{2.83}$$

where $B_1, B_2, ..., B_k$ are mutually exclusive events that make up the total sample space, S, and A is an event such that $P(A) \neq 0$.

Example 2.20

Three jars, 1, 2, and 3, each contain two marbles. Jar 1 contains two red marbles, jar 2 contains one red marble and one blue marble, and jar 3 contains two blue marbles. A marble is randomly selected from a randomly selected jar and it is red; what is the probability that the other marble in the same jar is also red?

Solution

The solution is the probability that jar 1 was selected given that a red marble was selected, $P(\text{jar 1} \backslash \text{red marble})$. Using Bayes' theorem, this can be expressed mathematically as

$$P(\text{jar 1} \backslash \text{red marble}) = \frac{P(\text{jar 1})P(\text{red marble} \backslash \text{jar 1})}{\sum_{i=1}^{3} P(\text{jar i})P(\text{red marble} \backslash \text{jar i})}$$

Since there are three jars, the probability of selecting each of the jars, $P(\text{jar i})$, is 1/3. Given that jar 1 is selected, the probability of selecting a red marble, $P(\text{red marble\jar 1})$, is 1. Given that jar 2 is selected, the probability of selecting a red marble, $P(\text{red marble\jar 2})$, is 1/2. Given that jar 3 is selected, the probability of selecting a red marble, $P(\text{red marble\jar 3})$, is 0. Thus,

$$P(\text{jar 1} \backslash \text{red marble}) = \frac{\left(\frac{1}{3}\right)(1)}{\left(\frac{1}{3}\right)(1) + \left(\frac{1}{3}\right)\left(\frac{1}{2}\right) + \left(\frac{1}{3}\right)(0)} = \frac{2}{3}$$

A more specific depiction of Bayes' theorem which is often used in reliability is

$$P(A \backslash B) = \frac{P(A \cap B)}{P(B)} \tag{2.84}$$

where A and B are events. If A and B represent the probability of survival to specific times, then the intersection of the two events is the probability of survival to the largest time. Rewriting this expression for reliability,

$$R(t_2 \backslash t_1) = \frac{R(t_2)}{R(t_1)}, t_2 > t_1 \tag{2.85}$$

Example 2.21

The probability of a system surviving until time = 25 is 0.68. The probability of the same system surviving until time = 50 is 0.39. At time = 25 the system is still functional. What is the probability the system will survive until time = 50?

Solution

The probability of survival until time = 50 given survival until time = 25 is

$$R(50 \backslash 25) = \frac{0.39}{0.68} = 0.5735$$

In the above examples, previously known information, such as the color of marbles in each box, was known. This information is known as a **prior** distribution. Bayesian analysis uses the prior distribution along with joint probability density functions to compute the **posterior** distribution, which yields probabilities of events occurring given that other events have occurred.

The posterior probability density function, given survival until t_1 is

$$f(t \backslash t_1) = \frac{f(t)}{R(t_1)}, \quad t_1 < t \tag{2.86}$$

The posterior reliability function, given survival until t_1 is

$$R(t \backslash t_1) = \frac{R(t)}{R(t_1)}, \quad t_1 < t \tag{2.87}$$

Suppose a component has a Weibull time to fail distribution with a shape parameter of 4 and a scale parameter of 100 weeks. Also, suppose that this component has survived for 85 weeks. The **prior** distribution is simply the Weibull distribution with a shape parameter of 4 and a scale parameter of 100 weeks. The **posterior** distribution is derived from the prior distribution given that the component has survived for 85 weeks. The prior and posterior probability density functions for this example are shown in Figure 2.27. The prior and posterior reliability functions for this example are shown in Figure 2.28.

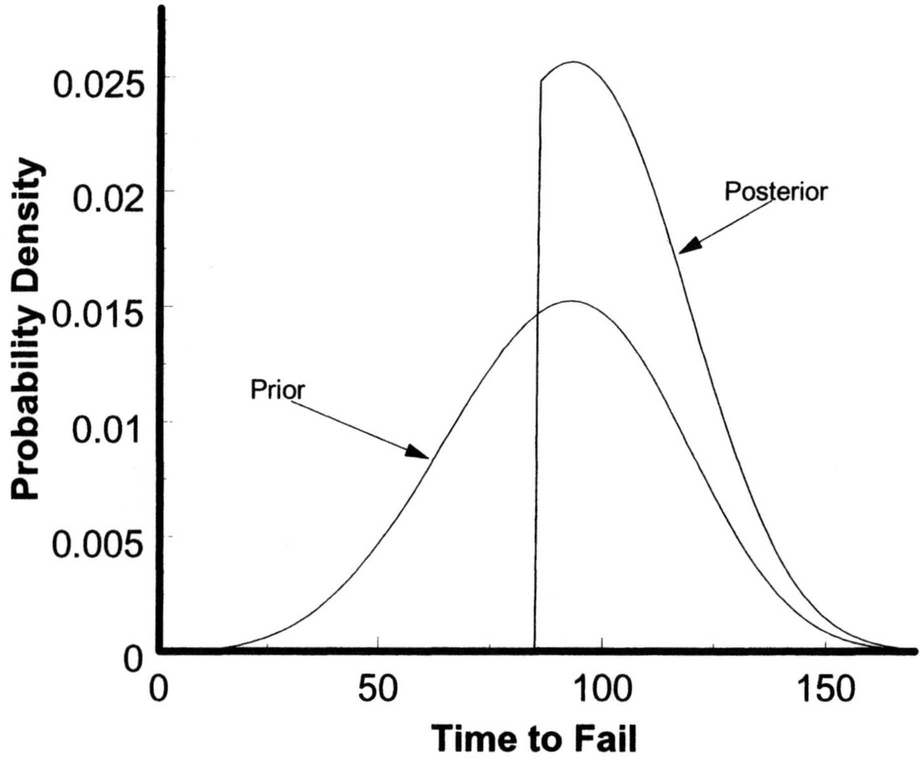

Figure 2.27. Prior and posterior density functions.

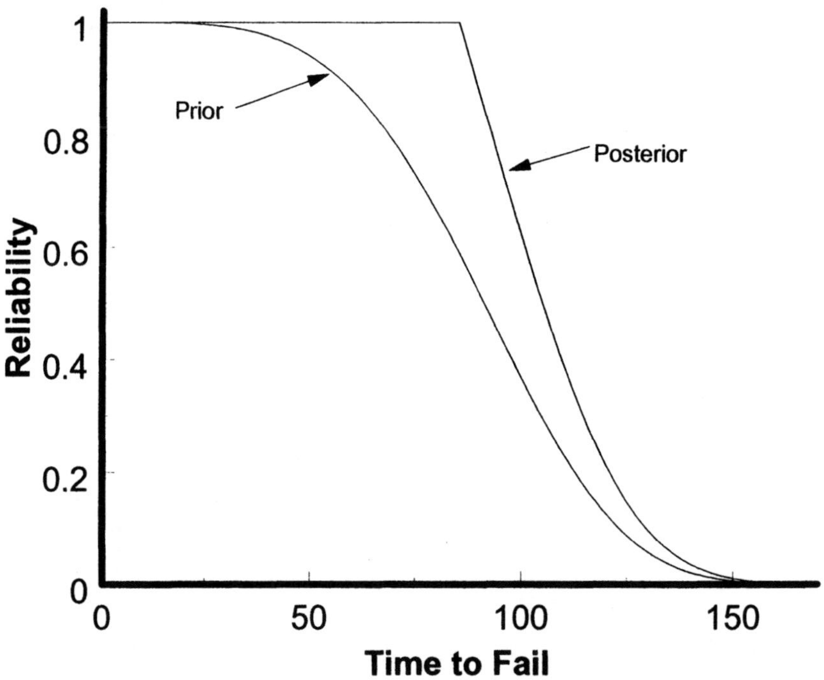

Figure 2.28. Prior and posterior reliability functions.

2.8 Propagation of errors

How can an engineer predict the reliability of the system being designed? A finished product may be a combination of many individual components. And even without the recent emphasis on reduced product development times, it may be infeasible to build prototypes for the purpose of reliability testing. Since statistical information may be available for individual components, the problem is to model the behavior of the system through statistical knowledge of its components. When a variable is a function of several other variables, $y = f(x_1, x_2, \ldots, x_n)$, the procedure of determining the variability of y from knowledge of the variables, $x_1, x_2, \ldots, x_n$, is called **propagation of error**.

This section describes methods for modeling combinations of random variables. The discussion is broken into three parts:

1. properties of means and variances of functions of several variables
2. combinations of random variables that form known distributions, and
3. transformation of random variables.

2.8.1 Properties of means and variances of functions of several variables

Table 2.1 gives the mean and variance for functions of several variables.

Table 2.1. Means and variances of functions of several variables.

Function	Mean	Variance
$y = ax$	$E(y) = aE(x)$	$V(y) = a^2V(x)$
$y = a + x$	$E(y) = a + E(x)$	$V(y) = V(x)$
$y = x_1 + x_2$	$E(y) = E(x_1) + E(x_2)$	$V(y) = V(x_1) + V(x_2)$ *
$y = x_1 - x_2$	$E(y) = E(x_1) - E(x_2)$	$V(y) = V(x_1) + V(x_2)$ *
$y = x^2$	$E(y) = [E(x)]^2 + V(x)$	$V(y) = E(x^4) - \left[E(x^2) + V(x)\right]^2$
$y = x_1x_2$	$E(y) = E(x_1)E(x_2)$ *	$V(y) = V(x_1)V(x_2) + V(x_2)[E(x_1)]^2 + V(x_1)[E(x_2)]^2$ *
$y = \dfrac{x_1}{x_2}$	$E(y) = E(x_1)E\left(\dfrac{1}{x_2}\right)$	$V(y) = \left[V(x_1) + E^2(x_1)\right]E\left(\dfrac{1}{x_2^2}\right) - E^2(x_1)E^2\left(\dfrac{1}{x_2}\right)$

* x_1 and x_2 are independent.

Example 2.22

Determine the expected area and the variance of the area of a rectangle with sides x and y. The variable x is normally distributed with a mean of 20 and a standard deviation of 4; the variable y is normally distributed with a mean of 60 and a standard deviation of 8.

Solution

From Table 2.1, the expected value of the area, $z = xy$ is

$$E(\text{area}) = E(x)E(y) = (20)(60) = 1200$$

The variance of the area is

$$V(\text{area}) = (4^2)(8^2) + (8^2)(20^2) + (4^2)(60^2) = 84,224$$

A useful approximation for the mean of a function of several variables is

$$E(y) \approx f(x_1, x_2, ..., x_n) + \frac{1}{2}\sum_{i=1}^{n} \frac{\partial^2 f(x_1, x_2, ..., x_n)}{\partial x_i^2} V(x_i) \tag{2.88}$$

The second term of this expression is often ignored, leaving the approximation as

$$E(y) \approx f(x_1, x_2, ..., x_n) \tag{2.89}$$

A useful approximation for the variance of a function of several variables is

$$V(y) \approx \sum_{i=1}^{n} \left(\frac{\partial f(x_x, x_2, ..., x_n)}{\partial x_i} \right)^2 V(x_i) \tag{2.90}$$

Example 2.23

The current in a circuit is defined as

$$I = \frac{V}{R}$$

where V is the voltage and R is the resistance. Find the mean current and the variance of the current if the mean voltage is 220 with a standard deviation of 5, and the mean resistance is 4 with a standard deviation of 0.2

Solution

From Equation 2.89,

$$E(I) \approx f(V, R) \approx \frac{220}{4} = 55$$

To utilize Equation 2.90, the partial derivatives of the expression for current are needed.

$$\frac{\partial f}{\partial V} = \frac{1}{R}$$

$$\frac{\partial f}{\partial R} = -\frac{V}{R^2}$$

The approximated variance is

$$V(I) = \left(\frac{1}{4}\right)^2 (5^2) + \left(-\frac{220}{4^2}\right)^2 (0.2^2) = 9.12$$

2.8.2 Combinations of random variables that form well-known distributions

The sum or difference of normally distributed variables is also normally distributed. If x_1 and x_2 are normally distributed with means μ_{x1} and μ_{x2}, and variances σ_{x1}^2 and σ_{x2}^2, then $y = x_1 + x_2$ is normally distributed with a mean of

$$E(y) = \mu_{x1} + \mu_{x2} \tag{2.91}$$

and a variance of

$$V(y) = \sigma_{x1}^2 + \sigma_{x2}^2 + 2\rho\sigma_{x1}\sigma_{x2} \tag{2.92}$$

where ρ is the correlation coefficient, and is defined as

$$\rho = \frac{\operatorname{cov}(x_1, x_2)}{\sigma_{x1}, \sigma_{x2}} \tag{2.93}$$

where $\operatorname{cov}(x_1, x_2)$ is the covariance between variables x_1 and x_2 and is defined as

$$\operatorname{cov}(x_1, x_2) = \int_{-\infty}^{\infty}\int_{-\infty}^{\infty} (x_1 - \mu_1)(x_2 - \mu_2) f(x_1, x_2)\, dx_1 dx_2 \tag{2.94}$$

The value, $y = x_1 - x_2$, is also normally distributed with a mean of

$$E(y) = \mu_{x1} - \mu_{x2} \tag{2.95}$$

and a variance of

$$V(y) = \sigma_{x1}^2 + \sigma_{x2}^2 - 2\rho\sigma_{x1}\sigma_{x2} \tag{2.96}$$

Note that Equations 2.92 and 2.96 are identical when the correlation coefficient is zero (when the variables are independent).

Example 2.24

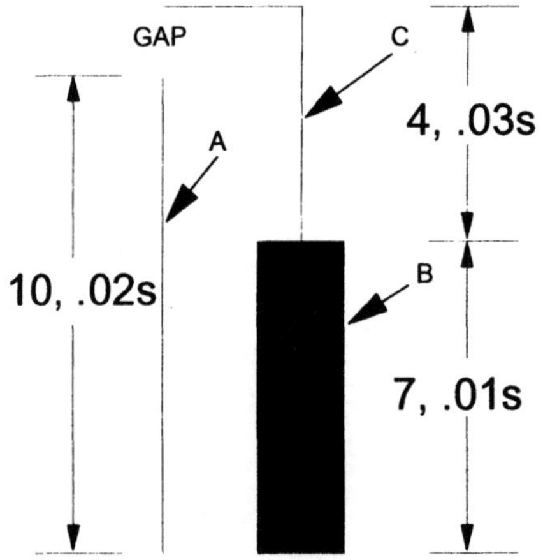

Figure 2.29. Sample fixture.

Referring to Figure 2.29, the sample fixture consists of three parts; the vertical bar to the left (part A), with a mean length of 10 and a standard deviation of 0.02, the solid block (part B), with a mean of length of 7 and a standard deviation of 0.01, and the vertical bar on top of the block (part C) with a mean length of 4 and a standard deviation of 0.03. Determine the expected length and the standard deviation of the identified gap.

Solution

The mean length of the solid block and the vertical bar on top of it is

$$E(\text{block and bar}) = 7 + 4 = 11$$

There is no reason to believe the length of these parts are not independent, thus, the standard deviation of the length of the solid block and the vertical bar is

$$V(\text{block and bar}) = (0.01)^2 + (0.03)^2 = 0.001$$

The mean length of the gap is

$$E(\text{gap}) = 11 - 10 = 1$$

The standard deviation of the gap is

$$s = \sqrt{V(\text{gap})} = \sqrt{0.001 + 0.02^2} = 0.0374$$

One of the most commonly used statistical principles is the theory of large numbers, also known as the central limit theorem. Simply stated, if

$$y = x_1 + x_2 + \ldots x_n$$

where $x_1, x_2, \ldots, x_n$ are independent random variables, then, regardless of the distributions of $x_1, x_2, \ldots, x_n$, y will become a normally distributed as n becomes large. In this case, the mean of y is

$$E(y) = E(x_1) + E(x_2) + \ldots + E(x_n) \tag{2.97}$$

If $x_1, x_2, \ldots, x_n$ are independent, the variance of y is

$$V(y) = V(x_1) + V(x_2) + \ldots + V(x_n) \tag{2.98}$$

Figures 2.30. 2.31, and 2.32 show the phenomenon of the sum of independent random variables tending to normal as the number of variables being summed increases. A distribution that has a shape much different from normal, the exponential distribution, becomes normal as n increases. Figure 2.30 is the sum of two independent, exponentially distributed variables.

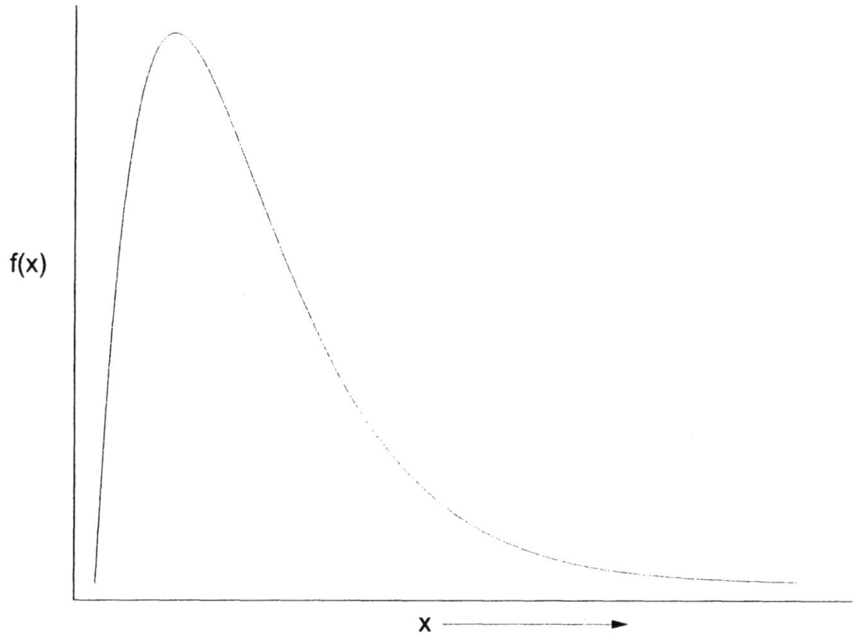

Figure 2.30. The sum of two exponentially distributed variables.

Figure 2.31 is the sum of five independent, exponentially distributed variables.

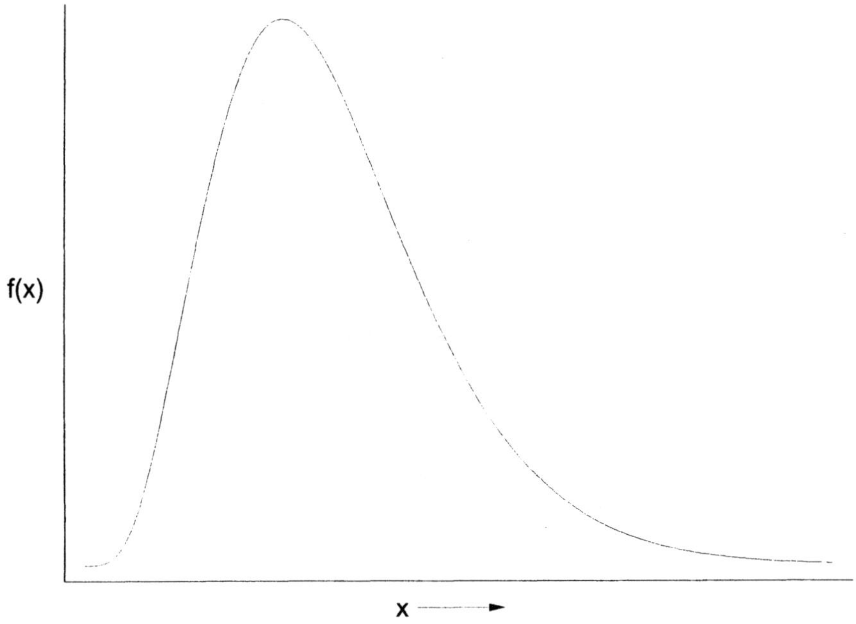

Figure 2.31. The sum of five exponentially distributed variables.

Figure 2.32 is the sum of thirty independent, exponentially distributed variables.

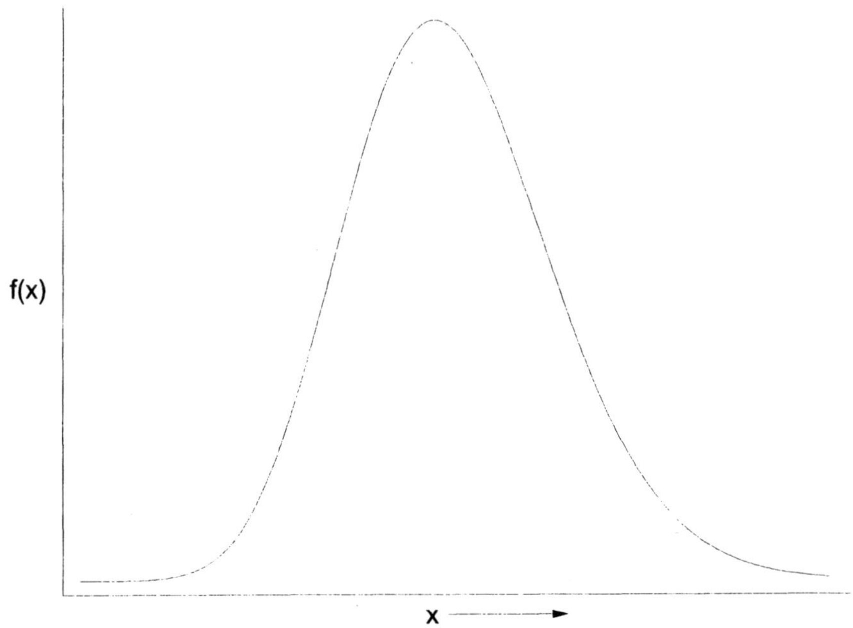

Figure 2.32. The sum of thirty exponentially distributed variables.

The distributions in the previous figures may look familiar. This is because the sum of exponentially distributed random variables is a gamma distributed random variable. Let x_1, x_2, ..., x_n be independent, exponentially distributed random variables with a failure rate of λ. Then $y = x_1, x_2, \ldots, x_n$ follows the gamma distribution with a shape parameter, $\eta = n$ and a scale parameter of λ.

The sum of gamma distributed random variables also follows a gamma distribution. Let $x_1, x_2, \ldots, x_n$ be independent, gamma distributed random variables with a shape parameter of η and a scale parameter of λ. Then $y = x_1, x_2, \ldots, x_n$ follows the gamma distribution with a shape parameter equal to $n(\eta)$and a scale parameter equal to λ.

Example 2.25

A system consisting of 12 components fails after all 12 components fail. The components are standby redundant, and the system has perfect switching. If the components are exponentially distributed with a failure rate of 30 units, what is the reliability function for this system? If the components follow a gamma distribution with a shape parameter of 2 and a scale parameter of 30 units, what is the reliability function for this system?

Solution

The random variable being modeled is

$$y = \sum_{i=1}^{12} x_i$$

Since the individual variables are exponentially distributed, the sum follows a gamma distribution with a shape parameter equal to the individual failure rate, 30, and a shape parameter equal to the number of components, 12. From Section 2.5.1, the gamma reliability function is

$$R(y) = \sum_{k=0}^{11} \frac{(30x)^k e^{-30x}}{k!}$$

If the components had a time to fail that followed a gamma distribution with $\eta = 2$, the parameters of the resulting distribution would also be gamma with a shape parameter equal to the sum of the individual shape parameters, which is equal to 24, and a scale parameter equal to the individual failure rate, 30. The resulting reliability function is

$$R(y) = \sum_{k=0}^{23} \frac{(30x)^k e^{-30x}}{k!}$$

The Poisson distribution is also reproductive. The sum of n independent Poisson distributed variables follows a Poisson distribution with a mean equal to the sum of the individual means. For example, if x, y, and z are Poisson distributed with means of 8, 14, and 16, $w = x + y + z$ is Poisson distributed with a mean of 38.

2.8.3 Transformation of random variables

When transforming random variables, if the resulting probability density function is not obvious, it must be computed. For the case of a single random variable, the probability density function for the transformed variable is

$$g(y) = f\left[w(y)\right]\left|\frac{dw}{dy}\right| \tag{2.99}$$

where: $f(x)$ is the probability density function for x,

$y = u(x)$ is the transformation function, and

$x = w(y)$ is the inverse of the transformation function, and has only one root.

Example 2.26

The random variable x is exponentially distributed with a failure rate λ. Determine the probability density function for the variable $y = 4x + 12$.

Solution

For this example, the transformation function is $y = 4x + 12$. Thus, the inverse transformation function is

$$x = \frac{y-12}{4}$$

The derivative of the inverse transformation function is

$$\frac{dx}{dy} = \frac{1}{4}$$

The exponential probability density function is

$$f(x) = \lambda e^{-\lambda x}\ ,\ x > 0$$

Substituting into Equation 2.99 gives

$$g(y) = \left(\lambda e^{-\lambda(y-12)/4}\right)\left(\frac{1}{4}\right)\ ,\ y > 12$$

The acceptable range for x is 0 to infinity, after transformation, when $x = 0$, $y = 12$, thus, the range of y is 12 to infinity.

If the inverse of the transformation function has more than one root, the probability density function for the transformed variable is

$$g(y)=\sum_{i=1}^{n} f\left[w_i(y)\right]\left|\frac{dw_i}{dy}\right| \tag{2.100}$$

where, n is the number of roots of the inverse transformation function.

Table 2.2 contains some useful transformations of functions involving one variable.

Table 2.2. Transformations involving one random variable.

Transformation Function	Resulting Probability Density Function
$y=a+bx$	$g(y)=\left\|\frac{1}{b}\right\| f\left(\frac{y-a}{b}\right)$
$y=\frac{1}{x}$	$g(y)=\frac{1}{y^2} f\left(\frac{1}{y}\right)$
$y=e^x$	$g(y)=\left\|\frac{1}{y}\right\| f(\ln y)$
$y=\ln x$	$g(y)=e^y f\left(e^y\right)$
$y=x^2$	$g(y)=\frac{1}{2\sqrt{y}}\left[f\left(\sqrt{y}\right)+f\left(-\sqrt{y}\right)\right]$
$y=\sqrt{x}$	$g(y)=2\|y\| f(y^2)\ ,y>0$
$y=x^{1/n}$	$g(y)=\left\|ny^{n-1}\right\| f(y^n)\ ,y>0$
$y=x^n$	$g(y)=\left\|\frac{1}{n}y^{(1/n)-1}\right\|\sum_{i=1}^{n} f(x_i)$ $x_1, x_2, \ldots, x_n$ are the roots of $x=y^{1/n}$.
$y=e^{ax}$	$g(y)=\left\|\frac{1}{ay}\right\| f\left[\ln\left(y^{1/a}\right)\right]$

$f(x)$ is the probability density function for x.

Now consider the case where the random variables x_1 and x_2 have a joint probability distribution of $f(x_1,x_2)$ and are transformed to two new random variables, $y_1=u_1(x_1,x_2)$ and $y_2=u_2(x_1,x_2)$. If the inverse of the transformation functions, $x_1=w_1(y_1,y_2)$ and $x_2=w_2(y_1,y_2)$ have single roots, the joint probability density function for y_1 and y_2 is

$$g(y_1,y_2)=f\left[w_1(y_1,y_2),w_2(y_1,y_2)\right]|J| \tag{2.101}$$

where J is the Jacobian and is defined as the determinant of the partial derivatives;

$$J = \begin{vmatrix} \partial x_1 / \partial y_1 & \partial x_1 / \partial y_2 \\ \partial x_2 / \partial y_1 & \partial x_2 / \partial y_2 \end{vmatrix}$$

In many cases, the transformation is to a single variable, y_1. In this case, simply choose a dummy variable for y_2. When choosing this dummy variable, be sure that y_1 and y_2 are independent; this will simplify the calculations because the joint probability density function for the two variables will be the product of the individual probability density functions.

Some useful transformations for functions of more than one random variable are given in Table 2.3.

Table 2.3. Transformations for functions of more than one random variable.

Transformation Function	Resulting Probability Density Function
$y = x_1 + x_2$	$g(y) = \int f_{x1}(y-z) f_{x2}(z)dz$
$y = x_1 x_2$	$g(y) = \int \left\vert \frac{1}{z} \right\vert f_{x1}\left(\frac{y}{z}\right) f_{x2}(z)dz$
$y = \frac{x_1}{x_2}$	$g(y) = \int \left\vert \frac{z}{y^2} \right\vert f_{x1}(z) f_{x2}\left(\frac{z}{y}\right)dz$
$y = x_1^2 + x_2^2$	$g(y) = \int \left\vert \frac{1}{2\sqrt{y-z^2}} \right\vert f_{x1}(z) f_{x2}\left(\sqrt{y-z^2}\right)dz$
$y = \sqrt{x_1^2 + x_2^2}$	$g(y) = \int \left\vert \frac{1}{2\sqrt{y^2-z^2}} \right\vert f_{x1}(z) f_{x2}\left(\sqrt{y^2-z^2}\right)dz$

f_{x1} and f_{x2} are the density functions for the random variables x_1 and x_2.

2.9 Exercises

1. Given that a population of components has a normal time to fail distribution with a mean of 400 hours and a standard deviation of 60 hours, graph the probability density function, the reliability function and the hazard function.
2. For the components in Exercise 1, if 200 components are tested for 430 hours, what is the expected number of survivors? What is the expected number of failures between time = 410 and time = 430?
3. For the components in Exercise 1, graphically show the probability of a failure before 425 hours of operation, 410 hours of operation, and from this graph, determine the

probability of a failure in the interval from 410 hours of operation to 425 hours of operation.

4. Repeat Exercises 1 & 2 with the components following a lognormal time to fail with the same mean and standard deviation.
5. Given the following probability density function, determine the mean and variance.

$$f(x) = \frac{1}{0.4x\sqrt{2\pi}} \exp\left[-\frac{1}{2}\left(\frac{\ln x - 2}{0.4}\right)^2\right], x > 0$$

6. A population of components has a constant failure rate of 0.003 per hour. What is the probability of failure before 200 hours of operation?
7. What is the value of the exponential probability density function at the intersection with the *y*-axis?
   a. 0
   b. the failure rate of the distribution
   c. the mean time to fail for the distribution
   d. the variance of the distribution
   e. infinity
   f. none of the above
8. Given the reliability function

$$R(t) = e^{-\frac{t}{\theta}}, t > 0,$$

determine the mean and variance.

9. An automobile manufacturer averages 4 defects per vehicle, what is the probability of an automobile containing (a)no defects, (b) 3 or more defects?
10. The failure rate for a carpet manufacturer is 3.7 per 1000 square yards. What is the probability of finding no defects in a random sample of 100 square yards? Use the Poisson distribution to compute the answer.
11. Repeat Exercise 10 using the binomial distribution to compute the answer.
12. A salesperson has a success rate of 15%. What is the probability of making two successful sales calls in a row.
13. For the sales person in Exercise 12, how many sales calls will have to be made in a day to have a 90% chance of having at least two successful calls?
14. A lot of 15 small military rockets was manufactured, and it is known that two of the rockets received faulty fuses. If four of the rockets are sampled at random, what is the probability that, (a) none of the four have faulty fuses, (b) one or less of the rockets have faulty fuses, and (c) two or less of the rockets have faulty fuses?

15. A certain military rocket has an 84% probability of successfully destroying its intended target. What is the probability that, (a) the first rocket fired at a specified target destroys the target, (b) the second rocket fired at a specified target destroys the target, and (c) the third rocket fired at a specified target destroys the target.

16. A component has the following reliability function

$$R(t) = \exp\left[-\left(\frac{t}{89}\right)^{3.7}\right], t > 0$$

Plot the hazard function.

17. What is the mean and variance of the time to fail for the component in Exercise 16?

18. What is the probability the component in Exercise 16 will fail before time = 80? What is the instantaneous failure rate at this time?

19. A system consists of four identical components, with three being standby redundant. The components have an exponential time-to-fail distribution with a failure rate of 1/235. Determine the probability density function for this system, and determine the system mean time to fail and variance.

20. The sides of a square have an expected length of 8.2 with a variance of 0.74. What is the expected area of this square, and what is the variance of the area?

21. A fixture consists of two pipes, A and B, welded end-to-end. Historically, the average length of pipe A has been 3.7 feet with a standard deviation of 1.2. Historically, the average length of pipe B has been 8.2 feet with a standard deviation of 2.7. What is the average length and standard deviation of the fixture?

22. An component is produced by extruding a shaft and removing a portion of the extrusion. The expected length of the extruded shaft is 23 inches with a standard deviation of 0.3 inches. The expected value for the length removed is 0.6 inches with a standard deviation of 0.09 inches. What is the expected length and standard deviation of the resulting component?

23. The diameter of a rod is normally distributed with mean μ, and standard deviation σ. What is the probability density function for the cross sectional area of the rod?

24. The variable x is exponentially distributed with a mean of 7. If $y = 1/x$, what is the probability density function of y?

25. Given the probability density function

$$f(x) = xe^{-x^2/2}, 0 < x < \infty,$$

(a) prove this is a valid density function, (b) find the reliability function, and (c) find hazard function.

26. Given the hazard function

$$h(x) = c, o < x < \infty,$$

determine the reliability and probability density functions.

27. A component has an exponential distribution with a mean time to fail of 500 hours. Given the component survives to 400 hours, what is the probability of the component surviving for an additional 200 hours?
28. A component has a Weibull time to fail distribution with a shape parameter of 4, a scale parameter of 50 and a location parameter of 10. Given the component survives until time = 60 and fails before time = 80, what is the probability of the component surviving until time = 75?

2.10 References

Crowder, M. J., A. C. Kimber, R. L. Smith, and T. J. Sweeting. 1991. ***Statistical Analysis of Reliability Data***, London: Chapman & Hall.

Dodson, Bryan, and Jack Kirkland. 1994. ***The Reliability & Maintenance Analyst***, Auburn Hills, MI: Espinoza Consulting.

Dodson, Bryan. 1994. ***Weibull Analysis (with Software)***. Milwaukee, WI: ASQ–Quality Press.

Dodson, B. L., and M. D. Mulcahy. 1992. ***Certified Reliability Engineer Examination Study Guide***. Tucson, AZ: Quality Publishing.

Dovich, R. A. 1990. ***Reliability Statistics***. Milwaukee, WI: ASQ–Quality Press.

Dovich, R. A. 1992. ***Quality Engineering Statistics***. Milwaukee, WI: ASQ–Quality Press.

Hahn, G. J., and S. S. Shapiro. 1967. ***Statistical Models in Engineering***. New York: John Wiley & Sons.

Johnson, N. L., and S. Kotz. 1970. ***Distributions in Statistics: Continuous Univariate Distributions-1***. New York: John Wiley & Sons.

Johnson, N. L., and S. Kotz. 1970. ***Distributions in Statistics: Continuous Univariate Distributions-2***. New York: John Wiley & Sons.

Juran, J. M., and F. M Gryna. 1988. ***Juran's Quality Control Handbook***. New York: McGraw Hill.

Kapur, K. C., and L. R. Lamberson. 1977. ***Reliability in Engineering Design***. New York: John Wiley & Sons.

Krishnamoorthi, K. S. 1992. ***Reliability Methods for Engineers.*** Milwaukee, WI: ASQC–Quality Press.

Lewis, E. E. 1987. ***Introduction To Reliability Engineering.*** New York: John Wiley & Sons.

O'Conner, P. D. T. 1993. ***Practical Reliability Engineering.*** New York: John Wiley & Sons.

Shooman, M. L. 1990. ***Probabilistic Reliability: An Engineering Approach.*** Malabar, FL: Robert E. Krieger.

Sundarajan, C. 1991. ***Guide To Reliability Engineering: Data, Analysis, Applications, Implementation, And Management.*** New York: Van Nostrand Reinhold.

Tobias, P. A., and D. C. Trindade. 1986. ***Applied Reliability.*** New York: Van Nostrand Reinhold.

Walpole, R. E., and R. H. Myers. 1989. ***Probability and Statistics for Engineers and Scientists.*** New York: Macmillan.

♦♦♦

CHAPTER

3

Parameter Estimation

There are many methods available for parameter estimation. In reliability engineering, the most popular methods are:

- Maximum likelihood
- Hazard plotting
- Probability plotting
- Moment estimation

It is desirable for a parameter estimator to have the following properties:

1. Lack of bias—if the expected value of the estimator is equal to the true value of the parameter, it is said to be unbiased.
2. Minimum variance—the smaller the variance of the estimate, the smaller the sample size required to obtain the level of accuracy desired, and the more efficient the estimator. The most efficient estimator is the estimator with minimum variance.
3. Consistency——as the sample size is increased, the value of the estimated parameter becomes closer to the true value of the parameter.
4. Sufficiency—the estimator uses all information available in the data set.

The method used to estimate parameters depends on the type of data or testing involved and the distribution of interest. In addition to the parameter estimation methods listed above, this section describes censored data and presents methods of parameter estimation for the exponential, normal, lognormal, Weibull, and gamma distribution. Methods are presented for complete and censored data.

Parameter estimation techniques are tedious, and computers are often employed to aid in data analysis. Dodson and Kirkland (1994) have developed a comprehensive software package for parameter estimation. The software is capable of maximum likelihood estimation, hazard plotting, probability plotting, and moment estimation. Available distributions are the exponential, Weibull, normal, lognormal, and gamma.

3.1 Censored data

A censored data set includes information on failed items and surviving items. Consider the data in Table 3.1. Eight items were placed on test stands. Three of the items failed, and five of the items were removed from testing without failing.

Table 3.1. Example of censored data.

Time to Fail	
35	60 +
45	60 +
55	60 +
60 +	60 +

Obviously, the sample average and the sample standard deviation for the three failed items cannot be used to estimate the parameters of the normal distribution in this case. The sample average is (35 + 45 + 55)/3 = 45. The time to fail for each of the remaining five items is greater than 60; the true sample average is considerably greater than 45.

Censoring is described in many ways.

Type I Censoring—A test is initiated with a number of items, and the test is terminated after a predetermined time. An example would be to place 100 transistors on test stands and terminate testing after 1,000 hours regardless of the number of failures. This is also known as **time censoring**. In this case, the test time is fixed and the number of failures is the random variable.

Type II Censoring—A test is initiated with a number of items, and the test is terminated after a predetermined number of failures. An example would be to place 100 transistors on test stands and terminate testing after 30 failures regardless of the elapsed time. This is also known as **failure censoring** or **truncation**. In this case, the number of failures is fixed and the testing time is the random variable.

These examples above involve **right** censoring. An item is censored on the right if the failure time is not known, but it is known that the item survived to a known time without failure. If an item is known to be in a failed condition at or before a specific time, but the exact failure time is not known, it is **left** censored.

Single censoring occurs when there is only one censoring point. If 100 transistors are placed on test stands and the test is terminated after 1,000 hours, there is a single censoring point at 1,000 hours. If 20 transistors were removed without failure after 1,000 hours of testing and another 15 transistors were removed without failure after 1,200 hours of testing, there are two censoring points, and the resulting data are **multiple** censored. If exact failure times are not known, but the number of failures in a time interval is recorded, this is **interval** or **grouped data**.

3.2 Maximum likelihood estimation

Maximum likelihood is the most widely used method generating estimators. It is based on the principle of determining the parameter(s) value(s) that maximize(s) the probability of obtaining the sample data.

The likelihood function for a given distribution is a representation of the probability of obtaining the sample data. Let $x_1, x_2, ..., x_n$ be independent, random variables from the probability density function $f(x,\theta)$, where θ is the single distribution parameter. Then

$$L(x_1, x_2, ..., x_n; \theta) = f(x_1, \theta) f(x_2, \theta) ... f(x_n, \theta) \tag{3.1}$$

is the joint distribution of the random variables, or the **likelihood function.** The maximum likelihood estimate, $\hat{\theta}$, maximizes the likelihood function. This estimate is asymptotically normal. Often the natural logarithm of the likelihood function is maximized to simplify computations.

The variances of the estimates can be found by inverting the matrix of the negative of the second partial derivatives of the likelihood function, also known as the local information matrix. These estimates are asymptotically normal and the variances obtained from the local information matrix are used to calculate confidence intervals.

3.3 Hazard plotting

Hazard plotting is a graphical method of parameter estimation. The cumulative hazard function is transformed to a linear expression, usually by a logarithmic transformation, and plotted. The slope and the intercept of the plot provide the information needed to estimate the parameters of the distribution of interest.

If manually constructing a hazard plot, distribution-specific hazard paper is required. By using hazard paper, the failure times and cumulative hazard function estimates can be plotted directly. With the power of personal computers and electronic spreadsheets, specialized graph paper is no longer needed, as the necessary transformations can be made quickly and easily.

3.4 Probability plotting

Probability plotting is a graphical method of parameter estimation. For the assumed distribution, the cumulative distribution function is transformed to a linear expression, usually by a logarithmic transformation, and plotted. If the plotted points form a straight line, the assumed distribution is acceptable, and the slope and the intercept of the plot provide the information needed to estimate the parameters of the distribution of interest. The median rank is usually used to estimate the cumulative distribution function, although there are several alternatives such as the mean rank and the Kaplan-Meier product limit estimator.

If manually constructing a probability plot, distribution-specific probability paper is required. By using probability paper, the failure times and cumulative distribution function estimates can be plotted directly. With the power of personal computers and electronic spreadsheets specialized graph paper is no longer needed, as the necessary transformations can be made quickly and easily.

3.5 Moment estimation

Moment estimation is based on the concept of matching the moments of the sample data with the moments defined by the distribution of interest and its parameters. The first moment about the origin

$$E(x)=\begin{cases}\sum_x xf(x) & \text{if } x \text{ is discrete}\\ \int_{-\infty}^{\infty} xf(x)\,dx & \text{if } x \text{ is continuous}\end{cases} \tag{3.2}$$

is equal to the distribution mean. The second moment about the mean is equal to the distribution variance.

$$E\left[(x-\mu)^2\right]=\begin{cases}\sum_x x^2 f(x)-\mu^2 & \text{if } x \text{ is discrete}\\ \int_{-\infty}^{\infty} x^2 f(x)\,dx-\mu^2 & \text{if } x \text{ is continuous}\end{cases} \tag{3.3}$$

where μ is the distribution mean.

For some distributions, such as the normal distribution, these moments provide direct estimates of the parameters. For other distributions, such as the Weibull, lognormal and gamma, distribution parameters are estimated by setting the sample moments equal to the theoretical moments and solving for the distribution parameters. The number of moments needed is determined by the number of parameters being estimated. If the distribution has one parameter, this parameter can be estimated by setting Equation 3.2 equal to the sample mean and solving for the parameter. If three parameters are being estimated, three equations are required. The distribution skewness, which is the third moment about the mean, is used with the two equations given above.

3.6 Exponential distribution

The simplest method of parameter estimation for the exponential distribution is the method of maximum likelihood. Maximum likelihood provides an unbiased estimate, but provides no indication of goodness of fit. Graphical methods, while more involved, provide a visual goodness of fit test. Often graphical methods will be used in conjunction with maximum likelihood estimation.

3.6.1 Maximum likelihood estimation

The exponential probability density function is

$$f(x)=\frac{1}{\theta}e^{-x/\theta},x\geq 0 \tag{3.4}$$

The maximum likelihood estimation for the parameter θ is

$$\hat{\theta}=\frac{\sum_{i=1}^{n}x_i}{r} \tag{3.5}$$

where: x_i is the *i*th data point; this may be a failure or a censoring point,
n is the total number of data points both censored and uncensored, and
r is the number of failures.

This estimate is unbiased and is the minimum variance estimator.

Example 3.1

The cycles to fail for seven springs are:

30,183	14,871	35,031	76,321
43,891	31,650	12,310	

Assuming an exponential time to fail distribution, estimate the mean time to fail and the mean failure rate.

Solution

The mean time to fail is

$$\hat{\theta} = \frac{30,183+14,871+35,031+76,321+43,891+31,650+12,310}{7}$$

$$= \frac{244,257}{7} = 34,893.9 \text{ cycles}$$

The mean failure rate is the inverse of the mean time to fail

$$\hat{\lambda} = \frac{1}{34,893.9} = 0.0000287 \text{ failures per cycle}$$

Example 3.2

Assume the data in Example 3.1 represents cycles to fail for seven springs, but an additional 10 springs were tested for 80,000 cycles without failure. Estimate the mean time to fail and the mean failure rate.

Solution

The mean time to fail is

$$\hat{\theta} = \frac{244,257+10(80,000)}{7} = 149,179.6 \text{ cycles}$$

The mean failure rate is

$$\lambda = \frac{1}{149,179.6} = 0.0000067 \text{ failures per cycle}$$

For a time truncated test a confidence interval for θ is

$$\frac{2\sum_{i=1}^{n} x_i}{\chi^2_{(\alpha/2,2r+2)}} \le \theta \le \frac{2\sum_{i=1}^{n} x_i}{\chi^2_{(1-\alpha/2,2r)}} . \tag{3.6}$$

Note that the χ^2 degrees of freedom differ for the upper and lower limits.

Example 3.3

Fifteen items were tested for 1000 hours. Failures occurred at times 120 hours, 190 hours, 560 hours, and 812 hours. Construct a 90% confidence interval for the mean time to fail and the failure rate.

Solution

This is a time truncated test. The mean life estimate is

$$\hat{\theta} = \frac{120 + 190 + 560 + 812 + 11(1000)}{4} = \frac{12,682}{4} = 3,170.5$$

For a 90% confidence interval, $\alpha = 0.1$. From Appendix H, $\chi^2_{(0.05,10)} = 18.307$, and $\chi^2_{(0.95,8)} = 2.733$. The 90% confidence interval for θ is

$$\frac{2(12,682)}{18.307} \le \theta \le \frac{2(12,682)}{2.733}$$

$$1,385.5 \le \theta \le 9,280.6$$

The confidence interval for the failure rate is the inverse of the confidence interval for the mean time to fail.

$$\frac{1}{9,280.6} \le \lambda \le \frac{1}{1,385.5}$$

$$0.0001077 \le \lambda \le 0.0007217$$

For a failure truncated test a confidence interval for θ is

$$\frac{2\sum_{i=1}^{n} x_i}{\chi^2_{(\alpha/2,2r)}} \le \theta \le \frac{2\sum_{i=1}^{n} x_i}{\chi^2_{(1-\alpha/2,2r)}} \tag{3.7}$$

Note that the χ^2 degrees of freedom are the same for the for the upper and lower limits.

Example 3.4

Twelve items were tested with failures occurring at times of 43 hours, 67 hours, 92 hours 94 hours, and 149 hours. At a time of 149 hours, the testing was stopped for the remaining seven items. Construct a 95% confidence interval for the mean time to fail.

Solution

This is a failure truncated test. The mean life estimate is

$$\hat{\theta} = \frac{43 + 67 + 92 + 94 + 149 + 7(149)}{5} = \frac{1488}{5} = 297.6$$

For a 95% confidence interval, $\alpha = 0.05$. From Appendix H, $\chi^2_{(0.025,10)} = 20.483$, and $\chi^2_{(0.975,10)} = 3.247$. The 95% confidence interval for θ is

$$\frac{2(1488)}{20.483} \leq \theta \leq \frac{2(1488)}{3.247}$$

$$145.3 \leq \theta \leq 916.5$$

For failure free testing the one sided lower confidence limit simplifies to

$$\frac{-nt}{\ln \alpha} \leq \theta \tag{3.8}$$

where: *t* is the testing time,

ln is the natural logarithm, and

α is the significance (α = 0.05 for a 95% limit).

Example 3.5

Twenty items are tested for 230 hours without failure. Determine a 90% lower confidence limit for θ.

Solution

$$\frac{-20(230)}{\ln(0.1)} = 1{,}997.8 \text{ hours}$$

A confidence interval for reliability is

$$e^{-\frac{x}{\theta_L}} \leq R(x) \leq e^{-\frac{x}{\theta_U}} \tag{3.9}$$

where: θ_L is the lower confidence limit for the mean time to fail, and

θ_U is the upper confidence limit for the mean time to fail.

A confidence interval for percentiles is

$$-\theta_L \ln(1-P) \leq x \leq -\theta_U \ln(1-P) \tag{3.10}$$

where: *P* is the probability of failure prior to time = ***x***.

3.6.2 Hazard Plotting

The exponential cumulative hazard function is

$$H(x) = \frac{x}{\theta} \tag{3.11}$$

If a data set is exponentially distributed, a plot of Equation 3.11 yields a linear fit with a zero intercept and a slope of $1/\theta$. To construct a hazard plot, an estimate for $H(t)$ is needed. The cumulative hazard function is estimated by the cumulative of the inverse of the reverse ranks. For a data set of n points ordered from smallest to largest, the first point has a rank of n, the second $n-1$, etc.

Example 3.6

Construct a hazard plot for the data in Example 3.4.

Solution

Table 3.2 give the calculations necessary for the hazard plot. The plot is shown in Figure 3.1. Note that censored data points are not plotted. The slope of best fit straight line through the data with an intercept of zero is 0.00333. The estimated mean is $\theta = 1/0.00333 = 300.3$.

Table 3.2. Tabulations for an exponential hazard plot.

Time to Fail	Reverse Rank	$h(t)$	$H(t)$
43	12	0.0833	0.0833
67	11	0.0909	0.1742
92	10	0.1000	0.2742
94	9	0.1111	0.3854
149	8	0.1250	0.5104
149 c	7		
149 c	6		
149 c	5		
149 c	4		
149 c	3		
149 c	2		
149 c	1		

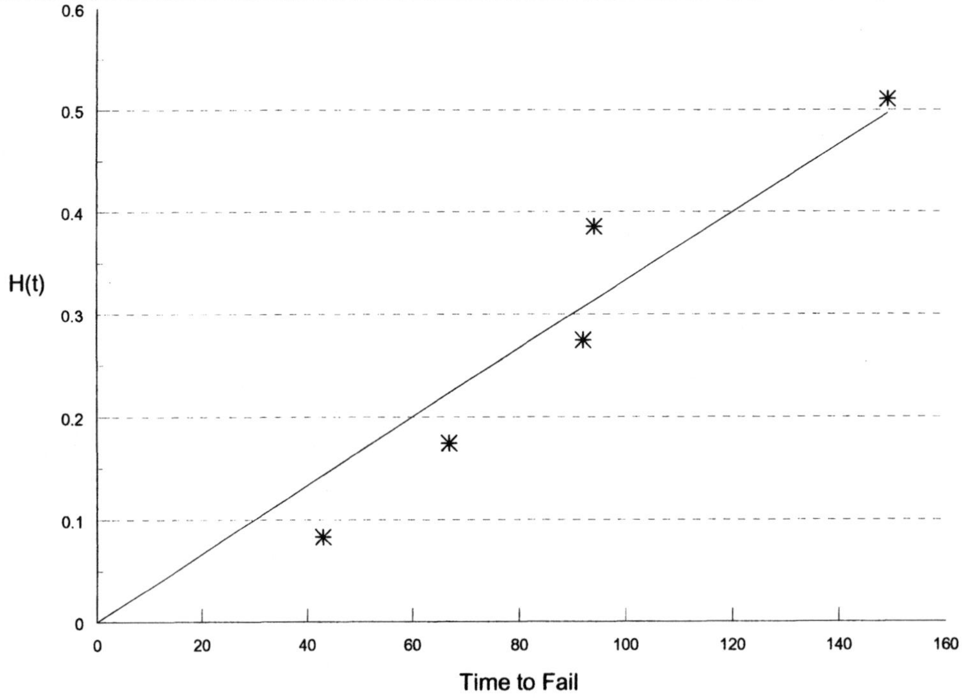

Figure 3.1. Hazard plot for exponential data.

3.6.3 Probability plotting

The exponential cumulative distribution function is

$$F(x) = 1 - e^{-\frac{x}{\theta}} \tag{3.12}$$

By manipulation this expression can be transformed to a linear expression

$$\ln \frac{1}{1 - F(x)} = \frac{x}{\theta} \tag{3.13}$$

If a data set follows an exponential distribution, a plot of

$$\ln \frac{1}{1 - F(x)}$$

versus x will be linear with a zero intercept and a slope of $1/\theta$. Before a plot can be constructed, an estimate for $F(x)$ is needed. The cumulative distribution function, $F(x)$, is usually estimated from the median rank, but other estimates such as the mean rank and the Kaplan-Meier product limit estimator are also used. The median rank estimate for $F(x)$ is

$$\hat{F}(x) = \frac{O_i - 0.3}{n + 0.4} \tag{3.14}$$

where O_i is the modified order of failure of the ith data point.

A modified order of failure is only needed if censored data is involved; if not the original order of failure, *i,* is equivalent to the modified order of failure. The logic for a modified order of failure is as follows. Consider three items; the first was tested for 3 hours and the test was stopped without failure, the second item was tested and failed after 4 hours, and the third item was tested and failed after 4.5 hours. For this data set the failure order is unclear. The first item could have been either the first failure, the second failure or the third failure; thus it is not certain that the first item to fail, the second item, is the first ordered failure. The modified order of failure is computed from the expression

$$O_i = O_{i-1} + I_i \tag{3.15}$$

where I_i is the increment for the *i*th failure, and is computed from the expression

$$I_i = \frac{(n+1) - O_p}{1+c} \tag{3.16}$$

where: n is the total number of points in the data set, both censored and uncensored,

c is the number of points remaining in the data set including the current point,

O_p is the order of the previous failure.

An alternative to plotting x versus

$$\ln \frac{1}{1-F(x)}$$

on conventional graph paper is to plot x versus $F(x)$ on specialized probability paper. The advantage of probability paper is that the values of

$$\ln \frac{1}{1-F(x)}$$

do not have to be computed. With electronic spreadsheets, the disadvantages of probability paper outweigh the advantages. Electronic spreadsheets with built-in statistical functions can quickly make data transformations, determine a best fit straight line using linear regression, and graph the results.

Example 3.7

Construct a probability plot for the data in Example 3.4.

Solution

Table 3.3 contains the calculations necessary for plotting. The probability plot is shown in Figure 3.2. The slope of the best fit straight line through the origin is 0.00304 which estimates the failure rate for the exponential distribution. The mean of the distribution is $\theta = 1/0.00304 = 328.9$.

Table 3.3. Tabulations for exponential probability plotting.

Time to Fail	I_i	O_i	Median Rank, $F(t)$	$\frac{1}{1-F(t)}$	$\ln\frac{1}{1-F(t)}$
43	1	1	0.0565	1.0598	0.0581
67	1	2	0.1371	1.1589	0.1475
92	1	3	0.2177	1.2784	0.2456
94	1	4	0.2984	1.4253	0.3544
149	1	5	0.3790	1.6104	0.4765
149 c					
149 c					
149 c					
149 c					
149 c					
149 c					
149 c					

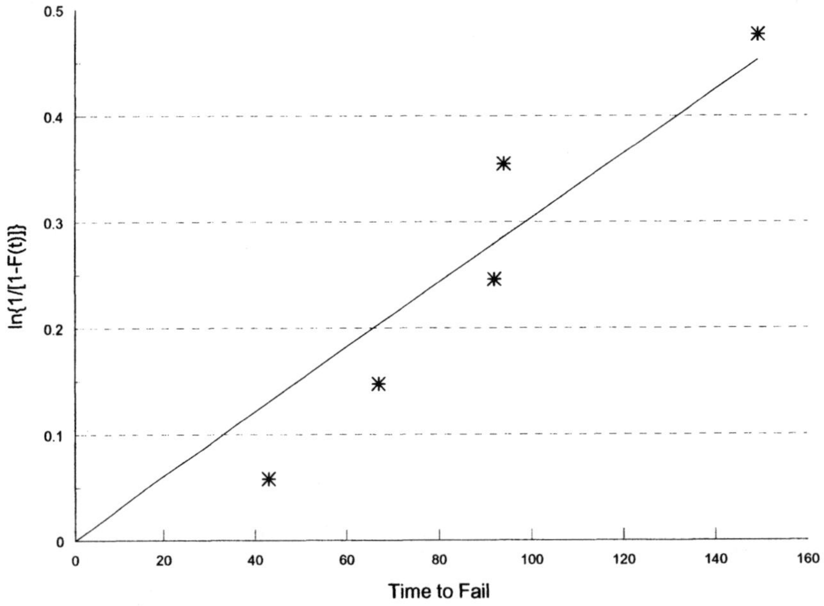

Figure 3.2. Exponential probability plot.

Often, reliability confidence limits are added to probability plots. Upper and lower confidence limits are approximated by 5% and 95% ranks. These ranks can be taken from the rank tables in Appendix C or calculated from the expressions

$$w_\alpha = \frac{\dfrac{j}{n-j+1}}{F_{1-\alpha,2(n-j+1),2j} + \dfrac{j}{n-j+1}}, \alpha \le 0.5 \tag{3.17}$$

$$w_\alpha = \frac{\dfrac{j}{n-j+1} F_{\alpha,2j,2(n-j+1)}}{1 + \dfrac{j}{n-j+1} F_{\alpha,2j,2(n-j+1)}}, \alpha > 0.5 \tag{3.18}$$

where w_α is the $100(1-\alpha)\%$ nonparametric confidence limit,
j is the failure order,
n is the total number of data points, both censored and uncensored, and
$F_{\alpha,v1,v2}$ is the critical value from the F distribution.

When multiple censored data are encountered, the modified failure orders will not be integers, and the rank values will have to be interpolated. The rank values are not plotted against the corresponding failure time. Any deviation of the failure time from the best-fit straight line through the data is considered sampling error, and the time the rank values are plotted against is found by moving parallel to the x-axis until the best-fit straight line is intersected. This plotting position is

$$x_i = \theta \ln\left(\frac{1}{1-F(x_i)}\right) \tag{3.19}$$

Example 3.8

Construct a probability plot for the previous example including confidence limits.

Solution

Table 3.4 shows the 5% and 95% ranks along with their plotting position for the previous example. A probability plot of this example including confidence limits is shown in Figure 3.3.

Table 3.4. Exponential probability plotting data including confidence limits.

Time to Fail	5% Rank	$\ln\frac{1}{1-5\%\text{Rank}}$	95% Rank	$\ln\frac{1}{1-95\%\text{Rank}}$	Plotting Time
43	0.0043	0.0043	0.2209	0.2496	19.1
67	0.0305	0.0310	0.3387	0.4135	48.5
92	0.0719	0.0746	0.4381	0.5764	80.8
94	0.1229	0.1311	0.5273	0.7493	116.6
149	0.1810	0.1997	0.6091	0.9393	156.7
149 c					
149 c					
149 c					
149 c					
149 c					
149 c					
149 c					

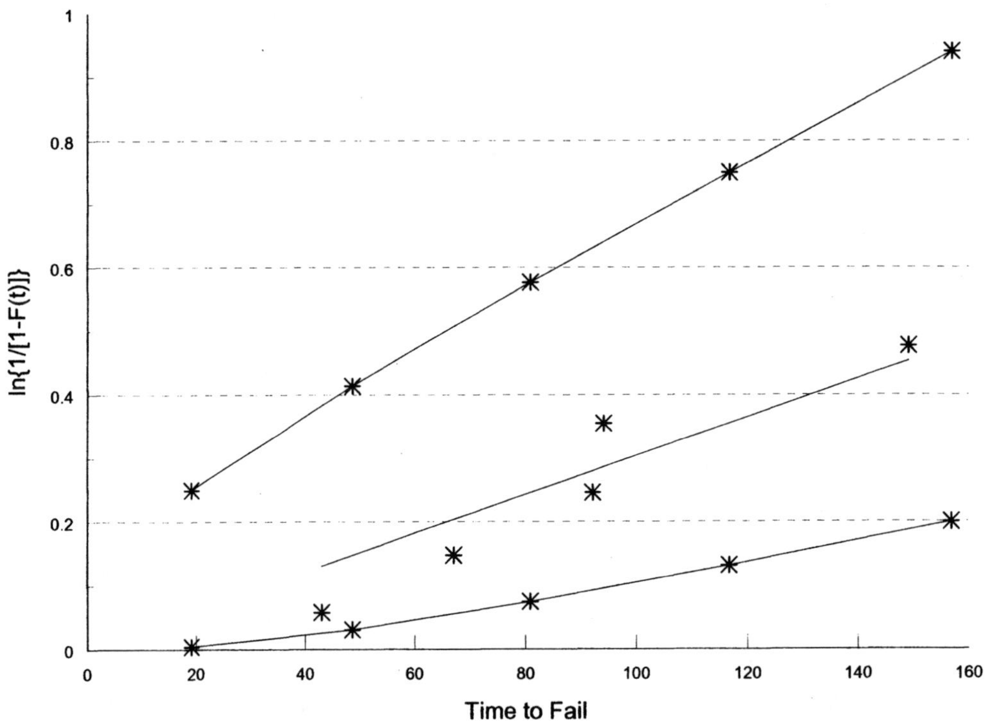

Figure 3.3. Exponential probability plot with confidence limits.

3.7 Normal distribution

The normal probability density function is

$$f(x) = \frac{1}{\sigma\sqrt{2\pi}} \exp\left[-\frac{1}{2}\left(\frac{x-\mu}{\sigma}\right)^2\right], \quad -\infty < x < \infty \tag{3.20}$$

where: μ is the distribution mean, and

σ is the distribution standard deviation.

If no censoring is involved, the distribution mean is estimated from the expression

$$\hat{\mu} = \bar{x} = \frac{\sum_{i=1}^{n} x_i}{n} \tag{3.21}$$

where: n is the sample size.

If no censoring is involved, the distribution standard deviation is estimated from the expression

$$\hat{\sigma} = \sqrt{\frac{n\sum_{i=1}^{n} x_i^2 - \left(\sum_{i=1}^{n} x_i\right)^2}{n(n-1)}} \tag{3.22}$$

However, when censored data are involved, parameter estimation becomes complicated. Three popular methods for parameter estimation for the normal distribution when censored data are encountered are (1) maximum likelihood estimation, (2) hazard plotting, and (3) probability plotting. The following sections present each of these alternatives.

3.7.1 Maximum likelihood estimation

The maximum likelihood equations for the normal distribution are

$$\frac{\partial L}{\partial \mu} = \frac{r}{\sigma}\left[\frac{\bar{x}-\mu}{\sigma} + \sum_{i=1}^{k} \frac{h(x_i)}{r}\right] = 0 \tag{3.23}$$

$$\frac{\partial L}{\partial \sigma} = \frac{r}{\sigma}\left[\frac{s^2 + (\bar{x}-\mu)^2}{\sigma^2} - 1 + \sum_{i=1}^{k} \frac{z(x_i)h(x_i)}{r}\right] = 0 \tag{3.24}$$

where: r is the number of failures,

k is the number of censored observations,

$\bar{x}$ is the sample mean of the failures,

s is the sample standard deviation for the failures,

$z(x_i)$ is the standard normal deviate

$$z(x_i) = \frac{x_i - \mu}{\sigma}, \text{ and}$$

$h(x_i)$ is the hazard function evaluated at the ith point

$$h(x_i) = \frac{\phi(z(x_i))}{\sigma[1-\Phi(z(x_i))]}$$

where: $\phi(z(x_i))$ is the standard normal probability density function evaluated at the *i*th point, and

$\Phi(z(x_i))$ is the standard normal cumulative distribution function evaluated at the *i*th point.

Note that if no censored data are involved these expressions reduce to the sample mean and the sample standard deviation.

Iterative techniques are necessary to solve Equations 3.23 and 3.24. A standard method based on Taylor series expansions involves repeatedly estimating the parameters until a desired level of accuracy is reached. Estimates of μ and σ are given by the expressions

$$\hat{\mu}_j = \hat{\mu}_{j-1} + h \tag{3.25}$$

$$\hat{\sigma}_j = \hat{\sigma}_{j-1} + k \tag{3.26}$$

where: *h* is a correction factor for the distribution mean, and

k is a correction factor for the distribution standard deviation.

For each iteration, the correction factors are estimated from the expressions

$$h\frac{\partial^2 L}{\partial \mu^2} + k\frac{\partial^2 L}{\partial \mu \partial \sigma} = -\frac{\partial L}{\partial \mu} \tag{3.27}$$

$$h\frac{\partial^2 L}{\partial \mu \partial \sigma} + k\frac{\partial^2 L}{\partial \sigma^2} = -\frac{\partial L}{\partial \sigma} \tag{3.28}$$

where,

$$\frac{\partial^2 L}{\partial \mu^2} = -\frac{r}{\sigma^2}\left[1 + \sum_{i=1}^{k}\frac{A_i}{r}\right] \tag{3.29}$$

$$\frac{\partial^2 L}{\partial \mu \partial \sigma} = -\frac{r}{\sigma^2}\left[\frac{2(\bar{x}-\mu)}{\sigma} + \sum_{i=1}^{k}\frac{B_i}{r}\right] \tag{3.30}$$

$$\frac{\partial^2 L}{\partial \sigma^2} = -\frac{r}{\sigma^2}\left[\frac{3\{s^2 + (\bar{x}-\mu)^2\}}{\sigma^2} - 1 + \sum_{i=1}^{k}\frac{C_i}{r}\right] \tag{3.31}$$

and

$$A_i = h(x_i)[h(x_i) - z(x_i)]$$

$$B_i = h(x_i) + z(x_i)A_i$$

$$C_i = z(x_i)[h(x_i) + B_i]$$

The estimated parameters are asymptotically normal. The variances of the estimates can be found by inverting the local information matrix.

$$F = \begin{bmatrix} -\dfrac{\partial^2 L}{\partial \mu^2} & -\dfrac{\partial^2 L}{\partial \mu \partial \sigma} \\ -\dfrac{\partial^2 L}{\partial \mu \partial \sigma} & -\dfrac{\partial^2 L}{\partial \sigma^2} \end{bmatrix} \tag{3.32}$$

After inversion, the variances are

$$F^{-1} = \begin{bmatrix} \mathrm{var}(\hat{\mu}) & \mathrm{cov}(\hat{\mu},\hat{\sigma}) \\ \mathrm{cov}(\hat{\mu},\hat{\sigma}) & \mathrm{var}(\hat{\sigma}) \end{bmatrix} \tag{3.33}$$

Approximate $(1-\alpha)100\%$ confidence intervals for the estimated parameters are

$$\hat{\mu} - K_{\alpha/2}\sqrt{\mathrm{var}(\hat{\mu})} \le \mu \le \hat{\mu} + K_{\alpha/2}\sqrt{\mathrm{var}(\hat{\mu})} \tag{3.34}$$

$$\frac{\hat{\sigma}}{\exp\left[\dfrac{K_{\alpha/2}\sqrt{\mathrm{var}(\hat{\sigma})}}{\hat{\sigma}}\right]} \le \sigma \le \hat{\sigma}\exp\left[\frac{K_{\alpha/2}\sqrt{\mathrm{var}(\hat{\sigma})}}{\hat{\sigma}}\right] \tag{3.35}$$

where $K_{\alpha/2}$ is the inverse of the standard normal probability density function.

These confidence intervals are approximate, but approach exactness as the sample size increases.

Confidence intervals for reliability can be found using the expressions

$$\mathrm{var}(\hat{z}) \approx \left(\frac{\mathrm{var}(\hat{\mu}) + \hat{z}^2\,\mathrm{var}(\hat{\sigma}) + 2\hat{z}\,\mathrm{cov}(\hat{\mu},\hat{\sigma})}{\hat{\sigma}^2}\right) \tag{3.36}$$

$$\hat{z} - K_{\alpha/2}\sqrt{\mathrm{var}(\hat{z})} \le z \le \hat{z} + K_{\alpha/2}\sqrt{\mathrm{var}(\hat{z})} \tag{3.37}$$

$$1 - \Phi\left(\hat{z} + K_{\alpha/2}\sqrt{\mathrm{var}(\hat{z})}\right) \le R(x) \le 1 - \Phi\left(\hat{z} - K_{\alpha/2}\sqrt{\mathrm{var}(\hat{z})}\right) \tag{3.38}$$

Confidence intervals for percentiles can be found using the expressions

$$\mathrm{var}(\hat{x}) \approx \mathrm{var}(\hat{\mu}) + \hat{z}^2\,\mathrm{var}(\hat{\sigma}) + 2\hat{z}\,\mathrm{cov}(\hat{\mu},\hat{\sigma}) \tag{3.39}$$

$$\hat{x} - K_{\alpha/2}\sqrt{\mathrm{var}(\hat{x})} \le x \le \hat{x} + K_{\alpha/2}\sqrt{\mathrm{var}(\hat{x})} \tag{3.40}$$

Maximum likelihood estimation is tedious and computer routines are often employed. Commercial software is available for these calculations, such as the software developed by Dodson and Kirkland (2000).

3.7.2 Hazard Plotting

The normal cumulative hazard function is

$$H(x) = -\ln\left[1 - \Phi\left(\frac{x-\mu}{\sigma}\right)\right] \tag{3.41}$$

where $\Phi(x)$ is the standard normal cumulative distribution function.

By rearranging Equation 3.41, the survival time can be represented as a function of the cumulative hazard function.

$$x = \mu + \sigma\Phi^{-1}\left[1 - e^{-H(x)}\right] \tag{3.42}$$

where $\Phi^{-1}(x)$ is the inverse of the standard normal cumulative distribution function.

By plotting x versus

$$\Phi^{-1}\left[1 - e^{-H(x)}\right]$$

the resulting y-intercept equals μ and the resulting slope equals σ. The hazard function, $h(x)$, is estimated from the inverse of the reverse rank of the ordered failures, and $H(x)$ is the cumulative of the values of $h(x)$. Censored data points are used to compute ranks, but are not included in hazard plots.

An alternative to plotting x versus

$$\Phi^{-1}\left[1 - e^{-H(x)}\right]$$

on conventional graph paper is to plot x versus $H(x)$ on specialized hazard paper. The advantage of hazard paper is that the values of

$$\Phi^{-1}\left[1 - e^{-H(x)}\right]^{*}$$

do not have to be computed. With sophisticated electronic spreadsheets, the disadvantages of hazard paper outweigh the advantages. Electronic spreadsheets with built-in statistical functions can quickly make data transformations, determine a best fit straight line using linear regression, and graph the results.

Example 3.9

Use hazard plotting to determine the parameters of the normal distribution given the following multiple censored data set. A "c" following an entry indicates the censoring.

Time to Fail		
150 c	183 c	234.9
157 c	209	235
167 c	216 c	248 c
179	217 c	257

* If using Lotus 123™, the inverse of the standard normal distribution is found with the @function @normal(x,0,1,1).

Solution

Table 3.5 is constructed to obtain the necessary plotting data.

Table 3.5. Tabulations for a hazard plot for the normal distribution.

Time to Fail	Reverse Rank	$h(t)$	$H(t)$	$1-e^{-H(t)}$	$\Phi^{-1}\left[1-e^{-H(t)}\right]$
150 c	12				
157 c	11				
167 c	10				
179	9	0.1111	0.1111	0.1052	–1.2527
183 c	8				
209	7	0.1429	0.2540	0.2243	–0.7578
216 c	6				
217 c	5				
235	4	0.2500	0.5040	0.3959	–0.2640
235	3	0.3333	0.8373	0.5671	0.1691
248 c	2				
257	1	1.0000	1.8373	0.8408	0.9976

Now the five failure times can be plotted against

$$\Phi^{-1}\left[1-e^{-H(t)}\right].$$

This is shown in Figure 3.4. The best fit straight line through the data points was found using linear regression. The *y*-intercept of the best fit straight line through the five points provides an estimate of μ–230.3, and the slope of the line provides an estimate of σ–32.9.

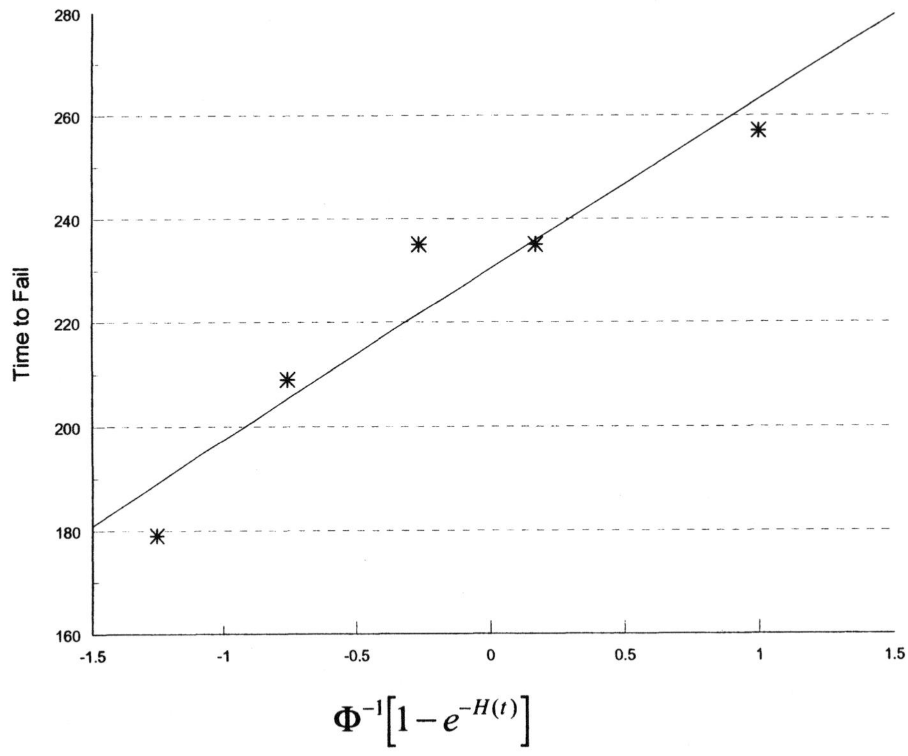

Figure 3.4. Hazard plot for the normal distribution.

3.7.3 Probability Plotting

By rearranging the normal cumulative distribution function, a linear expression can be obtained.

$$x = \mu + \sigma\Phi^{-1}\left[F(x)\right] \tag{3.43}$$

where: $F(x)$ is the normal cumulative distribution function, and

$\Phi^{-1}(x)$ is the inverse of the standard normal cumulative distribution function.

By plotting x versus

$$\Phi^{-1}\left[F(x)\right]$$

the resulting y-intercept equals μ and the resulting slope equals σ. The cumulative distribution function, $F(x)$, is usually estimated from the median rank, but other estimates such as the mean rank and the Kaplan-Meier product limit estimator are also used. Median ranks are estimated using Equations 3.14–3.16 in Section 3.7.6.3.

An alternative to plotting x versus

$$\Phi^{-1}\left[F(x)\right]$$

on conventional graph paper is to plot x versus $F(x)$ on specialized probability paper. The advantage of probability paper is that the values of

$$\Phi^{-1}\left[F(x)\right]$$

do not have to be computed. With sophisticated electronic spreadsheets, the disadvantages of hazard paper outweigh the advantages. Electronic spreadsheets with built-in statistical functions can quickly make data transformations, determine a best fit straight line using linear regression, and graph the results.

Example 3.10

Use probability plotting to determine the parameters of the normal distribution given the following data set. A "c" following an entry indicates the censoring.

Time to Fail

150 c	183 c	235
157 c	209	235
167 c	216 c	248 c
179	217 c	257

Solution

Table 3.6 is constructed to obtain the data for plotting. This data is plotted in Figure 3.5. The slope of this plot is 34.8, which is the estimated value of σ. The y-intercept is 235.3, which is the estimated value of μ.

Table 3.6. Tabulations for a probability plot for normal data.

Time to Fail	I_i	O_i	Median Rank, $F(x)$	$\Phi^{-1}\left[F(x)\right]$
150 c				
157 c				
167 c				
179	1.3000	1.3000	0.0806	–1.4008
183 c				
209	1.4625	2.7625	0.1986	–0.8467
216 c				
217 c				
235	2.0475	4.8100	0.3637	–0.3486
235	2.0475	6.8575	0.5288	0.0723
248 c				
257	3.0713	9.9288	0.7765	0.7605

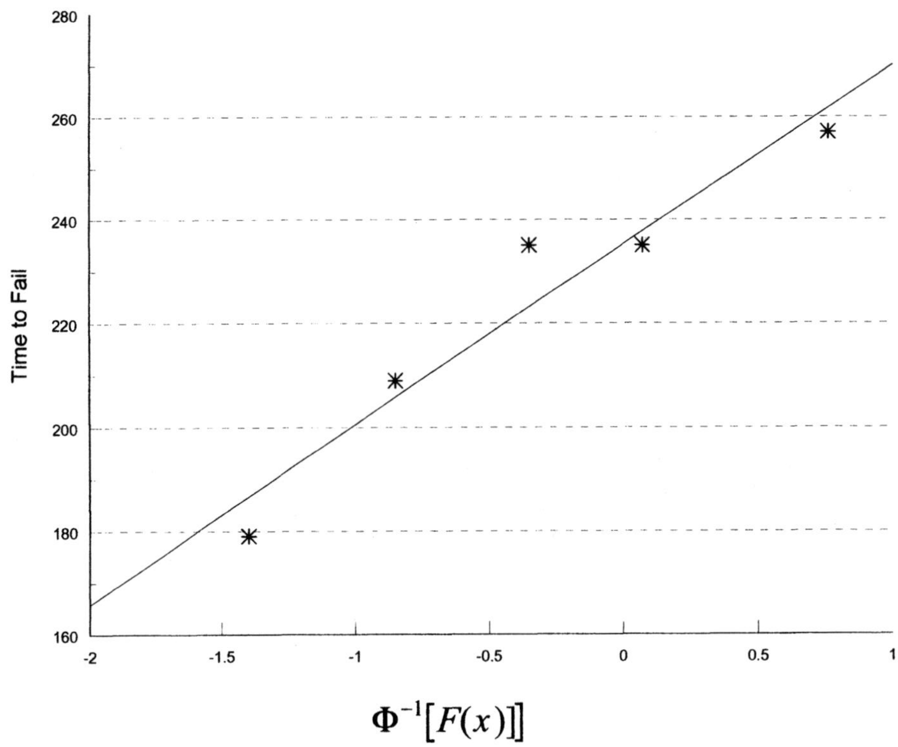

Figure 3.5. Normal probability plot.

Nonparametric confidence limits for reliability can be added to a normal probability plot using 5% and 95% ranks as explained in Section 3.7.6.3. The plotting position for the standard normal inverse of the 5% and 95% ranks is

$$x = \mu + \sigma\Phi^{-1}\left[F(x)\right] \tag{3.44}$$

Note that $F(x)$ is the median rank.

Example 3.11

Construct a probability plot including confidence limits for the previous example.

Solution

Table 3.7 contains the calculations to obtain the confidence limits and their plotting positions.

Table 3.7. Confidence limit calculations for a normal probability plot.

Time to Fail	Failure Order	Median Rank	$\Phi^{-1}[F(x)]$	5% Rank	Interpolated 5% Rank	Φ^{-1} 5% Rank	95% Rank	Interpolated 95% Rank	Φ^{-1} 95% Rank	Plotting Position
	1			0.0043			0.2209			
179	1.3	0.0806	−1.4008		0.0121	−2.2533		0.2562	−0.6550	186.55216
	2			0.0305			0.3387			
209	2.7625	0.1986	−0.8467		0.0620	−1.5379		0.4145	−0.2160	205.83484
	3			0.0719			0.4381			
	4			0.1229			0.5273			
235	4.81	0.3637	−0.3486		0.1700	−0.9543		0.5936	0.2368	223.16872
	5			0.1810			0.6091			
	6			0.2453			0.6848			
235	6.8575	0.5288	0.0723		0.3053	−0.5093		0.7447	0.6580	237.81604
	7			0.3152			0.7547			
	9			0.4727			0.8190			
257	9.9288	0.7765	0.7605		0.5555	0.1397		0.9204	1.4075	261.7654
	10			0.5619			0.9281			

A probability plot including confidence limits is shown in Figure 3.6.

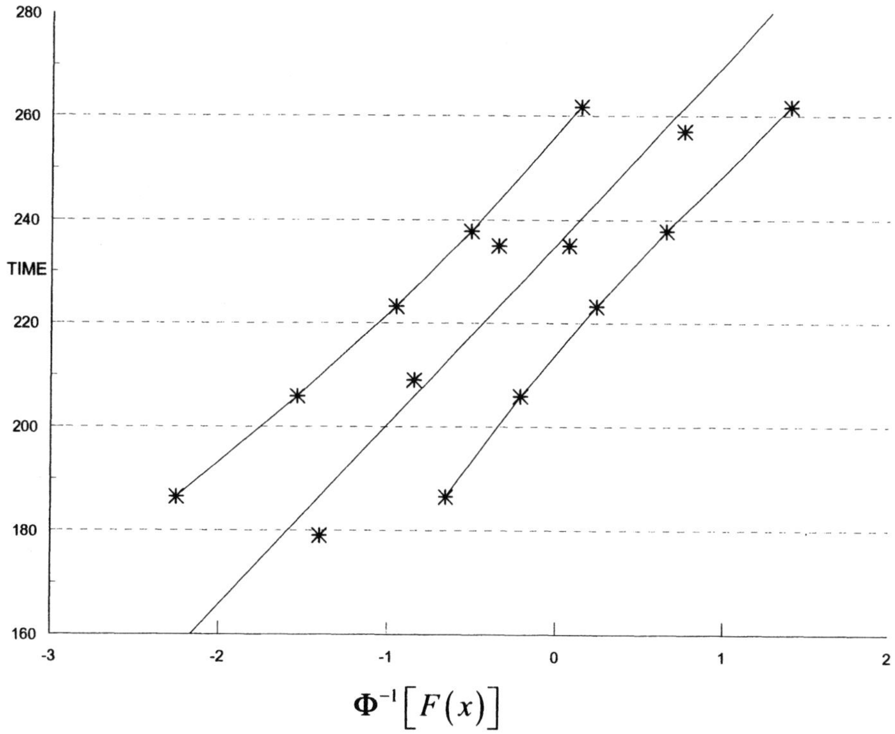

Figure 3.6. Normal probability plot with confidence limits.

3.8 Lognormal distribution

The lognormal probability density function is

$$f(x)=\frac{1}{\sigma x\sqrt{2\pi}}\exp\left[-\frac{1}{2}\left(\frac{\ln x-\mu}{\sigma}\right)^2\right],x>0 \tag{3.45}$$

where: μ is the location parameter, and
σ is the shape parameter.

If x is a lognormal distributed random variable, then y = ln(x) is a normally distributed random variable. The location parameter is equal to the mean of the logarithm of the data points, and the shape parameter is equal to the standard deviation of the logarithm of the data points. Thus, the lognormal distribution does not have to be dealt with as a separate distribution. By taking the logarithm of the data points, the techniques developed for the normal distribution discussed in the previous section can be used to estimate the parameters of the lognormal distribution.

3.9 Weibull distribution

The Weibull probability density function is

$$f(x)=\frac{\beta x^{\beta-1}}{\theta^{\beta}}\exp\left(\frac{x}{\theta}\right)^{\beta},x\geq 0 \tag{3.46}$$

where: β is the shape parameter, and
θ is the scale parameter.

In some cases, a three-parameter Weibull distribution provides a better fit than the two-parameter Weibull distribution. The difference in the two distributions is the location parameter, δ, which shifts the distribution along the x-axis. By definition, there is a zero probability of failure for x < δ. Although unusual, the location can be negative; this implies that items were failed prior to testing. The three parameter Weibull distribution is

$$f(x)=\frac{\beta(x-\delta)^{\beta-1}}{\theta^{\beta}}\exp\left(\frac{x-\delta}{\theta}\right)^{\beta},x\geq\delta \tag{3.47}$$

Four methods for estimating the parameters of the Weibull distribution are presented in this section: (1) maximum likelihood estimation, (2) hazard plotting, (3) probability plotting, and (4) matching moments.

3.9.1 Maximum likelihood estimation

The maximum likelihood equations for the Weibull distribution are

$$\frac{1}{r}\sum_{i=1}^{r}\ln(x_i)=\left[\sum_{i=1}^{n}x_i^{\beta}\ln(x_i)\right]\left[\sum_{i=1}^{n}x_i^{\beta}\right]^{-1}-\frac{1}{\beta} \tag{3.48}$$

$$\hat{\theta}=\left[\frac{1}{r}\sum_{i=1}^{n}x_i^{\hat{\beta}}\right]^{1/\hat{\beta}} \tag{3.49}$$

where: r is the number of failures, and

n is the total number of data points, both censored and uncensored.

Iterative techniques are required to solve Equation 3.48. The estimated parameters are asymptotically normal. The variances of the estimates can be found by inverting the local information matrix. is

$$F=\begin{bmatrix}-\frac{\partial^2 L}{\partial\beta^2} & -\frac{\partial^2 L}{\partial\beta\partial\theta}\\ -\frac{\partial^2 L}{\partial\beta\partial\theta} & -\frac{\partial^2 L}{\partial\theta^2}\end{bmatrix} \tag{3.50}$$

The second partial derivatives of the log-likelihood equation are

$$\frac{\partial^2 L}{\partial\beta^2}=\sum_r\left[-\frac{1}{\beta}-\left(\frac{x_i}{\theta}\right)^{\beta}\ln^2\left(\frac{x_i}{\theta}\right)\right]+\sum_k\left[-\left(\frac{x_i}{\theta}\right)^{\beta}\ln^2\left(\frac{x_i}{\theta}\right)\right] \tag{3.51}$$

$$\frac{\partial^2 L}{\partial\theta^2}=\sum_r\left[\frac{\beta}{\theta^2}-\left(\frac{x_i}{\theta}\right)^{\beta}\left(\frac{\beta}{\theta^2}\right)(\beta+1)\right]+\sum_k\left[-\left(\frac{x_i}{\theta}\right)^{\beta}\left(\frac{\beta}{\theta^2}\right)(\beta+1)\right] \tag{3.52}$$

$$\frac{\partial^2 L}{\partial\beta\partial\theta}=\sum_r\left\{-\frac{1}{\theta}+\left(\frac{x_i}{\theta}\right)^{\beta}\left(\frac{1}{\theta}\right)\left[\beta\ln\left(\frac{x_i}{\theta}\right)+1\right]\right\}$$
$$+\sum_k\left\{\left(\frac{x_i}{\theta}\right)^{\beta}\left(\frac{1}{\theta}\right)\left[\beta\ln\left(\frac{x_i}{\theta}\right)+1\right]\right\} \tag{3.53}$$

where: $\sum_r$ represents summation over all failures, and

$\sum_k$ represents summation over all censored points.

The variances of the estimated parameters are

$$F^{-1}=\begin{bmatrix}\operatorname{var}(\hat{\beta}) & \operatorname{cov}(\hat{\beta},\hat{\theta})\\ \operatorname{cov}(\hat{\beta},\hat{\theta}) & \operatorname{var}(\hat{\theta})\end{bmatrix} \tag{3.54}$$

Approximate $(1-\alpha)100\%$ confidence intervals for the estimated parameters are

$$\frac{\hat{\beta}}{\exp\left(\frac{K_{\alpha/2}\sqrt{\text{var}(\hat{\beta})}}{\hat{\beta}}\right)} \leq \beta \leq \hat{\mu} + \hat{\beta}\exp\left(\frac{K_{\alpha/2}\sqrt{\text{var}(\hat{\beta})}}{\hat{\beta}}\right) \tag{3.55}$$

$$\frac{\hat{\theta}}{\exp\left[\frac{K_{\alpha/2}\sqrt{\text{var}(\hat{\theta})}}{\hat{\theta}}\right]} \leq \theta \leq \hat{\theta}\exp\left[\frac{K_{\alpha/2}\sqrt{\text{var}(\hat{\theta})}}{\hat{\theta}}\right] \tag{3.56}$$

where $K_{\alpha/2}$ is the inverse of the standard normal probability density function.

These confidence intervals are approximate, but approach exactness as the sample size increases.

Confidence intervals for reliability can be found using the expressions

$$\begin{aligned}&\exp\left[-\exp\left(u + K_{a/2}\sqrt{\text{var}(u)}\right)\right] \\ &\quad \leq R(x) \leq \exp\left[-\exp\left(u - K_{a/2}\sqrt{\text{var}(u)}\right)\right]\end{aligned} \tag{3.57}$$

$$u = \beta\left[\ln(x) - \ln(\theta)\right] \tag{3.58}$$

$$\text{var}(\hat{u}) \approx \beta^2\left[\left(\frac{\text{var}(\hat{\theta})}{\hat{\theta}^2}\right) + \left(\frac{\hat{u}^2\,\text{var}(\hat{\beta})}{\beta^4}\right) - \left(\frac{2\hat{u}\,\text{cov}(\hat{\beta},\hat{\theta})}{\hat{\beta}^2\hat{\theta}}\right)\right] \tag{3.59}$$

Confidence intervals for percentiles can be found using the expressions

$$e^{y_L} \leq \hat{x} \leq e^{y_U} \tag{3.60}$$

$$\hat{x} = \hat{\theta}\left[-\ln(1-p)\right]^{1/\hat{\beta}} \tag{3.61}$$

$$y_L = \ln\hat{\theta} + \frac{\ln\left[-\ln(1-p)\right]}{\hat{\beta}} - K_\alpha\sqrt{\text{var}(\hat{y})} \tag{3.62}$$

$$y_U = \ln\hat{\theta} + \frac{\ln\left[-\ln(1-p)\right]}{\hat{\beta}} + K_\alpha\sqrt{\text{var}(\hat{y})} \tag{3.63}$$

$$\begin{aligned}\text{var}(\hat{y}) &= \frac{\text{var}(\hat{\theta})}{\hat{\theta}^2} + \frac{\left\{\ln\left[-\ln(1-p)\right]\right\}^2 \text{var}(\hat{\beta})}{\beta^4} \\ &\quad - \frac{2\left\{\ln\left[-\ln(1-p)\right]\right\}\text{cov}(\hat{\theta},\hat{\beta})}{\hat{\beta}^2\hat{\theta}}\end{aligned} \tag{3.64}$$

Maximum likelihood estimation is tedious and computer routines are often employed. Commercial software is available for these calculations, such as the software developed by Dodson and Kirkland (1994).

3.9.2 Hazard Plotting

The Weibull cumulative hazard function is

$$H(x) = -\ln[1 - F(x)] \tag{3.65}$$

Replacing $F(x)$ and rearranging gives a linear expression.

$$\ln H(x) = \beta \ln x - \beta \ln \theta \tag{3.66}$$

By plotting $\ln[H(x)]$ versus $\ln(x)$ the resulting slope (censored points are not plotted) provides an estimate of β. The y-intercept of this plot is an estimate of βlnθ. Thus, θ is estimated from the expression

$$\hat{\theta} = \exp\left(-\frac{y_0}{\hat{\beta}}\right) \quad (3.67$$

where, y_0 is the y-intercept of the hazard plot.

The hazard function $h(x)$ is estimated from the inverse of the reverse rank of the ordered failures; and the cumulative hazard function, $H(x)$, is the cumulative of the values of $h(x)$. An alternative to plotting $\ln[H(x)]$ versus $\ln x$ is to directly plot $H(x)$ versus x on specialized Weibull hazard paper. The advantage of hazard paper is that logarithmic transformations do not have to be computed. With electronic spreadsheets, these transformations are easily computed, and hazard paper is not needed. Electronic spreadsheets also make fitting a straight line through the data points and constructing the graph easier.

Example 3.12

Determine the parameters of the Weibull distribution using the multiple censored data in the table below. A "c" following an entry indicates censoring.

Time to Fail	
309 c	229
386	104 c
180	217 c
167 c	168
122	138

Solution

Table 3.8 is constructed to obtain the necessary plotting data.

Table 3.8. Tabulations for a hazard plot for the Weibull distribution.

Time to Fail	Reverse Rank	$h(t)$	$H(t)$	$\ln H(t)$	$\ln t$
104 c	10				
122	9	0.1111	0.1111	–2.1972	4.8040
138	8	0.1250	0.2361	–1.4435	4.9273
167 c	7				
168	6	0.1667	0.4028	–0.9094	5.1240
180	5	0.2000	0.6028	–0.5062	5.1930
217 c	4				
229	3	0.3333	0.9361	–0.0660	5.4337
309 c	2				
386	1	1.0000	1.9361	0.6607	5.9558

Now the final two columns of Table 3.8 can be plotted. This plot is shown in Figure 3.7. The slope of the best-fit straight line through the data (found by linear regression) is equal to 2.34 and provides an estimate of β. The y-intercept of the best-fit straight line through the data is
–13.004. The estimated scale parameter for the Weibull distribution is

$$\hat{\theta} = \exp\left(-\frac{-13.004}{2.34}\right) = 259.1$$

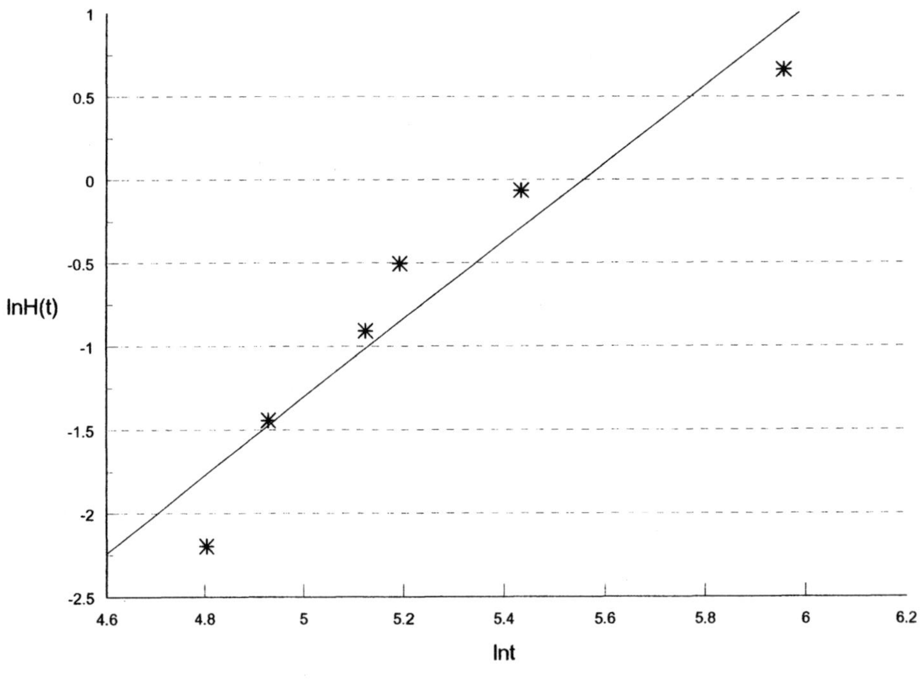

Figure 3.7. Weibull distribution Hazard plot.

3.9.3 Probability Plotting

By taking the logarithm of the Weibull cumulative distribution function twice and rearranging,

$$\ln\left[\ln\left(\frac{1}{1-F(x)}\right)\right] = \beta \ln x - \beta \ln \theta \tag{3.68}$$

By plotting

$$\ln\left[\ln\left(\frac{1}{1-F(x)}\right)\right]$$

versus $\ln(x)$, and fitting a straight line to the points, the parameters of the Weibull distribution can be estimated. The slope of the plot provides an estimate of β, and the y-intercept can be used to estimate θ.

$$\hat{\theta} = \exp\left(-\frac{y_0}{\hat{\beta}}\right) \tag{3.69}$$

The cumulative distribution function, $F(x)$, is usually estimated from the median rank, but other estimates such as the mean rank and the Kaplan-Meier product limit estimator are also used. Median ranks are estimated using Equations 3.14–3.16 in Section 3.7.6.3. Specialized probability paper is available for probability plotting. Using probability paper eliminates the need to transform the data prior to plotting, but this hardly seems necessary with the availability and speed of electronic spreadsheets.

Example 3.13

Determine the parameters of the Weibull distribution using probability plotting for the data in Example 3.12.

Solution

Table 3.9 is constructed to obtain the necessary plotting data.

Table 3.9. Tabulations for a probability plot for the Weibull distribution.

Time to Fail	I_i	O_i	Median Rank, $\hat{F}(t)$	$\ln\left[\ln\left(\frac{1}{1-F(t)}\right)\right]$	$\ln t$
104 c					
122	1.1000	1.1000	0.0769	–2.5252	4.8040
138	1.1000	2.2000	0.1827	–1.6008	4.9273
167 c					
168	1.2571	3.4571	0.3036	–1.0167	5.1240
180	1.2571	4.7143	0.4245	–0.5934	5.1930
217 c					
229	1.5714	6.2857	0.5755	–0.1544	5.4337
309 c					
386	2.3571	8.6429	0.8022	0.4827	5.9558

The last two columns of Table 3.9 are plotted in Figure 3.8. The slope of the best-fit straight line through the data (found using linear regression) is 2.41, which is the estimated value of β. The *y*-intercept of the best-fit straight line through the data is –13.55. The estimated shape parameter is

$$\hat{\theta} = \exp\left(-\frac{-13.55}{2.41}\right) = 276.6$$

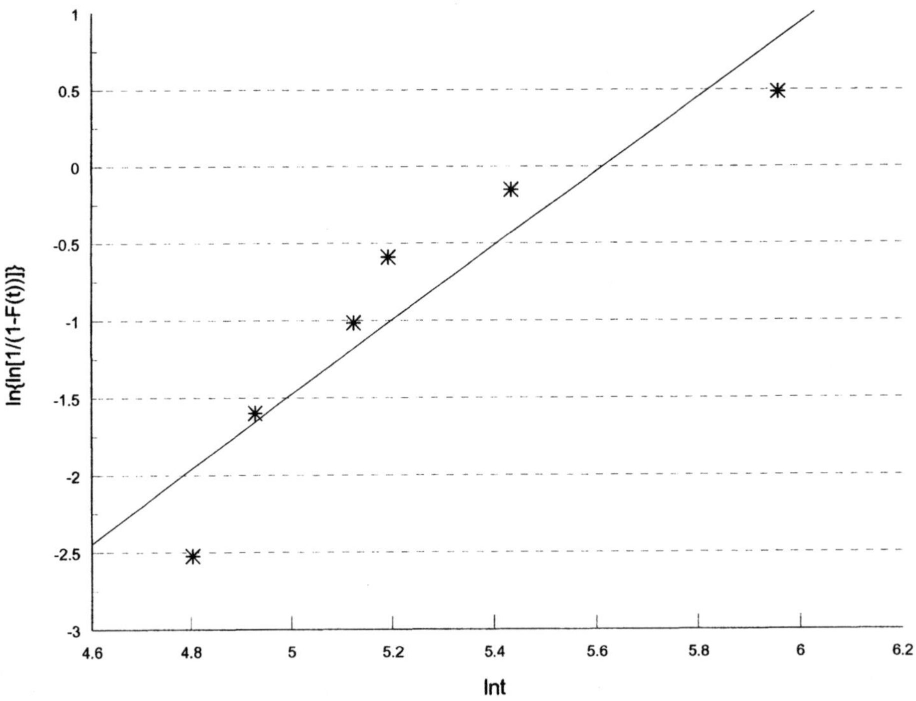

Figure 3.8. Weibull distribution Probability.

Confidence limits can be added to this plot using 5% and 95% ranks as described in Section 3.7.6.3. Plotting position for the 5% and 95% ranks are found from the expression

$$x_i = \theta\left[\ln\left(\frac{1}{1-F(x)}\right)\right]^{1/\beta} \tag{3.70}$$

3.9.4 Matching Moments

The mean of the Weibull distribution is defined as

$$\mu = \theta\Gamma\left(1+\frac{1}{\beta}\right) \tag{3.71}$$

The variance of the Weibull distribution is

$$\sigma^2 = \theta^2\left[\Gamma\left(1+\frac{1}{\beta}\right)-\Gamma^2\left(1+\frac{1}{\beta}\right)\right] \tag{3.72}$$

The parameters of the Weibull distribution can be found by equating the two equations above to the sample mean and sample variance of a data set.

Example 3.14

Determine the parameters of the Weibull distribution by matching moments for the following failure times; 34, 56, 71, 43, and 99.

Solution

The sample mean of the five failure times is 60.6, and the sample variance is 655.3. By matching moments, the following two expressions are obtained

$$60.6 = \theta\Gamma\left(1+\frac{1}{\beta}\right)$$

$$655.3 = \theta^2\Gamma\left[\Gamma\left(1+\frac{1}{\beta}\right)-\Gamma^2\left(1+\frac{1}{\beta}\right)\right]$$

Solving these two expressions simultaneously gives approximate values of 2.54 as an estimate of β and 68.3 as an estimate of θ.

3.10 Goodness of fit

Before applying a distribution, a test of fit should be applied. There are many goodness-of-fit tests. Some are distribution specific, and some are general. Two popular goodness-of-fit tests, the chi-square test and the Kolmogorov-Smirnov test are of limited value for reliability analysis. The chi-square test requires a large sample size ($n > 25$) and cannot be used with multiple censored data, and the Kolmogorov-Smirnov test generally has less power than distribution specific tests.

A general goodness-of-fit test that can be used with multiple censored data is the Hollander-Proschan test. This test is best illustrated with an example. Consider the data in Table 3.10. It is believed that this data follows a Weibull time to fail distribution, and maximum likelihood estimates for the shape and scale parameters are 2.974 and 203.3.

Table 3.10. Example data for goodness-of-fit.

Time to Fail			
42.1 +	105.9	151.3 +	195.6 +
77.8	117.0	157.3	207.0
83.3 +	126.9	163.8	215.3 +
88.7 +	138.7	177.2 +	217.4
101.8	148.9	194.3 +	258.8 +

A + following an observation indicates the item was removed from testing prior to failure.

To test if the Weibull distribution with shape and scale parameters of 2.974 and 203.3 adequately fit this data, follow the steps below.

1. For each **uncensored** observation, use the Kaplan-Meier product limit method to estimate reliability. The Kaplan-Meier reliability estimate is

$$\hat{R}(x_i) = \prod_{i=1}^{i} \left(\frac{n-i}{n-i+1} \right)^{Q_i} \tag{3.73}$$

where: Q_i = 1 for uncensored data and 0 for censored data, and

n is the sample size including both censored and uncensored observations.

2. For each observation and the final observation, uncensored or not, compute the jump of the Kaplan-Meier reliability estimate. The jump for censored observations is zero with the exception of the final observation. The Kaplan-Meier jump is

$$\hat{f}(x_i) = \frac{1}{n} \prod_{j=1}^{i-1} \left(\frac{n-j+1}{n-j} \right)^{1-Q_j} \tag{3.74}$$

With the exception of the final observation, if it is censored, the Kaplan-Meier jump can also be estimated from the expression

$$\hat{f}(x_i) = \hat{R}(x_p) - \hat{R}(x_i) \tag{3.75}$$

where $\hat{R}(x_p)$ is the Kaplan-Meier reliability estimate for the most previous uncensored point.

3. Calculate the reliability for every observation using the distribution being tested for goodness of fit. For this example, the reliability expression is

$$R_0(x) = e^{-\left(\frac{x}{203.3}\right)^{2.974}}$$

4. For every observation, compute the product of the Kaplan-Meier reliability estimate and the Kaplan-Meier jump.
5. For every observation, raise the reliability computed for the distribution being tested for goodness of fit to the fourth power.
6. For every observation, calculate the difference between reliability using the distribution being tested for goodness of fit for the previous point and the current point. For the first observation this is equal to one minus the reliability using the distribution being tested for goodness of fit.
7. For every observation, multiple the value obtained in step 6 by $n/(n-i+1)$.

The calculations in steps 1–7 are summarized in Table 3.11.

Table 3.11. Summary calculations for steps 1–7.

a	b		c	d	e	f	g	h	i	j
Order	x		$\frac{n-i}{n-i+1}$	$\hat{R}(x)$	Jump	$R_0(x)$	(e)*(f)	$R_0^4(x)$	$R_0^4(x_{i-1})$ - $R_0^4(x_i)$	$\frac{n-i}{n-i+1}*(i)$
1	42.1	c			0.0000	0.9908	0.0000	0.9637	0.0363	0.0363
2	77.8		0.9474	0.9474	0.0526	0.9442	0.0497	0.7947	0.1690	0.1779
3	83.3	c			0.0000	0.9320	0.0000	0.7546	0.0401	0.0445
4	88.7	c			0.0000	0.9186	0.0000	0.7122	0.0424	0.0499
5	101.8		0.9375	0.8882	0.0592	0.8800	0.0521	0.5997	0.1125	0.1406
6	105.9		0.9333	0.8289	0.0592	0.8661	0.0513	0.5627	0.0370	0.0494
7	117.0		0.9286	0.7697	0.0592	0.8242	0.0488	0.4614	0.1013	0.1447
8	126.9		0.9231	0.7105	0.0592	0.7818	0.0463	0.3735	0.0879	0.1352
9	138.7		0.9167	0.6513	0.0592	0.7256	0.0430	0.2772	0.0963	0.1605
10	148.9		0.9091	0.5921	0.0592	0.6729	0.0398	0.2051	0.0721	0.1312
11	151.3	c			0.0000	0.6601	0.0000	0.1899	0.0152	0.0305
12	157.3		0.8889	0.5263	0.0658	0.6273	0.0413	0.1549	0.0350	0.0778
13	163.8		0.8750	0.4605	0.0658	0.5910	0.0389	0.1220	0.0329	0.0822
14	177.2	c			0.0000	0.5145	0.0000	0.0701	0.0519	0.1483
15	194.3	c			0.0000	0.4173	0.0000	0.0303	0.0398	0.1325
16	195.6	c			0.0000	0.4100	0.0000	0.0283	0.0021	0.0082
17	207.0		0.7500	0.3454	0.1151	0.3482	0.0401	0.0147	0.0136	0.0679
18	215.3	c			0.0000	0.3055	0.0000	0.0087	0.0060	0.0399
19	217.4		0.5000	0.1727	0.1727	0.2950	0.0509	0.0076	0.0011	0.0113
20	258.8	c			0.1727	0.1287	0.0222	0.0003	0.0073	0.1460
							0.5244			1.8147

8. The sum of column *g* is designated as *C*, and the sum of column *j* is designated as *Y*. For this example, $C = 0.5244$ and $Y = 1.8147$.
9. Estimate the standard deviation of *C* using the expression

$$\hat{\sigma} = \sqrt{\frac{Y}{16}} \tag{3.76}$$

For this example,

$$\hat{\sigma} = \sqrt{\frac{1.8147}{16}} = 0.3368$$

Note that *C* is approximately standard normally distributed.

10. Calculate the statistic

$$C^* = \frac{\sqrt{n}(C-0.5)}{\hat{\sigma}} \tag{3.77}$$

where *n* is the total number of observations, censored and uncensored. For this example,

$$C^* = \frac{\sqrt{20}(0.5244-0.5)}{0.3368} = 0.3240$$

11. Determine if C^* falls in the acceptable region of the test. For a significance of α = 0.05, the acceptable region is $\pm z_{\alpha/2} = \pm 1.96$. For this example, the calculated statistic falls within the acceptable region, and the Weibull distribution with a shape parameter of 2.974 and a scale parameter of 203.3 is considered an adequate model for the data.

3.11 Exercises

1. Fifty items were tested for 100 hours, and 4 items failed at time = 2, time = 28, time = 59 and time = 81. Assuming an exponential distribution, determine the mean life.
2. Determine a 90% lower confidence limit for the failure rate for the data in Exercise 29.
3. Determine a 90% lower confidence for 99% reliability for the data in Exercise 1.
4. Determine a 90% lower confidence limit for percentiles at time = 40 for the data in Exercise 1.
5. For the data in Exercise 1, determine the parameters of the Weibull distribution using probability plotting.
6. For the data in Exercise 1, determine the parameters of the Weibull distribution using hazard plotting.

7. For the data in Exercise 1, determine the parameters of the Weibull distribution using maximum likelihood estimation.
8. Determine 95% confidence limits for the parameters, reliability at 90% and percentiles at time = 30, for the distribution found in Exercise 1.
9. For the data set below, determine the parameters of the Weibull distribution using probability plotting.

12 c	22 c	15
28	19	30
20 c	31 c	

10. Determine goodness of fit for the distribution found in Exercise 9 using the Hollander-Proschan test.
11. For the data given in Exercise 9, determine the parameters of the lognormal distribution using probability plotting. Include confidence limits on the probability plot using 5% and 95% ranks.
12. For the data given in Exercise 9, determine the parameters of the lognormal distribution using hazard plotting.
13. For the data given in Exercise 9, determine the parameters of the lognormal distribution using maximum likelihood estimation. Determine 95% confidence limits for the parameters.
14. For the data given in Exercise 9, determine the parameters of the normal distribution using maximum likelihood estimation. Determine a 90% confidence interval for 95% reliability.

3.12 References

Crowder, M. J., A. C. Kimber, R. L. Smith, and T. J. Sweeting. 1991. ***Statistical Analysis of Reliability Data***, London: Chapman & Hall.

Dodson, Bryan, and Jack Kirkland. 1994. ***The Reliability & Maintenance Analyst***, Auburn Hills, MI: Espinoza Consulting.

Dodson, Bryan. 1994. ***Weibull Analysis (with Software)***. Milwaukee, WI: ASQ–Quality Press.

Dodson, B. L., and M. D. Mulcahy. 1992. ***Certified Reliability Engineer Examination Study Guide***. Tucson, AZ: Quality Publishing.

Dovich, R. A. 1990. ***Reliability Statistics***. Milwaukee, WI: ASQ–Quality Press.

Dovich, R. A. 1992. ***Quality Engineering Statistics***. Milwaukee, WI: ASQ–Quality Press.

Hahn, G. J., and S. S. Shapiro. 1967. ***Statistical Models in Engineering.*** New York: John Wiley & Sons.

Kapur, K. C., and L. R. Lamberson. 1977. ***Reliability in Engineering Design.*** New York: John Wiley & Sons.

Krishnamoorthi, K. S. 1992. ***Reliability Methods for Engineers.*** Milwaukee, WI: ASQC–Quality Press.

Lewis, E. E. 1987. ***Introduction To Reliability Engineering.*** New York: John Wiley & Sons.

Shooman, M. L. 1990. ***Probabilistic Reliability: An Engineering Approach.*** Malabar, FL: Robert E. Krieger.

Sundarajan, C. 1991. ***Guide To Reliability Engineering: Data, Analysis, Applications, Implementation, And Management.*** New York: Van Nostrand Reinhold.

Tobias, P. A., and D. C. Trindade. 1986. ***Applied Reliability.*** New York: Van Nostrand Reinhold.

Walpole, R. E., and R. H. Myers. 1989. ***Probability and Statistics for Engineers and Scientists.*** New York: Macmillan.

♦♦♦

CHAPTER

4

Statistical Inference

Making a data based decisions is not always obvious. Consider a manufacturing process with a target of 10. The last 100 samples averaged 9.98. Should the process be adjusted to meet the target of 10? The process should be adjusted if 9.98 is statistically different than 10. If two samples of size n are taken from a population, the mean of these two samples will probably be different, and neither will be equal to the population mean. For example, a 100(1–α)% confidence interval contains the true population mean 100(1–α)% of the time. Confidence intervals and hypothesis tests are used to determine if a difference is real or if it is sampling error.

4.1 Confidence intervals for the mean and variance

For data that are normally distributed with unknown mean and variance, a 100(1–α)% confidence interval for the mean is

$$\bar{x} - t_{\alpha/2}\frac{s}{\sqrt{n}} < \mu < \bar{x} + t_{\alpha/2}\frac{s}{\sqrt{n}} \tag{4.1}$$

where: $\bar{x}$ is the sample mean,

s is the sample standard deviation, and

$t_{\alpha/2}$ is the critical value of the t-distribution with $\nu = n$-1 degrees of freedom.

The t-distribution is based on sampling from a normal distribution, but provides good results as long as the distribution is bell-shaped. The t-distribution is tabulated in Appendix F, but the function in Microsoft Excel is more convenient..

If the variance of the population is known or if the sample size is large ($n > 25$), the confidence interval for the population mean is

$$\bar{x} - z_{\alpha/2}\frac{\sigma}{\sqrt{n}} < \mu < \bar{x} + z_{\alpha/2}\frac{\sigma}{\sqrt{n}} \tag{4. 2}$$

where: $\bar{x}$ is the sample mean,

σ is the population standard deviation, and

$z_{\alpha/2}$ is the critical value of the standard normal distribution.

Example 4.1

Given the data below, construct a 90% confidence interval for the mean.

12	15	21
22	24	32

Solution

The sample mean is

$$\bar{x} = \frac{12+15+21+22+24+32}{6} = 21.0$$

The sample standard deviation is

$$s = \sqrt{\frac{n\sum_{i=1}^{n} x_i^2 - \left(\sum_{i=1}^{n} x_i\right)^2}{n(n-1)}} = \sqrt{\frac{6(2894)-(126)^2}{6(6-1)}} = 7.04$$

For a 90% confidence interval, $\alpha = 0.1$. From Appendix F, $t_{0.05}$ with 5 degrees of freedom is 2.015. The 90% confidence interval for the mean is

$$21 - 2.015\frac{7.04}{\sqrt{6}} < \mu < 21 + 2.015\frac{7.04}{\sqrt{6}}$$

$$15.21 < \mu < 26.79$$

Example 4.2

Twenty-eight samples are drawn from a population with a known variance of 3.4. The mean of the sample is 78.7. Construct a 95% confidence interval for the mean.

Solution

From Appendix A, $z_{\alpha/2} = 1.96$. The confidence interval is constructed using the expression

$$78.7 - 1.96\frac{\sqrt{3.4}}{\sqrt{28}} < \mu < 78.7 + 1.96\frac{\sqrt{3.4}}{\sqrt{28}}$$

$$78.0 < \mu < 79.4$$

For data that are normally distributed with sample variance s^2, a 100(1–α)% confidence interval for the variance is

$$\frac{(n-1)s^2}{\chi^2_{\alpha/2,n-1}} \le \sigma^2 \le \frac{(n-1)s^2}{\chi^2_{1-(\alpha/2),n-1}} \tag{4.3}$$

where, $\chi^2_{\alpha/2,n-1}$ is the critical value of the chi-square distribution with n–1 degrees of freedom.

Example 4.3

Ten boxes of cereal were randomly sampled and weighed. The sample mean was 12.1 ounces and the sample standard deviation was 0.85. Determine a 95% confidence interval for the variance of cereal box weight.

Solution

From Appendix H, the critical values of the chi-square distribution are

$$\chi^2_{0.025,9} = 19.023$$

$$\chi^2_{0.975,9} = 2.70$$

The confidence interval is

$$\frac{9(0.85)^2}{19.023} \le \sigma^2 \le \frac{9(0.85)^2}{2.7}$$

$$0.342 \le \sigma^2 \le 2.408$$

The chi-square distribution can also be used to construct confidence intervals for the number of occurrences. Given a sample of r occurrences, a (1-α)% interval for the average number of occurrences is

$$\frac{X^2_{1-\alpha/2,2r}}{2} < c < \frac{X^2_{\alpha/2,2r+2}}{2} \tag{4.4}$$

4.2 Confidence Intervals for proportions

A 100(1–α)% confidence interval for a proportion is

$$\frac{2r-1+z_{\alpha/2}^2-z_{\alpha/2}\sqrt{\frac{(2r-1)(2n-2r+1)}{n}+z_{\alpha/2}^2}}{2\left(n+z_{\alpha/2}^2\right)}$$

$$\leq\frac{r}{n}\leq\frac{2r+1+z_{\alpha/2}^2+z_{\alpha/2}\sqrt{\frac{(2r+1)(2n-2r-1)}{n}+z_{\alpha/2}^2}}{2\left(n+z_{\alpha/2}^2\right)} \tag{4.5}$$

where: r is the number of occurrences in the sample, and

n is the sample size.

If $\hat{p}$ is the proportion of successes in a random sample of size n, and $n\hat{p}\geq 5$ and $n(1-\hat{p})\geq 5$, then a 100(1–α)% confidence interval is

$$\hat{p}-z_{\alpha/2}\sqrt{\frac{\hat{p}(1-\hat{p})}{n}}\leq p\leq\hat{p}+z_{\alpha/2}\sqrt{\frac{\hat{p}(1-\hat{p})}{n}} \tag{4.6}$$

Example 4.4

In a sample of 500 items 19 were found to be deficient. Determine a 90% confidence interval for the proportion of deficient items.

Solution

From Appendix A, $z_{\alpha/2}$ = 1.645. The confidence interval is

$$\frac{19}{500}-1.645\sqrt{\frac{\frac{19}{500}\left(1-\frac{19}{500}\right)}{500}}\leq p\leq\frac{19}{500}+1.645\sqrt{\frac{\frac{19}{500}\left(1-\frac{19}{500}\right)}{500}}$$

$$0.024\leq p\leq .052$$

The beta distribution can also be used to determine a lower confidence limit for the number of occurrences. Appendix D provides the information for determination of a lower reliability confidence limit at confidence levels of 90%, 95%, 99% and 99.5%.

To determine the lower reliability confidence limit without the use of tables requires solving the equation below by trial and error.

$$\alpha=\sum_{i=0}^{r}\left(\frac{\begin{matrix}\prod_{n+2-i}^{n+1}(n+2-i) & i>0\\ 1 & i=0\end{matrix}}{i!}\right)(R_l)^{n+1-i}(1-R_l)^i \tag{4.7}$$

where: α is the level of significance ($\alpha = 0.05$ for a 95% limit),
r is the number of failures,
n is the number of trials (successes and failures), and
R_l is the lower reliability confidence limit.

This expression is easily solved with the solver routines included with electronic spreadsheets. Guess at the lower reliability confidence limit and assign this cell as adjustable. In another set of cells, use the guess for R_l to compute the right side of Equation 4.7. Make two constraint cells, with one being α – the computed right side of Equation 4.7, and the other being the computed right side of Equation 4.7 – α.

Example 4.5

Forty items were tested with 2 failures and 38 successes. Determine the lower reliability confidence limit at 90% confidence.

Solution

From Appendix D, with $n = 40$, $r = 2$, the lower reliability confidence limit at 90% confidence is 0.875.

A 100(1–α)% nonparametric confidence interval for reliability is

$$\frac{n-r}{n-r+(r+1)F_{\alpha/2,2(r+1),2(n-r)}} \leq R \leq \frac{(n-r+1)F_{\alpha/2,2(n-r+1),2r}}{r+(n-r+1)F_{\alpha/2,2(n-r+1),2r}} \tag{4.8}$$

where: n is the sample size,
r is the number of failures, and
$F_{\alpha,v1,v2}$ is the critical value of the F-distribution.

Example 4.6

Determine the 80% nonparametric confidence interval for reliability for the data in Example 4.5.

Solution

For a confidence level of 80%, $\alpha = 0.20$. For this example, $n = 40$ and $r = 2$. From Appendix E,

$$F_{0.1,6,76} = 1.8531$$

$$F_{0.1,78,4} = 3.783$$

The 80% confidence interval is

$$\frac{40-2}{40-2+(2+1)1.8531} \leq R \leq \frac{(40-2+1)3.783}{2+(40-2+1)3.783}$$

$$0.8724 \leq R \leq 0.9866$$

4.3 Tolerance Intervals

A **tolerance interval** contains (1–γ)% of the observations from a population with 100(1–α)% confidence. Tolerance intervals establish bounds on individual values as opposed to confidence intervals that establish bounds on parameters.

If the population is normally distributed with unknown mean and variance, a 100(1–α)% confidence interval containing (1–γ)% of the observations is

$$\bar{x} \pm ks \tag{4.9}$$

where: $\bar{x}$ is the sample mean,

s is the sample standard deviation, and

k is a factor given in Tables I.1a and I.1b in Appendix I.

Tolerance intervals can also be constructed nonparametrically, but the intervals will be wider. The extreme values of the sample contain 100(1–α)% of the population (1–γ)% of the time, where (1–γ) is obtained from Table I.2 in Appendix I. Table I.3 in Appendix I contains the sample size required to contain (1–γ)% of the population with 100(1–α)% confidence.

Example 4.7

A fuse-well is being drilled into a warhead, and the fuse-well depths from a random sample of 10 warheads are given below. Assuming a normal population, construct a 99% confidence interval that contains 95% of the population.

12.34	12.37	12.31	12.33	12.24
12.38	12.32	12.32	12.36	12.35

Solution

The sample mean is $\bar{x} = 12.332$, and the sample standard deviation is s = 0.0397. From Appendix I, Table I.1 for α = 0.01, γ = 0.05 and n = 10, k = 4.265. The tolerance interval is

$$12.332 \pm 4.265(0.0397)$$

It can be stated with 99% confidence that 95% of the fuse-well depths will fall in the range from 12.163 to 12.501

4.4 Hypothesis testing

Reliability engineers often make data based decisions about specific systems. For example an engineer might want to scientifically demonstrate that design A has a longer mean time to fail than design B. This can be done with statistical hypothesis testing. First, a null hypothesis is formed. This hypothesis is then tested and will either be rejected or fail to be rejected. If the null hypothesis is not rejected, it does not necessarily mean it is true, rather there is not sufficient evidence to reject it.

The hypothesis being tested is referred to as the **null hypothesis**, H_0. If the null hypothesis is rejected, the **alternative hypothesis**, H_1, is accepted. There is a probability of making incorrect decisions when hypothesis testing. If the null hypothesis is rejected when in reality it is true, this is referred to as **Type I error** or **α error**. If the null hypothesis is not rejected, when in reality it is false, this is referred to as **Type II error** or **β error**. This is shown in the diagram below.

	H_0: True	H_0: False
Accept H_0:	Correct Decision	Type II Error
Reject H_0:	Type I Error	Correct Decision

The power of a test is equal to the probability of rejecting the null hypothesis when the alternate hypothesis is true.

4.4.1 Hypothesis tests for means

When testing to determine if a population mean is equal to a specific value, μ_0, the null and alternative hypotheses are

$$H_0: \ \mu = \mu_0,$$
$$H_1: \ \mu \neq \mu_0 \ \text{or} \ \mu < \mu_0 \ \text{or} \ \mu > \mu_0$$

For a two-tailed test, assuming a normal sampling distribution, H_0 is rejected at significance α when the computed t-statistic does not fall in the range from $-t_{\alpha/2,n-1}$ to $t_{\alpha/2,n-1}$. The t-statistic is computed from the expression

$$t = \frac{\bar{x} - \mu_0}{s/\sqrt{n}} \tag{4.10}$$

where: n is the sample size and

s is the sample standard deviation.

If the standard deviation of the population is known, the statistic computed is

$$z = \frac{\bar{x} - \mu_0}{\sigma/\sqrt{n}} \tag{4.11}$$

and the acceptance region is $\pm z_{\alpha/2}$ for a two-tailed test.

When testing to determine if two population means are equal, the null and alternative hypotheses are

$$H_0: \ \mu_1 = \mu_2,$$
$$H_1: \ \mu_1 \neq \mu_2 \ \text{ or } \ \mu_1 < \mu_2 \ \text{ or } \ \mu_1 > \mu_2$$

Assuming the sampling distribution is normally distributed, but the variance is unknown, the statistic tested is

$$t' = \frac{\bar{x}_1 - \bar{x}_2}{\sqrt{s_1^2 / n_1 + s_2^2 / n_2}} \tag{4.12}$$

For a two-tailed test, H_0 is rejected at significance α when the computed t-statistic does not fall in the range from $-t_{\alpha/2}$ to $t_{\alpha/2}$. The degrees of freedom for the statistic are

$$\nu = \frac{\left(s_1^2 / n_1 + s_2^2 / n_2\right)^2}{\frac{\left(s_1^2 / n_1\right)^2}{n_1 - 1} + \frac{\left(s_2^2 / n_2\right)^2}{n_2 - 1}} \tag{4.13}$$

Assuming the standard deviations of the two populations are equal but unknown, the test statistic is

$$t = \frac{\bar{x}_1 - \bar{x}_2}{s_p \sqrt{1/n_1 + 1/n_2}} \tag{4.14}$$

where

$$s_p = \sqrt{\frac{s_1^2 (n_1 - 1) + s_2^2 (n_2 - 1)}{n_1 + n_2 - 2}} \tag{4.15}$$

For a two tailed test, H_0 is not rejected at significance α when

$$-t_{\alpha/2, n1+n2-2} < t < t_{\alpha/2, n1+n2-2}$$

When the standard deviation of the two populations is equal and known, the test statistic is

$$z = \frac{\bar{x}_1 - \bar{x}_2}{\sqrt{\sigma_1^2 / n_1 + \sigma_2^2 / n_2}} \tag{4.16}$$

where σ is the standard deviation of each of the populations.

For a two-tailed test, H_0 is not rejected at significance α when

$$-z_{\alpha/2} < z < z_{\alpha/2}$$

where z is the critical value of the standard normal distribution.

Example 4.8

A new beam is being designed and the design engineer wants to determine if an alloying agent affects strength. Ten beams of Design A, which includes the alloying agent, and 9 beams of design B, which does not include the alloying agent were tested to failure. The results follow. At 5% significance, can the engineer state the beams have equal strength?

Design A		Design B	
89	93	88	93
91	88	87	92
90	87	94	90
92	86	91	97
88	90	94	

Solution

There is no reason to assume the variances are equal, so the tested statistic is

$$t' = \frac{89.4 - 91.78}{\sqrt{2.22^2/10 + 3.15^2/9}} = -1.88$$

The degrees of freedom for the statistic are

$$v = \frac{\left(2.22^2/10 + 3.15^2/9\right)^2}{\frac{\left(2.22^2/10\right)^2}{10-1} + \frac{\left(3.15^2/9\right)^2}{9-1}} = 14.2$$

Rounding the above value to $v = 14$, at a significance of 2.5% (2.5% on each tail totals 5%) from Appendix F, the means of the two designs are accepted as equal if

$$-2.145 < t' < 2.145$$

Thus, the design engineer concludes the strengths are equal (the design engineer fails to reject the null hypothesis).

When testing to determine if two proportions are equal, the null and alternative hypotheses are

$$H_0: \; p_1 = p_2,$$

$$H_1: \; p_1 \neq p_2 \;\text{ or }\; p_1 < p_2 \;\text{ or }\; p_1 > p_2$$

The statistic for testing the difference between two proportions is

$$z = \frac{\hat{p}_1 - \hat{p}_2}{\sqrt{\hat{p}(1-\hat{p})(1/n_1 + 1/n_2)}} \tag{4.17}$$

where

$$\hat{p} = \frac{x_1 + x_2}{n_1 + n_2} \tag{4.18}$$

where: x_1 and x_2 are the number of successes in each of the two samples, and
z is the critical value of the standard normal distribution.

For a two-tailed test, H_0 is not rejected at significance α when

$$-z_{\alpha/2} < z < z_{\alpha/2}$$

4.4.2 Hypothesis tests for variance

If the population being sampled is normal, the chi-square distribution can be used to test if the variance of the distribution is equal to a specific value, σ_0. The hypotheses for this test are

$$H_0: \ \sigma = \sigma_0$$
$$H_1: \ \sigma \neq \sigma_0 \ \text{ or } \ \sigma < \sigma_0 \ \text{ or } \ \sigma > \sigma_0$$

The test statistic is

$$\chi^2_{\alpha,v} = \frac{(n-1)s^2}{\sigma_0^2} \tag{4.19}$$

where: n is the sample size
s is the sample standard deviation,
α is the significance, and
v is the degrees of freedom (n–1).

For a two-tailed test, H_0 is not rejected at a significance of α if

$$\chi^2_{1-\alpha/2,v} < \chi^2 < \chi^2_{\alpha/2,v}$$

Example 4.9

A specific type of computer monitor has a normally distributed mean life of 35,000 hours with a standard deviation of 3,000 hours. The manufacturer is testing a new design, and a random sample of 10 monitors has a standard deviation 2,800 hours. Does the new design have a lower standard deviation at 5% significance?

Solution

This is a one tailed test. The hypotheses are

$$H_0: \ \sigma = 3000$$
$$H_1: \ \sigma < 3000$$

From Appendix H, with 9 degrees of freedom, the critical value of the chi-square distribution is

$$\chi^2_{1-\alpha} = \chi^2_{0.95} = 3.325$$

The test statistic is

$$\chi^2 = \frac{(10-1)2800^2}{3000^2} = 7.84$$

Since $\chi^2 > \chi^2_{1-\alpha}$, the null hypothesis is not rejected, and it cannot be stated that the new design has a lower variance.

When testing to determine if there is a difference in the variance of two populations, assuming the populations are normally distributed, the test statistic is

$$F = \frac{s_1^2}{s_2^2} \tag{4.20}$$

For a two-tailed test, H_0 (the variances of the two populations are equal) is not rejected at a significance of α if

$$F_{1-\alpha,v1,v2} < F < F_{\alpha,v1,v2}$$

where: $v_1 = n_1 - 1$ and

$v_2 = n_2 - 1$

Example 4.10

A design engineer is considering two components. Ten Type A components were tested and the sample standard deviation of the strength was 0.092. Fifteen Type B components were tested and the sample standard deviation of the strength was 0.097. At a significance of 0.05, is the variance of the strength of component type A lower than that of component type B?

Solution

This is a one tailed test. The test statistic is

$$F = \frac{0.092^2}{0.097^2} = 0.8996$$

From Appendix E, the critical value of the F-distribution is

$$F_{0.95,9,14} = 0.331$$

Since the calculated statistic is greater than the critical value, it cannot be stated at 5% significance that the two variances are different.

4.5 Exercises

1. Given the following data, construct a 95% confidence interval for the mean.

45	70	52
78	61	82

2. Construct a 90% confidence interval for the mean of the data in Exercise 43.
3. A random sample of 50 items yielded 40 successes. Determine a 95% confidence interval for the proportion of successes.
4. How many successful launches must occur to demonstrate 95% reliability with 90% confidence with one failed launch?
5. Construct a 95% tolerance interval that contains 95% of the population for the data given in Exercise 43.
6. For the data given in Exercise 43, determine with 10% significance if the population mean is greater than 60.
7. For the data given in Exercise 43, determine with 10% significance if the population mean is less than 60.
8. For the data given in Exercise 43, determine with 10% significance if the population mean is different than 60.
9. For the data given in Exercise 43, determine with 10% significance if the population variance is different than 45.
10. Determine if the mean of population A and B are different given the data below at a significance level of 0.05.

A	B
25	34
33	38
43	39
45	41
51	50
52	58

11. Determine if the variances of populations A and B in Exercise 52 are different at a significance level of 0.1.

4.6 References

Dodson, Bryan, and Jack Kirkland. 1994. ***The Reliability & Maintenance Analyst***, Auburn Hills, MI: Espinoza Consulting.

Dodson, Bryan. 1994. ***Weibull Analysis (with Software)***. Milwaukee, WI: ASQ–Quality Press.

Dodson, B. L., and M. D. Mulcahy. 1992. ***Certified Reliability Engineer Examination Study Guide***. Tucson, AZ: Quality Publishing.

Dovich, R. A. 1990. ***Reliability Statistics***. Milwaukee, WI: ASQ–Quality Press.

Dovich, R. A. 1992. ***Quality Engineering Statistics***. Milwaukee, WI: ASQ–Quality Press.

Hahn, G. J., and S. S. Shapiro. 1967. ***Statistical Models in Engineering***. New York: John Wiley & Sons.

Johnson, N. L., and S. Kotz. 1970. ***Distributions in Statistics: Continuous Univariate Distributions-1***. New York: John Wiley & Sons.

Johnson, N. L., and S. Kotz. 1970. ***Distributions in Statistics: Continuous Univariate Distributions-2***. New York: John Wiley & Sons.

Juran, J. M., and F. M Gryna. 1988. ***Juran's Quality Control Handbook***. New York: McGraw Hill.

Kapur, K. C., and L. R. Lamberson. 1977. ***Reliability in Engineering Design***. New York: John Wiley & Sons.

Krishnamoorthi, K. S. 1992. ***Reliability Methods for Engineers***. Milwaukee, WI: ASQC–Quality Press.

Lewis, E. E. 1987. ***Introduction To Reliability Engineering***. New York: John Wiley & Sons.

O'Conner, P. D. T. 1993. ***Practical Reliability Engineering***. New York: John Wiley & Sons.

Shooman, M. L. 1990. ***Probabilistic Reliability: An Engineering Approach***. Malabar, FL: Robert E. Krieger.

Sundarajan, C. 1991. ***Guide To Reliability Engineering: Data, Analysis, Applications, Implementation, And Management***. New York: Van Nostrand Reinhold.

Tobias, P. A., and D. C. Trindade. 1986. ***Applied Reliability***. New York: Van Nostrand Reinhold.

Walpole, R. E., and R. H. Myers. 1989. ***Probability and Statistics for Engineers and Scientists***. New York: Macmillan.

♦♦♦

CHAPTER

5

Reliability Modeling

This chapter describes a variety of mathematical models for predicting the reliability of a system. The validity of these models is based on the accuracy of the reliability estimates of the components that compose the models. Methods are presented for obtaining reliability information from public databases and from military standards. After a system has been modeled, it is often desirable to optimize the system. Methods of reliability allocation are also discussed.

5.1 Models with standard data

Military Handbook 217 was created to establish and maintain consistent and uniform methods for estimating the inherent reliability of military electronic equipment systems. It provides a common basis for reliability predictions during acquisitions programs and establishes a common basis for comparing and evaluating reliability predictions of related or competitive designs.

When the standard was first released the reliability of most electronics, such as vacuum tubes, was very poor. The affect of manufacturing on the reliability of the electronic system was small with respect to the affect of the reliability of the individual components. This allowed relatively accurate predictions of system reliability to be made from component data for electronic systems. With the introduction of solid state electronics, the dramatic increase in component reliability has caused the system reliability for electronic systems to be impacted by manufacturing much more than component reliability. Finn Jensen has reported reliability differences of more than

300% for the same system being manufactured at more than one location. This has greatly reduced the ability to predict the reliability of system reliability from standard data. The accuracy of this method of reliability prediction has decreased so much that Patrick O'Connor and other reliability experts have criticized this approach.

Nevertheless, the approach described in Military Handbook - 217 continues to be used. This handbook has been enhanced by the *Reliability Analysis Center*; the enhanced version is titled "Consolidated Reliability Assessment Model (CRAM)". The handbooks contain two methods for predicting reliability, **Part Stress Analysis** and **Parts Count**. The Part Stress Analysis method requires a greater amount of detailed information and is applicable during the later design phase when actual hardware and circuits are being designed. The Parts Count method requires less information, generally part quantities, quality level and the applicable environment. The Parts Count method is applicable during the early design phase and during proposal formulation, and will generally result in a more conservative estimate. The following similar references have also been developed:

- Bellcore — used by telecom companies in USA
- CNET — French model
- BTT — British Telephone and Telecom
- NTT — Nippon Telephone and Telecom (Japan)
- SAE Reliance — Auto industry model
- IEC-1709 — European model

The Part Stress Analysis method is based on the expression

$$\lambda_p = \lambda_b \Pi_Q \Pi_E \ldots \tag{5.1}$$

where: λ_p is the part failure rate,

λ_b is the base failure rate usually expressed by a model relating the influence of electrical and temperature stresses on the part,

Π_Q is a modification factor for the quality level, and,

Π_E is a modification factor for environmental stresses.

All Π factors modify the base failure rate for the category application and other parameters that affect the part reliability. The Π_E and Π_Q factors are used in virtually all models. Most models also contain other Π factors. The base failure rate models are presented in each part section along with identification of the applicable model factors. All the part models include failure data from both catastrophic and permanent drift failures and are based on a constant failure rate, except for motors which show an increasing failure rate over time. Failures associated with connection of parts into circuit assemblies are not included within the part failure rate models. Information on connection reliability is provided in Sections 16 and 17 of the handbook.

The Parts Count method is based on the expression

$$\lambda_{\text{EQUIP}} = \sum_{i=1}^{n} N_i \left(\lambda_g \Pi_Q\right)_i \tag{5.2}$$

where: λ_{EQUIP} is the total equipment failure rate (failures per 1 million hours),

λ_g is the generic failure rate for the *i*th generic part (failures per 1 million hours),

Π_Q is the quality factor for the *i*th generic part,

N_i is the quantity of the *i*th generic part, and

n is the number of different generic part categories in the equipment.

5.2 Mathematical models

After component reliability information has been obtained, system reliability can be computed. Most systems can be modeled using reliability block diagrams. The model may be a series system, a parallel system, a standby redundant system, or a combination of the previous.

5.2.1 Series systems

For a series system to operate successfully, all components must operate successfully. A simple series system is shown in Figure 5.1.

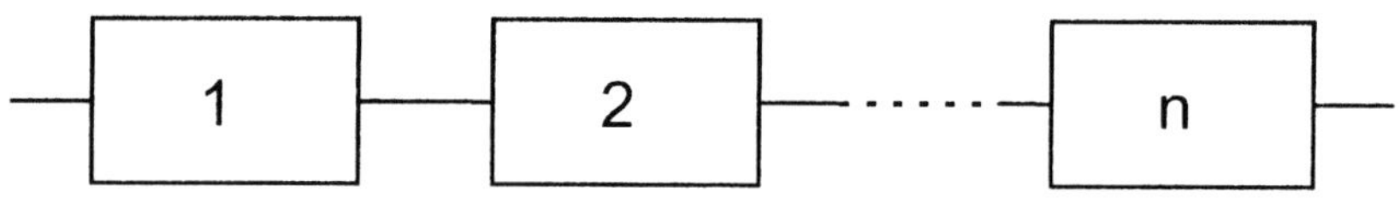

Figure 5.1. Series block diagram.

If each component in the system is independent, the reliability of a series system is

$$R_s = \prod_{i=1}^{n} R_i \tag{5.3}$$

The system hazard function is

$$h_s = \sum_{i=1}^{n} h_i \tag{5.4}$$

If the components in the system are exponentially distributed, the system mean time to fail is

$$MTTF_s = \frac{1}{\sum_{i=1}^{n} \lambda_i} \tag{5.5}$$

where λ_i is the failure rate of the *i*th component.

Example 5.1

Three components each with a reliability of 0.9 are placed in series. What is the reliability of the system?

Solution

The system reliability is the product of the component reliabilities.

$$R_s = 0.9^3 = 0.729$$

Example 5.2

The components in the system below are exponentially distributed with the indicated failure rates. Develop an expression for the reliability of the system. What is the system reliability at time = 100 hours?

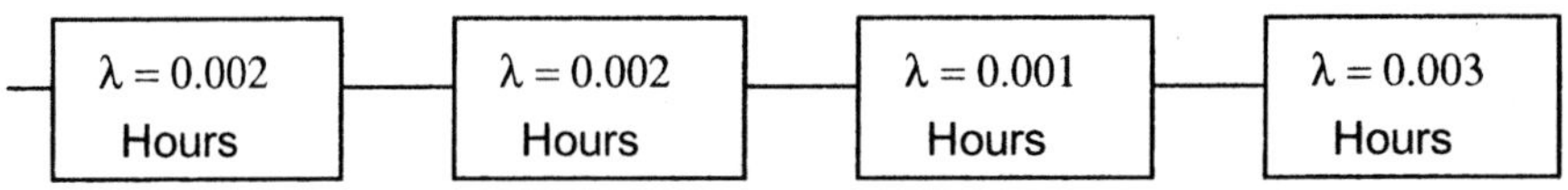

Solution

For the exponential distribution, reliability is defined as

$$R(t) = e^{-\lambda t}$$

The reliability of the system above is

$$R_s(t) = \left(e^{-0.002t}\right)\left(e^{-0.002t}\right)\left(e^{-0.001t}\right)\left(e^{-0.003t}\right) = e^{-0.008t} = 0.4493$$

An alternate solution method is to sum the individual values of the hazard function.

$$\lambda_s = 0.002 + 0.002 + 0.001 + 0.003 = 0.008$$

$$R_s = e^{-0.008t}$$

At time = 100 hours, the reliability is

$$R_s = e^{-0.008(100)} = 0.4493$$

5.2.2 Parallel systems

For a parallel system to operate successfully, **at least one** of the components in the system must operate successfully. A general parallel system is shown in Figure 5.2.

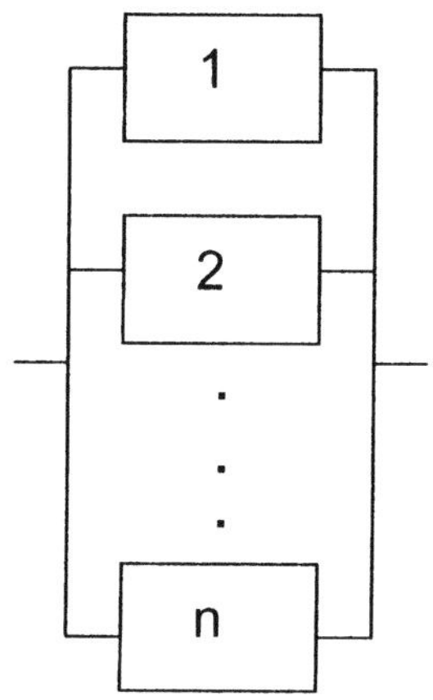

Figure 5.2. Parallel block diagram.

The system above is an **active** parallel system with n–1 redundancy. In an active parallel system all components are in operation when the system is in operation. An active parallel system may be a **pure** parallel system or a **shared** parallel system. In a pure parallel system, there is no change in the failure rate of the surviving components after the failure of a companion component. In a shared parallel system, the failure rates of the remaining components change when a companion component fails. An example is an automobile wheel being held on by four bolts. The load is being shared by four bolts, and when one of the bolts fails, the load on the three remaining bolts increases.

In a **passive** parallel system, also known as a **standby redundant** system, only one component functions at a given time. The remaining components do not come into service until required by the failure of other components. A switching device is usually used to transfer operation from one component to another. Passive parallel systems are discussed in Section 5.2.4.

The reliability of a pure, active parallel system is

$$R_s = 1 - \prod_{i=1}^{n} (1 - R_i) \tag{5.6}$$

A parallel system consisting of components with n identical components with an exponential time to failure distribution has a mean time to fail of

$$MTTF_s = \sum_{i=1}^{n} \frac{1}{i\lambda} \tag{5.7}$$

where λ is the failure rate of each of the components.

Example 5.3

A system consists of three components. At time = 0, all three components are activated, and as long as at least one of the three components is functional, the system is functional. Write the expression for system reliability if each of the components has an exponential time to fail distribution with mean times to fail of 40 hours, 80 hours and 85 hours. What is the reliability at time = 25 hours?

Solution

This is a system with three components—two in active redundancy The reliability of the system is

$$R_s(t) = 1 - \left(1 - e^{-t/40}\right)\left(1 - e^{-t/80}\right)\left(1 - e^{-t/85}\right)$$

At time = 25 hours, the reliability is

$$R_s(25) = 1 - \left(1 - e^{-25/40}\right)\left(1 - e^{-25/80}\right)\left(1 - e^{-25/85}\right) = 0.9682$$

5.2.3 r-out-of-n systems

Another form of active redundancy is a system consisting of n components in which r of the n components must function for the system to function. An example of this is a four engine aircraft that can maintain flight with only two of the four engines operating. If the reliability, R, of all components is equal, the reliability of an r-out-of-n system is

$$R_s = \sum_{x=r}^{n} \binom{n}{x} R^x (1 - R)^{n-x} \tag{5.8}$$

For the special case where the component failure rates are constant, the system mean time to fail is

$$\theta_s = \sum_{i=r}^{n} \frac{1}{i\lambda} \tag{5.9}$$

where: λ is the failure rate of each of the components.

Example 5.4

A system consists of four components. If more than two of the components fail, the system fails. If the components have an exponential time to fail distribution with a failure rate of 0.000388, what is the reliability of the system at time = 300? What is the system mean time to fail?

Solution

This is a 2-out-of-4 system; at least 2 of the 4 components must survive for the system to survive. The reliability of each component is

$$R = e^{-0.000388(300)} = 0.89$$

The reliability of the system is

$$R_s = \sum_{x=2}^{4} \binom{4}{x} R^x (1-R)^{4-x}$$

$$\begin{aligned} R_s &= \left[\frac{4!}{(2!)(4-2)!}\right](0.89)^2(1-0.89)^{4-2} \\ &+ \left[\frac{4!}{(3!)(4-3)!}\right](0.89)^3(1-0.89)^{4-3} \\ &+ \left[\frac{4!}{(4!)(4-4)!}\right](0.89)^4(1-0.89)^{4-4} \\ &= 0.995 \end{aligned}$$

The system mean time to fail is

$$\begin{aligned} \theta_s &= \sum_{i=2}^{4} \frac{1}{i(0.000388)} \\ &= \frac{1}{2(0.000388)} + \frac{1}{3(0.000388)} + \frac{1}{4(0.000388)} \\ &= 2792 \end{aligned}$$

5.2.4 Standby redundant systems

In a standby redundant system only one component is in operation at a given time, if the on-line component fails, another component is brought on-line. A switching mechanism is usually responsible for bringing a redundant component on-line. A standby redundant system is shown in Figure 5.3.

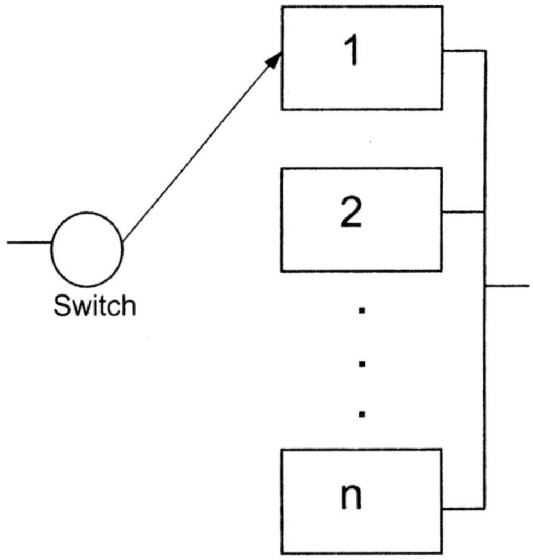

Figure 5.3. Standby redundant system.

In the case of perfect switching and exponential component life distributions, the standby redundant system is modeled by the gamma distribution. This is because the sum of n exponentially distributed random variables follows a gamma distribution. For the special case where all components have the same failure rate, λ, the reliability of a standby redundant system with n components is

$$R(x) = e^{-\lambda x} \sum_{i=0}^{n-1} \frac{(\lambda x)^i}{i!} \tag{5.10}$$

Example 5.5

Three power supplies are configured in a standby redundant system with perfect switching. The failure rate for each of the power supplies is constant with a mean time between failures of 20,000 hours. What is the probability of the system failing in less than 100,000 hours?

Solution

A constant failure rate indicates an exponential time to fail distribution. The system reliability is

$$R_s(t) = e^{-0.00005t} \sum_{i=0}^{3-1} \frac{(0.00005t)^i}{i!}$$

The reliability at 100,000 hours is

$$R_s(t) = e^{-5}\left[\frac{5^0}{0!} + \frac{5^1}{1!} + \frac{5^2}{2!}\right] = 0.1247$$

The probability of a failure in less than 100,000 hours of operation is

P(failure) = 1– 0.1247 = 0.8753

In many cases, the assumption of perfect switching is unrealistic. For the simple case of two components in standby redundancy with exponential time to fail distributions, the reliability of the system is

$$R_s(x) = e^{-\lambda_a x} + \frac{p\lambda_a}{\lambda_b - \lambda_a}\left(e^{-\lambda_a x} - e^{-\lambda_b x}\right) \tag{5.11}$$

where: p is the reliability of the switch,

λ_a is the failure rate for the primary component, and

λ_b is the failure rate of the standby component.

If the components have identical failure rates, this expression reduces to

$$R_s(x) = e^{-\lambda x}\left(1 + p\lambda x\right) \tag{5.12}$$

Example 5.6

In the figure below, components 1 and 2 have exponential time-to-fail distributions with mean failure times of 66,667 hours and 100,000 hours respectively. The reliability of the switch is 0.95. What is the reliability of the system at 10,000 hours?

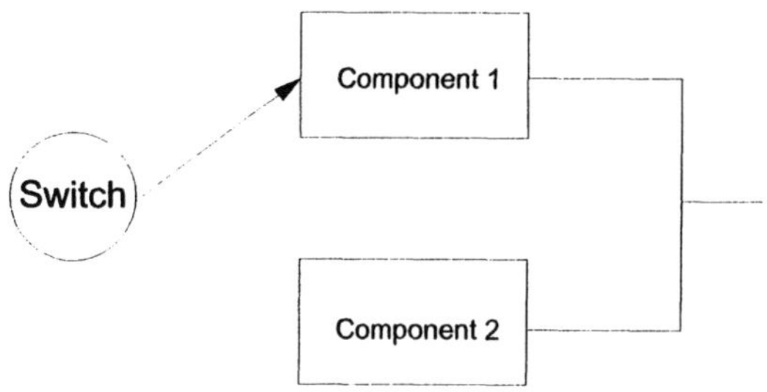

Solution

The system reliability is

$$R_s(t) = e^{-\frac{t}{66,667}} + \frac{0.95\left(\frac{1}{66,667}\right)}{\frac{1}{100,000} - \frac{1}{66,667}}\left(e^{-\frac{t}{66,667}} - e^{-\frac{t}{100,000}}\right)$$

At time = 10,000 hours, this expression is

$$R_s(10,000) = e^{-0.15} - 2.85\left(e^{-0.15} - e^{-0.1}\right) = .9865$$

Now consider the case where the switch is a complex piece of equipment with a constant failure rate, λ_S. For a two-component system, with one component in standby redundancy, the reliability of the system is

$$R_s(x) = e^{-\lambda x}\left[1 + \frac{\lambda}{\lambda_s}\left(1 - e^{-\lambda_s x}\right)\right] \tag{5.13}$$

where: λ is the failure rate of each of the two components, and

λ_s is the failure rate of the switch.

The reliability of a three-component system with identical failure rates is

$$R_s(t) = e^{-\lambda t}\left[1 + \frac{\lambda}{\lambda_s}\left(1 - e^{-\lambda_s t}\right)\right] + e^{-\lambda t}\left(\frac{\lambda}{\lambda_s}\right)^2\left[1 - e^{-\lambda_s t} - \lambda_s t e^{-\lambda_s t}\right] \tag{5.14}$$

Example 5.7

Two components with constant failure rates of 0.003 make up a standby redundant system. The device that transfers operation to the second component after failure of the first component has a constant failure rate of 0.0005. Determine the system reliability at time = 400.

Solution

The reliability of this system is

$$R_s(t) = e^{-0.003t}\left[1 + \frac{0.003}{0.0005}\left(1 - e^{-0.0005t}\right)\right]$$

At time = 400, the reliability is

$$R_s(t) = e^{-1.2}\left[1 + 6\left(1 - e^{-0.2}\right)\right] = 0.6288$$

In shared load parallel systems, as one component fails the failure rate of the remaining components increases. For a two-component system with both components having identical, constant failure rates, the system reliability is

$$R_s(x) = e^{-2\lambda_h x} + \frac{2\lambda_h}{2\lambda_h - \lambda_f}\left(e^{-\lambda_f x} - e^{-2\lambda_h x}\right) \tag{5.15}$$

where: λ_b is the half load failure rate (both components operating), and

λ_f is the full load failure rate.

For the special case where $2\lambda_h = \lambda_f$,

$$R_s(x) = e^{-2\lambda_h x} + 2\lambda_h x e^{-\lambda_f x} \tag{5.16}$$

Example 5.8

Two cables support the load of a crane. The failure rate of each of the cables is constant at 0.0007 if both cables are operating properly. The failure rate increases to 0.006 after one cable fails. Determine the reliability at time = 200.

Solution

The reliability of this system is

$$R_s(t) = e^{-2(0.0007)t} + \frac{2(0.0007)}{2(0.0007) - 0.006}\left(e^{-0.006t} - e^{-2(0.0007)t}\right)$$

At time = 200, the reliability is

$$R_s(t) = e^{-0.28} - 0.3043\left(e^{-1.2} - e^{-0.28}\right) = 0.8941$$

For the situations described above, there is no probability of failure of a component while it is in standby mode. Often this is not a valid assumption. The reliability of a system consisting of two components with one being standby redundant is

$$R_s(x) = e^{-\lambda_a x} + \frac{\lambda_a}{\lambda_a - \lambda_b + \lambda_{b,s}}\left[e^{-\lambda_b x} - e^{-(\lambda_a + \lambda_{b,s})x}\right] \tag{5.17}$$

where: λ_a is the failure rate for the primary component,

λ_b is the failure rate of the standby component when it is active, and

$\lambda_{b,s}$ is the failure rate of the standby unit when it is in standby mode.

For the special case where the failure rate of both components is the same in the active mode, the system reliability is

$$R_s(x) = \left(1 + \frac{\lambda}{\lambda_s}\right)e^{-\lambda x} - \frac{\lambda}{\lambda_s}e^{-(\lambda + \lambda_s)x} \tag{5.18}$$

where: λ is the failure rate of each of the components in active mode, and

λ_s is the failure rate of the standby component while in standby mode.

The system mean time to fail is

$$MTTF = \frac{1}{\lambda} + \frac{1}{\lambda_s} - \left(\frac{\lambda}{\lambda_s}\right)\left(\frac{1}{\lambda + \lambda_s}\right) \tag{5.19}$$

Example 5.9

A pump operates continuously with a mean time to fail of 200 hours that follows the exponential distribution. A second, identical pump is placed in standby redundancy, and the mean time to fail while the pump is inactive is 1,000 hours. The standby time to fail is also exponentially distributed. What is the mean time to fail for the system, and what is the system reliability at time = 300 hours?

Solution

The pump failure rate in the active mode is λ = 1/200 = 0.005. The pump failure rate in the standby mode is λ_s = 1/1000 = 0.001. The system mean time to fail is

$$MTTF = \frac{1}{0.005} + \frac{1}{0.001} - \left(\frac{0.005}{0.001}\right)\left(\frac{1}{0.005+0.001}\right) = 366.7$$

The system reliability at time = 300 hours is

$$R_s(300) = \left(1 + \frac{0.005}{0.001}\right)e^{-0.005(300)} - \frac{0.005}{0.001}e^{-(0.005+0.001)300} = 0.512$$

5.2.5 Complex systems

In reality many systems are not simple series, parallel or standby redundant systems. They are combinations of these systems. The reliability of these complex systems can be obtained by simplifying the reliability block diagram. Consider the system in Figure 5.4.

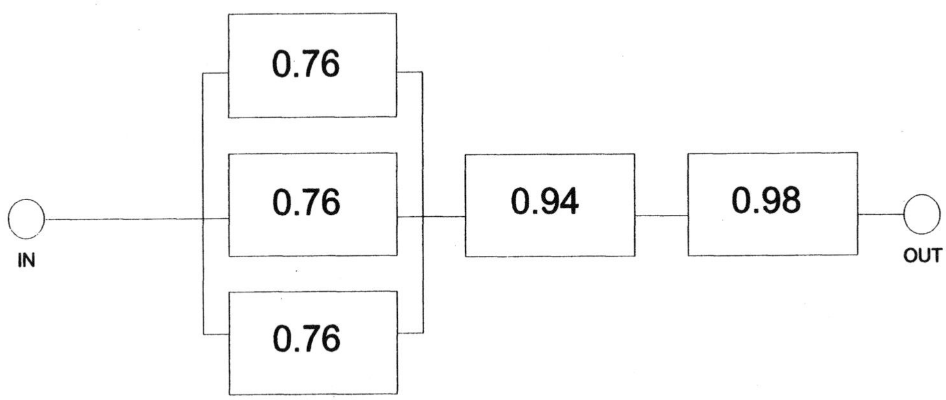

Figure 5.4. Parallel and series combination reliability block diagram.

The three components with reliability of 0.76 are a parallel system. This parallel system is in series with two more components. By reducing the parallel system, a simple series system can be obtained. The reliability of the parallel system is

$$R = 1-(1-0.76)(1-0.76)(1-0.76) = 0.9862$$

The system in Figure 5.4 can now be shown as

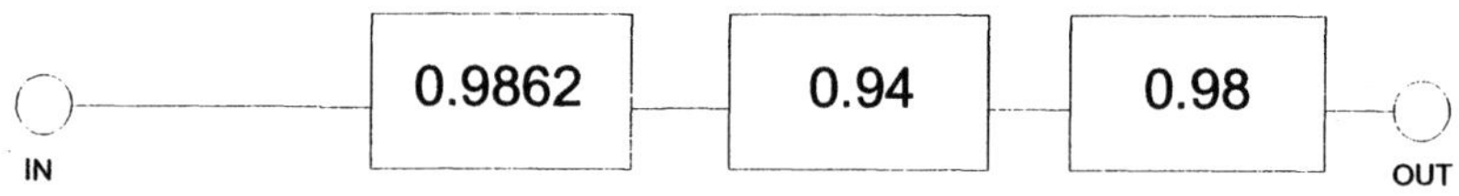

Figure 5.5. Figure 5.4 simplified.

The reliability of this system is

$$R = (0.9862)(0.94)(0.98) = 0.9085$$

Suppose only two of the three components with reliability of 0.76 in Figure 5.4 are required for the system to operate. Now a 2-out-of-3 system is in series with two other components. The reliability of the 2-out-of-3 system is

$$R = \sum_{x=2}^{3} \binom{3}{x} R^x (1-R)^{3-x}$$

$$R = \binom{3}{2} 0.76^2 (1-0.76)^1 + \binom{3}{3} 0.76^3 (1-0.76)^0$$

$$R = 0.415872 + 0.438976 = 0.8548$$

The system can now be shown as three components in series with reliabilities of 0.8548, 0.94 and 0.98. The system reliability is

$$R = (0.8548)(0.94)(0.98) = 0.7874$$

In some cases, the system cannot be simplified as easily. Consider the system in Figure 5.6.

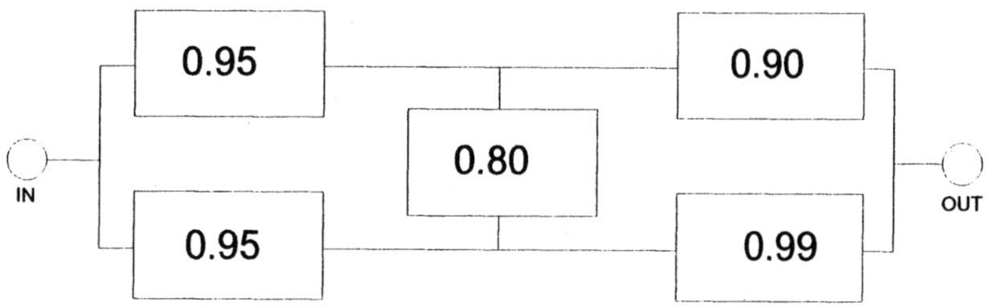

Figure 5.6. Complex reliability block diagram.

There is not an obvious way to simplify this system. By inspection, the component in the center of the diagram plays a key role. If the center component operates, then the system consists of two parallel systems in series. This is shown in Figure 5.7.

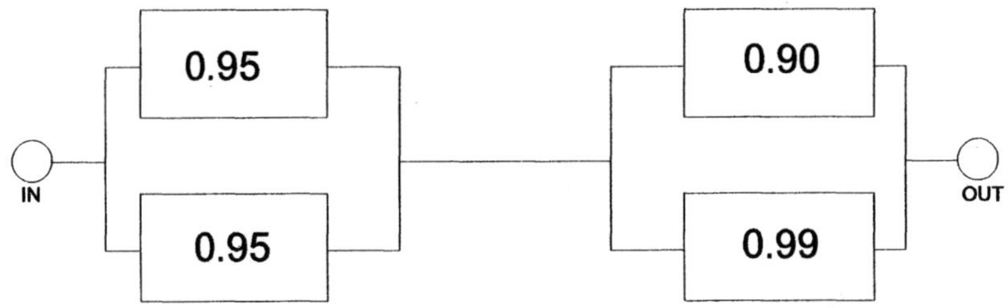

Figure 5.7. Figure 5.6 with the center component functioning.

The reliability of the components in parallel on the left side of Figure 5.7 is

$$R = 1 - (1 - 0.95)(1 - 0.95) = 0.9975$$

The reliability of the other set of parallel components is

$$R = 1 - (1 - 0.9)(1 - 0.99) = 0.999$$

The reliability of the system shown in Figure 5.7 is

$$R = (0.9975)(0.999) = 0.9965$$

Now assume the component in the center of Figure 5.6 fails. The resulting system is shown in Figure 5.8.

Figure 5.8. Figure 5.6 with center component failed.

Figure 5.8 can be further reduced by computing the reliabilities of the components in series. The reliability of the series system in the top half of Figure 5.8 is

$$R = (0.95)(0.90) = 0.855$$

The reliability of the series system in the bottom half of Figure 5.8 is

$$R = (0.95)(0.99) = 0.9405$$

The reliability of the system shown in Figure 5.8 is

$$R = 1 - (1 - 0.855)(1 - .9405) = 0.9914$$

So far it is known that if the component in the center of Figure 5.6 functions, the reliability of the system is 0.9965; and that if that component fails, the reliability of the system is 0.9914. The system reliability is the reliability of the system given the center component operates properly, multiplied by the probability of the center component operating properly, plus the reliability of the system given the center component fails, multiplied by the probability the center component fails. Since the reliability of the center component is 0.8, the reliability of the system is

$$R = (0.8)(0.9965) + (0.2)(0.9914) = 0.9955$$

5.2.6 Stress-strength systems

If a component has a strength, y, and is subjected to a stress, x, the component fails when $y < x$. If the variables y and x are random and can be described by a probability density function, then the probability of failure can be modeled. Consider Figure 5.9.

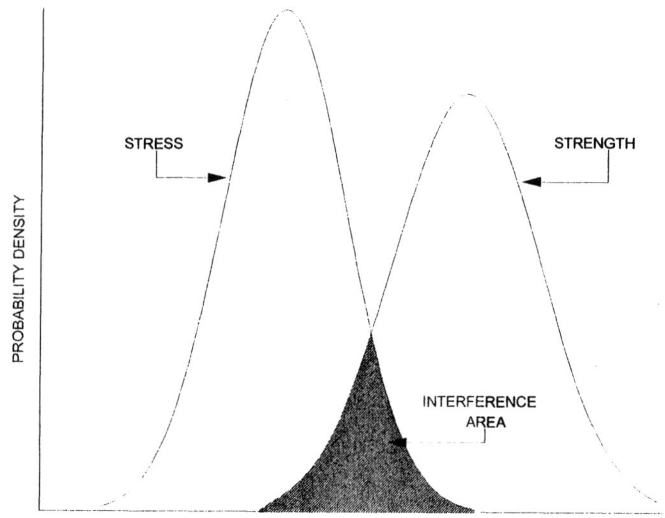

Figure 5.9. Stress-strength interference.

The shaded region of Figure 5.9 is known as the interference area. In the interference area, and only in the interference area, there is a probability of stress being greater than strength. Note that the area in the shaded region of Figure 5.9 is **not** equal to the probability of stress being greater than strength. The general expression for the reliability of a stress-strength system is

$$R = P(\text{stress} < \text{strength}) = \int_0^\infty \int_0^\infty g(y+x) f(x)\, dx dy \tag{5.20}$$

where: $g(y)$ is the strength probability density function, and

$f(x)$ is the stress probability density function.

If the probability density functions for stress and strength are both normally distributed and independent, the reliability (the probability that strength is greater than stress) is

$$R = \Phi\left(\frac{\mu_y - \mu_x}{\sqrt{\sigma_y^2 + \sigma_x^2}}\right) \tag{5.21}$$

where: $\Phi(x)$ is the standard normal cumulative distribution function,
μ_y is the mean strength,
σ_y is the strength standard deviation,
μ_x is the mean stress, and
σ_x is the stress standard deviation.

Example 5.10

The strength of an axle is normally distributed with a mean of 20,000 kPa and a standard deviation of 1,500 kPa. The axle is subjected to a stress that is normally distributed with a mean of 15,000 kPa and a standard deviation of 2,500 kPa. What is the reliability of the axle?

Solution

The axle reliability is

$$R = \Phi\left(\frac{20{,}000 - 15{,}000}{\sqrt{1500^2 + 2500^2}}\right) = \Phi(1.715)$$

From Appendix A, the area under the standard normal curve to the left of z = 1.715 is 0.9568.

Example 5.11

A component will be subjected to a normally distributed stress with a mean of 65 units and a standard deviation of 8 units. What must the mean strength of the component be if a reliability of 0.999 is required and the standard deviation of the strength is 3 units?

Solution

The first step is to find the inverse of the standard normal distribution function. From Appendix A,

$$0.999 = \Phi(3.0905)$$

From Equation 5.21,

$$3.0905 = \frac{\mu_y - 65}{\sqrt{8^2 + 3^2}}$$

The required mean strength is 91.41 units.

Example 5.12

The strength of a component is normally distributed with an unknown mean and a known variance of 25. Twenty components were randomly selected and tested. The average strength of these 20 components was 150. What is the maximum stress that these components can endure to be 90% confident of a 95% reliability?

Solution

A 90% lower confidence level for the mean strength is found from the expression

$$\mu_{90\%L} = 150 - \left(\frac{1.28}{\sqrt{20}}\right)\sqrt{25} = 148.57$$

The inverse of the standard normal cumulative distribution function, $\Phi(0.95)$, is 1.645. The maximum stress is found from the expression

$$1.645 = \frac{148.57 - x}{\sqrt{25}}$$

$$\mu_x = 140.35$$

5.3 Fault trees

Fault tree analysis provides a formal method of determining the combinations of primary events that result in the occurrence of a specified system level event. H. R. Watson conceived fault-tree analysis while working on safety evaluation for the Minuteman Launch Control System. Watson realized that the logic flow in data processing equipment could be used for analyzing the logic of system failures resulting from component failures.

Fault-trees provide a graphical representation of the system being studied, and their use often results in the discovery of failure combinations that would not have been obvious if other failure identification methods were used. Fault trees are useful for both qualitative and quantitative analyses. Some common uses for fault trees are:

- Root cause failure analysis
- Failure modes and effects analysis
- Preliminary hazard analysis
- Probability calculations

5.3.1 Fault-tree construction

Fault trees are built from gates and events. A list of fault-tree gates is provided in Table 5.1; a list of fault tree events is provided in Table 5.2.

Consider how an automobile might fail to complete a trip. The event under consideration is the failure of the automobile to complete the trip. In the fault tree, this will be the *Top* event, and will be represented by a conditional event.

Simplifying the problem, an automobile fails to complete a trip if

1. two of the five tires go flat (including the spare),
2. the engine fails, or
3. the transmission fails.

For this example, it is assumed all other systems work. This is depicted with a house event. The fault tree for this example is shown in Figure 5.10.

The *Top* event in this fault tree is the conditional event—automobile fails to complete the trip. Following the *Top* event is an *OR* gate. The *Top* event will take place if any of the four events below the *OR* gate take place. Two of these four events are primary events; engine fails and transmission fails. The conditional event, 4 tires not available, is dependent on the outcome of the *r-out-of-n* gate. The number "2" in the *r-out-of-n* gate signifies that two of the five events below the gate must transpire for the conditional event above the gate to occur. The house event is used to indicate if a component not detailed in the fault tree failed. If this house event is turned off, all of the automobile's components not detailed in the fault tree function properly. If turned on, the top event occurs. A logical analysis for this example would be to turn the house event off.

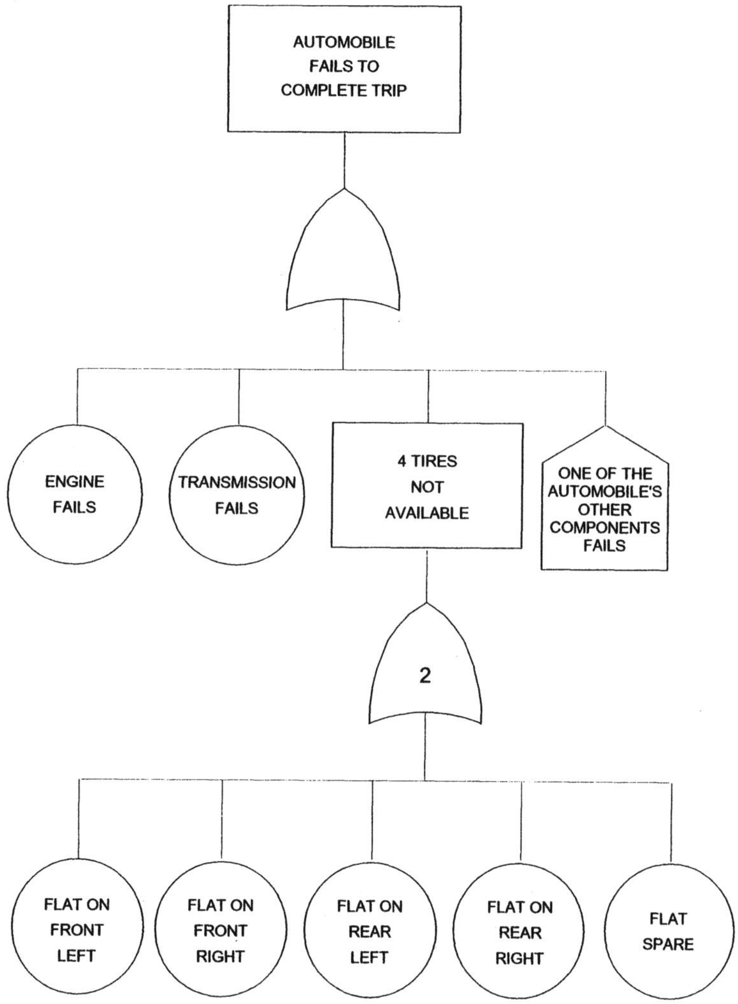

Figure 5.10. Automobile fault tree.

Table 5.1. Fault tree gates.

Symbol	Name	Description
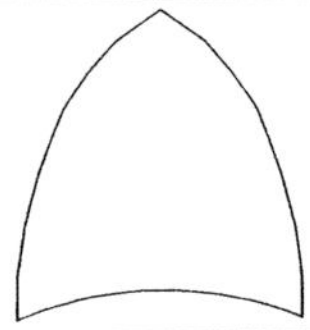	OR Gate	The event above this gate occurs if **any** of the events below this gate occur. An ***OR*** gate signifies a union of events.
	AND Gate	The event above this gate occurs if **all** the events below this gate occur. An ***AND*** gate signifies an intersection of events.
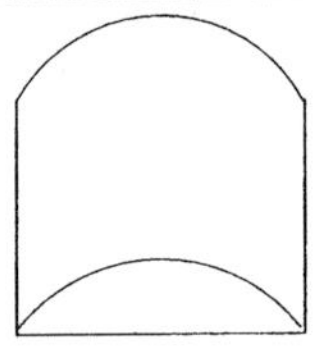	Exclusive OR Gate	The event above this gate occurs if one of the events below the gate occur and the others do not. The triggering event is specified within the gate.
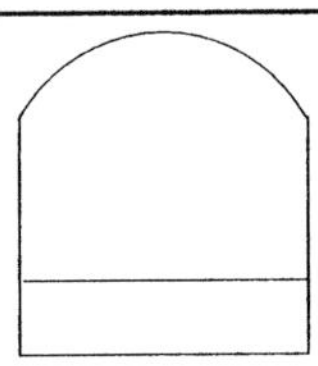	Priority AND Gate	The event above this gate occurs if all the events below the gate occur in the order specified. The order may be listed in the gate or in an attached oval. If no order is specified, the default order is left to right.
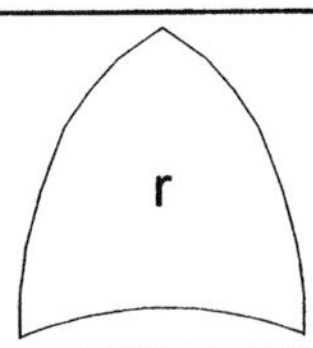	r-out-of-n Gate	The event above this gate occurs if ***r*** of the ***n*** events below the gate take place.
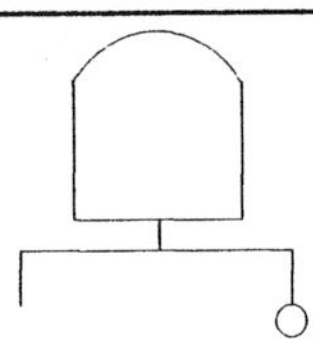	AND NOT Gate	The event above this gate occurs if all the events on a leg without a circle occur, and all the events on a leg with a circle **do not** occur.
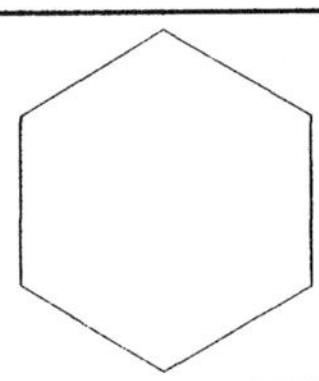	Inhibit Gate	A special type of ***AND*** gate. The event above this gate transpires if the event listed within the Inhibit gate is satisfied **and** the event below the gate takes place.

Table 5.2. Fault tree events.

Symbol	Name	Description
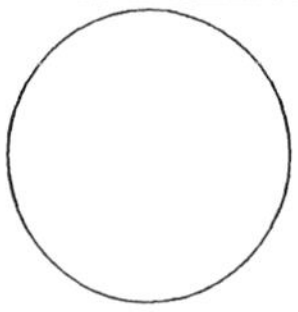	Primary event. Also known as a Basic event.	The lowest level of failure possible. Primary events are individual component failures, human errors and software errors.
	Conditional Event	Conditional events are dependent on Primary events. They are located above gates in the fault tree.
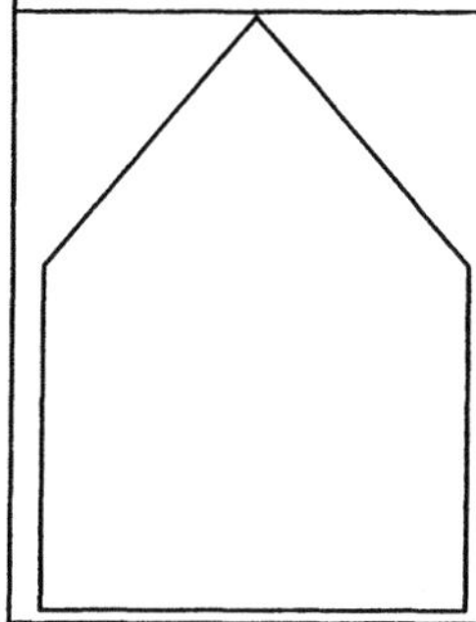	House Event	An event that can be turned on or off. If turned on, the probability of occurrence is 1; if turned off, the probability of occurrence is 0. Often used for scenario analysis.
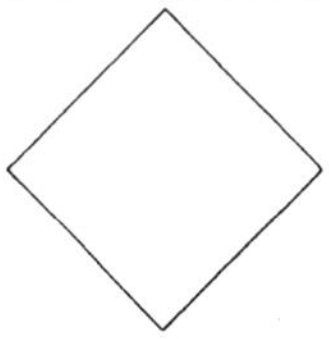	Undeveloped Event	This event has not been defined. Some reasons for not defining an event are; lack of interest, lack of information, or further development will not provide additional understanding.
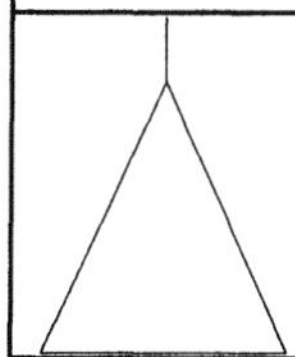	Triangle In	Use to repeat a portion of the tree or to continue the tree on another page.
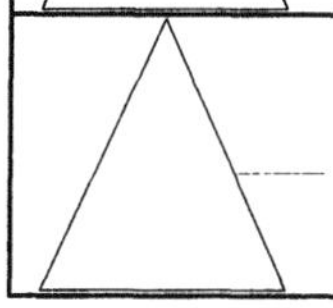	Triangle Out	Used in conjunction with the Triangle In block.

5.3.2 Cut sets and path sets

A **cut set** is a set of primary events that cause the *Top* event to transpire if they **all** transpire. A cut set may be a subset of another cut set. For the automobile trip example: if the engine fails, the transmission fails, there is a flat on the front left tire, there is a flat on the front right, and a flat spare tire, then the automobile will fail to make the trip. Thus, the occurrence of all of these events is a cut set. However, if there is a flat on the front left tire and the front right tire, the automobile also fails to make the trip. The set: flat on front left and flat on front right, is also a cut set.

A **minimum cut set** is a cut set that is no longer a cut set if any events are removed from the cut set. The first cut set mentioned above is not a minimum cut set. Several events were removed (engine fails, transmission fails, left, flat spare) and the set remained a cut set. The cut set: flat on front left and flat on front right, is a minimum cut set. If either of the two events: flat on front left or flat on front right, is removed from the cut set, the remaining event in the set is not sufficient to cause the *Top* event to occur. Thus: the set, flat on front left and flat on front right, is a minimum cut set. The remaining minimum cut sets for this problem are:

1. Engine fails
2. Transmission fails
3. House event turned on
4. Flat on front left and flat on rear left
5. Flat on front left and flat on rear right
6. Flat on front left and flat spare
7. Flat on front right and flat on rear left
8. Flat on front right and flat on right
9. Flat on front right and flat spare
10. Flat on rear left and flat on rear right
11. Flat on rear left and flat spare
12. Flat on rear right and flat spare

Minimum cut sets can be determined intuitively in small fault trees, however, for large trees, quantitative methods are needed. The four most common techniques for determining minimum cut sets are:

1. Use of Boolean algebra
2. The prime number method
3. The binary string bit method
4. The combination testing method

For more information concerning these methods consult Sundararajan (1991).

In some cases the interest is in what prevents the *Top* event from occurring, rather than what causes it to occur. This is done by identifying path sets; the opposite of cut sets. If **none** of the events in a **path set** occur, the *Top* event is prevented from occurring.

A path set, that is no longer a path set if any events are removed from the set, is a **minimum path set.** The minimum path sets for the previous example are (assuming the house event is turned off):

1. a) engine fails and
   b) transmission fails and
   c) flat on front left and
   d) flat on front right and
   e) flat on rear left and
   f) flat on rear right
2. a) engine fails and
   b) transmission fails and
   c) flat on front left and
   d) flat on front right and
   e) flat on rear left and
   f) flat spare
3. a) engine fails and
   b) transmission fails and
   c) flat on front left and
   d) flat on front right and
   e) flat spare rear left and
   f) flat on rear right
4. a) engine fails and
   b) transmission fails and
   c) flat on front left and
   d) flat spare and
   e) flat on rear left
   f) flat on rear right
5. a) engine fails and
   b) transmission fails and
   c) flat spare and
   d) flat on front right and

e) flat on rear left and

f) flat on rear right

The fault tree displayed in Figure 5.11 demonstrates the *AND* gate, the *AND NOT* gate, and the Inhibit gate. The *Top* Event is followed by an *AND* gate and takes place only if Event A and Event B take place. Event A is followed by an *AND NOT* gate and transpires if Event 1 occurs and Event 2 does not occur. Event B is followed by an Inhibit gate and occurs if Events 3 and 4 take place.

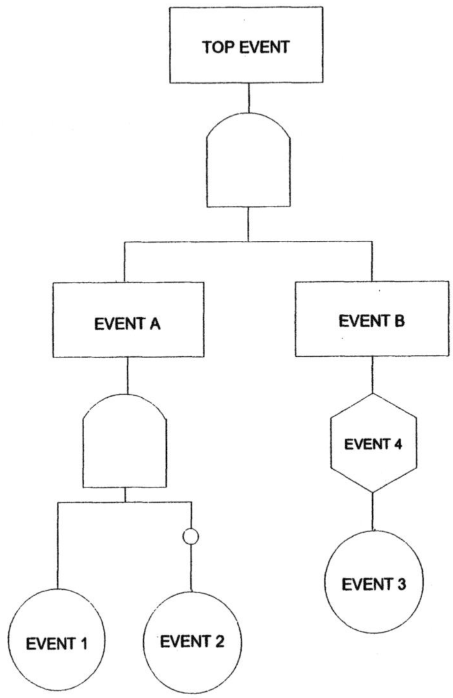

Figure 5.11. Example fault tree.

The fault tree presented in Figure 5.12 illustrates the Exclusive *OR* gate and the Priority *AND* gate. Being followed by an Exclusive *OR* gate, the *Top* Event transpires if either Event A or Event 3 occur, but does not transpire if both Event A and Event 3 occur. Event A takes place if Event 1 and Event 2 take place, but only if Event 1 occurs before Event 2.

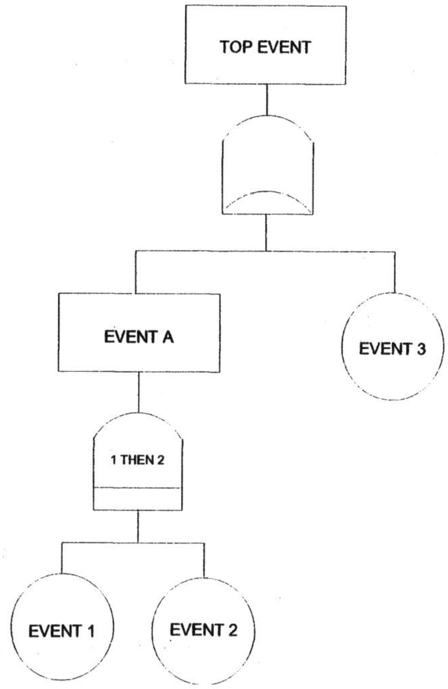

Figure 5.12. Example fault tree.

5.3.3 Quantitative fault tree analysis

If an event does not occur in more than one place in a fault tree, the probability of the *Top* event occurring can be determined by starting at the bottom of the tree and evaluating the probabilities of each branch of the tree. For example, compute the probability of the *Top* Event occurring in the fault tree in Figure 5.13, given the following probabilities for occurrence

1. probability of Event 1 = 0.3,
2. probability of Event 2 = 0.06,
3. probability of Event 3 = 0.12,
4. probability of Event 4 = 0.22, and
5. probability of Event 5 = 0.08.

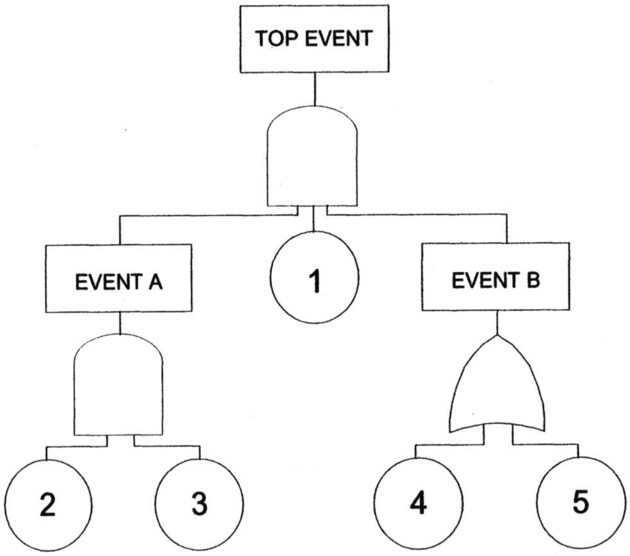

Figure 5.13. Quantitative fault tree example.

Solution

Event A occurs if Events 2 and 3 both occur; the probability of this happening is (0.06)(0.12) = 0.0072. Event B occurs if either event 4 or event 5 occur; this probability is [1–(1–0.22)(1–0.08)] = 0.2824. The ***Top*** Event occurs if events A, B and 1 all occur; this probability is (0.0072)(0.2824)(0.3) = 0.00061.

Now consider the fault tree in Figure 5.14. This tree is identical to the tree depicted in Figure 5.13, except for two changes; Event 5 has been replaced by event 1, and the gate immediately below the *Top* Event has been changed from an *AND* gate to an *OR gate*. The probability of Events 1–4 occurring is the same as in the previous example.

Because Event 1 appears in more than one location in the fault tree, the branches of the tree are no longer independent. Thus, the method used to solve the previous problem cannot be used.

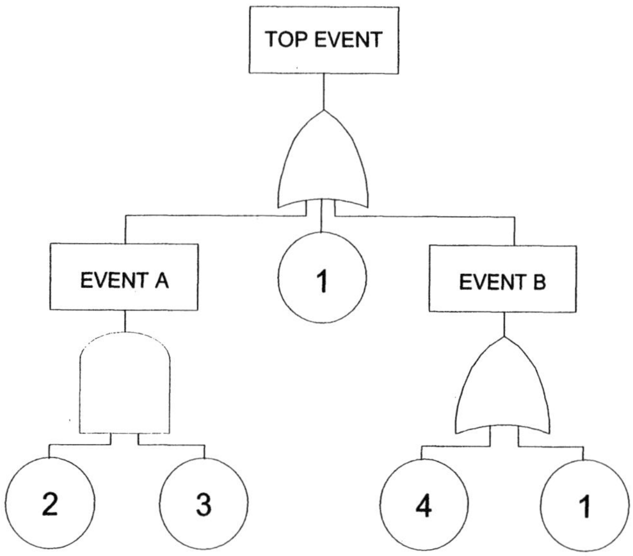

Figure 5.14. Quantitative fault tree example.

The probability of the *Top* Event occurring in a fault tree can be determined if the minimum cut sets for the tree are known. The expression for the probability of the *Top* Event occurring is

$$P_T = \sum_{i=1}^{N} P(C_i) - \sum_{i=1}^{N}\sum_{j=1}^{N} P(C_i \cap C_j) + \sum_{i=1}^{N}\sum_{j=1}^{N}\sum_{k=1}^{N} P(C_i \cap C_j \cap C_k) \ldots + (-1)^{N-1} P\left(\bigcap_{i=1}^{N} C_i\right) \tag{5.22}$$

where N is the number of minimum cut sets and $P(C_i)$ represents the probability of the minimum cut set i occurring.

Solution

The cut sets for this tree are; (1), (4), (1,2), (1,3), (1,4), (2,3), (2,4), (3,4), (1,2,3), (1,2,4), (1,3,4), (2,3,4), and (1,2,3,4). The minimum cut sets are (1), (4), and (2,3). The probability of each minimum cut set occurring is

Minimum Cut Set	Probability of Occurrence
Event 1	0.3
Event 4	0.22
Event 2 and Event 3	(0.06)(0.12)=0.0072

Using Equation 5.22, the probability of the *Top* Event occurring is:

$$P_T = 0.3 + 0.22 + 0.0072 - (0.3)(0.22) - (0.3)(0.0072) - (0.22)(0.0072) + (0.3)(0.22)(0.0072) = 0.4579$$

5.4 Monte Carlo Simulation

As seen in the previous sections, reliability modeling can be difficult mathematically. And in many cases, it is impossible to mathematically model the situation desired. Monte Carlo Simulation is a useful tool under these and many other circumstances, such as:

- Verifying analytical solutions
- Studying dynamic situations
- Gaining information about event sequences; often expected values and moments do not provide enough detail
- Determining the important components and variables in a complex system
- Determining the interaction among variables
- Studying the effects of changes without the risk, cost, and time constraints of experimenting on the real system
- Teaching

5.4.1 Random number generators

The heart of any simulation is the generation of random numbers. If a programming language such as, BASIC, C, or FORTRAN is used, random number generators will have to be created. If simulation languages such as, Siman, Slam, Simscript, or GPSS, are used, random number generating functions are included in the software.

Random numbers from specific distributions are generated by transforming random numbers from the unit, uniform distribution. Virtually all programming languages, as well as electronic spreadsheets, include a unit, uniform random number generator[2]. Technically, these unit, uniform random number generators are pseudo-random number generators, as the algorithms used to generate them takes away a small portion of the randomness. Nevertheless, these algorithms are extremely efficient and for all practical purposes the result is a set of truly random numbers.

A simple way to generate distribution-specific random numbers is to set the cumulative distribution function equal to a unit, random number and take the inverse. Consider the exponential distribution

$$F(x) = 1 - e^{-\lambda x} \tag{5.23}$$

---

[2] A unit, uniform random variable can be generated using the "RND" function in the BASIC programming language (Visual Basic, Quick Basic, GWBasic, and BASICA), and the "@RAND" function in Lotus™ 123. Microsoft™ Excel includes a random generating tool that allows random numbers to be generated from several distributions.

By setting a random variable uniformly distributed from 0 to 1, r, equal to $F(x)$ and inverting the function, an exponentially distributed random variable, x, with a failure rate of λ is created.

$$\begin{aligned} r &= 1 - e^{-\lambda x} \\ 1 - r &= e^{-\lambda x} \\ \ln(1-r) &= -\lambda x \\ x &= -\frac{\ln(1-r)}{\lambda} \end{aligned} \tag{5.24}$$

This expression can be further reduced; the term $1-r$ is also uniformly distributed from 0 to 1. The result is

$$x = -\frac{\ln r}{\lambda} \tag{5.25}$$

Table 5.3 contains some common random number generators.

Table 5.3. Random number generators.

Distribution	Probability Density Function	Random Number Generator
Uniform	$f(x) = \frac{1}{b-a},\ a \le x \le b$	$x = a + (b-a)r$
Exponential	$f(x) = \lambda e^{-\lambda x},\ 0 < x < \infty$	$x = -\frac{1}{\lambda}\ln r$
Normal	$f(x) = \frac{1}{\sigma\sqrt{2\Pi}}\exp\left[-\frac{1}{2}\left(\frac{x-\mu}{\sigma}\right)^2\right],$ $-\infty < x < \infty$	$x_1 = \left[\sqrt{-2\ln r_1}\cos(2\Pi r_2)\right]\sigma + \mu$ $x_2 = \left[\sqrt{-2\ln r_1}\sin(2\Pi r_2)\right]\sigma + \mu$ †
Lognormal	$f(x) = \frac{1}{\sigma x\sqrt{2\Pi}}\exp\left[-\frac{1}{2}\left(\frac{\ln x-\mu}{\sigma}\right)^2\right]$ $x > 0$	$x_1 = \exp\left[\sqrt{-2\ln r_1}\cos(2\Pi r_2)\right]\sigma + \mu$ $x_2 = \left[\sqrt{-2\ln r_1}\sin(2\Pi r_2)\right]\sigma + \mu$ †

Continued on next page...

Table 5.3—continued...

Distribution	Probability Density Function	Random Number Generator
Weibull	$f(x)=\frac{\beta x^{\beta-1}}{\theta^{\beta}}\exp\left(\frac{x}{\theta}\right)^{\beta}, x>0$	$x=\theta(-\ln r)^{1/\beta}$
Poisson	$f(x)=\frac{e^{-\lambda t}(\lambda t)^{x}}{x!}, x=0,1,2,...,\infty$	$x=\begin{cases}0, -\frac{1}{\lambda}\ln r>t \\ x, \sum_{i=1}^{x}-\frac{1}{\lambda}\ln r_i<t<\sum_{i=1}^{x+1}-\frac{1}{\lambda}\ln r_i\end{cases}$ ‡
Chi-square	$f(x)=\frac{1}{2^{\nu/2}\Gamma(\nu/2)}x^{(\nu/2-1)}e^{-x/2}, x>0$	$x=\sum_{i=1}^{\nu}z_i^2$, z_i is a standard normal random deviate.
Beta	$f(x)=\frac{1}{B(p,q)}x^{p-1}(1-x)^{q-1}$, $0\le x\le 1, p>0, q>0$	$x=\frac{r^{1/p}}{r^{1/p}+r^{1/q}}$
Gamma	$f(x)=\frac{\lambda^{\eta}}{\Gamma(\eta)}x^{(\eta-1)}e^{-\lambda x}$, $x\ge 0, \eta\ge 0, \lambda\ge 0$	1. η is a non-integer shape parameter. 2. Let η_1 = the truncated integer root of η. 3. Let $q=-\ln\prod_{j=1}^{\eta_1}r_j$. 4. Let $A=\eta-\eta_1$, and B = 1–A. 5. Generate a random number and let $y_1=r_i^{1/A}$. 6. Generate a random number and let $y_2=r_{i+1}^{1/B}$ 7. If $y_1+y_2\le 1$ go to 9. 8. Let i=i+2 and go to 5. 9. Let $z=y_1/(y_1+y_2)$. 10. Generate a random number, r_n. 11. Let $W=-\ln r_n$. 12. $x=(q+zW)\lambda$.

Continued on next page...

Table 5.3—continued...

Distribution	Probability Density Function	Random Number Generator
Binomial	$p(x)=\binom{n}{x}p^x(1-p)^{n-x}, x=0,1,\ldots,n$	$x=\sum_{i=1}^{n} y_i,\ y_i=\begin{cases}0, r_i>p\\1, r_i\le p\end{cases}$
Geometric	$p(x)=p(1-p)^{x-1}, x=1,2,3,\ldots$	$\frac{\ln(1-r)}{\ln(1-p)}\le x\le\frac{\ln(1-r)}{\ln(1-p)}+1$ ‡
Student's *t*	$f(x)=\frac{\Gamma[(\nu+1)/2]}{\Gamma(\nu/2)\sqrt{\Pi\nu}}\left(1+\frac{x^2}{\nu}\right)^{-(\nu+1)/2}$, $-\infty<x<\infty$	$x=\frac{z_1}{\left(\sum_{i=2}^{\nu+1}\frac{z_i^2}{\nu}\right)^{1/2}}$ z_1 is a standard normal random deviate.
F	$f(x)=\left(\frac{\Gamma[(\nu_1+\nu_2)/2](\nu_1/\nu_2)^{\nu_1/2}}{\Gamma(\nu_1/2)\Gamma(\nu_2/2)}\right) * \left(\frac{x^{\nu_1/2-1}}{(1+\nu_1x/\nu_2)^{(\nu_1+\nu_2)/2}}\right), x>0$	$x=\frac{\nu_2\sum_{i=1}^{\nu_1}z_i^2}{\nu_1\sum_{i=\nu_1+1}^{\nu_1+\nu_2}z_i^2}$ z_1 is a standard normal random deviate.

† Two uniform random numbers are generated, with the result being two normally distributed random numbers.

‡ Increase the value of ***x*** until the inequality is satisfied.

5.4.2 Simulation modeling

After the desired random number generator(s) have been constructed, the next step is to mathematically model the situation under study. After completing the model, it is important to **validate** and **verify** the model. A valid model is a reasonable representation of the situation being studied. A model is verified by determining that the mathematical and computer model created represents the intended conceptual model.

Enough iterations should be included in the simulation to provide a **steady-state** solution. A steady-state solution is reached when the output of the simulation from one iteration to the next changes negligibly. When calculating means and variances, 1,000 iterations is usually sufficient. If calculating confidence limits, many more iterations are required; after all, for 99% confidence limits the sample size for the number of random deviates exceeding the confidence limit is 1/100th the number of iterations.

Simulation can be used to determine the result of transformations of random variables. Suppose the mean and variance of the product of two normally distributed variables are needed for design purposes. The following BASIC code will produce the desired result. A flowchart for this code is shown in Figure 5.15.

```
REM simulation for the product of two normally distributed variables
e1=100
v1=7
e2=120
v2=16
DEFINT I–L
FOR i = 1 TO 5000
a = RND
b = RND
REM x is normal with mean=e1 and standard deviation=v1
x = (((–2 * LOG(a)) ^ .5) * COS(2 * 3.141592654# * b)) * v1 + e1
REM y is normal with mean=e2 and standard deviation=v2
y = (((–2 * LOG(a)) ^ .5) * SIN(2 * 3.141592654# * b)) * v2 + e2
z = x * y
ztot# = ztot# + z
zsq# = zsq# + z ^ 2
PRINT i
NEXT i
PRINT "ztot zsq"; ztot#; zsq#
PRINT "mean="; ztot# / 5000
zvar# = (5000 * zsq# – ztot# ^ 2) / (5000 * 4999)
PRINT "variance="; zvar#
```

Note: "RND" generates uniform random deviates on the interval from 0 to 1. The "LOG" function in BASIC represents the natural logarithm.

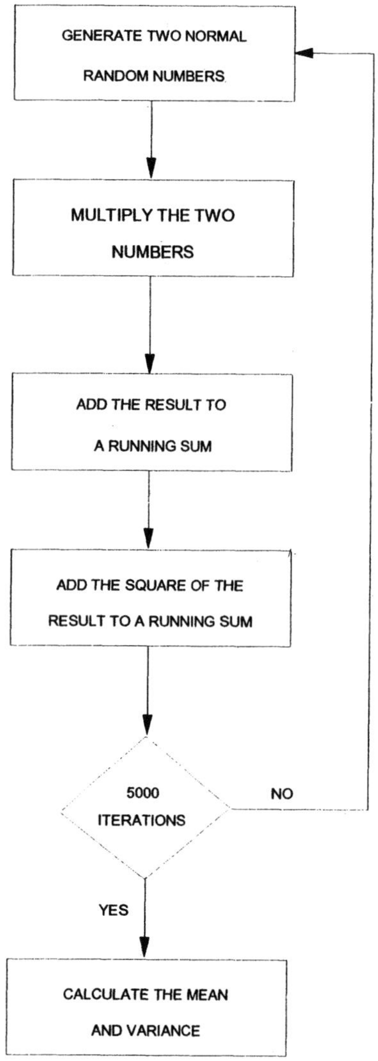

Figure 5.15. Flowchart for the simulation of two normally distributed normal variables.

In the above code, two normal random numbers are generated with the desired parameters and multiplied. This is repeated 5000 times, and the mean and variance of these 5000 random deviates are calculated. The result is a random variable with a mean of 12,009 and a variance of 3,307380.

For the above example, recall that the same result could have been obtained mathematically. A disadvantage of solving this problem mathematically is that there is no information regarding the shape of the resulting distribution.

With electronic spreadsheets, simulations no longer require computer code. The previous example is simulated using *Microsoft Excel*™ with the following steps[3].

1. In cell *A1* place the function =RAND()
2. In cell *B1* place the function =RAND()
3. In cell *C1* place the formula
   =7*((COS(2*PI()*A1))*(-2*LN(B1))^0.5)+100
4. In cell *A4* place the formula
   =7*((COS(2*PI()*A1))*(-2*LN(B1))^0.5)+100
5. In cell *E1* place the formula
   +C1*D1
6. Copy this row 5,000 times

In the above example, each row in the spreadsheet represents an iteration. The powerful @functions and graphics tools contained in the spreadsheet can then be used for analysis. Note that each change in the spreadsheet causes the random numbers to be changed. It may be helpful to convert the output from formulas to fixed numbers with the "Range Value" command.

Now consider a system consisting of four identical components which are exponentially distributed with a failure rate of 0.8. Three of the components are standby redundant with perfect switching. Information is needed regarding the shape of the resulting distribution. The following code produces four exponentially distributed random variables with a failure rate of 0.8, adds them, and writes the result to the file "c:\data."; this process is repeated 5,000 times. A flowchart for this problem is provided in Figure 5.16.

[3] Microsoft™ Excel can generate normal random variables directly.

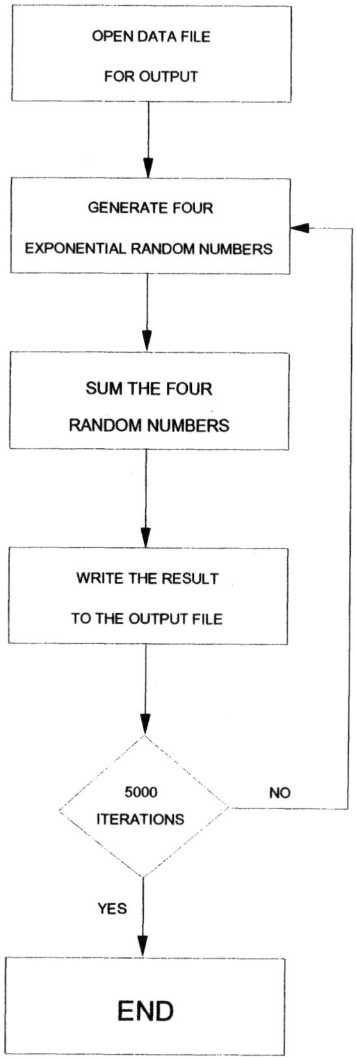

Figure 5.16. Simulation for the sum of exponential random variables.

```
REM simulation for the sum of four exponentially distributed variables
DEFINT I–L
OPEN "c:\data" FOR OUTPUT AS #1 LEN = 256
FOR i = 1 TO 5000
REM x1 is exponential with failure rate = 0.8
x1 = –(1 / .8) * LOG(RND)
REM x2 is exponential with failure rate = 0.8
x2 = –(1 / .8) * LOG(RND)
REM x3 is exponential with failure rate = 0.8
x3 = –(1 / .8) * LOG(RND)
REM x4 is exponential with failure rate = 0.8
x4 = –(1 / .8) * LOG(RND)
y = x1 + x2 + x3 + x4
PRINT #1, USING "##########.#####"; y
```

```
PRINT i
NEXT i
CLOSE
```

By importing the resulting data into an electronic spreadsheet or statistical program, a wide variety of analyses can be done on the data. A histogram of the data produced from the above code is shown in Figure 5.17.

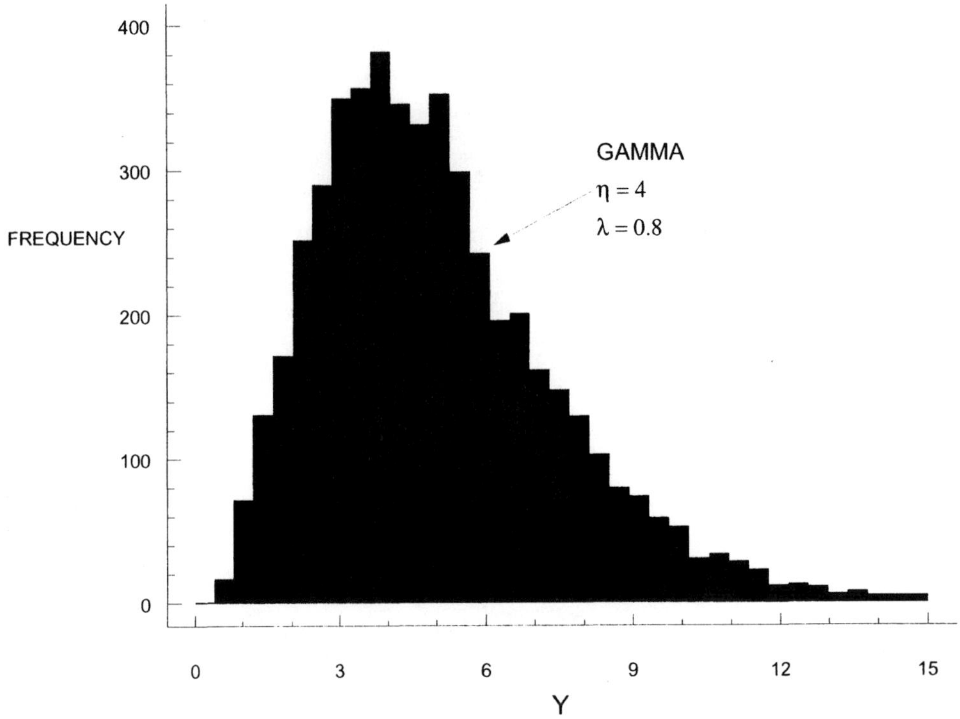

Figure 5.17. Histogram of the sum of four exponentially distributed random variables.

As seen from Figure 5.17, the sum of n exponentially distributed random variables with a failure rate of λ is a random variable that follows the gamma distribution with parameters $\eta = n$ and λ.

This problem is also easily simulated using an electronic spreadsheet. The steps required follow:

1. In cells *A1* through *D4*, place the formula
   =LN(RAND())*(-1/0.8)
2. In cell *E5* place the formula
   =SUM(A1..D4)
3. Copy the contents of row *A* 5000 times

Again, each row represents an iteration, and the spreadsheet can be used to obtain the desired simulation output.

Now consider a system consisting of three components in series. The components are Weibull distributed with parameters $\beta = 2$, $\theta = 300$; $\beta = 4$, $\theta = 100$; and, $\beta = 0.5$, $\theta = 200$. The code below depicts this situation. Figure 5.18 shows a flowchart for this simulation.

```
REM simulation three Weibull distributed variables in series
DEFINT I–L
DIM x(99)
OPEN "c:\data" FOR OUTPUT AS #1 LEN = 256
FOR i = 1 TO 5000
REM x(1) is Weibull shape parameter=2 and scale parameter=300
x(1) = 300 * (–LOG(RND)) ^ (1 / 2)
REM x(2) is Weibull shape parameter=4 and scale parameter=100
x(2) = 100 * (–LOG(RND)) ^ (1 / 4)
REM x(3) is Weibull shape parameter=0.5 and scale parameter=200
x(3) = 200 * (–LOG(RND)) ^ (1 / .5)
min = 999999999
FOR j = 1 TO 3
IF x(j) < min THEN min = x(j)
NEXT j
y = min
PRINT #1, USING "##########.#####"; y
PRINT i
NEXT i
CLOSE
```

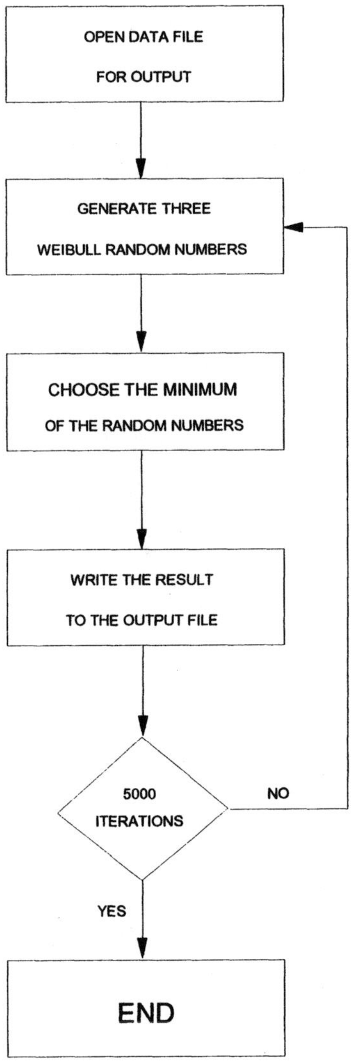

Figure 5.18. Simulation of a series system.

For a series system, the time to fail is the minimum of the times to fail of the components in series. A parallel system could be modeled by altering the above code to take the maximum time to fail of each of the components. Figure 5.19 is a histogram of the resulting data for the series system.

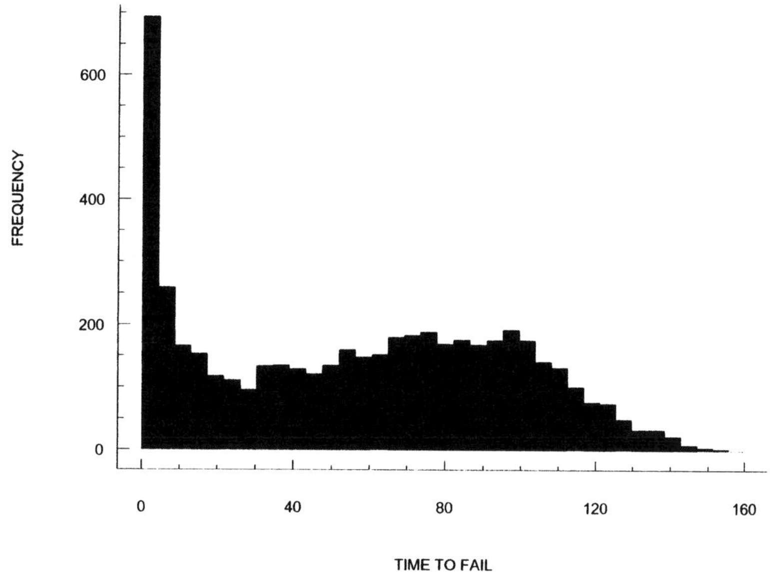

Figure 5.19. Histogram of a series system.

The large number of early failures are caused by the component with the high infant mortality rate ($\beta = 0.5$). The result is a distribution that does not appear to conform to any known distributions. However, with 5,000 points, a reliability function can be built empirically. The result is shown in Figure 5.20.

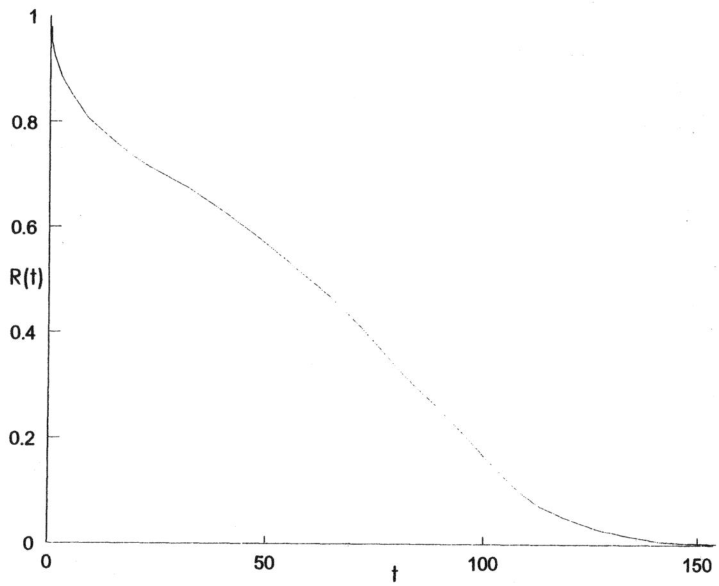

Figure 5.20. Reliability function for a series system.

The following steps are used to simulate the above problem using an electronic spreadsheet:

1. In cell *A1* place the formula
   =((-LN(RAND())^(1/2))*300)
2. In cell *B1* place the formula
   =((-LN(RAND())^(1/4))*300)
3. In cell *C1* place the formula
   =((-LN(RAND())^(1/0.5))*200)
4. In cell *D1* place the formula
   =MIN(A1..C1)
5. Copy the contents of row *A* 5,000 times.

Now consider a system with two Weibull distributed components, A and B. Component B is standby redundant, and the switching mechanism has a reliability of 95%. The parameters of component A are $\beta = 3$, $\theta = 85$. The parameters of component B are $\beta = 4.4$, $\theta = 95$. The code below models this system; a flowchart is given in Figure 5.21.

```
REM simulation of a switch for two Weibull distributed variables
DEFINT I–L
OPEN "c:\data" FOR OUTPUT AS #1 LEN = 256
FOR i = 1 TO 5000
REM x is Weibull shape parameter=3 and scale parameter=85
x = 85 * (–LOG(RND)) ^ (1 / 3)
REM y is Weibull shape parameter=4.4 and scale parameter=97
y = 97 * (–LOG(RND)) ^ (1 / 4.4)
s = RND
IF s >= .05 THEN swr = 1 ELSE swr = 0
IF swr = 1 THEN z = x + y ELSE z = x
PRINT #1, USING "##########.#####"; y
PRINT i
NEXT i
CLOSE
```

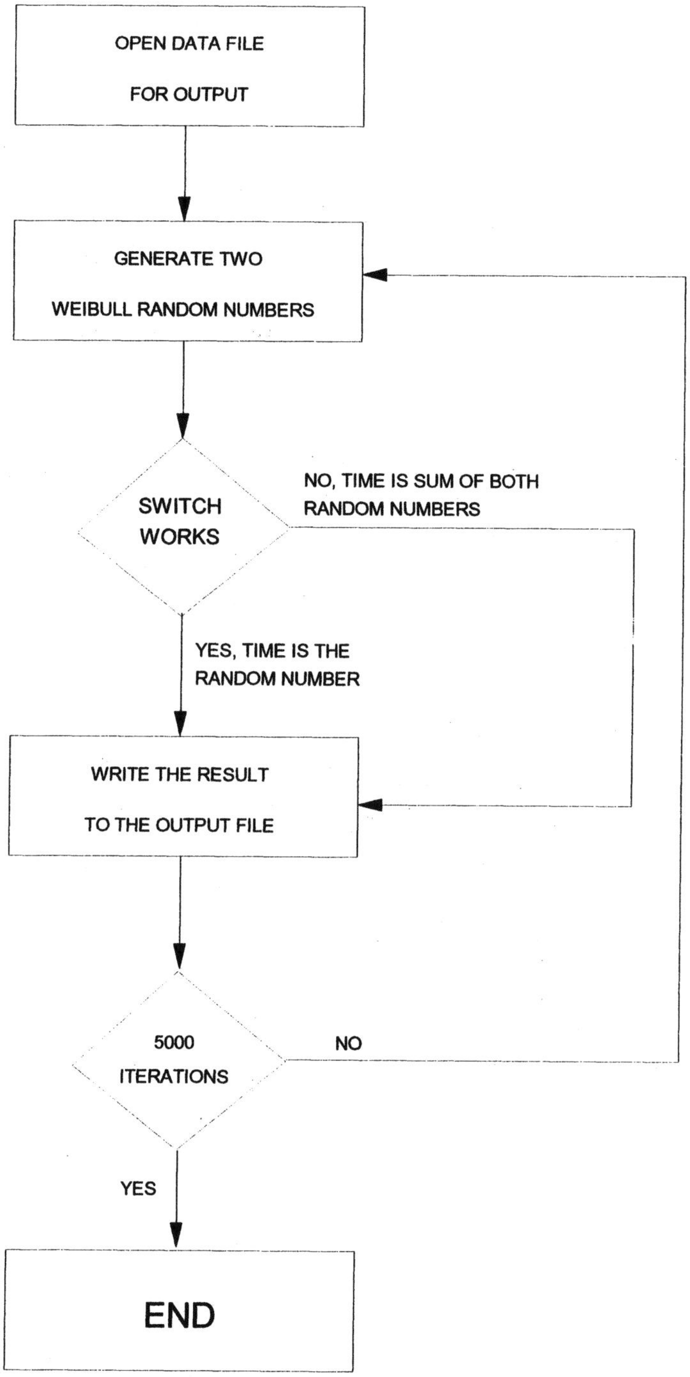

Figure 5.21. Simulation of a switching system.

A histogram of the 5,000 data points written to the data file is shown in Figure 5.22.

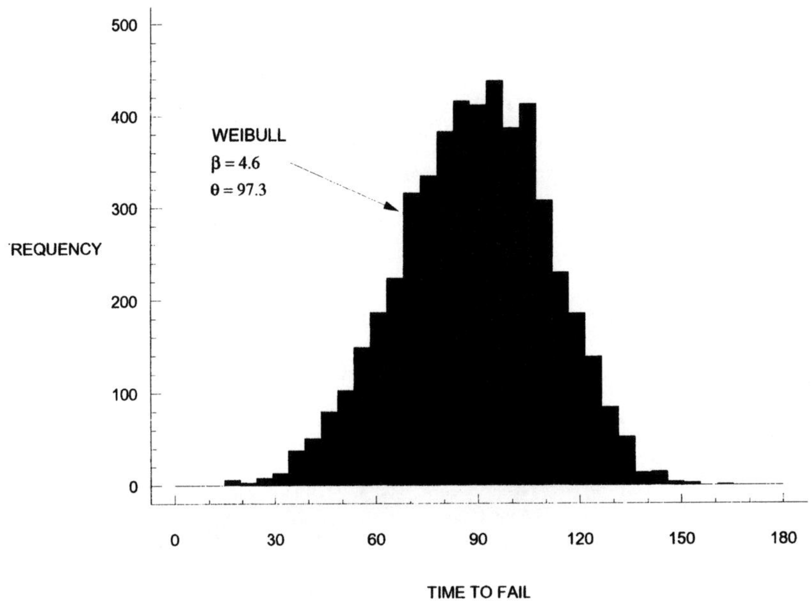

Figure 5.22. Histogram of a switching system.

The histogram shows the time to fail for the system follows a Weibull distribution. The reliability function for this system, built from the 5,000 simulation points, is shown in Figure 5.23.

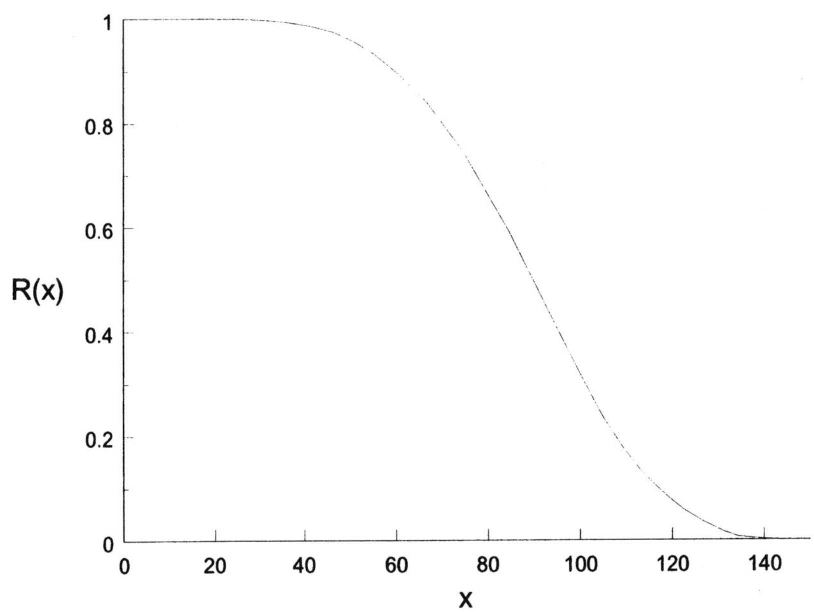

Figure 5.23. Reliability function for a switching system.

This situation can be simulated using an electronic spreadsheet. The required steps follow:

1. In cell *A1* place the formula
   =((-LN(RAND())^(1/3))*85)
2. In cell *B1* place the formula
   =((-LN(RAND())^(1/4.5))*97)
3. In cell *C1* place the formula
   =IF(RAND()<0.05,A1,A1+B1)
4. Copy the contents of row A 5,000 times.

In step 4 above, the reliability of the switch is tested using the unit, uniform random number generator. If the unit, uniform random number is less than 0.05, the switch fails, and the time to fail for the system is the time to fail for component *A* (the value in cell A1). If the switch operates, the system time to fail is the sum of the values in cells A1 and BA.

In summary, simulation is a powerful analytical tool that can be used to model virtually any system. For the above examples 5,000 iterations were used. The number of iterations used should be based on reaching a steady-state condition. Depending on the problem more or less iterations may be needed.

Once simulation is mastered, a danger is that it is overused because of the difficulty involved with mathematical models. Do not be tempted to use simulation before exploring other options. When manipulating models, mathematical models lend themselves to optimization whereas simulation models require trial and error for optimization.

5.5 Markov analysis

Markov analysis is used to model a system that can be defined as a set of **states**. For example, a system consisting of two components in parallel, A and B, can be defined as having the following four states:

1. Components A and B operating
2. Component A failed and component B operating
3. Component A operating and component B failed
4. Components A and B failed

An important feature of Markov models is that the probability of transition from one state to another in a given time interval is independent of all previous states. Also, the probability of more than one transition in a given interval is assumed to be zero. If the transition rates for the model are constant, the model is homogeneous. If the transition functions are time dependent, the model is non-homogeneous.

Using Markov analysis, the probability of the system being in a specific state at a given time can be determined. This information can be used to determine system performance measures, such as reliability and availability. The probability of a system being in state *i* at a given time, $P_i(t)$ is found by solving a set of differential equations. The differential equations are derived from the **state transition diagram** or the **transition rate matrix**. The state transition diagram is a graphical representation of the rate of transition from one state to another. For the two-state system described above, assuming component *A* has a constant failure rate of λ_a and component *B* has a constant failure rate of λ_b, the state transition diagram is shown in Figure 5.24.

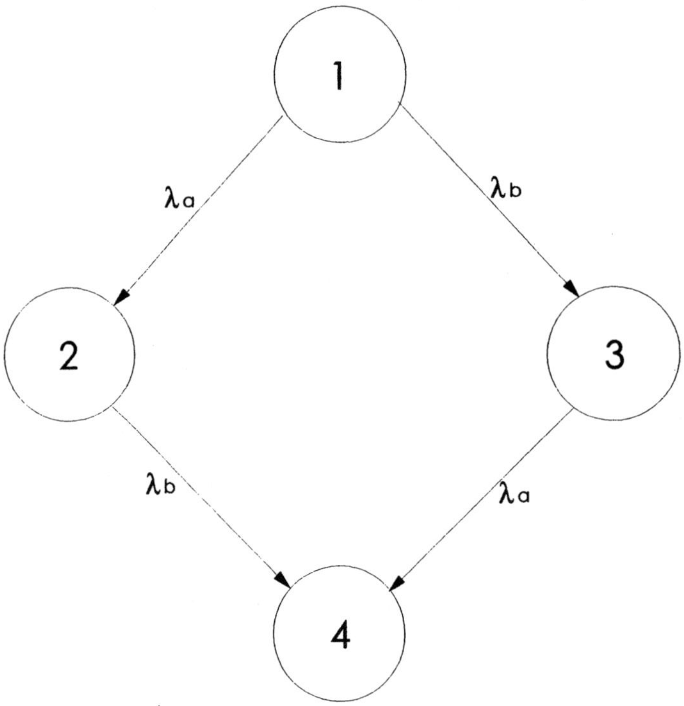

Figure 5.24. Parallel system state transition diagram.

The differential equations needed to describe this system can be determined from the State transition diagram. State 1 has no possibility of being entered, and it is exited with rates of λ_a and λ_b. The differential equation describing this is

$$\frac{dP_1(t)}{dt} = -\lambda_a P_1(t) - \lambda_b P_1(t) \tag{5.26}$$

State 2 is entered from State 1 at a rate of λ_a and exited from with a rate of λ_b. The describing equation is

$$\frac{dP_2(t)}{dt} = \lambda_a P_1(t) - \lambda_b P_2(t) \tag{5.27}$$

State 3 is entered from State 1 at a rate of λ_b and exited from with a rate of λ_a. The describing equation is

$$\frac{dP_3(t)}{dt} = \lambda_b P_1(t) - \lambda_a P_3(t) \tag{5.28}$$

State 4 is entered from State 2 at a rate of λ_b and is entered from State 3 at a rate of λ_a. Since there is no transition from State 4, it is called an **absorbing state**. The differential equation describing State 4 is

$$\frac{dP_4(t)}{dt} = \lambda_b P_2(t) + \lambda_a P_3(t) \tag{5.29}$$

With the exception of the equation describing $P_1(t)$, the equations above are coupled. A drawback of Markov analysis is the difficulty in solving the resulting system of differential equations. However, this problem has been lessened with modern symbolic software such as *Maple* and *MathCad*. Modern symbolic processors can quickly solve large systems of differential equations.

Solving the above system of equations yields

$$P_1(t) = e^{-(\lambda_a+\lambda_b)t} \tag{5.30}$$

$$P_2(t) = e^{-\lambda_b t} - e^{-(\lambda_a+\lambda_b)t} \tag{5.31}$$

$$P_3(t) = e^{-\lambda_a t} - e^{-(\lambda_a+\lambda_b)t} \tag{5.32}$$

$$P_4(t) = 1 - e^{-\lambda_a t} - e^{-\lambda_b t} - e^{-(\lambda_a+\lambda_b)t} \tag{5.33}$$

The transition rate matrix displays the same information as the state transition diagram in the form of a matrix. Each value in the matrix, r_{ij}, represents transition rate from state i to state j. For the system described in Figure 5.24, the transition rate matrix is

$$\Lambda = \begin{bmatrix} 0 & \lambda_a & \lambda_b & 0 \\ 0 & 0 & 0 & \lambda_b \\ 0 & 0 & 0 & \lambda_a \\ 0 & 0 & 0 & 0 \end{bmatrix} \tag{5.34}$$

The derivative of the probability of the system being in a given state at a specific time is

$$\frac{dP_i(t)}{dt} = \sum_{k=1}^{n} r_{ki} P_k(t) - \sum_{j=1}^{n} r_{ij} P_j(t) \tag{5.35}$$

The reliability of the system is found with the knowledge of which states are operating states and which states are failure states. The system reliability is probability the system is in an operating state

$$R_s(t) = \sum_{i=1}^{n} U_i P_i(t) \tag{5.36}$$

where: U_i equals 1 if the system operates in state i, and 0 if the system fails in state i.

Stated another way, reliability is the probability the system is not in a failed state

$$R_s(t) = 1 - \sum_{i=1}^{n} V_i P_i(t) \tag{5.37}$$

where: V_i equals 0 if the system operates in state i, and 1 if the system fails in state i.

For the parallel system described earlier, States 1, 2, and 3 represent operating states, and State 4 represents a failed state.

Markov analysis is useful for repairable systems. Consider two components, A and B, in series with failure rates of λ_a and λ_b. When a component fails, it is repaired. The component repair rates are λ_a and λ_b. The system can be defined as having three states:

1. Both components operating
2. Component A failed, component B operating
3. Component A operating, component B failed

The state transition diagram for this system is shown in Figure 5.25.

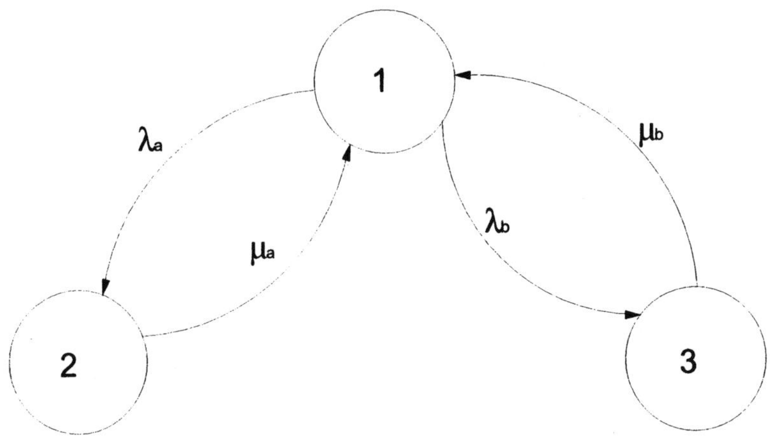

Figure 5.25. Transition state diagram for a repairable system.

The transition rate matrix for this system is

$$\Lambda = \begin{bmatrix} 0 & \lambda_a & \lambda_b \\ \mu_a & 0 & 0 \\ \mu_b & 0 & 0 \end{bmatrix} \tag{5.38}$$

The set of differential equations describing this system is

$$\frac{dP_1(t)}{dt} = -\lambda_a P_1(t) - \lambda_b P_1(t) + \mu_a P_2(t) + \mu_b P_3(t) \tag{5.39}$$

$$\frac{dP_2(t)}{dt} = \lambda_a P_1(t) - \mu_a P_2(t) \tag{5.40}$$

$$\frac{dP_3(t)}{dt} = \lambda_b P_1(t) - \mu_b P_3(t) \tag{5.41}$$

With the above set of equations, the availability of the system can be computed. The system is available if it is in State 1; and it is unavailable if it is in States 2 or 3.

In summary, Markov analysis is a powerful modeling tool. It is limited by the difficulty of solving systems of differential equations, but the power of modern symbolic software and numerical routines overcomes the drawback of dealing with systems of differential equations.

5.6 Reliability allocation

When designing, the required system reliability must be translated into sub-system and component reliabilities. This process is called **reliability allocation**. Reliability allocation involves solving the following inequality.

$$f(R_1, R_2, ..., R_n) \geq R_s \tag{5.42}$$

where: R_i is the reliability allocated to the ith sub-system or component,
f is the functional relationship between the components and the system, and
R_s is the required system reliability.

The functional relationship can be derived from reliability modeling, such as block diagrams or fault trees. For complex designs, it may be impossible to determine the functional relationship mathematically. In this case simulation may be used for allocation. Beware that using simulation requires a trial and error type of analysis, and can become time consuming.

When allocating sub-system and component reliabilities, the goal is system optimization. Two common allocation requirements are to:

1. Maximize system reliability while meeting a cost objective
2. Minimize cost while meeting a system reliability objective

Other common constraints when allocating for reliability are limits on weight, temperature, and physical space.

Several algorithms for reliability allocation have been developed, including commercial software, such as SHARPE. Three simple algorithms, the ARINC method, the AGREE method, and a method for repairable systems are presented in this section.

For more complicated allocation problems, dynamic programming may be used. Dynamic programming is a mathematical method for multi-stage decision making. If each component or sub-system is considered a stage, then an apportionment decision can be made at each stage. Dynamic programming is the most general method of reliability allocation, but obtaining solutions is very difficult manually. Computers are usually used to solve dynamic programming problems, and there are several commercial packages available. For further information concerning dynamic programming consult Puterman, 1979.

5.6.1 The ARINC apportionment technique

The ARINC apportionment method assumes all sub-systems are in series and have constant failure rates. The method allocates improved sub-system failure rates based on weighting factors, derived from the present allocation of sub-systems, that satisfy the expression

$$\sum_{i=1}^{n} \lambda_i^* \leq \lambda^* \qquad (5.43)$$

where: n is the number of sub-systems,

λ_i^* is the failure rate allocated to sub-system i, and

λ^* is the required system failure rate.

The steps of the algorithm are:

1. Determine the present sub-system failure rates.
2. Compute a weighting factor for each sub-system using the expression

$$w_i = \frac{\lambda_i}{\sum_{i=1}^{n} \lambda_i} \qquad (5.44)$$

3. Allocate the new sub-system failure rates using the expression

(5.45)

Example 5.13

A system which consists of 4 subsystems in series is required to have a reliability of 0.99 for a mission time of 50 hours. Allocate reliability requirements to each of the sub-systems given the following estimates for present sub-system failure rates.

$\lambda_1 = 0.004$ failures/hour

$\lambda_2 = 0.007$ failures/hour

$\lambda_3 = 0.008$ failures/hour

$\lambda_4 = 0.005$ failures/hour

Solution

The weighting factors are

$$w_1 = \frac{0.004}{0.004 + 0.007 + 0.008 + 0.005} = 0.1667$$

$$w_2 = \frac{0.007}{0.004 + 0.007 + 0.008 + 0.005} = 0.2917$$

$$w_3 = \frac{0.008}{0.004 + 0.007 + 0.008 + 0.005} = 0.3333$$

$$w_4 = \frac{0.005}{0.004 + 0.007 + 0.008 + 0.005} = 0.2083$$

Using the exponential reliability function,

$$R^*(0.99) = e^{-\lambda^* 50}$$

Thus, λ^* = 0.000201 failures/hour.

The allocated sub-system failure rates are

$\lambda_1^* = (0.1667)(0.000201) = 0.0000335$ failures/hour

$\lambda_2^* = (0.2917)(0.000201) = 0.0000586$ failures/hour

$\lambda_3^* = (0.3333)(0.000201) = 0.0000669$ failures/hour

$\lambda_4^* = (0.2083)(0.000201) = 0.0000418$ failures/hour

5.6.2 The AGREE allocation method

The AGREE allocation method determines a minimum acceptable mean life for each component in order to fulfill a minimum acceptable system mean life. It is assumed that sub-systems are in series and have constant failure rates. The method is based on sub-system complexity and considers the relationship between sub-system and system failure. The complexity of a sub-system is defined in terms of modules and their associated circuitry. Modules can be transistors, electron tubes, magnetic amplifiers, etc. Diodes represent half modules, and for computers, where module counts are high, the module count is reduced to compensate for the lower failure rates of digital parts in respect to radio-radar type parts. A sub-system's importance is assigned on a scale from 0 to 1, in proportion to the probability of the system failing if the sub-system fails. An importance factor of 0 indicates the sub-system's failure has no affect on system failure, and an importance factor of 1 indicates certain system failure if the sub-system fails.

Since $e^{-x} \approx 1 - x$ when x is small, the AGREE method allocates failure rates to sub-components using the expression

$$\lambda_i = \frac{n_i\left[-\ln R^*(T)\right]}{\left(\sum_{i=1}^{k} n_i\right) E_i t_i}, \quad i = 1, 2, \ldots, n \tag{5.46}$$

where: n_i is the number of modules in the ith sub-system,
$R^*(T)$ is the required system reliability,
T is the required mission time for the system,
t_i is the required operating time for the ith component ,
E_i is the importance factor for the ith sub-system, and
k is the number of sub-systems in the system.

The corresponding allocated component reliability is

$$R_i(t_i) = 1 - \frac{1 - \left[R^*(T)\right]^{n_i/N}}{E_i} \tag{5.47}$$

where: N is the total number of modules in the system ($N = \sum_{i=1}^{k} n_i$)

Example 5.14

A system of 3 sub-systems has a reliability requirement of 0.98 at time = 100 hours. Sub-system 1 is only required to operate for 80 hours and has an importance factor of 0.95. Sub-systems 2 and 3 are required to operate for 100 hours and have importance factors of 1.0. Sub-system 1 consists of 15 modules, Sub-system 2 consists of 30 modules, and Sub-system 3 consists of 20 modules. Use the AGREE method to allocate reliabilities to each of the subsystems.

Solution

From Equation 5.46,

$$\lambda_1 = \frac{15[-\ln 0.98]}{65(0.95)(80)} = 0.0000613$$

$$\lambda_2 = \frac{30[-\ln 0.98]}{65(1.0)(100)} = 0.0000932$$

$$\lambda_3 = \frac{20[-\ln 0.98]}{65(1.0)(100)} = 0.0000622$$

From Equation 4.47,

$$R_1(80) = 1 - \frac{1-(0.98)^{15/65}}{0.95} = 0.9951$$

$$R_2(100) = 1 - \frac{1-(0.98)^{30/65}}{1.0} = 0.9907$$

$$R_3(100) = 1 - \frac{1-(0.98)^{20/65}}{1.0} = 0.9938$$

5.6.3 Repairable systems allocation

The following method allocates failure rates to sub-systems in series with constant failure rates. The allocated sub-system failure rates allow the system to meet an availability objective. It is assumed that all sub-systems are identical and have a constant repair rate of μ. The following steps describe the procedure.

1. Determine the steady-state availability for each sub-system using the expression

$$A_s = \left(A^*\right)^{1/k} \tag{5.48}$$

where: A^* is the required system availability, and
$\boldsymbol{k}$ is the number of sub-systems.

2. Determine the ratio of the allocated failure rate to the repair rate for each sub-system using the expression

$$\theta = \left(\frac{1}{A_s} - 1\right) \tag{5.49}$$

3. Determine the failure rate allocated to each sub-system using the expression

$$\lambda = \theta\mu \tag{5.50}$$

where: μ is the repair rate for each of the subsystems.

Example 5.15

Consider a series system composed of 5 independent, identical sub-systems. The repair rate for each of the sub-systems is 0.4, and the required system availability is 0.99. What is the required failure rate for each of the sub-systems?

Solution

The steady-state availability for each of the sub-systems is

$$A_s = (0.99)^{1/5} = 0.998$$

The ratio of the allocated failure rate to the repair rate for each sub-system is

$$\theta = \left(\frac{1}{0.998} - 1\right) = 0.002$$

The failure rate allocated to each sub-system is

$$\lambda = 0.002(0.4) = 0.0008$$

5.7 Software reliability

The information processing industry is growing rapidly. The declining cost of computing hardware has allowed more tasks to be handled by software. Modern products are commonly a combination of hardware and software. Also, more and more computer systems operate in real time, such as machinery controls and reservation systems.

The increase in use of computing systems is matched by an increase in the size and complexity of the computing code used to drive the systems. The reliability of a product or process is a function of the reliability of the system's hardware and its software. Thus, the field of software reliability is becoming increasingly important.

5.7.1 Differences in software and hardware reliability

While the body of knowledge for hardware reliability is well developed, the field of software reliability is relatively new and is still developing. The definition of software reliability is the same as the definition of hardware reliability—the probability of operation for a specified time in a specified environment with no failures. There are several differences in hardware and software reliability:

- Software reliability is not a function of manufacturing. Each set of computer code is identical; there is no manufacturing variability.
- Software does not degrade over time. Mechanical components are subject to wear and degrade with use. Regardless of how much a computer program has been used, it is exactly the same as when it was new, and performs accordingly.
- The **external** environment has no affect on software. Vibration, temperature and other environmental factors may affect computer hardware, but software performance is independent of the external environment. The definition of software reliability refers to the **internal** environment, such as the type of transaction being processed or the hardware the software is operating on.
- All software failures are the result of design errors.
- Software can only be repaired by redesign. Thus, the term maintainability for software is different from the meaning for hardware. Maintainability is a function of the speed and ease with which a computer program can be redesigned.
- Software availability is not used as a performance measure, since most software is not idled during repair.

The techniques for improving software reliability are also different from the conventional techniques used for hardware. Obviously, software reliability cannot be improved with redundancy; any error in the primary code is also contained in the redundant code. Techniques for improving software reliability are:

- Using structured programming
- Using modular programming
- Programming for fault tolerance

Structured programming constrains the programmer to control structures with a single entry point and a single exit point. Thus, the use of GOTO statements are prohibited. Modular programming involves breaking the program into several smaller programs, each as independent as possible. These modules are easier to test and verify. Programs should also be able to tolerate faults. For example, if a user attempts to print a report, but fails to turn on the printer, the program should be able to detect that the printer is not available and notify the user, rather than crash.

5.7.2 Software reliability models

Several models for software reliability have been developed. The two most widely used models are the basic execution time model and the logarithmic Poisson execution time model. Both models have an execution time component and a calendar time component. Only the execution time component is discussed since execution time has been shown to be superior to calendar time for reliability modeling.

5.7.2.1 Basic execution time model

This model assumes failure occurrence is a non-homogeneous Poisson process. That is, at each point in time the occurrence of failures follows a Poisson distribution, and the parameters of the Poisson distribution vary with time. The failure occurrence rate varies with time because faults are being introduced and removed as time passes.

The failure intensity is modeled as a function of the number of failures experienced, and is defined as

$$\lambda(\mu) = \lambda_0\left(1 - \frac{\mu}{\nu_0}\right) \tag{5.51}$$

where: μ is the expected number of failures at a given point in time,

λ_0 is the initial failure intensity, and

ν_0 is the total number of failures that would occur in infinite time.

The rate of change in the failure intensity is

$$\frac{d\lambda}{d\mu} = -\frac{\lambda_0}{\nu_0} \tag{5.52}$$

The failure rate intensity as a function of time is

$$\lambda(t) = \lambda_0 e^{-\frac{\lambda_0 t}{\nu_0}} \tag{5.53}$$

The number of failures experienced as a function of execution time is

$$\mu(t) = \nu_0\left(1 - e^{-\frac{\lambda_0 t}{\nu_0}}\right) \tag{5.54}$$

Example 5.16

The initial failure intensity for a program was 15 failures per CPU hour, and the program has now experienced 50 failures. Assuming the program will experience 120 failures in an infinite period of time, what is the current failure intensity? What is the rate of change in the failure intensity?

Solution

The current failure intensity is

$$\lambda(\mu) = 15\left(1 - \frac{50}{120}\right) = 8.75 \text{ failures per CPU hour}$$

The rate of change in the failure intensity is

$$\frac{d\lambda}{d\mu} = -\frac{15}{120} = -0.125 \text{ per CPU hour}$$

Example 5.17

For the system in Example 5.16, what is the expected number of failures after 30 hours of execution? What is the expected failure intensity after 30 hours of execution?

Solution

The expected number of failures is

$$\mu(30) = 120\left(1 - e^{-\frac{15(30)}{120}}\right) = 117.2$$

The expected failure intensity is

$$\lambda(30) = 15e^{-\frac{15(30)}{120}} = 0.0353 \text{ failures per CPU hour}$$

5.7.2.2 Logarithmic Poisson execution time model

The basic execution time model implies a uniform operational profile. The change in failure rate intensity is constant. The logarithmic Poisson execution time model includes a failure intensity decay parameter that allows a non-uniform operational profile. The failure intensity at a given number of failures is

$$\lambda_i = \lambda_{i-1}(1-\theta) \tag{5.55}$$

where: λ_i is the failure intensity at the *i*th failure, and

θ is the failure decay parameter.

The logarithmic Poisson execution time model also assumes failure occurrence is a non-homogeneous Poisson process. The failure intensity as a function of the number of failures experienced is

$$\lambda(\mu) = \lambda_0 e^{-\theta\mu} \tag{5.56}$$

where: μ is the expected number of failures at a given point in time,

λ_0 is the initial failure intensity,

v_0 is the total number of failures that would occur in infinite time, and

θ is the failure intensity decay parameter.

The failure intensity rate of change is

$$\frac{d\lambda}{d\mu} = -\lambda_0 \theta e^{-\theta\mu} = -\lambda\theta \tag{5.57}$$

The failure rate intensity as a function of time is

$$\lambda(t) = \frac{\lambda_0}{\lambda_0 \theta t + 1} \tag{5.58}$$

The number of failures experienced as a function of execution time is

$$\mu(t) = \frac{\ln(\lambda_0 \theta t + 1)}{\theta} \tag{5.59}$$

Example 5.18

The initial failure intensity for a program was 15 failures per CPU hour, and the program has now experienced 50 failures. The failure intensity decay parameter is 0.03 per failure. Assuming the program will experience 120 failures in an infinite period of time, what is the current failure intensity? What is the rate of change in the failure intensity?

Solution

The current failure intensity is

$$\lambda(\mu) = 15e^{-0.03(50)} = 3.347 \text{ failures per CPU hour}$$

The rate of change in the failure intensity is

$$\frac{d\lambda}{d\mu} = -15(0.03)e^{-0.03(50)} = -0.1004$$

Example 5.19

For the system in Example 5.18, what is the expected number of failures after 30 hours of execution? What is the expected failure intensity after 30 hours of execution?

Solution

The expected number of failures is

$$\mu(t) = \frac{\ln[15(0.03)(30) + 1]}{0.03} = 89.1$$

The expected failure rate intensity is

$$\lambda(t) = \frac{15}{15(0.03)(30) + 1} = 1.03 \text{ failures per CPU hour}$$

Examples 5.16 and 5.18 are the same with the exception of the failure intensity decay parameter. Figure 5.26 demonstrates the difference in the failure intensity as a function of execution time of the two models.

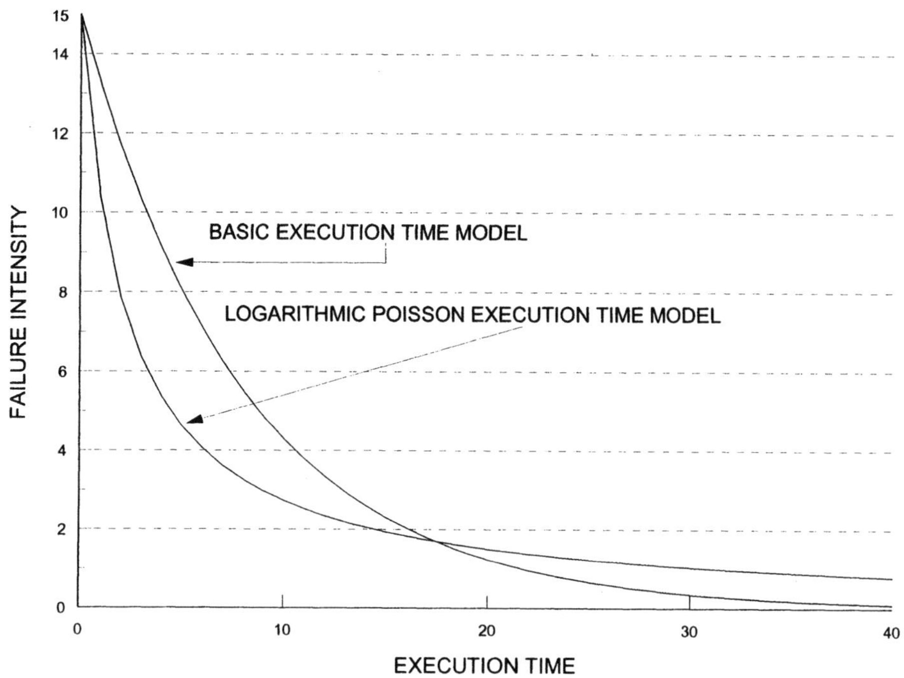

Figure 5.26. Failure intensity as a function of execution time.

Figure 5.27 demonstrates the difference in the expected number of failures as a function of execution time for the two models.

5.8 Reliability growth

During the design phase a product's reliability usually improves as design changes are made. Also, after a new piece of equipment is installed, its reliability tends to improve as modifications are made and the equipment operation is better understood. Reliability growth is a method of modeling the improvement of reliability over time. Reliability growth can be used to estimate the amount of time required to obtain a reliability goal, or to predict reliability. Note that the software reliability models presented in the previous section are growth models.

There are several reliability growth models; the two discussed in this section are the Duane model and the AMSAA model.

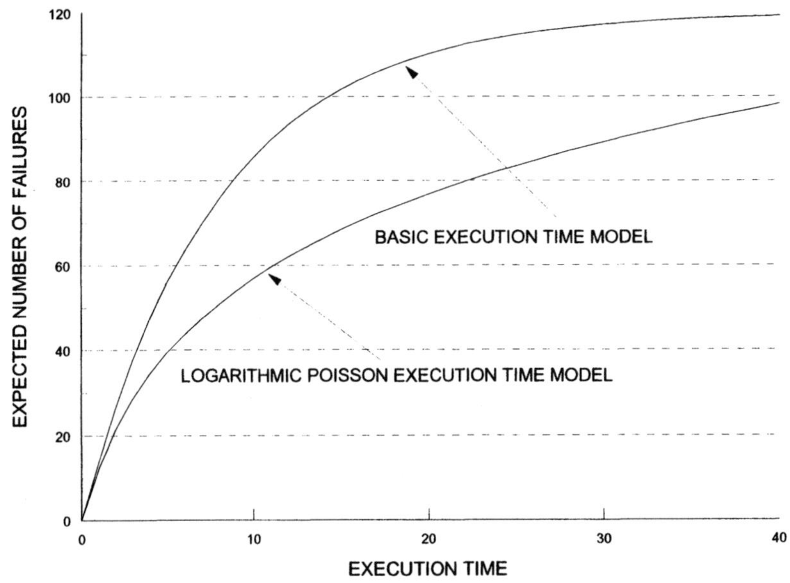

Figure 5.27. Expected cumulative number of failures as a function of execution time.

5.8.1 The Duane model

J. T. Duane was one of the first to introduce a reliability growth model. He empirically derived the expression

$$\theta_C = \alpha t^{\beta} \tag{5.60}$$

where: θ_C is the cumulative mean time between failures, and
β is the slope parameter, and
α is the initial mean time to fail.

By taking the logarithm of Equation 5.60 the following linear expression is found.

$$\ln(\theta_C) = \ln\alpha + \beta \ln t \tag{5.61}$$

The parameters of the Duane model can be found by plotting Equation 4.61.

Example 5.20

Five electronic switches were each tested for 20,000 cycles. After each failure, the root cause was determined, and design changes were made on all 5 switches. The failure data is given in Table 5.4. Determine the Duane model parameters.

Table 5.4. Example data.

Cumulative Cycles	**Cumulative Failures**
20,000	10
40,000	18

60,000	25
80,000	32
100,000	37

Solution

The parameters are determined by plotting the logarithm of the cumulative failures versus the logarithm of cumulative test cycles. Most authors recommend plotting the cumulative failures versus the cumulative test cycles on log-log paper. This is unnecessary, as the log-transformation can be easily handled with electronic spreadsheets. The electronic spreadsheet can also easily determine the best-fit straight line to the data with built-in regression routines. Table 5.5 contains the information required for the graph.

Table 5.5. Duane model plotting data.

Cumulative Cycles	Cumulative Failures	Cumulative Mean Cycles to Fail	Logarithm of Cumulative Cycles	Logarithm of Cumulative Mean Cycles to Fail
20,000	10	2000	9.9035	7.6009
40,000	18	2222	10.5966	7.7063
60,000	25	2400	11.0021	7.7832
80,000	32	2500	11.2898	7.8240
100,000	37	2703	11.5129	7.9020

A graph of the last two columns of Table 5.5 are shown in Figure 5.28.

The slope of the straight line in Figure 5.28 is $\beta = 0.179$. This can be computed from the graph, or the value can be obtained from an electronic spreadsheet. The *y*-intercept of the graph is 5.81 (again, this can be determined electronically). The initial mean time to fail is

$$\alpha = e^{\text{y intercept}}$$

$$\alpha = e^{5.814} = 335$$

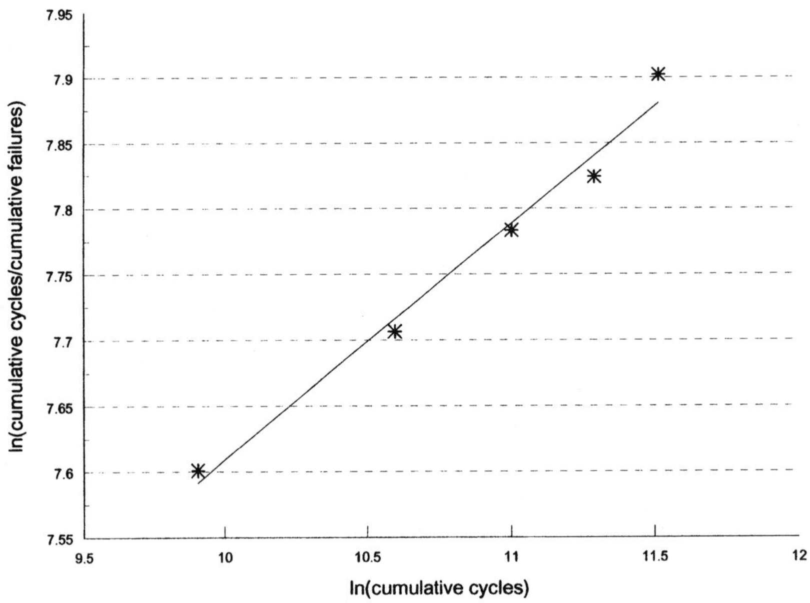

Figure 5.28. Duane reliability growth plot.

O'Conner (1993) quantifies the type of reliability improvement program required to achieve a desired slope for the Duane model. They are as follows:

β	Reliability Effort
0–0.2	No priority is given to reliability improvement. Corrective action is only taken for important modes, and it is done with low priority.
0.2	There is routine attention given to reliability improvement. Corrective action is taken for important failure modes.
0.3–0.4	Reliability improvement is a priority. There is analysis and corrective action for all important failure modes.
0.4–0.6	There is a program dedicated to failure elimination. It has top priority, and there is analysis and corrective action for all failures.

The current or instantaneous mean time to fail is

$$\theta_i(t) = \frac{\alpha t^{\beta}}{(1-\beta)} \tag{5.62}$$

Example 5.21

For the switch in Example 5.20, how long would testing have to continue to obtain a mean cycles to fail of 4,000?

Solution

The Duane reliability growth model for the instantaneous mean cycles to fail for this switch is

$$\theta_i(t) = \frac{335t^{0.179}}{(1-0.179)}$$

Solving the above expression for time

$$4000 = \frac{335t^{0.179}}{(1-0.179)}$$

$$t = 345{,}394 \text{ cycles}$$

Using the Duane model, it is estimated that 345,395 cumulative cycles of testing would be required.

5.8.2 The AMSAA model

The AMSAA (Army Material Systems Analysis Activity) model also assumes a non-homogeneous Poisson process. The instantaneous mean time to fail is

$$\theta_i(t) = \frac{\alpha}{\beta}\left(\frac{t}{\alpha}\right)^{1-\beta} \tag{5.63}$$

where: β is the shape parameter, and

α is the scale parameter.

These parameters can be estimated graphically using the expression

$$\ln\left[N(t)\right] = \beta \ln t - \beta \ln \alpha v \tag{5.64}$$

where: $N(t)$ is the cumulative number of failures at time = *t*.

By plotting $\ln\left[N(t)\right]$ on the *y*-axis and ln*t* on the *x*-axis, the slope of the resulting best-fit straight line through the data is the estimated shape parameter, β. The scale parameter is determined using the *y*-intercept.

$$\alpha = e^{-\frac{\text{y intercept}}{\beta}} \tag{5.65}$$

Example 5.22

For the data given in Example 5.20, determine the AMSAA growth model parameters, and the number of cycles of testing required to achieve a mean time to fail of 4,000 cycles.

Solution

The data required for plotting is contained in Table 5.6

Table 5.6. AMSAA reliability growth plotting data.

Cumulative Cycles	Cumulative Failures	Logarithm of Cumulative Cycles	Logarithm of Cumulative Failures
20,000	10	9.9035	2.3026
40,000	18	10.5966	2.8904
60,000	25	11.0021	3.2189
80,000	32	11.2898	3.4657
100,000	37	11.5129	3.6109

A plot of the data in Table 5.6 is shown in Figure 5.29.

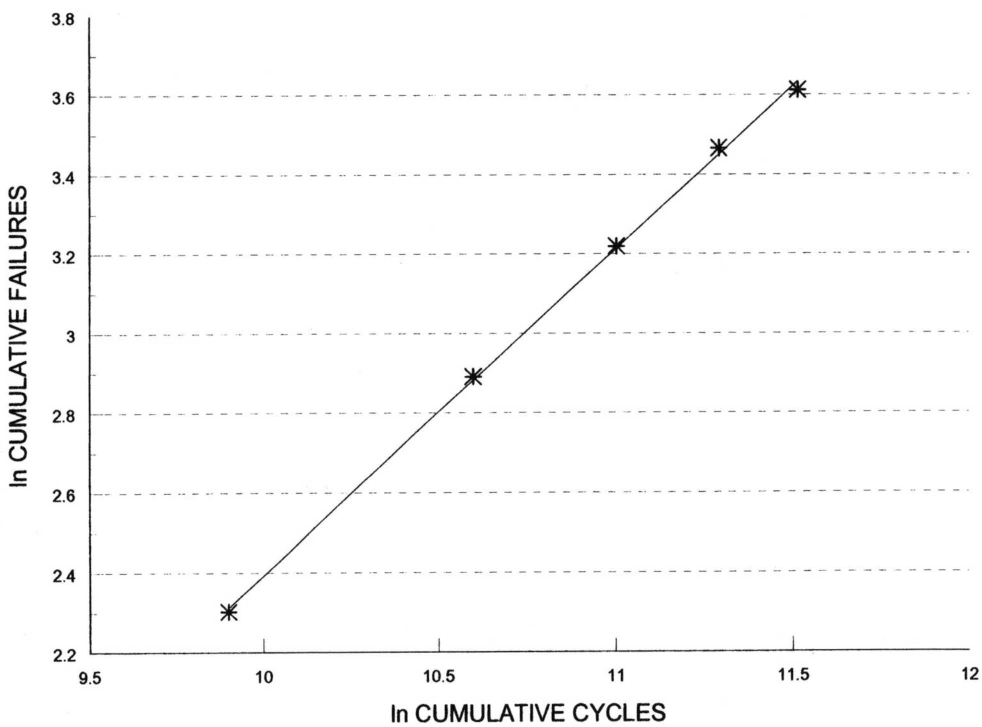

Figure 5.29. AMSAA reliability growth plot.

The slope of the straight line in Figure 5.29 is 0.821, which is equal to the shape parameter, β. The *y*-intercept of the straight line in Figure 5.29 is -5.815. The scale parameter is

$$\alpha = e^{-\frac{-5.81}{0.821}} = 1{,}184$$

The instantaneous mean time to fail is

$$\theta_i(t) = \frac{1184}{0.821}\left(\frac{t}{1184}\right)^{1-0.821}$$

Solving for $\theta_i(t)$ = 4,000 gives

t = 353,590 cycles

5.9 Sneak analysis

System failures can occur even if no component fails. Sneak analysis is a design analysis tool that looks for certain types of system failures in the absence of component failures. Sneak analysis was developed by Boeing in 1967 after the Apollo-1 capsule fire.

Sneak analysis is a tedious and time consuming job if performed manually. Software tools have been developed to aid in this analysis. Most of these tools are primitive in nature. It is expected that sneak analysis will become more useful and popular as more matured and thorough analytical tools are available. There are two types of sneak analyses: hardware sneak analysis and software sneak analysis.

5.9.1 Hardware sneak analysis

Hardware sneak analysis is also known as sneak circuit analysis (SCA). Sneak circuits are latent circuit paths that can cause unwanted functions to occur or inhibit desired functions from occurring. These circuits are characterized by their ability to escape detection during normal analysis and testing. Physical testing alone is inadequate at the system level since it becomes too complex to verify all combination of switches, modules and wiring harnesses. This alone justifies the need for an SCA.

Sneak circuits are inadvertently caused by poor system engineering, poor configuration management, system complexity, changes in the baseline design, and the involvement of many design groups and lack of communication between them. SCA is primarily used for analog circuits and power distribution and control circuits. Other specialized techniques are used for digital circuits that look for logic errors, timing races, intermittent outputs and unintended switching patterns.

SoHar Inc, based in Beverly Hills, California has developed a simple software tool to aid an SCA. This tool is called SCAT (*sneak circuit analysis tool*) and it identifies potential sneak conditions. These conditions should be validated by test or detailed analysis before a design action is taken.

The circuit below is the famous auto industry sneak scenario. The sneak path is highlighted. When the hazard switch is on and brake pedal pressed, a sneak path exists turning radio on (even if the ignition switch is off).

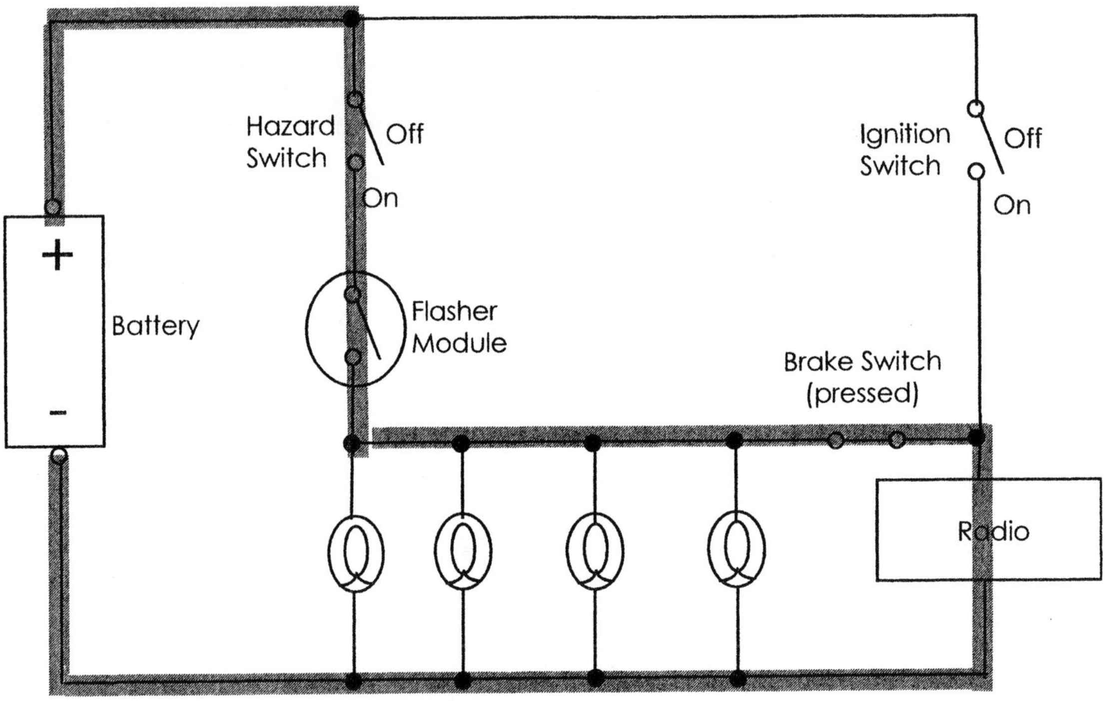

Figure 5.30. Hardware sneak example.

There are 5 categories of hardware sneak.

1. **Sneak Paths**: Current flows along an unexpected route.
2. **Sneak Opens**: Current does not flow along an expected route.
3. **Sneak Timing**: Current flows at the incorrect time or does not flow at the correct time.
4. **Sneak Indications**: False or ambiguous indications causing undesired action by the operator.
5. **Sneak Labels**: Incorrect or ambiguous labels causing undesired action by the operator.

To perform a hardware sneak analysis, transform the circuit schematic into a structure called a *network tree*. These network trees are a condensed depiction of the system under analysis. Power sources are shown at the top and ground returns at the bottom.

Identify any network trees with one of the special circuit patterns that have caused a subject to sneaks. There are total of five such basic circuit patterns:

1. single line,
2. ground dome,
3. power dome,
4. combination dome,
5. H pattern.

Apply previously learned, pattern specific, sneak clues to determine the possibility of sneak. For examples: bi-directional current flow through the crossbar is a clue for the most common sneak pattern, the *H* pattern. Of course, this is not the only clue for *H* pattern, analysts have enlisted over 100 such clues. Similarly grounds at different potential is one of the sneak clues for *ground dome* pattern.

SCAT software described previously does not require the generation of network trees and hence no identification with the patterns is required. It looks for *H* pattern and highlights bi-directional current paths as source of a potential sneak circuit. This is done for the normal operation as well as for circuit configurations resulting from blown fuses and disconnected ground paths (fuse pull-ground lift).

5.9.2 Software sneak analysis

By definition software sneak is the logic control path which causes an unwanted operation to occur or which bypasses a desired operation. These bugs can be inadvertently coded into the software. Software sneak analysis looks for these scenarios. This analysis does not require execution of the software itself to detect problems. Software utilities exist for performing a software sneak analysis.

The steps involved in a software sneak analysis are very similar to a hardware SCA steps. The software package to be analyzed must be reduced to a standardized *topological format pattern*, the basic pattern identified and appropriate problem clue applied. The first step is usually computerized. This is because it may be difficult to manually see a specific pattern in a program of reasonable size. The problem clues are the lessons learned from historical software sneak conditions. These clues are in the form of questions that the analyst must answer. These questions are designed to aid in the identification of the sneak paths latent in the software that can produce undesired output.

Like SCA, topological pattern recognition is the keystone of this type of analysis. It has been found that a total of six such basic patterns exists. These are:

1. single line,
2. return dome,
3. iteration or loop circuit,
4. parallel line,
5. entry dome, and
6. trap circuit.

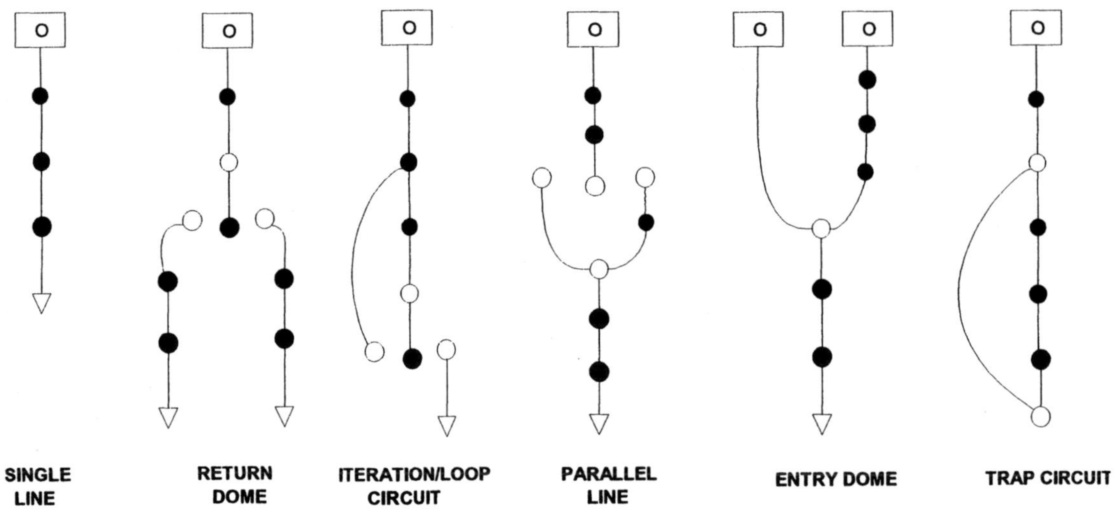

Figure 5.31. Software sneak classifications.

Although at first glance, a software program may appear to be more complex than these basic patterns, closer inspection will reveal that the code is actually composed of these basic structures in combination. Also, note that most of these sneak patterns are related to branching instructions (GOTO, IF-THEN-ELSE etc.) and improper subroutine calls.

Software sneaks are classified into four basic types:

1. **Sneak Output**: The wrong output is generated.
2. **Sneak Inhibit**: Undesired inhibition of an input or output.
3. **Sneak Timing**: The wrong output is generated as a result of its timing or incorrect input timing.

4. **Sneak Message**: A program message incorrectly reports the state of the system.

5.10 Exercises

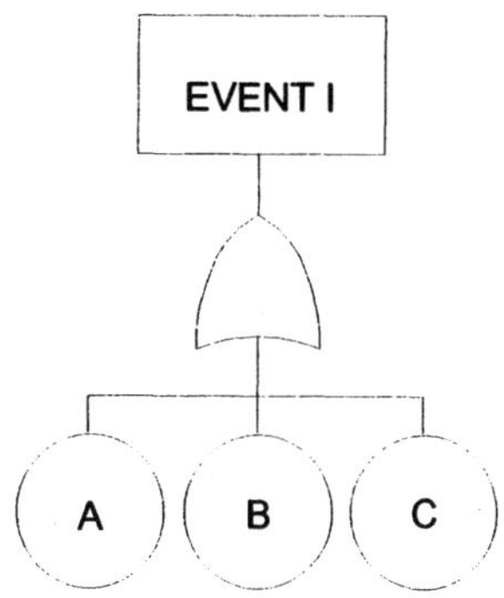

1. What is the probability of Event I occurring in the fault tree above? The probability of A, B, and C occurring is 0.2, 0.4, and 0.3 respectively.
2. For the fault tree in Exercise 1, determine a) all cut sets, b) all minimum cut sets, c) all path sets, and d) all minimum path sets.
3. For the fault tree in Exercise 1, replace the *OR* gate with an *AND* gate and rework the problem.
4. For the fault tree below, determine the minimum cut sets.

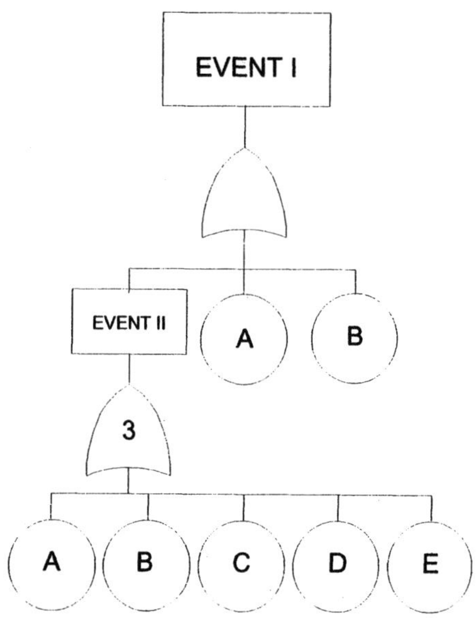

5. For the fault tree to in Exercise 4, determine the probability of Event I occurring given P(A) = P(B) = P(C) = P(D) = P(E) = 0.15.

6. For the fault tree below, indicate which of the scenarios below cause Event I to take place.

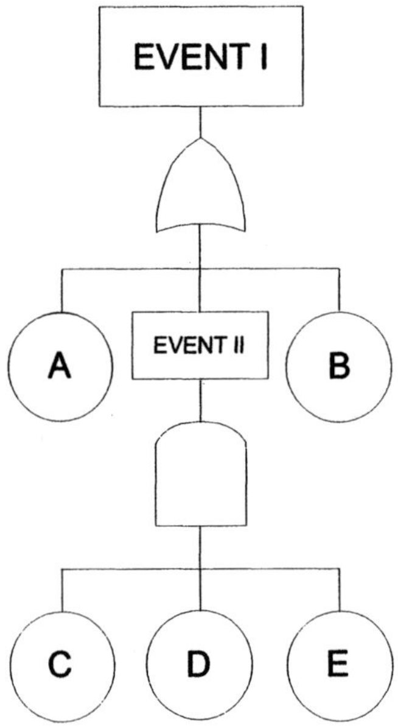

   a. Event A transpires
   b. Event B transpires
   c. Event C transpires
   d. Event D transpires
   e. Event E transpires
   f. Events A and B transpire
   g. Events A and C transpire
   h. Events A and D transpire
   f. Events C, D, and E transpire

7. For the fault tree given in Exercise 6, what is the probability of Event I occurring if P(A) = 0.3, P(B)= 0.09, P(C) = 0.08, P(D) = 0.2, and P(E) = 0.01?

8. Replace the ***OR*** gate in the fault tree in Exercise 6 with an ***AND*** gate and repeat Exercise 7.

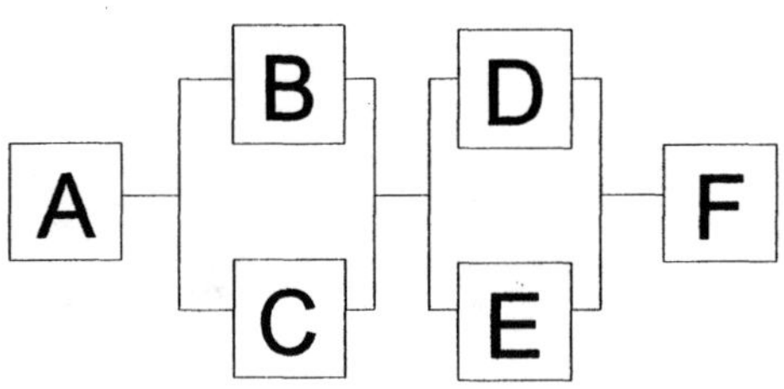

9. Create a fault tree to describe the system above.
10. Determine the minimum cut sets for the fault tree below.

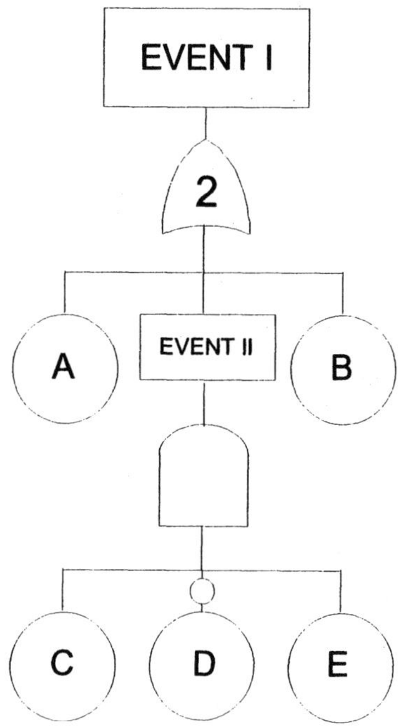

11. Given the following probabilities

P(A) = 0.001
P(B) = 0.02
P(C) = 0.08
P(D) = 0.01
P(E) = 0.06

What is the probability of Event I occurring for the fault tree in Exercise 10?

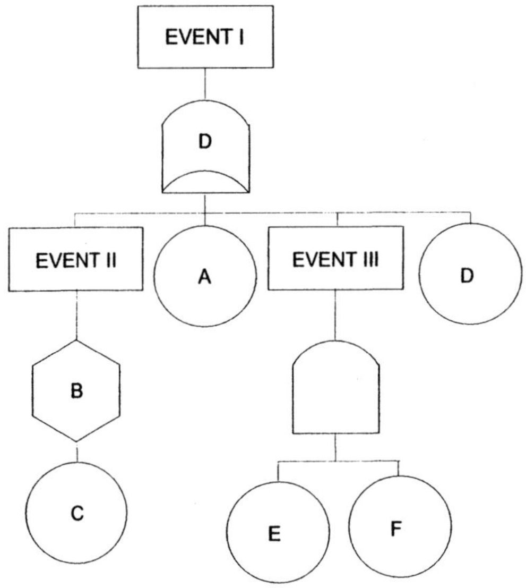

12. For the fault tree above, determine the probability of Event I occurring given

$$P(A) = 0.2$$
$$P(B) = 0.3$$
$$P(C) = 0.003$$
$$P(D) = 0.07$$
$$P(E) = 0.08$$
$$P(F) = 0.001$$

13. Explain why simulation might be used to model a reliability problem.
14. Give some instances when other techniques are more appropriate than simulation
15. Write a simulation to determine the mean time to fail for a system consisting of five components in series, each with exponential times to fail with a mean of 17.
16. For Exercise 15, modify the simulation to find a 95% lower confidence limit for the time to fail.
17. Repeat Exercise 15 with components that follow a Weibull distribution with a shape parameter of 4 and a scale parameter of 90.
18. Identify some public sources for reliability data.
19. What are some limitations of using public reliability data bases?
20. Components A, B, and C are exponentially distributed with mean times to fail of 40 hours, 60 hours, and 75 hours. If these components are in series, what is the reliability of the system at time = 30 hours?
21. What is the mean time to fail for the system described in Exercise 20?
22. If the components in Exercise 20 are in parallel, what is the system reliability at time = 50?
23. What is the system hazard function for the system in described in Exercise 20?

24. Three components with Weibull time to fail distributions are placed in parallel. The Weibull distribution parameters are:

Component	Shape parameter	Scale parameter
A	4	120
B	2	80
C	3.5	100

What is the system reliability at time = 200?

25. A system consists of 4 components with constant failure rates of 0.02. When the system is activated, all 4 components are activated. The system operates as long as at least 2 of the 4 components operate. What is the system reliability at time = 130? What is the system mean time to fail?
26. A system consists of three components. When the system is activated, only one component comes on-line. When the on-line component fails, another component comes on-line. The failure rate for each of the three components is 0.05. What is the system reliability at time = 80?
27. For the system described in Exercise 26, what is the reliability of the system at time = 80 if the mechanism that transfers operation from the on-line component to the standby component has a failure rate of 0.01?
28. Component A has a constant failure rate of 0.02, and component B has a constant failure rate of 0.03. If component A fails, a switch brings component B on-line. The switch has an 8% chance of failing. What is the reliability of this system at time = 100?
29. Two identical components with constant failure rates of 0.09 are activated simultaneously. If one of the components fails, the failure rate of the remaining component increases to 0.14. What is the reliability of this system at time = 20?
30. Three identical components with constant failure rates of 0.09 are activated simultaneously. If one of the components fails, the failure rate of the remaining two components increases to 0.14. If one of the remaining two components fails, the failure rate of the remaining component increases to 0.2. What is the reliability of this system at time = 110?
31. A system consists of two components, A and B, with exponentially distributed mean times to fail of 95 and 85 when they are operating. Component B is brought on-line only if component A fails. Component B has an exponentially distributed mean time to fail of 30 when it is in standby mode. What is the system reliability at time = 150?
32. A system consists of two identical components with a constant failure rate of 0.05 while operating and 0.02 while in standby. When the system in initialized, only one

component is operating. When the on-line component fails, the second component is brought on-line. What is the system mean time to fail?

33. What is the reliability of the system below?

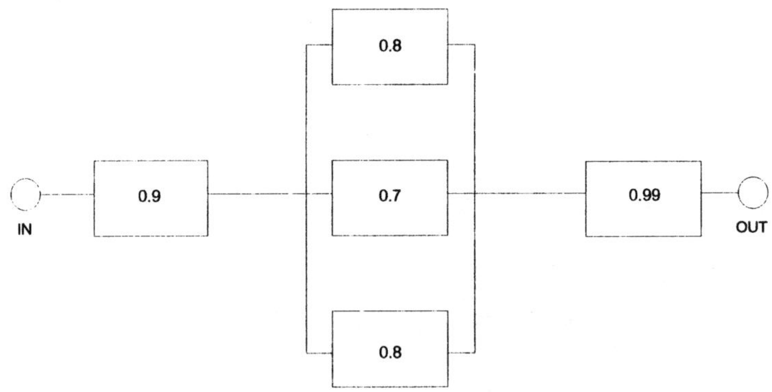

34. What is the reliability of the system below at time = 300?

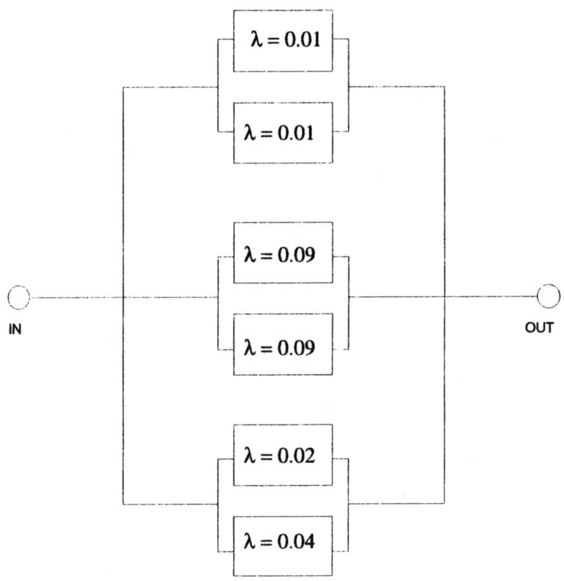

35. In the block diagram below, components A, B, and C have reliability of 0.9, 0.95, and 0.85. What is the reliability of the system?

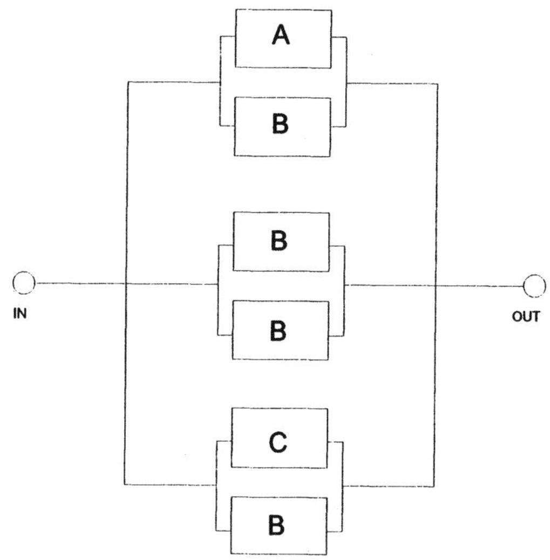

36.A type of beam is normally distributed with a mean strength of 40 and a variance of 0.8. This beam is subjected to a stress which is normally distributed with an average of 38 and a standard deviation of 1.1. What percentage of the beams fail?

37. With manufacturing changes, it is possible to reduce variance of the strength of the beam in the previous exercise. What must this variance be reduced to if only 7% of the beams are allowed to fail?

38. A company is developing a new fiber, and tested 12 specimens to fail. The failure strengths are:

14.2	14.7	15.1
14.4	14.8	15.3
14.6	14.8	15.5
14.6	14.9	15.8

What is the maximum stress that these fibers can withstand to be 90% confident of 95% reliability?

39. Write a flowchart to simulate Exercise 1.
40. Simulate Exercise 1 using an electronic spreadsheet.
41. Write a flowchart to simulate Exercise 4.
42. Simulate Exercise 4 using an electronic spreadsheet, given the probability of Events A, B, C, D and E occurring is 0.01, 0.02, 0.03, 0.04 and 0.05 respectively.
43. Develop the transition rate matrix for the system described in Exercise 20.
44. Develop the set of differential equations to describe the system in Exercise 20.
45. Construct the transition rate diagram and the transition rate matrix for the system described in Exercise 22.
46. A system consists of three sub-systems in series. The system reliability must be at least 0.99 at 80 hours. The three sub-systems have constant failure rates with mean times to

fail of 800, 1,000 and 1,200. Allocate reliability requirements to each of the sub-systems using the ARINC method.

47. Given that a computer program will experience 300 failures if run for an infinite period of time, the initial failure intensity is 10 failures per CPU hour, and the program has experienced 30 failures, use the basic execution time model to estimate the current failure intensity, and the rate of change in the failure intensity.
48. For the computer program in Exercise 47, what is the expected number of failures after 50 hours of execution time? What is the failure intensity after 50 hours of execution time?
49. How does software reliability differ form hardware reliability?
50. A pressure valve was tested during development and the following data obtained.

Failure #	Cumulative Time
1	83
2	201
3	385
4	619
5	888
6	1353
7	1776

Determine the parameters of the Duane growth model. What is the expected instantaneous mean time to fail after 1,000 hours of testing?

51. Repeat Exercise 50 using the AMSAA model.
52. The reliability requirement for a system is 0.995 at time = 190. The system is made up of 4 sub-systems with the following characteristics:

Sub-System	Importance Factor	Required Operating Time	Number of Modules
A	1.0	190	30
B	0.8	80	30
C	1.0	15	80
D	0.5	100	45

Use the AGREE method to allocate reliabilities to each sub-system.

53. A system consists of 4 identical components in series with a constant repair rate of 0.09. What is the required mean time to fail for each of the components if a system availability of 0.98 is required?

5.11 References

Bush, S. H. 1985. Statistics of Pressure Vessel and Piping Failures. In ***Pressure Vessel and Piping Technology—A Decade of Progress,*** edited by C. Sundararajan. New York: American Society of Mechanical Engineers.

Crowder, M. J., A. C. Kimber, R, L. Smith, and T. J. Sweeting. 1991. ***Statistical Analysis of Reliability Data.*** London: Chapman & Hall.

Dhillon, B. S. 1985. ***Quality Control, Reliability, and Engineering Design.*** New York: Marcel Dekker.

Dodson, B. L., and M. D. Mulcahy. 1992. ***Certified Reliability Engineer Examination Study Guide.*** Tucson, AZ: Quality Publishing.

Dovich, R. A. 1990. ***Reliability Statistics.*** Milwaukee, WI: ASQC–Quality Press.

Hahn, G. J., and S. S. Shapiro. 1967. ***Statistical Models in Engineering.*** New York: John Wiley & Sons.

Ireson, G. W., and C. F.Coombs. 1988. ***Handbook of Reliability Engineering and Management.*** New York: McGraw-Hill.

Kapur, K. C., and L. R. Lamberson. 1977. ***Reliability in Engineering Design.*** New York: John Wiley & Sons.

Kececioglu, D. 1991. ***Reliability Engineering Handbook, Volume 2.*** Englewood Cliffs, NJ: Prentice Hall.

Krishnamoorthi, K. S. 1992. ***Reliability Methods for Engineers.*** Milwaukee, WI: ASQC–Quality Press.

Lewis, E. E. 1987. ***Introduction To Reliability Engineering.*** New York: John Wiley & Sons.

Musa, J., A. Iannino, and K. Okumoto. 1987. ***Software Reliability: Measurement, Prediction, Application.*** New York: McGraw-Hill.

O'Conner, P. D. T. 1993. ***Practical Reliability Engineering.*** New York: John Wiley & Sons.

Puterman, M. L. 1979. ***Dynamic Programming and Its Applications.*** New York: Academic Press.

Shooman, M. L. 1990. ***Probabilistic Reliability: An Engineering Approach.*** Malabar, FL: Robert E. Krieger.

Sundarajan, C. 1991. ***Guide To Reliability Engineering: Data, Analysis, Applications, Implementation, And Management.*** New York: Van Nostrand Reinhold.

Tobias, P. A., and D. C. Trindade. 1986. ***Applied Reliability.*** New York: Van Nostrand Reinhold.

US Department of Defense. ***MIL-HDBK-189: Reliability Growth Management.*** Philadelphia: Naval Publications and Forms Center.

US Department of Defense. ***MIL-HDBK-217: Reliability Prediction of Electronic Equipment.*** Philadelphia: Naval Publications and Forms Center.

US Department of Defense. ***MIL-HDBK-338: Electronic Reliability Design Handbook.*** Philadelphia: Naval Publications and Forms Center.

US Department of Defense. ***MIL-HDBK-756: Reliability Modeling and Prediction.*** Philadelphia: Naval Publications and Forms Center.

US Department of Defense. ***MIL-STD-1556: Government Industry Data Exchange Program (GIDEP).*** Philadelphia: Naval Publications and Forms Center.

US Department of Defense. ***MIL-STD-1635: Reliability Growth Testing.*** Philadelphia: Naval Publications and Forms Center.

♦♦♦

CHAPTER

6

Reliability Testing

Products are put through many inspections and tests to ensure conformance to requirements and fitness for use. One or more of these evaluations is usually a reliability test. Unlike many quality tests, reliability testing can be time consuming, and is often destructive. There are several types of reliability tests, each with different purposes. Some common reliability tests are defined below.

Reliability Growth Testing is used at the development stage to model reliability gains. As a product is developed, it may not meet reliability requirements. Some reliability improvements can be made by fine tuning the existing design, but often a design change is required. The new design often results in a lower initial reliability which is improved and eventually exceeds the reliability of the previous design. This process is repeated until reliability requirements are met. The mathematics of reliability growth modeling is discussed in Chapter 4.

Environmental Stress Screening is used during the production phase to eliminate infant mortality failures. Items are subjected to environmental conditions designed to detect manufacturing flaws. The most common stresses include temperature, humidity, vibration and moisture. The stresses are usually cycled during testing, for example, the temperature may be raised from freezing to boiling and back again..

Reliability Qualification Testing is used to verify that an item meets its designed reliability. This test is usually an interim step of development: as each subassembly is qualified on the basis of reliability. Sequential testing, truncated testing and accelerated testing can be used to verify minimum mean time to failure requirements. Qualification testing is also known as design approval testing and demonstration testing.

Acceptance Testing is used to determine if an item is acceptable for use. This test may be conducted by the manufacturer, the purchaser, or both. Sequential testing, truncated testing and accelerated testing can be used to verify minimum mean time to failure requirements.

6.1 Accelerated testing

When items with very high reliability are encountered, the time required to gather failure data can become unreasonable. The time to fail can be reduced by increasing the stress on a component. When this is done, parameters cannot be directly estimated from the test data, and models are required to transform the data or results to normal operating conditions.

While any transformation function could be used to model acceleration, a linear transformation of the time scale is almost always used. Under this assumption, the time to fail under normal operating conditions is

$$t_o = \varepsilon t_\varepsilon \tag{6.1}$$

where: ε is the acceleration factor, and

t_ε is the time to fail under increased stress conditions.

If $f(t)$ represents the probability density function under accelerated conditions, then the probability density function under normal operating conditions is

$$f_o(t) = \frac{1}{\varepsilon} f\left(\frac{t}{\varepsilon}\right) \tag{6.2}$$

The reliability function under normal operating conditions is

$$R_o(t) = R\left(\frac{t}{\varepsilon}\right) \tag{6.3}$$

The hazard function under normal operating conditions is

$$h_o(t) = \frac{1}{\varepsilon} h\left(\frac{t}{\varepsilon}\right) \tag{6.4}$$

Table 6.1 gives transformed reliability functions for several distributions using an acceleration factor of ε.

Table 6.1. Reliability functions with an acceleration factor of ε.

Distribution	Density Function Determined From Accelerated Testing	Reliability Function Under Normal Operating Conditions
Weibull	$f(t)=\frac{\beta t^{\beta-1}}{\theta^{\beta}}\exp\left[-\left(\frac{t}{\theta}\right)^{\beta}\right]$	$R(t)=\exp\left[-\left(\frac{t}{\varepsilon\theta}\right)^{\beta}\right]$
lognormal	$f(t)=\frac{1}{\sigma t\sqrt{2\pi}}\exp\left[-\frac{1}{2}\left(\frac{\ln t-\mu}{\sigma}\right)^{2}\right]$	$R(t)=1-\Phi\left(\frac{\ln t-\ln\varepsilon-\mu}{\sigma}\right)$
normal	$f(t)=\frac{1}{\sigma\sqrt{2\pi}}\exp\left[-\frac{1}{2}\left(\frac{t-\mu}{\sigma}\right)^{2}\right]$	$R(t)=1-\Phi\left(\frac{t-\varepsilon\mu}{\varepsilon\sigma}\right)$
exponential	$f(t)=\lambda e^{-\lambda t}$	$R(t)=e^{-\frac{\lambda t}{\varepsilon}}$

Example 6.1

When tested at 180°F, the time to fail for a component was found to follow a Weibull distribution with a shape parameter of 3.5 and a scale parameter of 19.3. The normal operating temperature for the component is 110°F, and the acceleration factor between these two temperatures is known to be 12. Determine the reliability of the component under normal operating conditions at time = 200.

Solution

From Table 6.1 (or Equation 6.3), the expression for reliability under normal operating conditions is

$$R(t)=\exp\left[-\left(\frac{t}{12(19.3)}\right)^{3.5}\right]$$

At time = 200, the component reliability is

$$R(t)=\exp\left[-\left(\frac{200}{12(19.3)}\right)^{3.5}\right]=0.550$$

With linear acceleration, the shape of the failure distribution does not change due to acceleration. If the time to fail under accelerated conditions follows a

Weibull distribution, the time to fail under normal operation conditions will follow a Weibull distribution with the same shape parameter. This is easily demonstrated with probability plotting. The slope of the probability plot estimates the shape parameter for Weibull probability plots and lognormal probability plots. Thus, if failure data collected under different levels of stress are plotted on the same probability plot, the lines should be parallel (the lines will never be exactly parallel because of sampling error). If parallel lines do not result, either the acceleration is not linear, or the chosen probability density function does not adequately model the data.

Example 6.2

On the same graph, construct a Weibull probability plot for the data obtained under both levels of stress in Table 6.2.

Table 6.2. Time to fail at two levels of stress.

250°F	50°F
3.9	39.2
4.6	48.6
6.1	59.1
7.8	65.5
8.7	70.9
9.5	71.0
9.6	73.7
10.0	76.9
10.1	88.1
10.2	95.8
10.6	99.1
11.8	100.2
12.5	102.4
12.8	107.6
12.9	111.1
13.0	118.6
13.1	123.8
13.2	124.2
13.4	128.9
13.4	131.4
13.5	131.6
13.9	131.9
15.0	139.0
15.1	139.2
15.3	139.8
15.4	144.6

16.3	149.8
16.4	154.8
16.6	155.6
16.8	178.0

Solution

Figure 6.1 shows probability plots for both sets of data. The data set on the right of the figure was obtained under the lower level of stress (50°F). The figure demonstrates linear acceleration; the slope of the two lines is nearly identical.

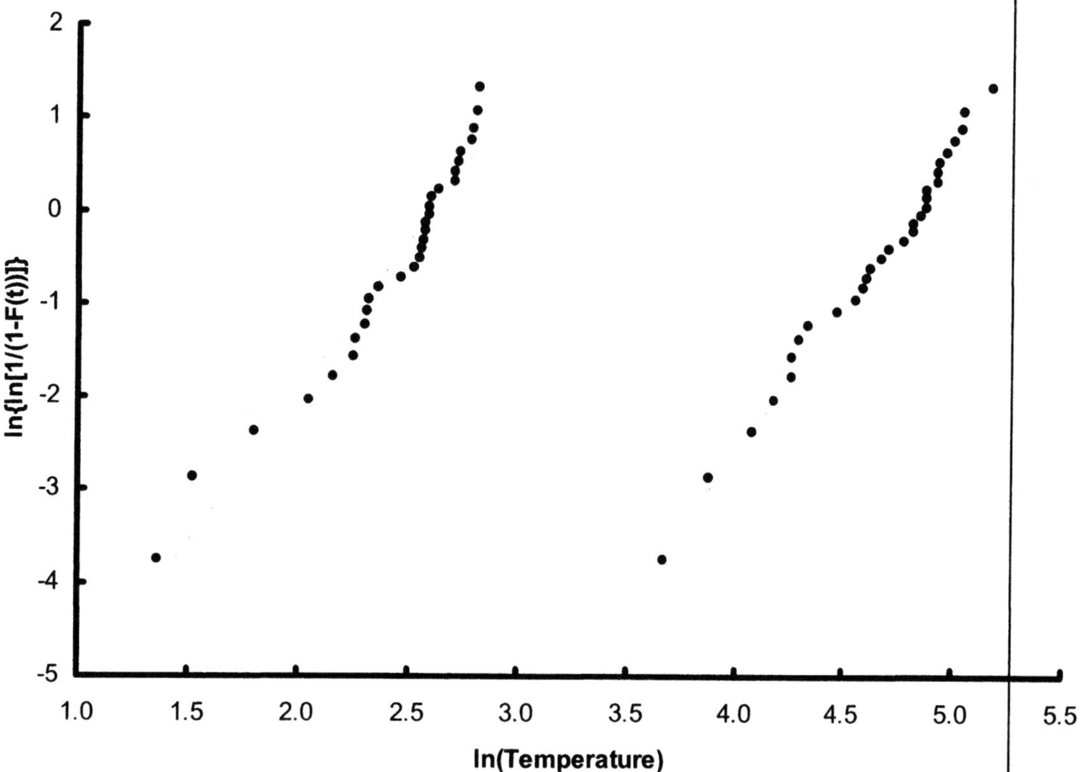

Figure 6.1. Weibull probability plots under two levels of stress.

6.1.1. The Arrhenius model

For the data given in Example 6.2, the acceleration factor can be estimated by taking the ratio of the scale parameters estimated at the two stress levels. If an acceleration factor is needed for another level of stress, testing at that stress level would be required unless there is a way to extrapolate the existing data. This is required in many cases because it is not feasible to test under normal operating conditions due to time constraints.

The Arrhenius model has been shown to accurately model accelerated test data when only thermal stresses are significant. The Arrhenius model is

$$t = ke^{c/T} \tag{6.5}$$

where: t is the time at which a specified portion of the population fails,
k is a constant,
c is a constant, and
T is the absolute temperature (measured in degrees Kelvin or Rankine)*.

By making a logarithmic transformation, the Arrhenius model can be written as

$$\ln t = \ln k + \frac{c}{T} \tag{6.6}$$

By plotting this expression, the values of the constants in the Arrhenius model can be determined. Kielpinski and Nelson (1975) have developed optimum test plans for the Arrhenius model. These plans estimate the 10th percentile of the time to failure distribution at the normal operating stress with the greatest possible large sample precision. These plans involve testing at only two temperatures, and Meeker and Hahn (1985) caution their use because the plans do not allow any degrees of freedom for accessing goodness of fit. Meeker and Hahn make the following recommendations for developing an accelerated test plan using the Arrhenius model:

- Restrict testing to a range of temperatures over which there is a good chance that the Arrhenius model adequately represents the data.
- Select a second temperature reasonably removed from the highest temperature.
- Select a low temperature that is as close as possible to the design temperature.
- Apportion more of the available test units to the lower levels of stress.

Example 6.3

A component was tested at four different temperatures, and the estimated Weibull distribution parameters are given below.

* Degrees Kelvin is equal to degrees Celsius + 273.16. Degrees Rankine is equal to degrees Fahrenheit + 459.69.

Temperature	30°C	50°C	100°C	150°C
β	3.16	3.08	3.14	3.10
θ	912.2	408.9	197.3	104.6

Determine the constants of the Arrhenius model, and assuming 20°C is the normal operating temperature, estimate the acceleration factor for a temperature of 70°C.

Solution

The first step is to estimate a percentile to plot against the absolute temperature. While any percentile is acceptable, for this example, the 10th percentile will be used*. This can be done nonparametrically using the rank distribution or a parametric model can be used. Since Weibull parameters have been estimated, the Weibull distribution will be used for this example. The time at which 10% of the population fails is

$$t = \theta\left[-\ln(0.9)\right]^{1/\beta} \tag{6.7}$$

The 10th percentiles are

Temperature	30°C	50°C	100°C	150°C
10th Percentile	447.52	196.93	96.34	50.61

The data required to plot the Arrhenius model are summarized in Table 6.3.

Table 6.3. Arrhenius model plotting data.

10th Percentile	ln(10th Percentile)	Temperature °C	Absolute Temperature °K	Inverse Absolute Temperature
447.52	6.1037	30	303.16	0.0032986
196.93	5.2828	50	323.16	0.0030944
96.34	4.5679	100	373.16	0.0026798
50.61	3.9241	150	423.16	0.0023632

The next step is to plot $1/T$ on the x-axis, $\ln t$ on the y-axis and fit a straight line to the data. This is shown in Figure 6.2. The straight line was fit to the data using least squares regression. The slope of the line is 2207.8 and the y-intercept is −1.3425. Thus, the constants of the Arrhenius model are

$$k = e^{y\text{-intercept}} = e^{-1.3425} = 0.2612$$

* It is also valid to use the scale parameter in place of the percentile. For example, the scale parameter of the Weibull distribution or the mean of the exponential distribution. See Tobias and Trindade (1986) for more information.

$c =$ slope $=2207.8$

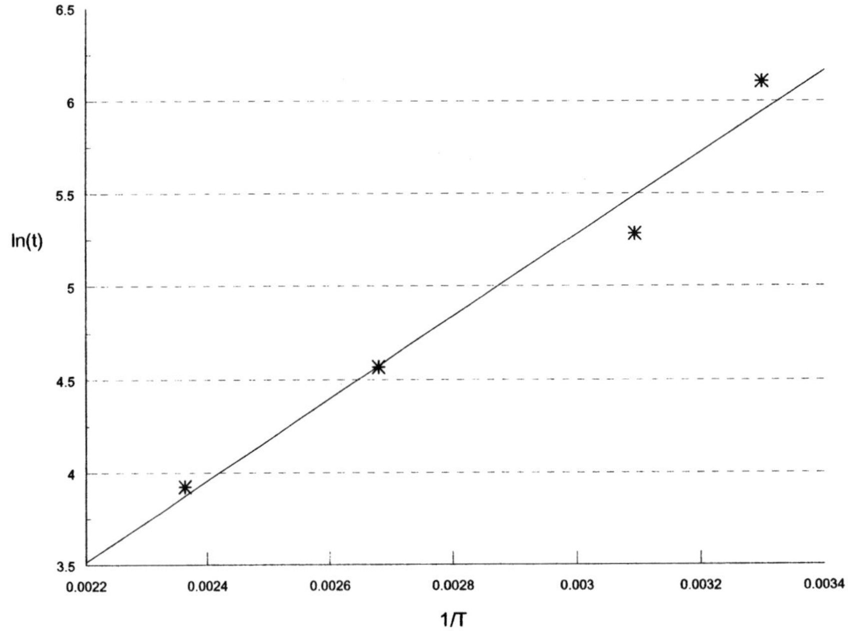

Figure 6.2. Plot of Arrhenius Model.

Assuming the normal operating temperature is 20°C, the acceleration factor at a temperature of 70°C is equal to the ratio of the percentiles at 20°C and 70°C.

$$\varepsilon = \frac{t_{20^0 C}}{t_{70^0 C}}$$

The estimated 10th percentile at 20°C is

$$t_{20^0 C} = 0.2612 e^{2207.8/(20+273.16)} = 487.15$$

The estimated 10th percentile at 70°C is

$$t_{70^0 C} = 0.2612 e^{2207.8/(70+273.16)} = 162.60$$

The acceleration factor at 70°C is

$$\varepsilon_{70^0 C} = \frac{487.15}{162.60} = 2.996$$

The Arrhenius model has been proven to be effective for modeling many types of stresses other than thermal, and unless chemical reactions are involved absolute temperatures are not necessary when thermal stress is used as an acceleration factor.

Example 6.4

A component was tested at three different vibration levels, and the estimated Weibull distribution parameters are given below.

Vibration(GRMS)	**5**	**20**	**50**
β	2.42	2.48	2.41
θ	391.7	179.3	87.6

Determine the constants of the Arrhenius model.

Solution

The first step is to estimate a percentile to plot against the absolute temperature. Since any percentile is acceptable, the 63.2% percentile will be used because this percentile is defined by the shape parameter of the Weibull distribution. The next step is to plot 1/GRMS on the x-axis, ln θ on the y-axis and fit a straight line to the data. The data to be plotted is shown in Table 6.4, and the plot is shown in Figure 6.3.

Table 6.4. Arrhenius model plotting data.

Vibration (GRMS)	**θ**	**1/GRMS**	**ln θ**
5	391.7	0.20	5.970
20	179.3	0.05	5.189
50	87.6	0.02	4.473

The straight line was fit to the data using least squares regression. The slope of the line is 7.32 and the y-intercept is 4.552. Thus, the constants of the Arrhenius model are

$$k = e^{y\text{-intercept}} = e^{4.552} = 94.8$$

$$c = \text{slope} = 7.32$$

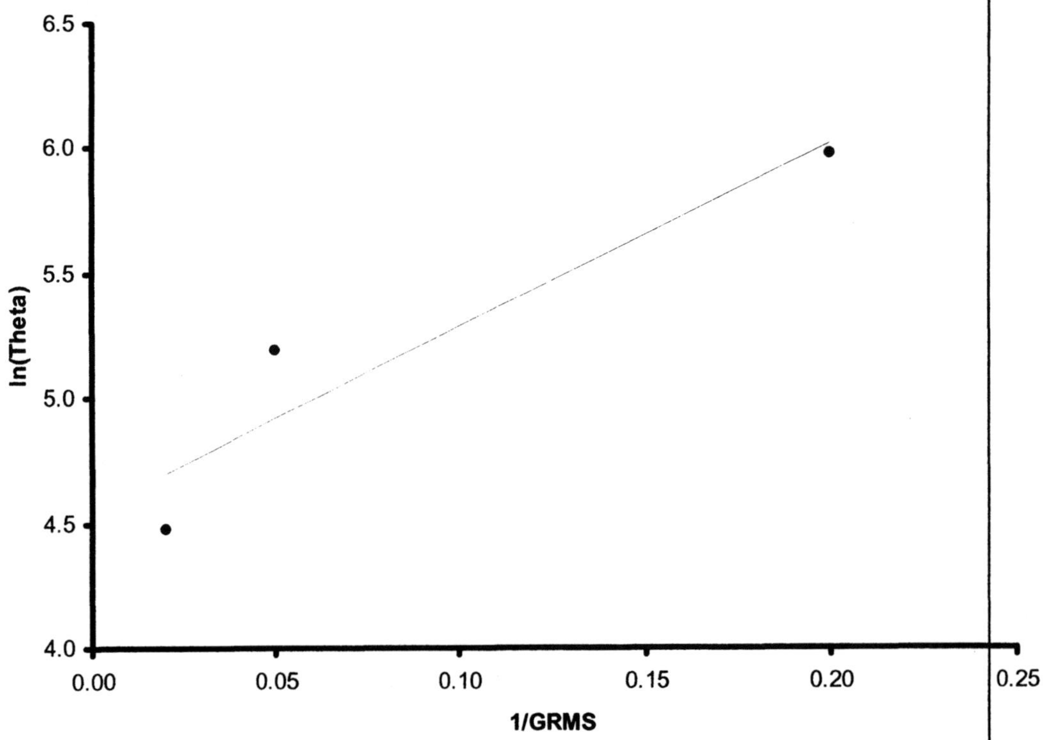

Figure 6.3. Plot of Arrhenius Model.

6.1.2. Other acceleration models

The Arrhenius model is limited to one type of stress, temperature. When there is more than one type of stress acting on a component, a more complex model is required. Eyring derived the following model based on chemical reaction rates and quantum mechanics.

$$t = aT^{\alpha} e^{b/T} e^{[c+(d/T)]S_1} \tag{6.8}$$

where: t is the time at which a specified portion of the population fails,

T is the absolute temperature (measured in degrees Kelvin or Rankine,

α, a, b, c, and d are constants, and

S_1 is the level of a second stress factor.

With five constants to be estimated for two stress factors in the Eyring model, testing must be conducted under at least five distinct stress combinations. Ideally, more than five testing conditions should be used to demonstrate goodness of fit. Determining the constants is not straight forward, and is best done with mathematical software, such as Mathcad®.

More stress factors may be added to the Eyring model. For example if three stress factors are require, the model would be

$$t = aT^{\alpha}e^{b/T}e^{[c+(d/T)]S_1}e^{[e+(f/T)]S_2} \tag{6.9}$$

Note that for every stress factor added to the model, the values of two additional constants must be estimated. A simpler form of the Eyring model for temperature and voltage has been proposed. The constants α and d are assumed to be zero, and $S_1 = \ln V$, where V is voltage. This simplified model is

$$t = ae^{b/T}V^{-c} \tag{6.10}$$

With this version of the Eyring model, only three constants require estimation, and this can be done with multiple linear regression by linearizing the model.

$$\ln t = \ln a + \frac{b}{T} + c(-\ln V) \tag{6.11}$$

The Eyring model has been proven to be effective for modeling many types of stresses other than thermal, and unless chemical reactions are involved absolute temperatures are not necessary when thermal stress is used as an acceleration factor.

Example 6.5

Testing was conducted on a component at two temperatures and two levels of voltage. The time for 10% of the components to fail is given below. Determine the parameters of the simplified Eyring model.

10th Percentile	Temperature °C	Voltage
253	40	60
148	80	60
97	40	120
61	80	120

Solution

The following table contains the data required for multiple least squares regression.

t	ln(t)	T (°C)	T (°K)	V	1/T (°K)	–ln(V)
253	5.5334	40	313.16	60	0.00319	–4.0943

148	4.9972	80	353.16	60	0.00283	–4.0943
97	4.5747	40	313.16	120	0.00319	–4.7875
61	4.1109	80	353.16	120	0.00283	–4.7875

Using ln(t) as the dependent variable, and $1/T$ and –ln(V) as independent variables, multiple least squares regression yields 6.5499 as the y-intercept, 1382 as the coefficient for $1/T$ and 1.3309 as the coefficient for –ln(V). The Eyring model parameters are

$$a = e^{6.5499} = 699.2$$

$$b = 1{,}382$$

$$c = 1.3309$$

There are many empirically derived acceleration models for voltage, humidity, vibration and other acceleration factors. The following empirical relationship has been show to hold true for accelerating failure by changing voltage levels.

$$\frac{t}{t_\varepsilon} = \left(\frac{V_\varepsilon}{V}\right)^3 \tag{6.12}$$

where: t is the time at which a specified portion of the population fails under normal operating conditions,

t_ε is the time at which a specified portion of the population fails under accelerated operating conditions,

V is the normal voltage, and

V_ε is the accelerated level of voltage.

6.2 Sequential testing

There are many occasions when reliability testing is required: at various stages of development, before shipment by the manufacturer, and before acceptance by the customer. Reliability testing may be conducted on individual components, sub-systems, or complete systems. Because of the high cost of reliability testing, in time and materials, sequential testing is usually used to minimize the number of units tested*. The truncated tests explained in Chapter 3 may also be used to verify minimum mean time to failure, but they usually require greater sample sizes.

* For information on fixed length testing (to a predetermined length of time, or a predetermined number of failures, consult the Parameter Estimation Chapter. Complete details, including confidence limits and failure free testing are explained.

Consider a test of the null hypothesis

$$H_0 : \theta = \theta_0$$

against the alternate hypothesis

$$H_1 : \theta = \theta_1$$

Sequential testing utilizes the available data on a continuous basis to make a decision. Three decisions are possible: (1) accept the null hypothesis, (2) reject the null hypothesis, or (3) continue testing. Decisions are based on the likelihood ratio, $L_{1,n} / L_{0,n}$, where

$$L_{k,n} = \prod_{i=1}^{n} f(x_i \backslash \theta_k) \tag{6.13}$$

where: $f(x)$ is the probability density function of the variable x, and
θ is the parameter being tested.

Sequential test plans have a Type I error (the probability of rejecting the null hypothesis when it is true) of α when $\theta = \theta_0$, and a Type II error (the probability of accepting the null hypothesis when it is not true) that is less than or equal to β when $\theta_1 \leq \theta$.

6.2.1 The binomial distribution

In many instances, the outcome of a reliability test is pass or fail for each of the individual items being tested. This situation can be modeled by the binomial distribution, and for an acceptance test, the hypotheses are

$$H_0 : p \leq p_0$$

$$H_1 : p > p_0$$

where: p is the probability of failure for an item, and
p_0 is the probability of failure such the probability of accepting H_0 is $1-\alpha$, where α is the **Type I error**, or **producer's risk**.

As each item is tested, the decision to continue testing, accept the null hypothesis, or reject the null hypothesis is determined from the expression

$$z = \left(\frac{p_1}{p_0}\right)^{y} \left(\frac{1-p_1}{1-p_0}\right)^{n-y} \tag{6.14}$$

where: p_1 is the probability of failure such the probability of accepting H_0 is β,
β is **Type II error** or **consumer's risk**,
n is the total number of trials, and

y is the total number of failures.

The test is continued if $A < z < B$, where

$$A = \frac{\beta}{1-\alpha} \qquad (6.15)$$

$$B = \frac{1-\beta}{\alpha}$$

The null hypothesis is accepted if $z < A$ and rejected if $z > B$. If neither of these conditions are met, testing continues.

Example 6.6

A sequential test was conducted to test the hypotheses

$H_0: p \leq 0.02$

$H_1: p > 0.02$

The level of producer's risk for the test was 0.05; the level of consumer's risk was 0.1; and $p_1 = 0.1$. The first 3 items tested were successful, and the 4th and 5th items tested failed. Determine the status of the test after the results of each trial were known.

Solution

The values of A and B are

$$A = \frac{0.1}{1-0.05} = 0.10526$$

$$B = \frac{1-0.1}{0.05} = 18.0$$

The values of z after each trial, and the resulting decisions, are given below.

Trial	z	Decision
1	0.9184	Continue
2	0.8434	Continue
3	0.7746	Continue
4	3.873	Continue
5	19.364	Reject H_0

It is often useful to represent acceptance tests graphically. The test described above can be defined as:

accept H_0 if $c < y$

reject H_0 if $d > y$

continue testing if $c < y < d$

where

$$c = \frac{n \ln\left(\frac{1-p_0}{1-p_1}\right)}{\ln\left[\left(\frac{p_1}{p_0}\right)\left(\frac{1-p_0}{1-p_1}\right)\right]} - \frac{\ln\left(\frac{1-\alpha}{\beta}\right)}{\ln\left[\left(\frac{p_1}{p_0}\right)\left(\frac{1-p_0}{1-p_1}\right)\right]} \quad (6.16)$$

and

$$d = \frac{n \ln\left(\frac{1-p_0}{1-p_1}\right)}{\ln\left[\left(\frac{p_1}{p_0}\right)\left(\frac{1-p_0}{1-p_1}\right)\right]} + \frac{\ln\left(\frac{1-\beta}{\alpha}\right)}{\ln\left[\left(\frac{p_1}{p_0}\right)\left(\frac{1-p_0}{1-p_1}\right)\right]} \quad (6.17)$$

Example 6.7

For the test described in Example 6.6 suppose 81 items were tested with the 4th, 18th, 37th, and 73rd items failing while all other items were successful. Show this test graphically.

Solution

This sequential test is shown in Figure 6.4.

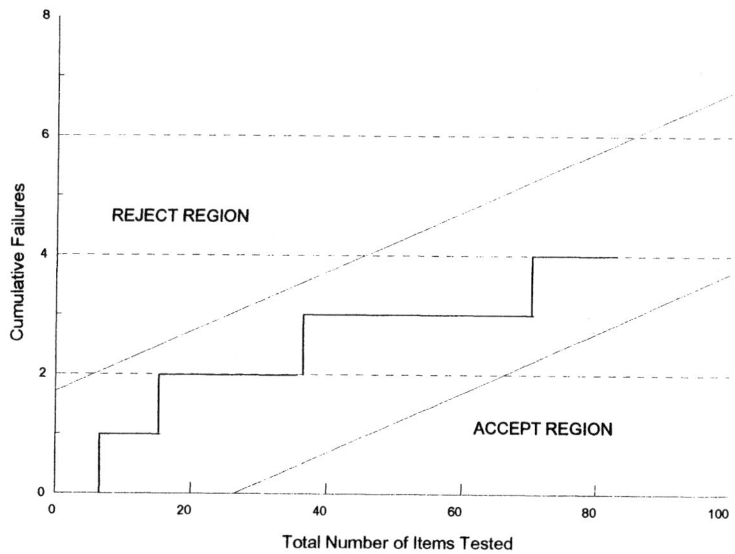

Figure 6.4. Sequential testing with the binomial distribution.

From Figure 6.4 it can be seen that the fewest number of trials required to reject the null hypothesis is 2 (two consecutive failures with no successes). It can also be seen from the figure that the fewest number of trials (all trials being successful) required to accept the null hypothesis is 27. Mathematically, the fewest number of trials required to reject the null hypothesis is

$$f_R = \frac{\ln \beta}{\ln\left[\left(\frac{p_1}{p_0}\right)\left(\frac{1-p_0}{1-p_1}\right)\right] - \ln\left(\frac{1-p_0}{1-p_1}\right)} \tag{6.18}$$

The fewest number of trials required to accept the null hypothesis is

$$f_A = \frac{\ln A}{\ln\left(\frac{1-p_1}{1-p_0}\right)} \tag{6.19}$$

The discriminatory power of a sampling plan is shown by its **operating characteristic curve** (OC curve). An OC curve displays the probability of accepting the null hypothesis versus the fraction defective. For binomial sequential sampling, the probability of accepting the null hypothesis is

$$P_a(p) = \frac{B^h - 1}{B^h - A^h} \tag{6.20}$$

where: p is the true fraction defective, and

h is an arbitrary term used to compute p (h usually ranges from –3 to 3) using the expression

$$p = \frac{1 - \left(\frac{1-p_1}{1-p_0}\right)^h}{\left(\frac{p_1}{p_0}\right)^h - \left(\frac{1-p_1}{1-p_0}\right)^h} \tag{6.21}$$

Example 6.8

Construct the OC curve for the sampling plan described in Example 6.6.

Solution

First, a value of –3 is arbitrarily assigned to h, and p is computed. The resulting value of p is then used to compute P_a. The value of h is then increased, and the procedure is repeated until enough values are obtained to construct a graph. Table 6.5 contains the information required to construct an OC curve.

Table 6.5. OC curve computations.

h	p	P_a
–3.0	0.2269	0.001166
–2.6	0.2011	0.002869
–2.2	0.1751	0.007051
–1.8	0.1492	0.017288
–1.4	0.1239	0.042058
–1.0	0.1000	0.100000
–0.6	0.0780	0.223531
–0.2	0.0587	0.435651
0.2	0.0426	0.683413
0.6	0.0297	0.862924
1.0	0.0200	0.950000
1.4	0.0130	0.983252
1.8	0.0082	0.994593
2.2	0.0051	0.998281
2.6	0.0031	0.999457
3.0	0.0018	0.999829

This OC curve is shown in Figure 6.5.

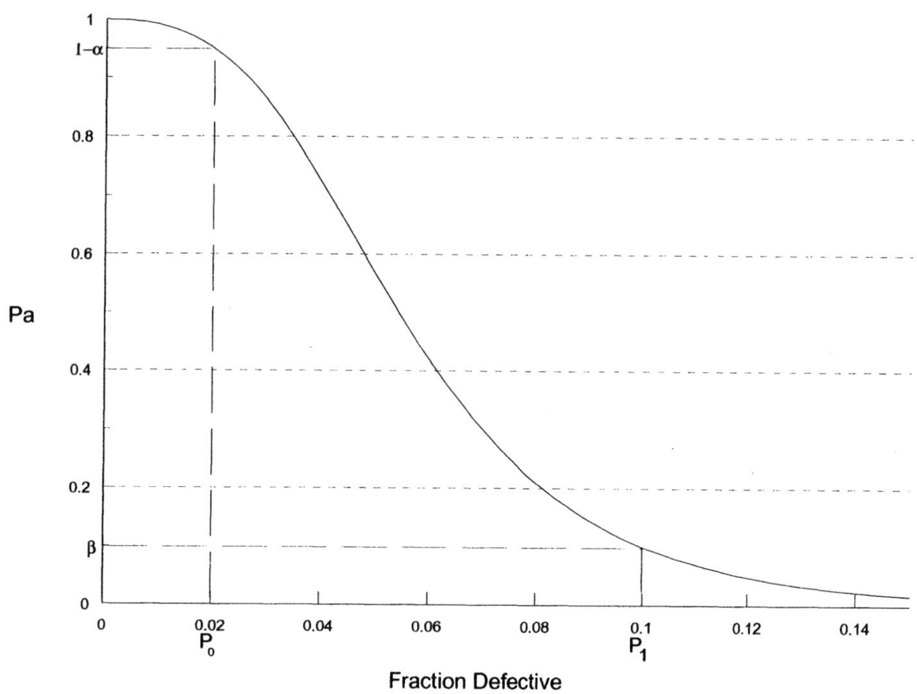

Figure 6.5. Sequential sampling OC curve.

Another characteristic that is valuable when planning a sequential test is the expected number of trials required to reach a decision. This value is a function of the true fraction defective, and is

$$E(p)=\frac{P_a(p)\ln A+\left[1-P_a(p)\right]\ln B}{p\ln\left(\frac{p_1}{p_0}\right)+(1-p)\ln\left(\frac{1-p_1}{1-p_0}\right)} \tag{6.22}$$

This function reaches a maximum between p_0 and p_1, and is relatively flat in this region.

6.2.2 The exponential distribution

In many instances, the time to fail for individual items is available, which provides more information than a simple pass or fail designation and reduces testing time. Assuming an exponential time to fail distribution, a sequential test can be conducted on the cumulative time to fail,

$$T=\sum_{i=1}^{n}t_i \tag{6.23}$$

where: t_i is the time to fail for the ith item, and
n is the total number of failures.

The test hypotheses are

$$H_0:\ \lambda<\lambda_0$$
$$H_1:\ \lambda>\lambda_0$$

where: λ is the failure rate of the exponential distribution.

The null hypothesis is accepted if

$$T>\frac{n\ln\left(\frac{\lambda_1}{\lambda_0}\right)+\ln\left(\frac{\alpha}{1-\beta}\right)}{\lambda_1-\lambda_0} \tag{6.24}$$

where: λ_0 is the failure rate such that the probability of accepting H_0 is 1–α, and
λ_1 is the failure rate such that the probability of accepting H_0 is β.

The null hypothesis is rejected if

$$T<\frac{n\ln\left(\frac{\lambda_1}{\lambda_0}\right)-\ln\left(\frac{\beta}{1-\alpha}\right)}{\lambda_1-\lambda_0} \tag{6.25}$$

Testing continues if

$$\frac{n\ln\left(\frac{\lambda_1}{\lambda_0}\right)-\ln\left(\frac{\beta}{1-\alpha}\right)}{\lambda_1-\lambda_0}<T<\frac{n\ln\left(\frac{\lambda_1}{\lambda_0}\right)+\ln\left(\frac{\alpha}{1-\beta}\right)}{\lambda_1-\lambda_0} \tag{6.26}$$

Example 6.9

Graphically construct a sequential test plan to test the null hypothesis, $\lambda > 0.02$ units against the alternate hypothesis, $\lambda < 0.02$ units, given $\alpha = 0.05$, $\beta = 0.1$ and $\lambda_1 = 0.025$ units.

Solution

Table 6.6 contains the computations required to construct a graph. The graph is shown in Figure 6.6.

Table 6.6. Exponential sequential test.

Cumulative Time	Failures Reject	Accept Time
0	450	–578
3	584	–444
6	718	–310
9	852	–176
12	986	–43
15	1,120	91
18	1,254	225
21	1,387	359
24	1,521	493
27	1,655	627
30	1,789	761

It can be seen from Figure 6.6 that the fastest the null hypothesis can be accepted is to test for 450 units without failure. The fastest the null hypothesis can be rejected is to have 14 consecutive, instantaneous failures.

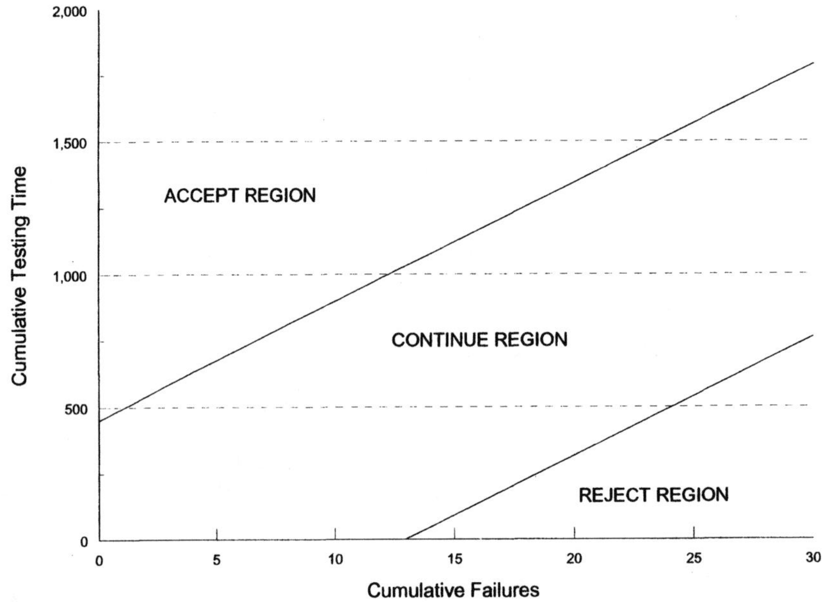

Figure 6.6. Exponential sequential test.

The OC curve for sequential testing using the exponential distribution is defined by the expression

$$P_a(\lambda) = \frac{B^h - 1}{B^h - A^h} \tag{6.27}$$

where: $P_a(\lambda)$ is the probability of accepting the null hypothesis when λ is the true failure rate,

$B = \frac{1-\beta}{\alpha}$,

$A = \frac{\beta}{1-\alpha}$, and,

λ is determined by arbitrarily selecting a value of h, and utilizing the expression

$$\lambda = \frac{h(\lambda_1 - \lambda_0)}{\left(\frac{\lambda_1}{\lambda_0}\right)^h - 1} \tag{6.28}$$

A typically range for the value of h is from –4 to 4.

Example 6.10

Sketch the OC curve for the test plan described in Example 6.9.

Solution

Table 6.7 contains the computations necessary to sketch the curve. The OC curve is shown in Figure 6.7.

Table 6.7. OC curve computations for an exponential, sequential test.

h	λ	$P_a(\lambda)$
–4.0	0.03388	0.000123
–3.4	0.03197	0.000474
–2.8	0.03013	0.001829
–2.2	0.02836	0.007051
–1.6	0.02664	0.027007
–1.0	0.02500	0.100000
–0.4	0.02342	0.319314
0.2	0.02191	0.683413
0.8	0.02047	0.915945
1.4	0.01909	0.983252
2.0	0.01778	0.996948
2.6	0.01653	0.999457
3.2	0.01535	0.999904

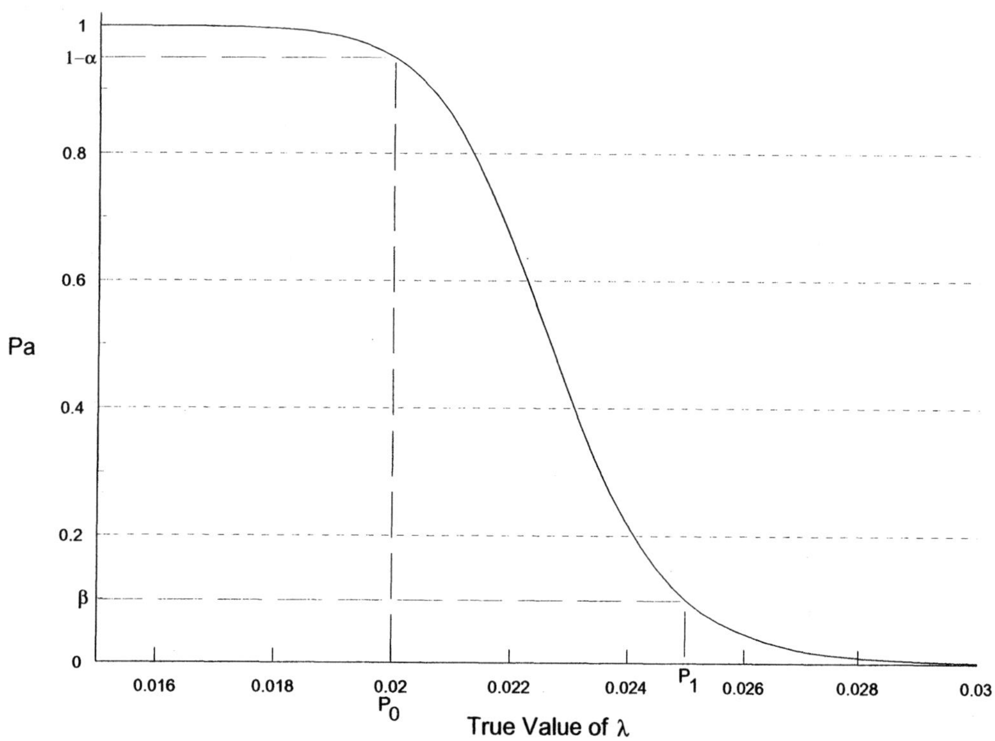

Figure 6.7. OC curve for sequential sampling with the exponential distribution.

The expected number of failures required to reach a decision is a function of the true failure rate, and is

$$E_r(\lambda)=\frac{-c-(d-c)P_a(\lambda)}{s-\frac{1}{\lambda}},\quad s\neq\frac{1}{\lambda} \tag{6.29}$$

$$E_r(\lambda)=\frac{-cd}{s^2},\quad s=\frac{1}{\lambda} \tag{6.30}$$

where

$$c=\frac{\ln\left(\frac{\alpha}{1-\beta}\right)}{\lambda_1-\lambda_0} \tag{6.31}$$

$$d=\frac{\ln\left(\frac{1-\alpha}{\beta}\right)}{\lambda_1-\lambda_0} \tag{6.32}$$

$$s=\frac{\ln\left(\frac{\lambda_1}{\lambda_2}\right)}{\lambda_1-\lambda_0} \tag{6.33}$$

The expected time to reach a decision is

$$E_T(\lambda)=\frac{E_r(\lambda)}{\lambda} \tag{6.34}$$

This time represents the cumulative test time for all items. If ***n*** items are being tested simultaneously with replacement, the elapsed calendar time required to make a decision is $E_t(\lambda)/n$.

Example 6.11

For the test described in Example 6.9, given the true failure rate is 0.02, determine the expected number of failures to reach a decision, and the expected cumulative testing time to reach a decision.

Solution

The values of c, d and s are

$$c=\frac{\ln\left(\frac{0.05}{1-0.1}\right)}{0.025-0.02}=-578.074$$

$$d=\frac{\ln\left(\frac{1-0.05}{0.1}\right)}{0.025-0.02}=450.258$$

$$s=\frac{\ln\left(\frac{0.025}{0.02}\right)}{0.025-0.02}=44.629$$

From the OC curve, $P_a(0.02)=0.95$. The expected number of failures to reach a decision is

$$E_r(0.02)=\frac{-(-578.074)-(450.258+578.074)0.95}{44.629-\frac{1}{0.02}}=74.4$$

The total expected testing time to reach a decision is

$$E_T(\lambda)=\frac{74.4}{0.02}=3720.0$$

6.2.3 The Weibull distribution

There are many instances when it is known that the exponential distribution does not model the time to fail for the testing situation under consideration. If the shape parameter, β, is known, the Weibull distribution can be used in sequential testing. The test hypotheses are

$$H_0:\ \theta>\theta_0$$
$$H_1:\ \theta<\theta_0$$

where: θ is the scale parameter of the Weibull distribution, and

θ_0 is the scale parameter such that the probability of accepting H_0 is $1-\alpha$.

The test region is defined as

$$nb - c \leq \sum_{i=1}^{n} t_i^{\beta} \leq nb + d \tag{6.35}$$

where

$$a = \frac{1}{\theta_1^{\beta}} - \frac{1}{\theta_0^{\beta}} \tag{6.36}$$

$$b = \frac{(\beta - 1)\ln\left(\frac{\theta_0}{\theta_1}\right)}{a} \tag{6.37}$$

$$c = \frac{\ln\left(\frac{1-\tau}{\alpha}\right)}{a} \tag{6.38}$$

$$d = \frac{\ln\left(\frac{1-\alpha}{\tau}\right)}{a} \tag{6.39}$$

and θ_1 is the scale parameter such that the probability of accepting H_0 is τ. The term τ is used to represent consumer's risk because the standard term, β, is used to represent the shape parameter of the Weibull distribution.

Example 6.12

Determine the sequential testing region using the Weibull distribution assuming a shape parameter of 3.0, given θ_0 = 500 hours, θ_1 = 400 hours, α =0.05, and τ = 0.1.

Solution

The calculated values of ***a***, ***b***, ***c***, and ***d*** are given below.

$$a = \frac{1}{400^3} - \frac{1}{500^3} = 7.625(10^{-9})$$

$$b = \frac{(3-1)\ln\left(\frac{500}{400}\right)}{7.625(10^{-9})} = 5.8529(10^{7})$$

$$c = \frac{\ln\left(\frac{1-0.1}{0.05}\right)}{7.625(10^{-9})} = 3.7907(10^{8})$$

$$d = \frac{\ln\left(\frac{1-0.05}{0.1}\right)}{7.625(10^{-9})} = 2.9525(10^8)$$

The test region is shown graphically in Figure 6.8.

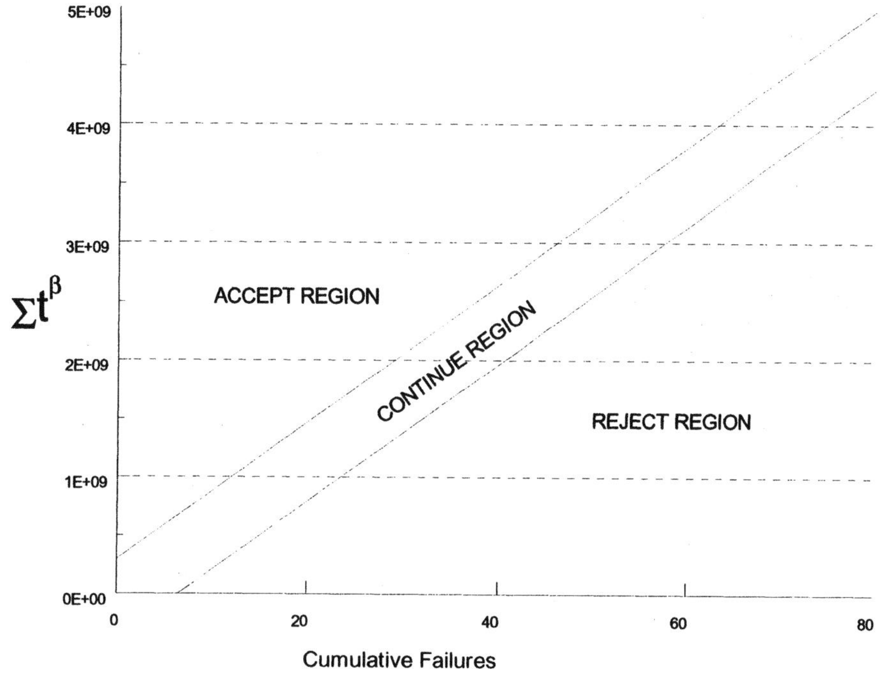

Figure 6.8. Sequential testing with the Weibull Distribution.

Example 6.13

For the test described in Example 6.12, determine the testing time required to accept if there are no failures.

Solution

Using the values of ***a***, ***b***, ***c***, and ***d*** from the previous example, the testing region is defined as

$$5.8528x10^7 n - 3.7907x10^8 \leq \sum_{i=1}^{n} t_i^\beta \leq 5.8528x10^7 n + 2.9525x10^8$$

With no failures, $n = 0$, and the null hypothesis is accepted when

$$t^\beta > 2.9525x10^8$$

The testing time required to accept is

$$t = \left[2.9525x\left(10^8\right)\right]^{1/3} = 665.9 \text{ hours.}$$

6.2.4 Bayesian testing with the Weibull distribution

In Example 6.13, note that this is the test time for a single unit. If two units were tested, the total test time would be

$$2\left(t^{3.0}\right) = 2.9525(10^8)$$

$$t = \left(\frac{2.9525(10^8)}{2}\right)^{1/3.0} = 528.5$$

When using Bayesian analysis with the Weibull distribution, a data transformation can be made, and the result is data that behaves exponentially. The transformation is

$$t^* = t^\beta \tag{6.40}$$

Assuming the value of the shape parameter is equal to β, the maximum likelihood estimate of the scale parameter is

$$\hat{\theta} = \left(\frac{\sum_{i=1}^{n} t_i^\beta}{r}\right)^{1/\beta} \tag{6.41}$$

where: t_i is the time to fail or the time of censoring for the *i*th unit,
n is the number of units tested, and
r is the number of failures.

The lower confidence limit for the estimated scale parameter is

$$\theta_{L,\alpha} = \left(\frac{2\sum_{i=1}^{n} t_i^\beta}{\chi^2_{\alpha,2r+2}}\right)^{1/\beta} \tag{6.42}$$

where: α is the level of significance (α = 0.1 for 90% confidence), and
$\chi^2_{\alpha,v}$ is the critical value of the chi-square distribution with significance α, and ν degrees of freedom

This expression assumes testing is discontinued after a predetermined amount of time. If testing is discontinued after a predetermined number failures, the degrees of freedom for the chi-square statistic is $2r$.

The estimated reliability at time t is

$$\hat{R}(t) = e^{\left[-\left(\frac{t}{\hat{\theta}}\right)^{\beta}\right]} \tag{6.43}$$

The lower confidence limit for reliability at time *t* is

$$R_{L,\alpha}(t) = e^{\left[-\left(\frac{t}{\theta_{L,\alpha}}\right)^{\beta}\right]} \tag{6.44}$$

Using the above expressions, it can be seen that the test time required to demonstrate a specified reliability, at time *t*, with confidence 1–α, and assuming no units fail, is

$$T = \theta\left(\frac{-\ln\alpha}{n}\right)^{1/\beta} \tag{6.45}$$

where θ is the characteristic life corresponding to the required reliability at time = *t*, and is

$$\theta = \frac{t}{\left\{-\ln\left[R(t)\right]\right\}^{1/\beta}} \tag{6.46}$$

From Equation 6.45, it is obvious that the duration of the test time is not directly inversely proportional to the number of units tested. Figure 6.9 shows the relationship between test duration and sample size for a verification test requiring 95% reliability with 90% confidence at 100,000 miles, assuming a shape parameter of 3.

From Figure 6.9, it can be seen that the total time on test (test duration X number of units tested), is increasing with the number of units tested. This is shown in Figure 6.10.

The behavior displayed in Figures 6.9 and 6.10 is unique to the case of a shape parameter greater than 1.0. If the shape parameter is equal to 1, the total time on test required to achieve a specified reliability with a given confidence is independent of the number of units tested. If the shape parameter is less than 1, the total time on test required to achieve a specified reliability with a given confidence decreases as the of the number of units tested increases.

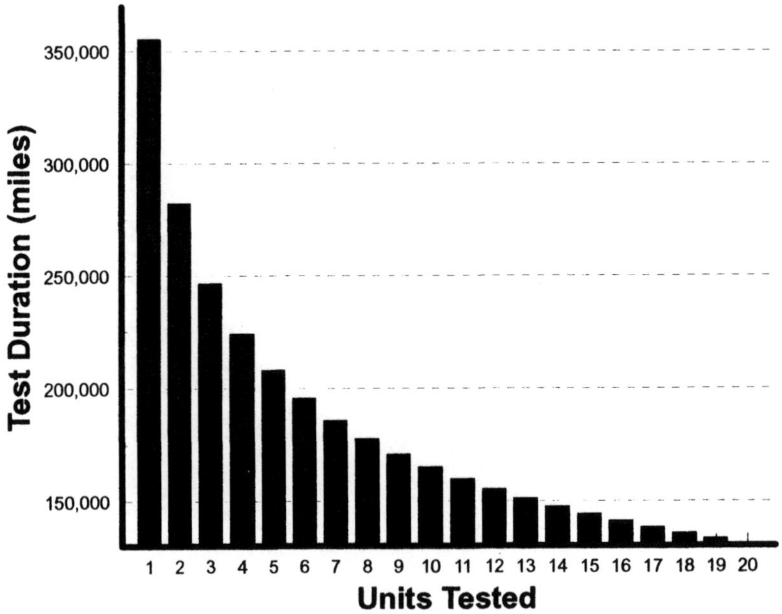

Figure 6.9. Test duration versus number of units tested to ensure 95% reliability with 90% confidence at 100,000 miles, given a shape parameter of 3, and no failures.

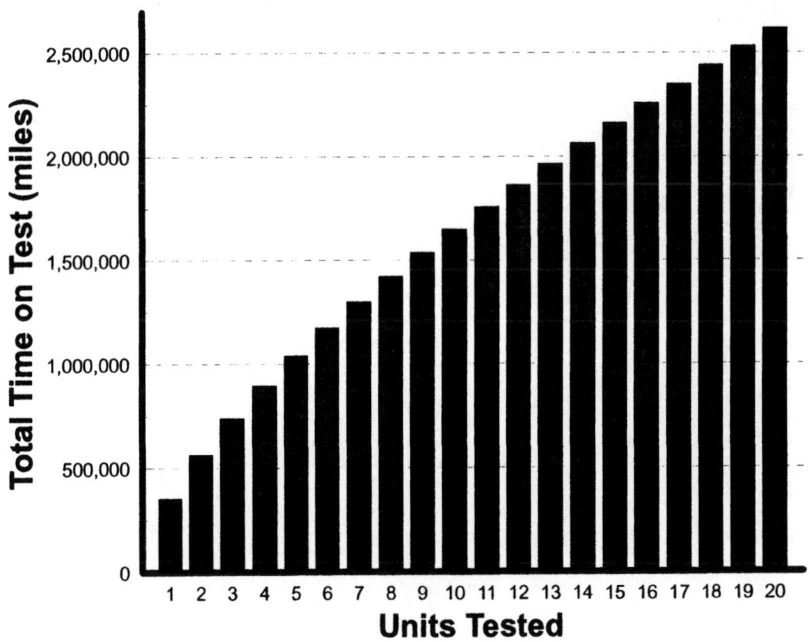

Figure 6.10. Total time on test versus number of units tested to ensure 95% reliability with 90% confidence at 100,000 miles, given a shape parameter of 3, and no failures.

6.3 Exercises

1. Explain the difference between environmental stress testing and acceptance testing.
2. Explain the difference between acceptance testing and qualification testing.
3. Given the density function obtained under accelerated conditions

$$f(t) = \frac{1}{30}\exp\left(-\frac{t}{30}\right)$$

   Determine the density function under normal operating conditions, given an acceleration factor of 5.
4. Determine the reliability function under normal operating conditions for the data given in Exercise 3.
5. Determine the hazard function under normal operating conditions for the data given in Exercise 3.
6. What is the reliability of the system given in Exercise 3 under normal operating conditions at time = 50?
7. Graph the reliability of the system given in Exercise 3 under normal operating conditions and accelerated conditions.
8. The acceleration factor for increasing vibration from 50 units to 200 units is 6.4. At a vibration level of 200 units, the time to fail is lognormal with a scale parameter of 0.8 and a location parameter of 3.2. What is the reliability function under normal operating conditions?
9. The time to fail for a component is normally distributed with a mean of 310 and a standard deviation of 42 at a temperature of 130°F. The normal operating temperature is 75°F. What is the reliability function under normal operating conditions given an acceleration factor of 1.8 between temperatures 75°F and 130°F?
10. How can linear acceleration be verified?
11. Given the following temperatures and Weibull distribution parameters, determine the constants of the Arrhenius model.

Temperature	β	θ
70°F	2.84	1021
98°F	2.82	954
23°F	2.86	759
175°F	2.83	631

12. For the data given in Exercise 11, determine the acceleration factor at a temperature of 100°F if the normal operating temperature is 60°F.

13. Given the following data determine the parameters of the Eyring model.

50th Percentile	Temperature	Voltage
45	30°C	20V
38	60°C	20V
35	30°C	40V
29	60°C	40V
33	45°C	30V

14. A sequential test plan is graphically shown in the figure below. What is the fewest number of trials required to reject the null hypothesis?

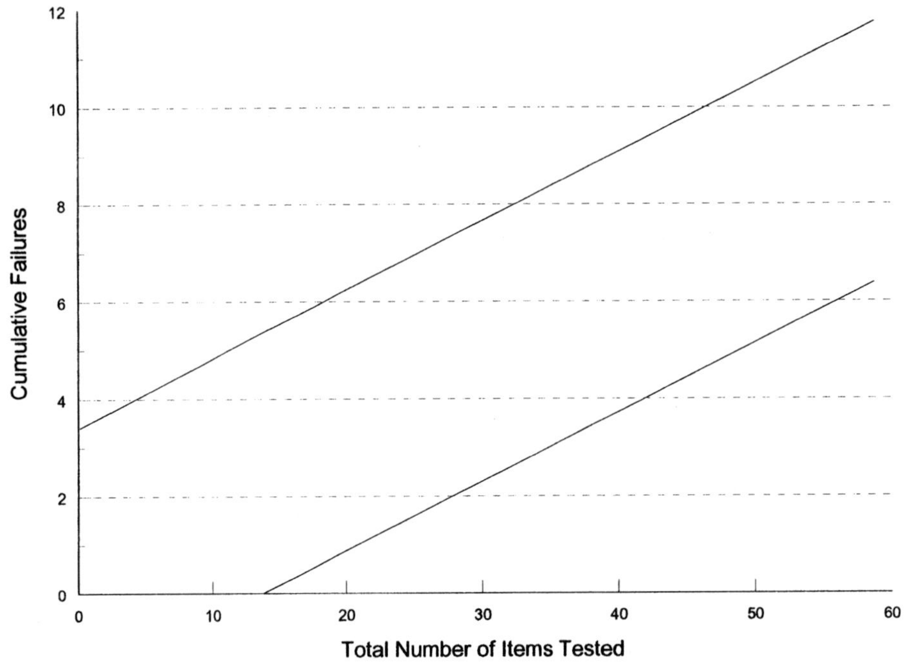

15. For the test plan shown in Exercise 14, what is the fewest number of trials required to accept the null hypothesis?
16. For the test plan shown in Exercise 14, if the first 3 items tested fail, and the next 18 succeed, what is the status of the test?
17. Construct a sequential test plan to test the null hypothesis $p \leq 0.05$ against the alternate hypothesis $p \geq 0.05$, given a producer's risk of 0.05, a consumer's risk of 0.1, and $p_1 = 0.1$.
18. Construct the OC curve for the test plan developed in Exercise 17.
19. For the test plan in Exercise 19, what is the fewest number of trials required to accept the null hypothesis?
20. For the test plan in Exercise 17, what is the fewest number of trials required to reject the null hypothesis?

21. Construct a sequential test plan to test the null hypothesis $\lambda \geq 0.05$ against the alternate hypothesis $\lambda \leq 0.05$, given a producer's risk of 0.05, a consumer's risk of 0.1, and $\lambda_1 = 0.07$.
22. Construct the OC curve for the test plan developed in Exercise 23.
23. What is the expected number of trials required to reach a decision for the test plan developed in Exercise 23?
24. Given a shape parameter of 1.7 for the Weibull distribution, and assuming no failures, how long will 20 items have to be tested to give a reliability of 95% at t = 100,000 with 90% confidence.
25. Five items were placed on test stands, one item failed after 42 hours, and the remaining four items were removed after 50 hours. It is known that the time to fail for these items follows a Weibull time to fail with a shape parameter of 2.2. Determine 90% symmetrical limits for the characteristic life.

6.4 References

Dhillon, B. S. 1985. ***Quality Control, Reliability, and Engineering Design.*** New York: Marcel Dekker.

Dodson, Bryan. 1994. ***Weibull Analysis (with Software).*** Milwaukee, WI: ASQC–Quality Press.

Dodson, B. L., and M. D. Mulcahy. 1992. ***Certified Reliability Engineer Examination Study Guide.*** Tucson, AZ: Quality Publishing.

Dovich, R. A. 1990. ***Reliability Statistics.*** Milwaukee, WI: ASQC–Quality Press.

Ireson, G. W., and C. F.Coombs. 1996. ***Handbook of Reliability Engineering and Management.*** New York: McGraw-Hill.

Kapur, K. C., and L. R. Lamberson. 1977. ***Reliability in Engineering Design.*** New York: John Wiley & Sons.

Kielpinski, T. J., and W. Nelson. 1975. Optimum censored accelerated life tests for normal and lognormal life distributions. ***IEEE Transactions on Reliability*** R-24:310–320.

Krishnamoorthi, K. S. 1992. ***Reliability Methods for Engineers.*** Milwaukee, WI: ASQ–Quality Press.

Lewis, E. E. 1995. ***Introduction To Reliability Engineering.*** New York: John Wiley & Sons.

Meeker, W. Q., and G. J. Hahn. 1985. ***How to Plan an Accelerated Life Test—Some Practical Guidelines.*** Milwaukee, WI: ASQ–Quality Press.

Meeker, W. and Nelson, W. (1976). Optimum accelerated life tests for Weibull and extreme value distributions. ***IEEE Transactions on Reliability*** R-25:20–24.

Nelson, W. 1990. ***Accelerated Testing: Statistical Models, Test Plans and Data Analysis.*** New York: John Wiley & Sons.

Shooman, M. L. 1990. ***Probabilistic Reliability: An Engineering Approach.*** Malabar, FL: Robert E. Krieger.

Tobias, P. A., and D. C. Trindade. 1995. ***Applied Reliability.*** New York: Van Nostrand Reinhold.

US Department of Defense. ***MIL-HDBK-189: Reliability Growth Management.*** Philadelphia: Naval Publications and Forms Center.

US Department of Defense. ***MIL-HDBK-781: Reliability Test Methods, Plans and Environments for Engineering Development, Qualification and Production.*** Philadelphia: Naval Publications and Forms Center.

US Department of Defense. ***MIL-STD-810: Environmental Test Methods and Engineering Guidelines.*** Philadelphia: Naval Publications and Forms Center.

US Department of Defense. ***MIL-STD-1635: Reliability Growth Testing.*** Philadelphia: Naval Publications and Forms Center.

♦♦♦

CHAPTER

7

Maintainability and Availability

Many systems are repairable; when the system fails, it is repaired. Examples are automobiles, dishwashers, most production equipment, etc. Maintainability is a measure of the difficulty to repair the system. More specifically, maintainability is:

The measure of the ability of a system to be retained in, or restored to, a specified condition when maintenance is performed by personnel having specified skill levels, using prescribed procedures and resources, at each prescribed level of maintenance and repair.

The six components of maintainability are shown in Figure 7.1.

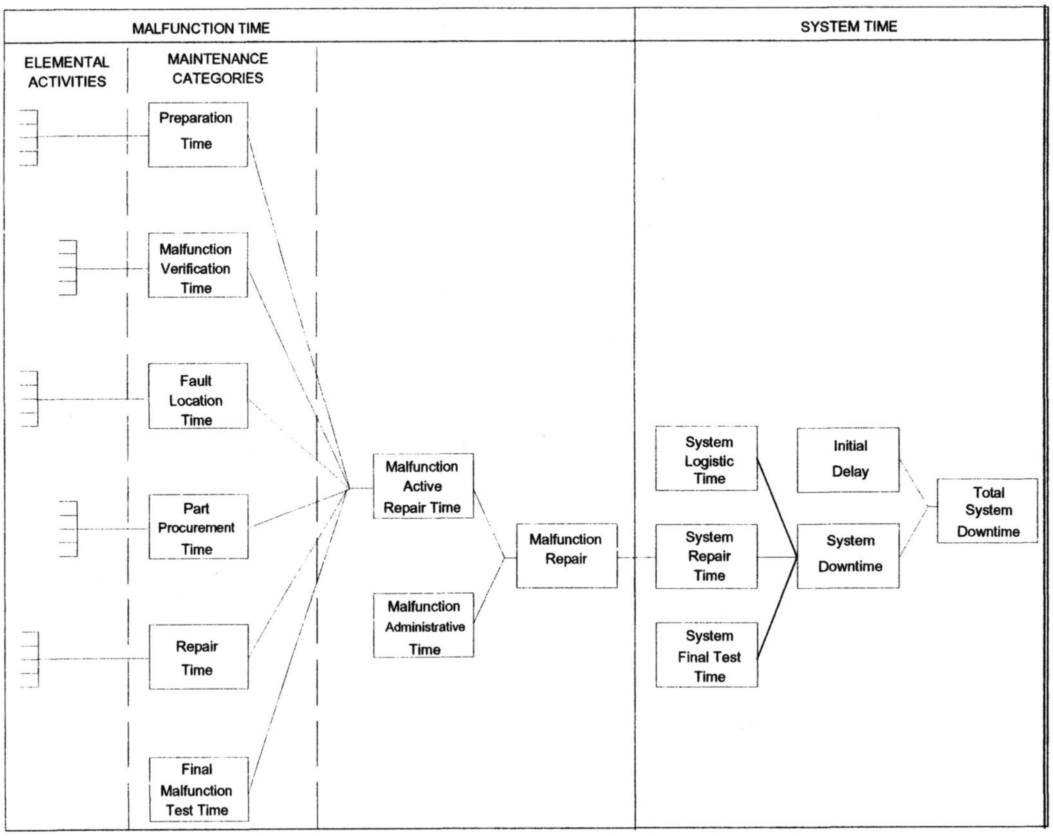

Figure 7.1. Structure of maintainability.

The six components shown in Figure 7.1 are discussed below.

1. **Elemental Activities** are simple maintenance actions of short duration and relatively small variance that do not vary appreciably from one system to another. An example of an elemental activity is the opening and shutting of a door.
2. **Malfunction Active Repair Time** consists of:
   a. Preparation time
   b. Malfunction verification time
   c. Fault location time
   d. Part procurement time
   e. Repair time
   f. Final malfunction test time

Items *a–f* above are composed of elemental activities.

3. **Malfunction Repair Time** consists of:
   a. Malfunction active repair time
   b. Malfunction administrative time

4. **System Repair Time** is the product of malfunction repair time and the number of malfunctions.
5. **System Downtime** includes:
   a. System logistic time
   b. System repair time
   c. System final test time
6. **Total System Downtime** is a combination of the distributions of
   a. Initial delay
   b. System downtime

A prediction for maintainability can be based on the structure described above. The philosophy of the procedure is based on the principles of synthesis and transferability. The synthesis principle involves a buildup of downtimes, step-by-step, progressing from the distribution of downtimes of elemental activities through various stages culminating finally with the distribution of system downtime. The transferability principle embodies the concept that data applicable to one type of system can be applied to similar systems, under like conditions of use and environment, to predict system maintainability.

Availability is a measure of the readiness of a system. More specifically, availability is:

> A measure of the degree to which a system is in an operable and committable state at the start of a mission when the mission is called for at a random time.

There are three categories of availability.

1. **Inherent Availability** is the ideal state for analyzing availability. It is a function only of the mean time to fail, MTBF, and the mean time to repair, MTTR; preventive maintenance is not considered. Inherent availability is defined as

$$A_I = \frac{MTBF}{MTBF + MTTR} \tag{7.1}$$

2. **Achieved Availability** includes preventive maintenance as well as corrective maintenance. It is a function of the mean time between maintenance actions, MTMA, and the mean maintenance time, MMT. Achieved availability is defined as

$$A_A = \frac{MTMA}{MTMA + MMT} \tag{7.2}$$

3. **Operational Availability** includes preventive maintenance, corrective maintenance, and delay time before maintenance begins, such as waiting for parts

or personnel. It is a function of the mean time between maintenance actions and the mean down time, MDT, and is defined as

$$A_O = \frac{MTMA}{MTMA + MDT} \tag{7.3}$$

It is important to note that the type of availability being described is often not distinguished. Many authors simply refer to "availability," MTTR may be the equivalent of MMT or MDT, and MTBF may be the equivalent MTMA.

7.1 Quantifying maintainability and availability

Maintainability is defined as

$$M(t) = P(T < t) = \int_0^t f(x)\,dx \tag{7.4}$$

where: $P(T < t)$ is the probability of completing the repairs in time < T, and

$f(x)$ is the repair time probability density function.

It has been shown that the lognormal distribution can often be used to model repair times. There are times when repairs are completed quickly, but it is rather uncommon to achieve repair times much shorter than normal. However, in some instances, problems arise causing repairs to take much longer than normal. The skewness of the lognormal distribution models this situation well.

Perhaps the most widely used measure of maintainability is the mean-time-to-repair (MTTR).

$$MTTR = \frac{\sum_{i=1}^{n} \lambda_i \tau_i}{\sum_{i=1}^{n} \lambda_i} \tag{7.5}$$

where: n is the number of sub-systems,

λ_i is the failure rate of the ith sub-system, and

τ_i is the time to repair of the ith unit.

Another useful measure is the median corrective maintenance time (MCMT). This is the time in which half of the maintenance actions can be accomplished; the median of the time to repair distribution. For normally distributed repair times, MTTR = MCMT.

Maximum Corrective Maintenance Time (CMAX%) is the maximum time required to complete a given percentile of maintenance tasks. Unless otherwise specified, CMAX% refers to 95% of all maintenance tasks.

Example 7.1

The time to repair for a system follows a lognormal distribution with a scale parameter of 1.2 and a location parameter of 2.2 Plot the maintainability function for this system, and determine the probability of completing a system repair in less than time = 50.

Solution

The maintainability function is

$$M(t) = \int_0^t f(x)\,dx$$

In Chapter 2, the lognormal probability density function is given as

$$f(x) = \frac{1}{\sigma x\sqrt{2\pi}} \exp\left[-\frac{1}{2}\left(\frac{\ln x - \mu}{\sigma}\right)^2\right], \; x>0$$

where: μ is the location parameter, and

σ is the scale parameter.

The integral of this function from 0 to *t* can be expressed as

$$M(t) = \Phi\left(\frac{\ln t - \mu}{\sigma}\right)$$

where $\Phi(z)$ is the area under the standard normal probability density function to the left of *z*. This value is tabulated in Appendix A.

The maintainability function is shown in Figure 7.2.

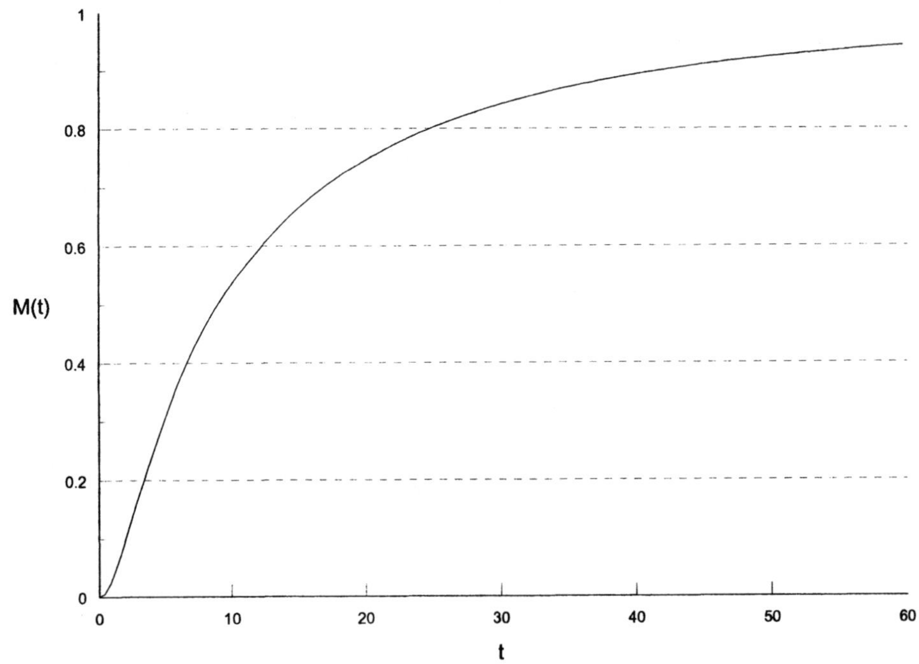

Figure 7.2. Maintainability example.

The probability of completing a repair in less than time = 50 is

$$M(50) = \Phi\left(\frac{\ln 50 - 2.2}{1.2}\right) = \Phi(1.4267) = 0.9232$$

In the special case where the time to fail and the time to repair follow exponential distributions, availability is

$$A(t) = \frac{\mu}{\mu+\lambda} + \frac{\lambda}{\mu+\lambda} e^{-(\mu+\lambda)t} \tag{7.6}$$

where: μ is the mean repair rate, and

λ is the mean failure rate.

It can be seen from Equation 7.6 that as time increases, availability reaches a steady state of

$$A = \frac{\mu}{\mu+\lambda} \tag{7.7}$$

This expression is equivalent to

$$A = \frac{\theta}{\theta+\alpha} \tag{7.8}$$

where: θ is the mean time to fail, and

α is the mean time to repair.

The steady-state availability is independent of the time to fail and time to repair distributions.

The average availability for the period from time = 0 to time = T is equal to the average uptime, which is

$$U(t)=\frac{1}{t}\int_0^t A(s)ds \tag{7.9}$$

For the special case of exponentially distributed times to fail and repair, this reduces to

$$U(t)=\frac{\mu}{\lambda+\mu}+\frac{\mu}{(\lambda+\mu)^2 t}-\frac{\mu}{(\lambda+\mu)^2 t}e^{-(\lambda+\mu)t} \tag{7.10}$$

Note that the steady-state uptime is equal to the steady-state availability.

Example 7.2

The time to fail for a component is exponentially distributed with a mean of 25. The time to repair for this component is also exponentially distributed with a mean of 40. Plot availability as a function of time. What is the steady-state availability?

Solution

The mean failure rate is

$$\lambda=\frac{1}{25}=0.04$$

The mean repair rate is

$$\mu=\frac{1}{40}=0.025$$

Availability is

$$A(t)=\frac{0.025}{0.025+0.04}+\frac{0.04}{0.025+0.04}e^{-(0.025+0.04)t}$$

Availability is plotted as a function of time in Figure 7.3.

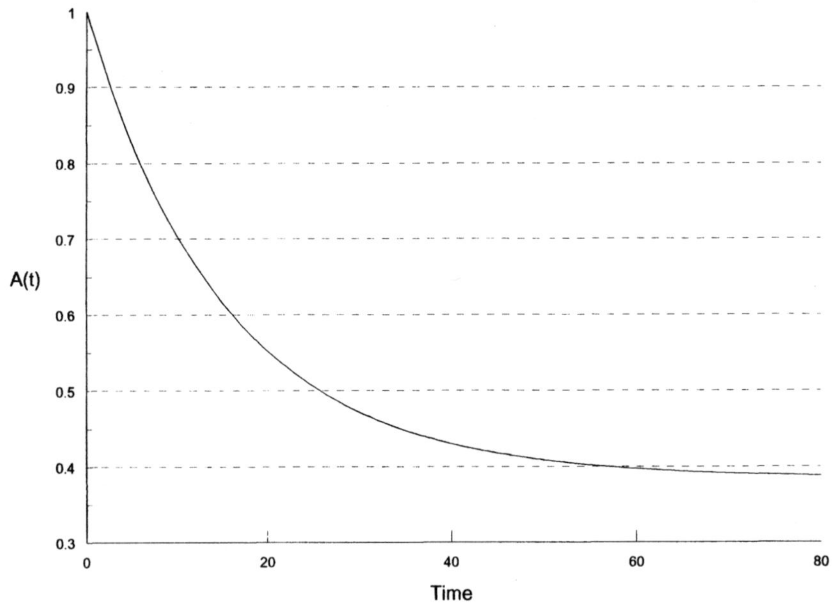

Figure 7.3. Availability as a function of time.

The steady-state availability is

$$A = \frac{25}{25+40} = 0.3846$$

This is verified by Figure 7.3.

To determine system availability, it is important to specify the operating assumptions. With the trivial assumption that all components in the system remain operating while failed components are repaired, system availability for a series system consisting of n components is

$$A_s(t) = \prod_{i=1}^{n} A_i(t) \tag{7.11}$$

This assumption is obviously not valid in many circumstances. Consider a six-stand rolling mill. It is impossible to operate five stands while repairing the other stand. Equation 7.11 also requires that n repairmen be available. If the n–1 operating components continue to run while the failed component is being repaired, it is possible to have all n components in a failed state.

Assuming all components remain running while the failed component is being repaired makes more sense for a parallel system. The purpose of redundant components is to prevent system failure or downtime. Using this assumption, and assuming n repairmen are available, the availability of a system with n systems in parallel is

$$A_s(t) = 1 - \prod_{i=1}^{n} \left[1 - A_i(t)\right] \tag{7.12}$$

Example 7.3

Five components in series each have a mean time to fail that is normally distributed, with a mean of 60 hours, and a standard deviation of 10. The mean time to repair of each of these components is exponentially distributed with a mean of 30 hours. There are five, repairmen and components are only removed from service during repair time (if the system is down because component number 1 failed, components 2–5 continue to operate). What is the system steady-state availability?

Solution

The steady-state availability of each component is

$$A = \frac{60}{60+30} = 0.667$$

The system steady-state availability is

$$A_s = (0.667)^5 = 0.132$$

Example 7.4

Rework Example 7.3 assuming the five components are in parallel.

Solution

The system steady-state availability is

$$A_s = 1 - (1 - 0.6667)^5 = 0.9959$$

As stated earlier, the assumption that all components in a series system continue to operate even though the failure of one component causes a system failure does not often apply to series systems. More often, all components are shut down until the failed item is repaired. Then, the entire system is brought on-line. In this case, system availability can be computed from the system mean time to fail and the system mean time to repair.

Example 7.5

Assuming exponential distributions for the mean time between failures and the mean time to repair for all components, compute the steady-state system availability of the series system shown below. All components are idled when the system fails.

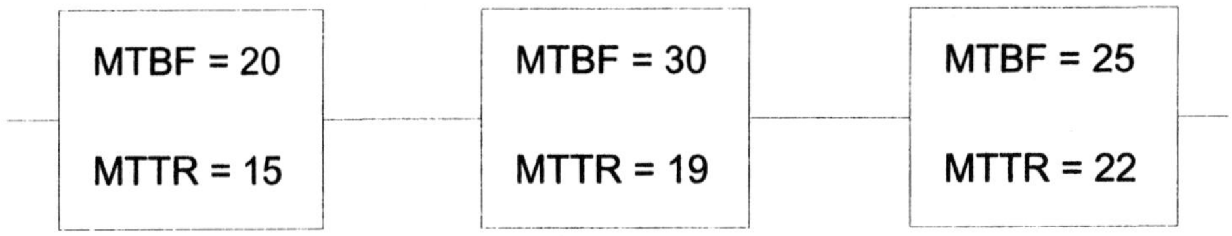

Solution

The system mean time between failures is

$$MTBF_s = \frac{1}{\frac{1}{20}+\frac{1}{30}+\frac{1}{25}} = 8.1081$$

The system mean time to repair is found by computing the weighted average of each component's mean repair time and the percentage of system failures caused by that component.

Component	MTBF	Failure Rate	MTTR	Weighted Average
1	20	0.05	15	0.75
2	30	0.0333	19	0.6333
3	25	0.04	22	0.88
Total		**0.1233**		**2.2633**

The system mean time to repair is

$$MTTR_s = \frac{2.2633}{0.1233} = 18.3560$$

The steady-state system availability is

$$A_s = \frac{8.1081}{8.1081+18.356} = 0.3061$$

Assuming all components continued to operate when the system fails, the system availability would be found by computing the product of the component availabilities.

Component	MTBF	MTTR	Availability
1	20	15	0.5714
2	30	19	0.6122
3	25	22	0.5319

System availability in this case is

$$A_s = (0.5714)(0.6122)(0.5319) = 0.1861$$

which is substantially less than the availability achieved by shutting the remaining components down during a system failure.

7.2 Preventive maintenance

In some cases, it is possible to prevent failures with preventive maintenance. The question is to determine if preventive maintenance is applicable, and if so, how often should it be scheduled. Referring to Figure 7.4, failures can be grouped into three categories based on the behavior of the failure rate. **Infant mortality** failures are characterized by a decreasing failure rate. Recall from Chapter 3, that the hazard function (failure rate) of the Weibull distribution is decreasing if the shape parameter, β, is less than 1. **Random**[*] failures exhibit a constant failure rate; the shape parameter of the Weibull distribution is equal to 1. **Wear-out** failures have an increasing failure rate; the shape parameter of the Weibull distribution is greater than 1.

[*] This term can be confusing, since all failures occur randomly. This portion of the bathtub curve is also referred to as the "useful life". This term is also confusing since many products exhibit a decreasing or increasing failure rate for their entire life.

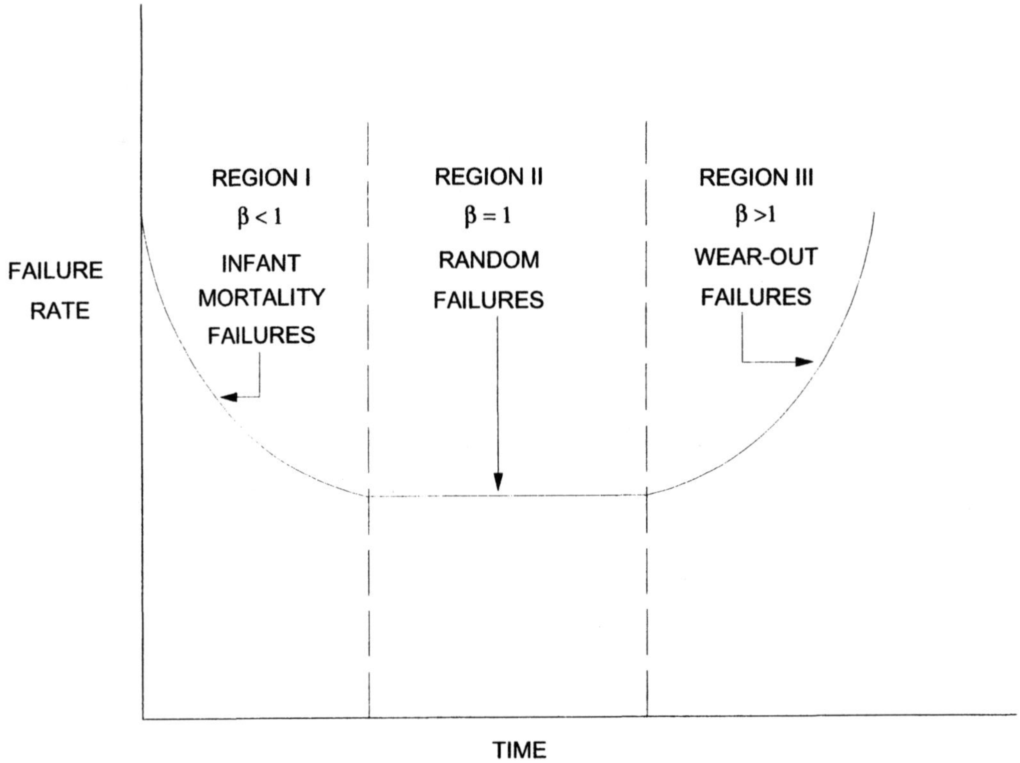

Figure 7.4. The bathtub curve.

Infant mortality failures are premature failures that can often be prevented by management. If infant mortality failures cannot be prevented, a burn-in procedure can be implemented to eliminate the failures before the product is shipped. Preventive maintenance is not applicable for an item with a decreasing failure rate. Performing preventive maintenance restores the system to its initial state which has a higher failure rate; preventive maintenance increases the number of failures.

Some causes of infant mortality failures are:

- Improper use
- Inadequate materials
- Over-stressed components
- Improper setup
- Improper installation
- Poor quality control
- Power surges
- Handling damage

Random failures cannot be prevented with preventive maintenance. The failure rate is constant, so preventive maintenance has no affect on failures. Reliability can be increased by redesigning the item, or in some cases, by implementing an inspection program.

Wear-out failures can be prevented with preventive maintenance. The failure rate is increasing with time, so preventive maintenance restores the system to a state with a lower failure rate. The question is how often should preventive maintenance be scheduled.

The time to fail for an item is variable, and can be represented by a probability distribution, $f(x)$. Referring to Figure 7.5, the cost of failures per unit time decreases as preventive maintenance is done more often, but the cost of preventive maintenance per unit time increases. There exists a point where the total cost of failures and preventive maintenance per unit time is at a minimum; the optimum schedule for preventive maintenance.

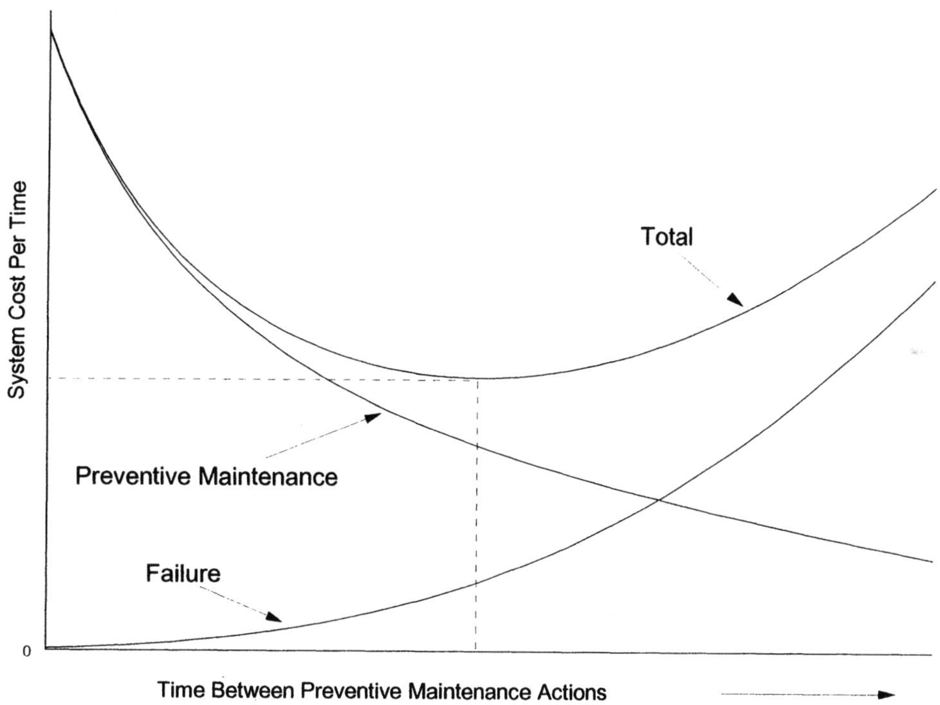

Figure 7.5. Optimum schedule for preventive maintenance.

The optimum time between maintenance actions is found by minimizing the total cost per unit time.

$$C_T = \frac{C_p \int_T^{\infty} f(t)dt + C_f \int_0^T f(t)dt}{T \int_T^{\infty} f(t)dt + \int_0^T tf(t)dt} \tag{7.13}$$

where: C_p is the cost of preventive maintenance,

C_f is the cost of a failure, and

T is the time between preventive maintenance actions.

Minimizing Equation 7.13 is tedious, and numerical routines are usually required. Dodson (1994), developed a tabular solution for this problem given the following assumptions.

1. The time to fail follows a Weibull distribution.
2. Preventive maintenance is performed on an item at time T, at a cost of C_p.
3. If the item fails before time = T, a failure cost of C_f is incurred.
4. Each time preventive maintenance is performed, the item is returned to its initial state; that is, the item is "as good as new."

The optimum time between preventive maintenance actions is

$$T = m\theta + \delta \tag{7.14}$$

where: m is a function of the ratio of the failure cost to the preventive maintenance cost, and the value of the shape parameter, and is given in Table 7.1.

θ is the scale parameter of the Weibull distribution, and

δ is the location parameter of the Weibull distribution.

Dodson and Kirkland (2000) have developed a software package, "The Reliability and Maintenance Analyst," that determines optimum preventive maintenance schedules as well as optimum inspection schedules. This package also determines the parameters of the Weibull distribution. See the accompanying CD for a demo version of "The Reliability and Maintenance Analyst".

Table 7.1. Values of m.

C_f/C_p	ß 1.5	2.0	2.5	3.0	4.0	5.0	7.0	10.0
2.0	2.229	1.091	0.883	0.810	0.766	0.761	0.775	0.803
2.2	1.830	0.981	0.816	0.760	0.731	0.733	0.755	0.788
2.4	1.579	0.899	0.764	0.720	0.702	0.711	0.738	0.777
2.6	1.401	0.834	0.722	0.688	0.679	0.692	0.725	0.766
2.8	1.265	0.782	0.687	0.660	0.659	0.675	0.713	0.758
3.0	1.158	0.738	0.657	0.637	0.642	0.661	0.702	0.749
3.3	1.033	0.684	0.620	0.607	0.619	0.642	0.687	0.739
3.6	0.937	0.641	0.589	0.582	0.600	0.627	0.676	0.730
4.0	0.839	0.594	0.555	0.554	0.579	0.609	0.662	0.719
4.5	0.746	0.547	0.521	0.526	0.557	0.591	0.648	0.708
5	0.676	0.511	0.493	0.503	0.538	0.575	0.635	0.699
6	0.574	0.455	0.450	0.466	0.509	0.550	0.615	0.683
7	0.503	0.414	0.418	0.438	0.486	0.530	0.600	0.671
8	0.451	0.382	0.392	0.416	0.468	0.514	0.587	0.661
9	0.411	0.358	0.372	0.398	0.452	0.500	0.575	0.652
10	0.378	0.337	0.355	0.382	0.439	0.488	0.566	0.645
12	0.329	0.304	0.327	0.357	0.417	0.469	0.550	0.632
14	0.293	0.279	0.306	0.338	0.400	0.454	0.537	0.621
16	0.266	0.260	0.288	0.323	0.386	0.441	0.526	0.613
18	0.244	0.244	0.274	0.309	0.374	0.430	0.517	0.605
20	0.226	0.230	0.263	0.298	0.364	0.421	0.508	0.598
25	0.193	0.205	0.239	0.275	0.343	0.402	0.492	0.584
30	0.170	0.186	0.222	0.258	0.328	0.387	0.478	0.573
35	0.152	0.172	0.207	0.245	0.315	0.374	0.468	0.564
40	0.139	0.160	0.197	0.234	0.304	0.364	0.459	0.557
45	0.128	0.151	0.187	0.225	0.295	0.356	0.451	0.550
50	0.119	0.143	0.179	0.217	0.288	0.348	0.444	0.544
60	0.105	0.130	0.167	0.204	0.274	0.335	0.432	0.534
70	0.095	0.120	0.157	0.193	0.264	0.325	0.422	0.526
80	0.087	0.112	0.148	0.185	0.255	0.316	0.415	0.518
90	0.080	0.106	0.141	0.177	0.248	0.309	0.407	0.513
100	0.074	0.101	0.135	0.172	0.241	0.303	0.402	0.507
150	0.057	0.082	0.115	0.150	0.217	0.278	0.379	0.487
200	0.047	0.071	0.103	0.136	0.203	0.263	0.363	0.472
300	0.035	0.058	0.087	0.119	0.182	0.243	0.343	0.454
500	0.025	0.045	0.071	0.100	0.161	0.219	0.319	0.431
1000	0.016	0.032	0.054	0.079	0.135	0.190	0.288	0.403

Example 7.6

The cost of failure for an item is $1,000. The cost of preventive maintenance for this item is $25. The following Weibull distribution parameters were determined from time to fail data: β =2.5, θ = 181 days, δ = 0. How often should preventive maintenance be done?

Solution

The ratio of failure cost to PM cost is

$$\frac{C_f}{C_p} = \frac{1000}{25} = 40$$

Entering Table 7.1 with this ratio and a shape parameter of 2.5, give 0.197 for the value of ***m***.

A PM should be done every

$$T = (0.197)(181) + 0 = 35.657 \text{ days}$$

This is shown graphically in Figure 7.6.

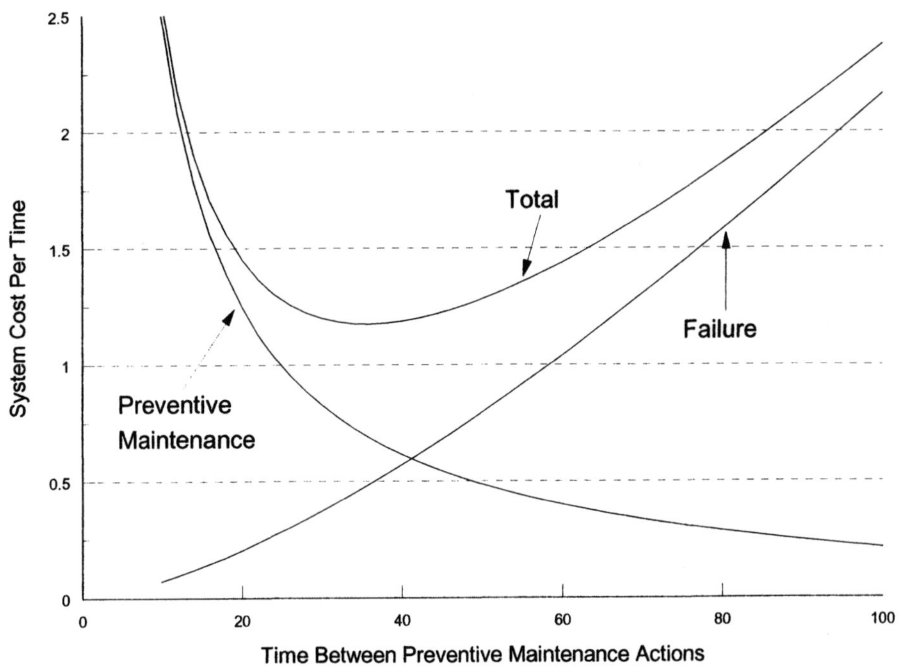

Figure 7.6. Solution to Example 7.6.

Example 7.7

Repeat Example 6.6 with the following Weibull distribution parameters: β =0.8, θ = 181 days, δ = 0.

Solution

When the shape parameter of the Weibull distribution is less than or equal to 1.0, the optimum solution is to do no preventive maintenance.

7.3 Designing for maintainability

Maintenance may be divided into the following tasks:

1. **Disassembly** — Equipment disassembly to the extent necessary, to gain access to the item that is to be replaced. How difficult is it to access the failed part? How much time does it take?
2. **Interchange**— Removing the defective item and installing the replacement. How difficult is it to exchange the defective part? How much time does it take?
3. **Reassembly** — Closing and reassembly of the equipment after the replacement has been made. Again difficulty and time should be a consideration.

Considering that the ease of maintenance may be measured in terms of man-hours, material, and money. Savings are attributable to the fact that designing for ease of maintenance is considered to be a design enhancement because it provides for the recognition and elimination of areas of poor maintainability during the early stages of the design life cycle. Otherwise, this area of poor maintainability would only become apparent during demonstration testing or actual use. After which time, correction of design deficiencies would be costly and would unduly delay schedules and missions. The ease with which corrective action may be taken is the key to designing for maintainability.

Moss (1985) gives eight guidelines that should be considered when designing for maintainability. Following these guidelines will help produce a design that is easily repaired, thus increasing maintainability and availability.

1. **Standardization** — Ensure compatibility among mating parts, and minimize the number of different parts in the system. Instead of using a 4mm bolt in one area and a 5mm bolt in another area, use a 5mm bolt in both areas. This will reduce spare parts requirements.
2. **Modularization** — Enforce conformance of assembly configurations to standards based on modular units with standard sizes, shapes, and interface locations. This allows use of standardized assembly and disassembly procedures.

3. **Functional packaging** — Expedite repair of faulty systems by containing all components of an item performing a function in a package. Replacing the faulty package repairs the unit.
4. **Interchangeability** — Control dimensional and functional tolerances to assure that items can be replaced in the field without physical rework.
5. **Accessibility** — Control spatial arrangements to allow accessibility. Items should be arranged to allow room for a worker to be comfortable when making repairs. All items should be accessible. Un-failed items should not have to be removed to gain access to a failed item.
6. **Malfunction annunciation**— Provide a means of notifying the operator when a malfunction is present. The oil pressure and temperature gauges on an automobile are examples.
7. **Fault isolation** — Assure that a malfunction can be traced. A simple method is to provide connectors that can be connected to external test equipment to determine if a component is operating properly. Often built-in diagnostics are provided with a system.
8. **Identification** — Provide a method for unique identification of all components as well as a method for recording preventive and corrective maintenance.

7.4 System effectiveness

The Weapons System Effectiveness Industry Advisory Committee (WSEIAC) developed a performance measurement for repairable systems that incorporates availability, capability and reliability. The measurement is **system effectiveness**. In order for a system to be effective, the system must be available when called on, the system must have the capability to complete the stated task, and the system must achieve the mission (i.e. be reliable).

The ARINC Research Corporation has also developed a measure of system effectiveness based on mission reliability, operational readiness and design adequacy. This relationship is shown in Figure 7.7.

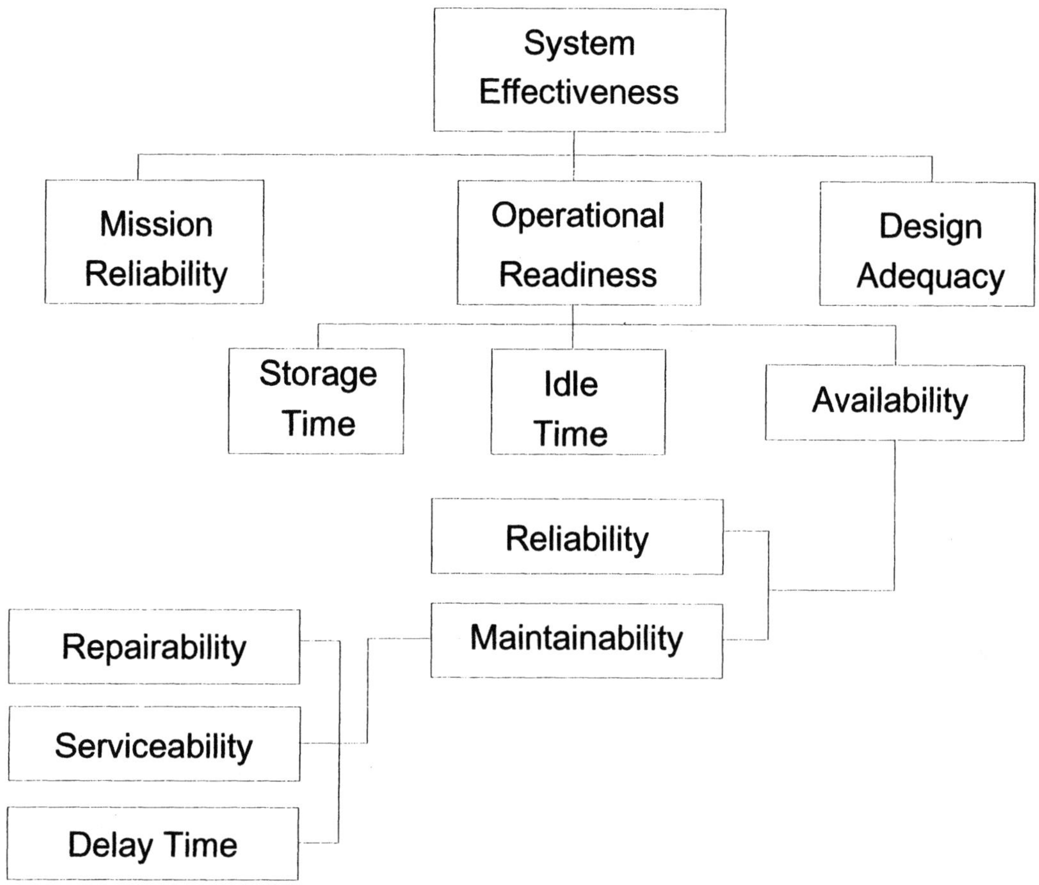

Figure 7.7. System effectiveness.

System effectiveness can be quantified as

$$SE = (MR)(OR)(DA) \tag{7.15}$$

where: MR is mission reliability,
OR is operational readiness, and
DA is design adequacy.

Example 7.8

A missile system not originally designed to intercept Scud missiles has been employed to do so. Assuming this missile's design adequacy for intercepting Scud missiles is 0.9. Determine the system effectiveness if operational readiness is 0.98 and mission reliability is 0.87.

Solution

System effectiveness is

$$SE = (0.87)(0.98)(0.9) = 0.7673$$

7.5 Exercises

1. Describe the 6 components of maintainability and their relationship.
2. Describe the differences in inherent availability, achieved availability, and operational availability.
3. Given the time to repair follows a Weibull distribution with a shape parameter of 4, a scale parameter of 130 and a location parameter of 10, graph maintainability as a function of time.
4. For the system described in Exercise 3, what is the maintainability at time = 5, time = 25, time = 200?
5. Explain why availability at time = 0 is equal to 1.0.
6. An item has a constant failure rate of 50, and a constant repair rate of 22, what is the steady-state availability?
7. For the system in Exercise 6, graph availability as a function of time.
8. For the system in Exercise 6, determine the uptime for the interval (0,5).
9. For the system in Exercise 6, plot uptime as a function of time.
10. Three components in series have mean times to fail of 50, 60, and 70 respectively. The mean repair times are 30, 40, and 50 respectively. What is the steady-state availability, assuming all non-failed components remain operating during system downtime?
11. Repeat Exercise 10 assuming the 3 components are in parallel.
12. Repeat Exercise 10 assuming all components cease operating during system downtime.
13. Repeat Exercise 10 assuming all components cease operating during system downtime, and the 3 components are in parallel.
14. Explain what is meant by infant mortality failures.
15. Explain what is meant by wear-out failures.
16. A component has a time to fail that follows a Weibull distribution with a shape parameter of 3.5 and a scale parameter of 67. The component can be replaced at a cost of $200; but if the component fails, the cost is $20,000. How often should the component be replaced?
17. Repeat Exercise 16 with Weibull parameters of $\beta = 0.89$ and $\theta = 190$.
18. What is the required operational readiness if a system effectiveness of 0.997 is required, and mission reliability is 0.999, and design adequacy is 0.9999?

7.6 References

Dhillon, B. S. 1985. ***Quality Control, Reliability, and Engineering Design.*** New York: Marcel Dekker.

Dodson, Bryan, and Jack Kirkland. 2000. ***The Reliability & Maintenance Analyst.*** Rochester, MI: Engineered Software (www.engineeredsoftware.com).

Dodson, Bryan. 1994. Determining the optimum schedule for preventive maintenance. ***Quality Engineering*** 6(4):667–679.

Dodson, Bryan. 1994. Using statistics to schedule maintenance. In ***Proceedings of the 2nd Annual Reliability and Maintainability Symposium.*** Dallas, TX.

Dodson, Bryan. 1993. Using the Weibull distribution to schedule preventive maintenance. In ***Proceedings of the 31st Annual Reliability Engineering and Management Institute.*** Tucson, AZ.

Dodson, Bryan. 1993. Optimizing schedules for equipment monitoring. In ***Proceedings of the 1993 CSI User Conference.*** Louisville, KY.

Dodson, B. L., and M. D. Mulcahy. 1992. ***Certified Reliability Engineer Examination Study Guide.*** Tucson, AZ: Quality Publishing.

Dovich, R. A. 1990. ***Reliability Statistics.*** Milwaukee, WI: ASQ–Quality Press.

Ireson, G. W., and C. F. Coombs. 1988. ***Handbook of Reliability Engineering and Management.*** New York: McGraw-Hill.

Kapur, K. C., and L. R. Lamberson. 1977. ***Reliability in Engineering Design.*** New York: John Wiley & Sons.

Krishnamoorthi, K. S. 1992. ***Reliability Methods for Engineers.*** Milwaukee, WI: ASQC–Quality Press.

Lewis, E. E. 1987. ***Introduction To Reliability Engineering.*** New York: John Wiley & Sons.

Moss, M. A. 1985. ***Designing for Minimal Maintenance Expense.*** New York: Marcel Dekker.

O'Conner, P. D. T. 1993. ***Practical Reliability Engineering.*** New York: John Wiley & Sons.

US Department of Defense. ***MIL-HDBK-472: Maintainability Prediction.*** Philadelphia: Naval Publications and Forms Center.

US Department of Defense. ***MIL-STD-470: Maintainability Program for Systems and Equipment.*** Philadelphia: Naval Publications and Forms Center.

US Department of Defense. ***MIL-STD-471: Maintainability Verification, Demonstration and Evaluation*** Philadelphia: Naval Publications and Forms Center.

♦♦♦

CHAPTER

8

Failure Mode, Effects and Criticality Analysis

Failure Mode, Effects and Criticality Analysis (FMECA) is a powerful design analysis tool that is used to increase system reliability. It can be applied during the initial design phase or to existing equipment. To be more effective, the FMECA should relate to the nature of the design process itself. In either case, it considers overall design, operating and service problems, while at the same time addressing process and safety problems.

If used as a design tool, the benefit of FMECA depends upon the timeliness in which information is communicated in the early design phase. Timeliness is probably the most important factor in differentiating between effective and ineffective implementation of the FMECA. The efforts and sophistication of the approach used much depends on the requirements of each individual program. In any case, the FMECA should contribute to the overall program decision.

8.1 Criticality analysis

The objective of an FMECA is to identify all failure modes in a system design. Its purpose is to identify all catastrophic and critical failure probabilities so they can be minimized as early as possible. Therefore, the FMECA should be started as soon as preliminary design information is available and extended as more information becomes available in suspected problem areas.

The effects of all failure modes are not equal with respect to the total impact on the system concerning safety and overall system performance. The designer, faced with this dilemma, needed a tool that would rank the significance of each potential failure for each component in the system's design alternatives. Because of the need for such justification, the criticality analysis function was added to the Failure Mode and Effects Analysis (FMEA) process, thus creating Failure Mode, Effects and Criticality Analysis (FMECA).

This tool has been used extensively by the military in the last three decades. In recent years, more commercial industries have been requiring the FMECA to evaluate new designs and even more recently to improve the reliability of existing equipment.

8.1.1 The qualitative approach to FMECA

This approach should be used when specific failure rate data is not available. Failure modes identified by the FMECA process are assessed by their probability of occurrence. To establish qualitative measures of occurrence, severity and detection, criteria must be established that subjectively relates to the overall effect on the process. Examples are offered in Tables 8.1, 8.2, and 8.3, to serve as guides in establishing qualitative measures. The product of the measures of occurrence, severity and detection is called the Risk Priority Number (RPN). Tables 8.1, 8.2, and 8.3 are for example only. The numbers or criteria assigned to any particular ranking system are at the discretion of the user. Detailed instructions on how to use this criterion on the FMECA form is explained in Section 8.1.4.

Table 8.1. Occurrence probabilities.

Rank	Occurrence Criteria	Occurrence Rates (cycles, hrs, etc.)
1	Unlikely. Unreasonable to expect this failure mode to occur.	—
2	Isolated. Based on similar designs having a low number of failures.	1/10,000
3	Sporadic. Based on similar designs that have experienced occasional failures.	1/1,000
4	Conceivable. Based on similar designs that have caused problems.	1/100
5	Recurrent. Certain that failures will ensue.	1/10

NOTE: The ranking criteria selected must be consistent throughout the FMECA.

Table 8.2. Severity Probabilities.

Rank	Severity Criteria
1	Minor. No noticeable effect. Unable to realize that a failure has occurred.
2	Marginal. Annoying. No system degradation.
3	Moderate. Causing dissatisfaction. Some system degradation.
4	Critical. Causing a high degree of dissatisfaction. Loss of system function.
5	Catastrophic. A failure which may cause death or injury. Extended repair outages.

NOTE: The ranking criteria selected must be consistent throughout the FMECA.

Table 8.3. Detection probabilities.

Rank	Detection Criteria	Probability
1	Very High probability of detecting the failure before it occurs. Almost always preceded by a warning.	80%–100%
2	High probability of detecting the failure before it occurs. Preceded by a warning most of the time.	60%–80%
3	Moderate probability of detecting the failure before it occurs. About a 50% chance of getting a warning.	40%–60%
4	Low probability of detecting the failure before it occurs. Always comes with little or no warning.	20%–40%
5	Remote probability of detecting the failure before it occurs. Always without a warning.	0%–20%

NOTE: The ranking criteria selected must be consistent throughout the FMECA.

8.1.2 FMECA quantitative approach

Figure 8.1 provides the worksheet for the quantitative FMECA process.

The failure mode and criticality number (C_m) is the portion of the criticality number for the item due to a particular failure mode. This criticality number replaces the RPN number used in the qualitative method described in the previous section. The C_m for a failure mode is determined by the expression

$$C_m = \beta \alpha \lambda_p t \quad (8.1)$$

where: β = conditional probability of loss of function,

α = failure mode ratio,

λ_p = part failure rate, and

t = duration or operating time.

System ______ Date ______
Indenture Level ______ Sheet ______ of ______
Reference Drawing ______ Compiled by ______
Mission ______ Approved by ______

ID #	Item Identification	Function	Failure Modes & Causes	Mission Operational Mode	Severity Class	Failure Probability & data Source	Failure Effect Probability	Failure Mode Ratio	Failure Rate	Operating Time	Failure Mode Criticality	Item Criticality	Remarks

Figure 8.1. MIL-STD-1629 worksheet for method 102.

The β values represent the analyst's judgment as to the conditional probability that the loss will occur and should be quantified in general accordance with the following:

Failure effect	Probability of loss of function
Actual loss	$\beta=1$
Probable loss	$0.10<\beta<1.00$
Possible loss	$0<\beta<0.10$
No effect	$\beta=0$

The failure mode ratio, α, is the probability that the part or item will fail. If all potential failure modes of a particular part or item are listed, the sum of the α values for that part or item will equal one. Individual failure mode multipliers may be derived from failure rate source data or from test and operational data. If failure mode data are not available, the α values should represent the analyst's judgment based upon an analysis of the item's functions.

Part failure rates, λ_p, are derived from appropriate reliability prediction methods using mean-time-between-failure (MTBF) data or possibly other data obtained from handbooks or reference material. Manufacturers often supply failure data; however, it is important that the environment the item will be subjected to is similar to the environment the manufacturer used when obtaining the failure data.

The operating time, t, is usually expressed in hours or the number of operating cycles of the item being analyzed.

The α and β values are often subjective, thus making the supposed quantitative method somewhat qualitative. All things considered, it is generally understood that the FMECA process is a qualitative method of analysis.

8.1.3 FMECA process

The process outlined here demonstrates a method, often used throughout industry today, that incorporates the basic FMECA and a qualitative Criticality Analysis on the same form. Exact forms, as defined by various standards, are usually modified by different users to fit individual conditions.

The FMECA process is a combination of group discussions, team meetings, and individual studies. The meetings are typically led by the design engineer or an expert in Failure Mode, Effects and Criticality Analysis.

Entire systems are much too complex to be covered by one FMECA. The system should be defined and separated into sub-systems, also known as the infrastructure, early in the process. It should be a consensus as to how and why the system is separated, since addressing a single part is usually not an efficient use of resources. The FMECA should be done at some major level. For example, for an automobile you might list the d*rive train, frame, body, and support equipment* separately.

The team is led into discussion about the individual functions of each sub-system and how they interact with the overall system. At this time, team members should be supplied with pertinent technical data including layouts, functional descriptions, prints, etc., of the system being analyzed. After the initial discussions and examination, the infrastructure is separated into infrastructure sub-systems. For example, an automobile drive train might be divided into the engine, the transmission, the drive gearing, and support equipment. This should be done for each infrastructure.

At this time, the engineer or team leader prompts the team to identify the process or machine functions. The functions of each infrastructure sub-system is then discussed, and the data is input on the FMECA form. After all functions are established, *FAILURE MODES, EFFECTS, CAUSES, PROBABILITIES, and RECOMMENDED CORRECTIVE ACTIONS* are then input on the form.

SYSTEM:					INFRASTRUCTURE:				
COMPANY/LOCATION:					PROJECT ENGINEER:				
OUTSIDE SUPPLIER(S):					DATE:				
INFRASTRUCTURE SUB-SYSTEM	**SUB-SYSTEM FUNCTION**	**FAILURE MODE**	**EFFECTS OF FAILURE**	**CAUSE(S) OF FAILURE**	**OCC**	**SEV**	**DET**	**RPN**	**RECOMMENDED CORRECTIVE ACTION & STATUS**
WHAT DOES IT DO?	WHAT FUNCTION CAN BE LOST?	HOW **COULD** IT FAIL TO PERFORM THIS FUNCTION? 1. 2. 3.	WHAT HAPPENS IF IT FAILS?	WHAT CAUSES IT TO FAIL?	LIKELIHOOD	CONSEQUENCES	PREDICTABILITY	O x S x D	HOW CAN THIS FAILURE BE PREVENTED?

Figure 8.2. Sample FMECA form for existing equipment.

8.1.4. Steps for completing the form

Refer to Figure 8.2 for the remainder of this section.

First Column—Infrastructure sub-system

What does this sub-system do? Describe briefly what the equipment is designed to do. Be as concise in the description as possible. For example, hold temperature below 200° F during continuous operation.

Second Column—Sub-system function

What function is lost if the sub-system fails? For example, temperature regulation is lost.

Third Column—Failure mode

Anticipation of failures is the goal. The question here is how could it fail, not will it fail. Try to determine all possibilities for the system failing to meet objectives. Some distinctive failure modes are:

Over/undersize	Cracked	Breaks
Torn or Split	Open	Mis-aligned
Rough	Out-of-balance	Pitting
Eccentric	Leaking	Deformed

NOTE: Try to be specific. For example, say, "gas line leaking and deformed" instead of "gas line fails." Always find a word to replace "fails or failed."

Fourth Column—Effects of failure

Describe what will happen as a result of each failure mode. Will the equipment be totally inoperative, inefficient, inaccurate, etc.?

Fifth Column—Cause of failure

Again, try to anticipate the cause of the failure mode described. Ask what might cause this particular failure mode. Fault-tree analysis is sometimes used at this point to determine root causes (refer to Section 8.3). If time does not permit the use of a fault tree, list all conceivable causes for each failure mode. Examples of causes might be:

Worn or broken tooling	Improper part location
Worn bearings	Overload
Poor handling	Improper weld
Contamination	Interference
Inefficient cooling	Over pressure
Heat treat shrinkage	Inadequate lubrication

NOTE: In columns 7, 8, and 9, it is important to remember that the ratings apply to the failure mode and not to the cause of the failure.

Sixth Column—Occurrence

Estimate the likelihood that the failure will occur using a previously set evaluation scale. Table 8.1 (Section 8.1) gives an example of an evaluation scale for the likelihood of a failure. The lowest number of the scale must indicate the lowest probability of occurrence, and the highest number must indicate the highest probability of occurrence.

Seventh Column—Severity

Estimate the consequences of the failure using a predetermined scale with the lowest number of the scale indicating minor concerns and the highest number indicating major consequences. Table 8.2 (Section 8.1) gives an example of an evaluation scale for the consequences of a failure.

Eighth Column—Detection

Estimate the ability to predict the failure. Table 8.3 (Section 8.1) gives an example of an evaluation scale for the ability to predict a failure. Use a predetermined scale with the lowest number indicating a very high probability that the failure mode will be detected before reaching the customer.

Ninth Column—RPN

Estimate the Risk Priority Number (RPN). The product of the occurrence, severity, and detection provides a relative priority to each failure mode. From this number critical summaries can be developed for prioritizing follow-up actions.

Tenth Column—Recommended corrective action & status

Corrective actions are most usually found to be design changes, either to the equipment or the process. Changes should be focused on reducing failure occurrences, reducing the consequences of failures, and increasing the ability to predict failures. Remember the idea is to reduce failures, not just detect them.

8.1.5. FMECA follow-up report

Follow-up is critical to the FMECA's success. The basic recommendations that the FMECA and the follow-up provide should set a solid foundation for corrective action.

A good tool to use for the follow-up report is a matrix. The first thing to do is establish the matrix criteria (see Table 8.4). Keep in mind that this information will be used to present recommendations to management for approval. The end result should provide answers to all major concerns, such as, why are these problems being worked on, how much will this project cost, what is the time frame, etc.?

Table 8.4. Sample matrix report criteria.

Equipment	Difficulty	% Solved	Cost ($)	Time Required	Actions
Failure mode & rank #	1= Very easy 2 = easy 3 = medium 4 = hard 5 = Very hard	Percent of total ranked order	a = 0 B = Very low c = low d = medium e = High	1 = 1–5 units 2 = 6–10 units 3 = 11–15 units 4 = 15–20 units 5 = 21–30 units 6 = 31–50 units	Recommended actions

The next step is to compile the RPNs from the highest number to the lowest (see Table 8.5). This example shows that the number one problem identified by the FMECA is the conveyor system's guide rollers breaking. Also note that the redesigning of these rollers will eliminate three other failure modes; guide rollers seizing, tracking bars bending and bent track rips belt.

Table 8.5. Ranked order chart.

Rank	RPN	Component	Failure mode
1	125	Conveyor system	Guide rollers break
2	100	Drying chamber	Plugged drains
3	90	Air compressor valves	Corrosion
4	80	Ventilation	Louvers stick
5	60	Conveyor system	Guide rollers seize
6	40	Conveyor system	Tracking bars bend
7	20	Conveyor system	Bent track rips belt
8	5	Drying chamber	High humidity

Now total the RPN numbers for a number representing 100% of the problems in the sub-system. In the example in Table 8.5, this number is 520.

If the RPNs for just the CONVEYOR SYSTEM are totaled, the result is 245. Accepting that 245 is 47% of 520, with 520 representing at, or near, 100% of the total population of failure modes, it can be reasonably said that by taking actions to eliminate the cause of the top failure mode, 47% of the overall problem can be solved. This is shown on the matrix in Table 8.6. The cost, complexity, time, etc., of any potential changes should be discussed as a team and a consensus drawn. Table 8.7 is an example of how a safety concern of the FMECA process could be handled on a matrix.

Table 8.6. Completed sample matrix for equipment.

Equipment	Difficulty	% Solved	Cost ($)	Time Required	Actions
Conveyor System; Rank = 1; RPN = 125; Other ranked items effected 5,6, & 7	Hard	47%	low	Category 6, 45 days	Installed self-lubricating double-guide rollers. Redesigned track bars. Redesigned retainers. Redesigned pins.

NOTE: This Table is based on Tables 8.1, 8.2, and 8.3, where 5 is the highest possible value in ranking criteria. Thus, 125 would be the highest possible RPN (5x5x5).

Table 8.7. Sample matrix report criteria.

Equipment	Difficulty	% Solved	Cost ($)	Time Required	Actions
Inline plugs	Hard	50%	Very low	5 weeks	Installed inline plugs in discharge

8.1.6. What it takes to do a successful FMECA

The keys to a successful FMECA are the same when applied to the initial design or to existing designs. Management support and the diversity of the team are the most important considerations, followed by training for the facilitator and the team members.

Steps must be taken to assure management's support for the project from beginning to end. First, management must understand the benefits of the FMECA. After completing the FMECA, a quantifying report reflecting the total impact of the FMECA's suggested corrective actions and implementation plans should be presented to management.

A good FMECA team will include personnel that are directly involved with the process to be analyzed. Examples include, but certainly not limited to: electrical and mechanical maintenance technicians, machine operators, technical specialists, design engineers, and even a representative from the equipment supplier if possible.

The FMECA facilitator should be trained to ask appropriate key questions and be prepared to probe the team for answers to describe the equipment's functions, failure modes and their effects. A good facilitator should also be able to assist the team with establishing criteria for the equipment's ranking indexes and organize information as it comes from the team members.

Some time should be allotted for new FMECA team members to become familiar with the FMECA process prior to beginning the actual FMECA. The team should have a good understanding of the purpose and benefits of the FMECA as well as an understanding of the team's objectives. If time allows, training should include an actual FMECA on some less complex piece of equipment such as a flashlight or cigarette lighter. On the basis of past experiences, this training can be accomplished in about four hours.

The use of a computer with the ability to project the screen during the FMECA team meetings is highly recommended. Software containing a standard FMECA blank form should be filled in as data is received from the team members. A copy of the form should be distributed after each meeting to enable the team to review the data and recommend any changes at the next meeting.

Although the use of the computer and software are not an absolute necessity for the successful FMECA, this practice is not only a time-saver, but helps keep the team's ideas organized and on track.

8.2 Environmental requirements and influence

When the FMECA team members are properly selected, the FMECA process will address all problems concerning design, manufacturing, process, safety and environment. The consequences of these problems are addressed during the study of the Failure Mode during the FMECA. Preventive measures for failures where an individual's well-being in the workplace is concerned should always be given top priority. According to John Moubray (1991), "A failure mode has safety consequences if it causes a loss of function or other damage which could hurt or kill someone."

The overall design must consider how the environment influences the system during testing, storage, installing and operating. When considering the failure mode effects, think about the impact on society in general if the failure happened. Society may, in some cases, be considered as a customer because if the product is not accepted due to environmental effects, it may not be successful. "A failure mode has environmental consequences if it causes a loss of function or other damage which could lead to the breach of any known environmental standard or regulation," (Moubray, 1991).

8.3 Differences between part function, part failure mechanism, failure effect, failure detection, and failure mode

The definitions which follow form the fundamentals of FMECAs. These definitions will enable the reader to understand the differences between Part Function, Part Failure Mechanism, Failure Effect, Failure Detection, and Failure Modes.

1. **Failure Mode and Effects Analysis (FMEA)** — A procedure by which each potential failure mode in a system is analyzed to determine the

results, or effects thereof, on the system and to classify each potential failure mode according to its severity.

2. **Failure mode** — The manner by which a failure is observed. Generally describes the way the failure occurs and its impact on equipment operation.
3. **Failure effect** — The consequence(s) a failure mode has on the operation, function, or status of an item. Failure effects are usually classified according to how the entire system is impacted.
4. **Failure cause** — The physical or chemical process, design defects, part misapplication, quality defects, or other processes that are the basic reason for failure or which initiate the physical process by which deterioration proceeds to failure.
5. **Environments** — The conditions, circumstances, influences, stresses and combinations thereof, surrounding and affecting systems or equipment during storage, handling, transportation, testing, installation, and use in standby status and operations.
6. **Detection mechanism** — The means or methods by which a failure can be discovered by an operator under normal system operation or can be discovered by the maintenance crew by some diagnostic action.
7. **Severity** — The consequences of a failure as a result of a particular failure mode. Severity considers the worst potential consequence of a failure, determined by the degree of injury, property damage, or system damage that could ultimately occur.
8. **Criticality** — A relative measure of the consequences of a failure mode and its frequency of occurrence.
9. **Criticality analysis (CA)** — A procedure by which each potential failure mode is ranked according to the combined influence of severity and probability of occurrence.
10. **FMECA-Maintainability information** — A procedure by which each potential failure is analyzed to determine how the failure is detected and the actions to be taken to repair the failure.
11. **Single failure point** — The failure of an item that would result in failure of the system and is not compensated for by redundancy or alternative operational procedure.

12. **Undetectable failure** — A postulated failure mode in the FMEA for which there is no failure detection method by which the operator is made aware of the failure.
13. **FMECA planning** — Planning the FMECA work involves the contractor's procedures for implementing their specified requirements. Planning should include updating to reflect design changes and analysis results. Worksheet formats, ground rules, assumptions, identification of the level of analysis, failure definitions, and identification of coincident use of the FMECA by the contractor and other organizational elements should also be considered.
14. **Ground rules and assumptions** — The ground rules identify the FMECA approach (e.g., hardware, functional or combination), the lowest level to be analyzed, and include statements of what might constitute a failure in terms of performance criteria. Every effort should be made to identify and record all ground rules and analysis assumptions prior to initiation of the analysis; however, ground rules and analysis assumptions may be adjusted as requirements change.
15. **Analysis approach** — Variations in design complexity and available data will generally dictate the analysis approach to be used. There are two primary approaches for the FMECA. One is the hardware approach that lists individual hardware items and analyzes their possible failure modes. The other is the functional approach that recognizes that every item is designed to perform a number of outputs. The outputs are listed and their failures analyzed. For more complex systems, a combination of the functional and hardware approaches may be considered.
16. **Hardware approach** — The hardware approach is normally used when hardware items can be uniquely identified from schematics, drawings, and other engineering and design data. This approach is recommended for use in a part level up approach often referred to as the bottom-up approach.
17. **Functional approach** — The functional approach is normally used when hardware items cannot be uniquely identified or when system complexity requires analysis from the top down.
18. **Level of analysis** — The level of analysis applies to the system hardware or functional level at which failures are postulated. In other words, how the system being analyzed is segregated. (e.g., a section of the system, component, sub-component, etc.)

19. **Failure definition** — This is a general statement of what constitutes a failure of the item in terms of performance parameters and allowable limits for each specified output.
20. **Trade-off study reports** — These reports should identify areas of marginal and state-of-the-art design and explain any design compromises and operating restraints agreed upon. This information will aid in determining the possible and most probable failure modes and causes in the system.
21. **Design data and drawings** — Design data and drawings identify each item and the item configuration that perform each of the system functions. System design data and drawings will usually describe the system's internal and interface functions beginning at system level and progressing to the lowest indenture level of the system. Design data will usually include either functional block diagrams or schematics that will facilitate construction of reliability block diagrams.
22. **Block diagrams** — Block diagrams that illustrate the operation, interrelationships, and interdependencies of the functions of a system are required to show the sequence and the series dependence or independence of functions and operations. Block diagrams may be constructed in conjunction with, or after defining the system and shall present the system breakdown of its major functions. More than one block diagram is sometimes required to represent alternative modes of operation, depending upon the definition established for the system.
23. **Functional block diagrams** — Functional block diagrams illustrate the operation and interrelationships between functional entities of a system as defined in engineering data and schematics. An example of a functional block diagram is shown in Figure 8.3.
24. **Reliability block diagrams** — Reliability block diagrams define the series dependence, or independence, of all functions of a system or functional group for each life-cycle event. An example of a reliability block diagram is shown in Figure 8.4.

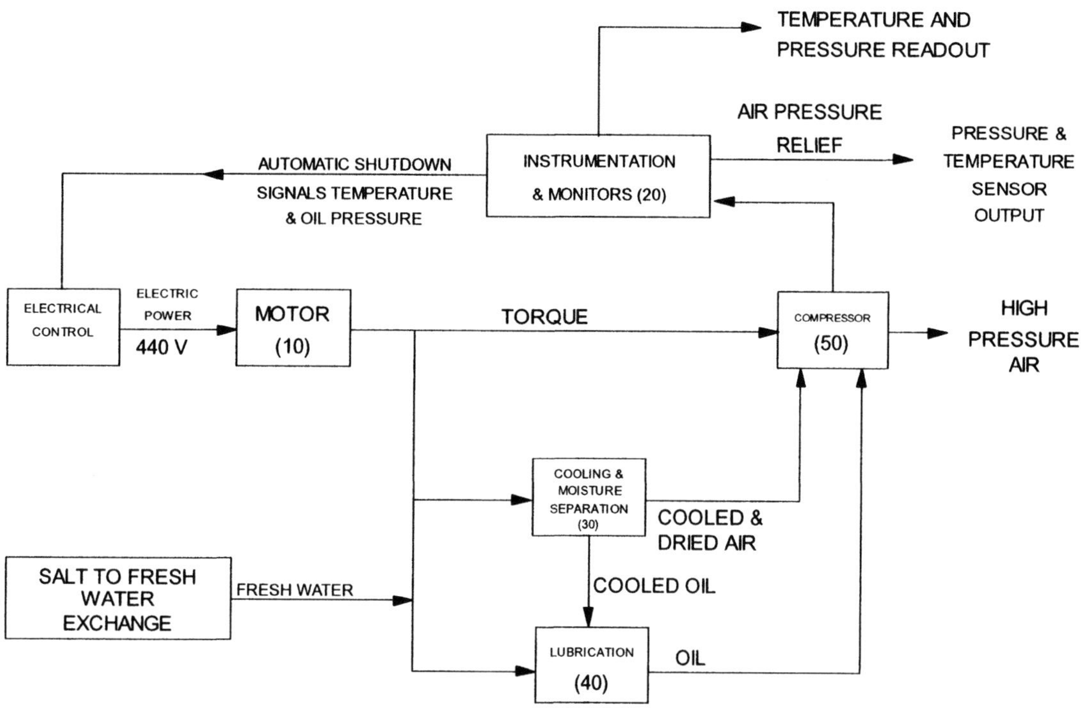

Figure 8.3. Functional block diagram.

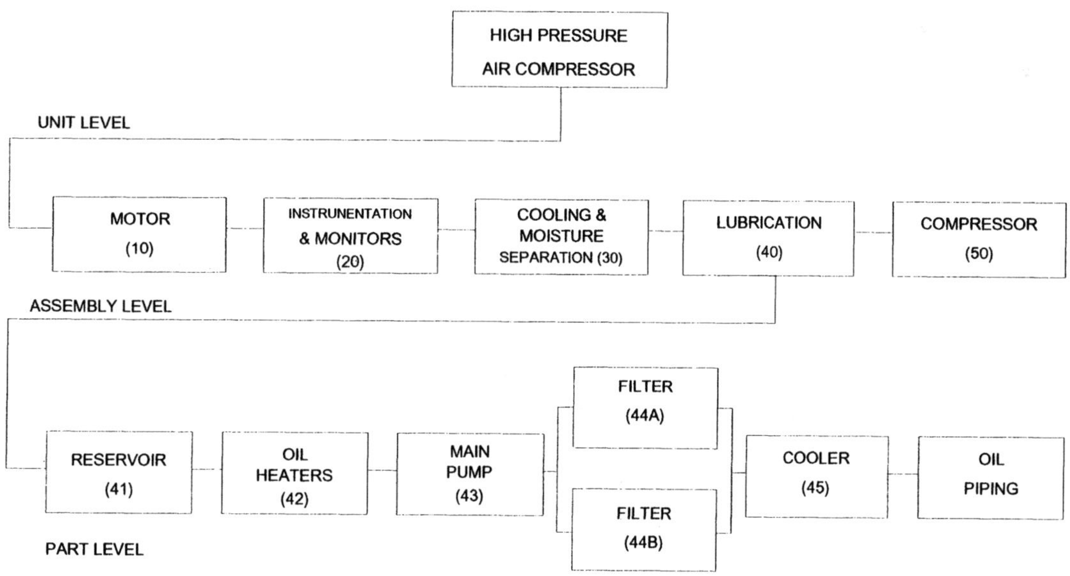

Figure 8.4. Reliability block diagram.

25. **Severity classification** — This classification is assigned to provide a qualitative measure of the worst potential consequences resulting from

design error or item failure. Classifications should be assigned to each identified failure mode and each item analyzed in accordance with the loss statements below. It may not be possible to identify an item or a failure mode according to the loss statements in the four categories below, but similar loss statements based on various inputs and outputs can be developed and included in the ground rules for the FMECA activity. Severity classification categories are defined as follows:

- **Category I–Catastrophic** — A failure that may cause injury or death.
- **Category II–Critical** — A failure which may cause severe injury, major property damage, or major system damage that will result in major downtime or production loss.
- **Category III–Marginal** — A failure which may cause minor injury, minor property damage, or minor system damage which will result in delay or loss of system availability or degradation.
- **Category IV–Minor** — A failure not serious enough to cause injury, property damage or system damage, but will result in unscheduled maintenance or repair.

These categories can be attached to a cost or any other factor, but when used in the established criteria, should be consistent throughout the analysis.

26. **FMECA report** — The results of the FMECA and other related analyses should be included in a report that identifies the level of the analysis, documents the data sources and techniques used in performing the analysis, and gives the system definition.

8.4 Exercises

1. On the FMECA form shown in Figure 8.5 (following page), list five possible failure modes for the conveyor.
2. For each failure mode listed in Exercise 1, input two failure effects in the example FMECA form.
3. For each failure mode listed in Exercise 1, list two failure causes.

SYSTEM: **Tipple Coal Mine**					INFRASTRUCTURE: **Separators**				
COMPANY/LOCATION: **Black Star Coal of America**					PROJECT ENGINEER:				
OUTSIDE SUPPLIER(S):					DATE:				
INFRASTRUCTURE SUB-SYSTEM	**SUB-SYSTEM FUNCTION**	**FAILURE MODE**	**EFFECTS OF FAILURE**	**CAUSE(S) OF FAILURE**	**OCC**	**SEV**	**DET**	**RPN**	**RECOMMENDED CORRECTIVE ACTION & STATUS**
Conveyor A	**Transport coal from Hopper #1 to Hopper #2 at 100 fpm ± 10 fpm at a maximum handling capacity of 50 tons ± 5 tons.**								

Figure 8.5. Example problem FMECA form.

4.Calculate the RPN for each failure mode in Exercise 1, given

Failure mode 1—OCC=4; SEV=5; DET=1
Failure mode 2—OCC=2; SEV=3; DET=3
Failure mode 3—OCC=3; SEV=3; DET=5
Failure mode 4—OCC=3; SEV=5; DET=5
Failure mode 5—OCC=3; SEV=3; DET=4

5. Input recommended actions for the item with the highest RPN in Exercise 4. What portion of the "total" problem does this item represent?
6. What does the failure mode and criticality number (Cm) replace in the most common qualitative methods?
   a. Severity number
   b. Risk Priority Number
   c. Failure Effects Number
   d. Failure Mode Number
7. The objective of an FMECA is to
   a. Identify all failure modes
   b. Identify failure probabilities
   c. Increase product reliability
   d. All of the above
8. A good reference for FMECA is
   a. MIL-STD-781

b. MIL-STD-414

c. MIL-STD-1629

d. MIL-HDBK-217

9.FMECA team meetings are typically led by a

a. Consultant

b. Engineer

c. Facilitator

d. Contractor

10. When identifying failure causes it is sometimes necessary to use

a. Root cause analysis

b. Fault tree analysis

c. Historical data

d. Failure test data

11. The FMECA process will address all problems concerning

a. Design

b. Process

c. Safety

d. All of the above

12. A failure is thought to have environmental consequences if it

a. Causes a safety problem in the workplace

b. Causes a hazard in the community

c. Breaches any environmental standard or regulation

d. All of the above

13. The FMECA eliminates failure modes in the design phase focusing attention on

a. Engineering standards

b. Company policies

c. Critical parts

d. Unreliable parts

14. FMECA classifies each failure mode according to

a. Probability

b. Criticality

c. Severity

d. Unreliability

15. The means or methods in which an operator detects a failure is called the

a. Alert mechanism

b. Alarm mechanism
c. Detection mechanism
d. Reliable mechanism

16. A relative measure of the consequences of a failure mode and its frequency is called the
    a. Severity
    b. Frequency
    c. Relative measure
    d. Criticality
17. Outside influences, stresses, circumstances and/or a combination thereof, effecting systems relates to the
    a. Designer
    b. Environment
    c. Operator
    d. None of the above

8.5 References

Dodson, B. L., and M. D. Mulcahy. 1992. ***Certified Reliability Engineer Examination Study Guide.*** Tucson, AZ: Quality Publishing.

Ireson, G. W., and C. F.Coombs. 1988. ***Handbook of Reliability Engineering and Management.*** New York: McGraw-Hill.

Kececioglu, D. 1991. ***Reliability Engineering Handbook, Volume 2.*** Englewood Cliffs, NJ: Prentice Hall.

Lewis, E. E. 1987. ***Introduction To Reliability Engineering.*** New York: John Wiley & Sons.

Moubray, John. 1991. ***Reliability-Centered Maintenance.*** Oxford: Butterworth-Heinemann.

Sundarajan, C. 1991. ***Guide To Reliability Engineering: Data, Analysis, Applications, Implementation, And Management.*** New York: Van Nostrand Reinhold.

US Department of Defense. 1980. ***MIL-STD-1629: Procedures for Performing a Failure Mode, Effects and Criticality Analysis.*** Philadelphia: Naval Publications and Forms Center.

♦♦♦

CHAPTER

9

Part Selection and Derating

Conducting an aggressive parts control and application program increases the probability of achieving and maintaining inherent equipment reliability, and minimizes parts proliferation, logistics support costs, and system life cycle costs. The added investment required for a vigorous program which controls parts selection and application can be offset by reduced system life cycle costs for reparable systems and by overall system effectiveness for non-reparable systems. In some cases the use of higher quality parts can even lower item acquisition cost through reduction in the amount of assembly line rework as well as eliminate additional costs for drawings and test data required when using non-standard parts.

Failure rates of parts are correlated to the stress and strength relationship of the design parameters. Using parts that are rated to withstand stresses above the design strength parameters is called **derating**. Derating parts will ultimately increase reliability by decreasing their failure rates.

9.1 Evaluation of parts for reliability

Parts and components are the basic items comprising higher level assemblies, which in turn ultimately constitute the system, where the "system" may be a radio, an automobile, or an airplane. Significant contributions toward system optimization can be realized by applying attention and resources to parts application, selection, and control starting early in the design phase and continuing throughout the life of the system.

An effective parts program requires the use of knowledgeable parts engineers by supporting activities. Logisticians should always be consulted for their inputs, because they will be required to support the system operationally. The investment in parts programs generally pays handsome dividends in terms of reduced operational costs and improved system operation effectiveness.

9.1.1 Application and specification

The decision as to the depth and extent of the parts program designed for a particular item acquisition should be made based on considerations of factors such as: operation criticality, parts essentiality (to assure successful operation and reduced frequency of maintenance), the maintenance concept, production quantity, parts availability, amount or degree of new design, and parts standardization status. A comprehensive parts program can be just as essential for a relatively simple device (e.g., radio) with a large projected inventory as for a single, complex device (e.g., space shuttle).

A comprehensive parts program will consist of the following elements:

a. Parts standardization,
b. Parts application (derating) guidelines,
c. Parts testing, screening, or validation, and
d. Government/Industry Data Exchange Program (GIDEP) participation as applicable.

If such a program is implemented, it is imperative that qualified personnel are assigned, because the dynamic nature of parts and component technology can quickly render existing knowledge and experience obsolete.

9.1.2 Program basic objectives

The basic objective of the parts program is to control the selection and use of standard and nonstandard parts. Occasionally, depending on the system, it becomes necessary to propose the use of a nonstandard part. Proposals for use of nonstandard parts should be made and accepted only after other options that use standard parts have been investigated. Before choosing an application, a judgment should be made on whether the part in question meets enough of the criteria to make it a standard part.

Parts programs should be initiated early in design and continued throughout the life of the system. Before the beginning of full-scale development, the procuring activity should clearly describe the part quality level preference order for use in the system. In addition, the procuring activity should identify prohibited part types. For certain applications, special tests of standard parts may have to be used to obtain acceptable parts which are then unique and must be identified as such. It is most important to emphasize, however, that special testing, identification, and selection inhibits standardization, since those processes produce a nonstandard part that may not be readily available to support the system through its life.

9.1.3 Program activities

Parts program activities are interrelated with analyses performed by other disciplines such as safety, quality engineering, maintainability, survivability and vulnerability. In some cases, upgraded or unique parts are needed to meet system requirements. In other cases, standard or readily available parts can be used to minimize life cycle costs and ensure supportability.

9.2 Reliability testing for parts

The reliability test program should serve three objectives in the following priorities:

1. Disclose deficiencies in item design, materiel and workmanship.
2. Provide measured reliability data as input for estimates of operational readiness, mission success, maintenance manpower cost, and logistics support cost.
3. Determine compliance with quantitative reliability requirements.

Cost and schedule investment in reliability testing should conform to these priorities to ensure that the overall reliability program is both effective and efficient. Four types of reliability testing that should be considered are:

1. Environmental Stress Screening
    a. A test, or a series of tests, specifically designed to disclose weak parts and workmanship defects.
    b. It should be applied to parts or equipment, as appropriate and cost-effective, to remove defects that would otherwise cause failures during higher-level testing or early field service.

2. Reliability Development or Growth Test
   a. A planned, pre-qualification, test-analyze-and-fix process, in which parts or equipment are tested under actual, simulated, or accelerated environments to disclose design deficiencies and defects.
   b. This testing is intended to provide a basis for early incorporation of corrective actions, and verification of their effectiveness, thereby promoting reliability growth.
3. Reliability Qualification Test
   a. A test intended to provide assurance that minimum acceptable reliability requirements have been met before items are committed to production.
   b. The test must be operationally realistic, and must provide estimates of demonstrated reliability.
4. Production Reliability Acceptance Test
   a. A test that is intended to simulate in-service evaluation of the delivered item or production lot.
   b. The statistical test plan must define criteria for compliance which limit the probability that the item tested, and the lot it represents, may have a true reliability less than the minimum acceptable reliability, and these criteria must be tailored for cost and schedule efficiency.

ESS and RDGT are reliability engineering tests. Program plans should emphasize early investment in these tests to avoid subsequent cost and schedule delays. Reliability qualification tests (RQT and PRAT) are reliability accounting tests. They should be tailored for effectiveness and efficiency (maximum return on cost and schedule investment) in terms of the management information they provide. A properly balanced reliability program will emphasize ESS and RDGT, and limit RQT and PRAT. However, be reminded that testing does not improve reliability. Only corrective actions that prevent the recurrence of failures in the operational inventory actually improve reliability.

9.2.1 Integrated testing

Performance, reliability and environmental stress testing should be combined, and environmental stress types should be combined insofar as practical. These tests should be drawn together into an integrated, effective and efficient test program. For example, mechanical, hydraulic, pneumatic and electrical equipment is usually subjected to three qualification tests: performance, environmental, and endurance (durability). The integration of these separate tests into a more comprehensive reliability test program can avoid costly duplication and ensure that deficiencies are not overlooked as they often are in the fragmented approach.

9.2.2 The performance test

Performance tests should be conducted as soon as items are fabricated. They should be brief, and should provide the immediate basis for correction of any deficiencies they disclose. However, an item that has passed its performance test must not be considered compliant until it has shown that it will perform reliability under realistic conditions.

9.2.3 The environmental test

Environmental tests should be considered an early portion of RDGT. They must be conducted early in development, to ensure that time and resources are available to correct the deficiencies they disclose, and the corrections must be verified under stress. Such information must be included in the Failure Reporting, Analysis and Corrective Action System (FRACAS) as an integral aspect of the reliability program.

9.2.4 The endurance test

Endurance (durability) testing usually consists of a normal test, an overload test, and a mission profile cycling test that duplicates or approximates the conditions expected in service. Failures must be evaluated, and corrective actions must be incorporated in the test items. The test must then be rerun or completed and an additional run conducted to show the problems have been corrected. This information must also be included in the FRACAS. An integrated test program will combine reliability testing and durability testing.

9.2.5 Test realism

A test is realistic to the degree that test conditions and procedures simulate the operational life, mission, and environmental profile of a production item. Realistic testing can disclose deficiencies and defects that otherwise would be discovered only after an item is put into service, and it can reduce the disparity between laboratory and operational reliability values. Therefore, test realism should be a primary consideration in every reliability test. A test that only discloses a small fraction of the operational failures it is supposed to disclose is a waste of time and resources. Conversely, a test that induces failures that will not occur in service forces unnecessary expenditures of time and resources to correct those failures. Finally, the degree to which any reliability test must simulate field service depends on the purpose of the test.

Low test realism is often due to omission of a relevant stress, or incomplete definition of a type of stress. For example, failures that are caused by vibrations are seldom found by tests that apply no vibration, or by tests that ignore the relevant combinations of vibration frequency, amplitude and duration of exposure. Establishment of realistic test conditions and procedures requires a knowledge of the life profile from factory to final expenditure, to include the micro-environments an item will experience during each phase of its life profile, based on measurement of the actual stresses experienced by similar items.

It is appropriate to apply stress levels greater that those expected in service, if the purpose of the test is to disclose deficiencies, and if test conditions do not induce failures that will not occur in service. On the other hand, both over-stress and under-stress make reliability estimates inaccurate and distort test results used to determine compliance. Therefore, over-stress (and step-stress) testing may be applied during ESS and the early portion of RDGT, but the final portion of RDGT, and both RQT and PRAT, should simulate the operational life profile insofar as it is practical and cost-effective.

Precise simulation of the operational life profile would expose each item and each part of each item to the exact stress types, levels, and durations they would experience in service. Such idealistic testing is seldom practical or cost-effective. Some stress types cannot be combined in the same test facility, and some may cost more to reproduce in the laboratory than they are worth in terms of the failures they cause in service. Stress types may be applied in series for ESS and the early portion of PRAT. Total test time may be compressed by reducing the amount of time spent in simulating less stressful phases of the life profile. Overstress is a valid way to accelerate the discovery of deficiencies and defects, but it is not a valid means of compressing test time when reliability is to be measured.

9.3 Vendor evaluation of part reliability

Vendors should be required to provide methods and procedures for control and testing of all reliability critical items. Reliability critical items are those items whose failure can significantly affect system safety, availability, mission success, or total maintenance support cost. Critical items should include, but not be limited to, those identified by reliability analysis and Failure Mode, Effects and Criticality Analysis (FMECA). High-value items should always be considered reliability critical for life cycle cost. When reliability critical items are specified, the vendor's procedures should include engineering support of the item during development or field testing, which include provisions for confirming failures that may occur, expediting failure cause determination, and determining and incorporating, or verifying, the necessary corrective action.

9.4 Derating principles

The strength of an item is a relative measure of the stress that can be applied to the item. The failure rate of an item increases as nominal stress approaches and exceeds the nominal strength. Failure rates generally decrease when stresses are reduced to levels below designed strengths. Derating may thus be defined as a method of assuring that stresses, either environmental or operational, are applied below rated values to enhance reliability by decreasing failure rates. For example, a capacitor rated at 400V assigned to a 200V application, or a structural load calculated at 25 tons, designed to support 125 tons.

Derating is defined as:

a. Using an item in such a way that applied stresses are below rated values, or

b. The lowering of the rating of an item in one stress field to allow an increase in another stress field.

The effect of failure rates on increased temperatures is an effect called the Arrhenius model. This model is covered in detail in Chapter 6. The model is used for predicting failure rates for electronic systems and is based on the Arrhenius equation. There is a "rule of thumb" for the Arrhenius effect which states that for every 10°C increase in temperature, the chemical reactions for electronic and electrical components double (Dovich, 1990). The following examples show this "rule of thumb" to be unreliable in some cases and should be used with caution.

Example 9.1

A MIL-C-81 capacitor, variable, ceramic, has a failure rate of 0.072 per 10^6 hours at a 50% stress level, at 50°C (T=125°C Max Rated). Using the table in Figure 9.1; what is the failure rate at 60°C with the same stress level?

Solution

Following down the left column under T_A(°C) to 50, and moving right to the point where 0.5 is intersected, a failure rate of 0.072 per 10^6 hours is found. Across from 60 the failure rate is found to be to be 0.077 per 10^6 hours. (NOTE: A 10° increase in temperature resulted in a failure rate increase of only 7% — not near double.)

Example 9.2

Referring to Figure 9.1, holding the temperature constant at 50°C, what would the failure rate be at a stress level of 30%?

Base Failure Rate - λ_b
(T = 125°C Max Rated)
(MIL-C-81 Styles CV 35, 36)

T_A (°C)	Stress .1	.3	.5	.7	.9
0	.0028	.015	.061	.16	.35
10	.0028	.015	.062	.17	.35
20	.0029	.016	.064	.17	.36
30	.0030	.016	.066	.18	.37
40	.0031	.017	.068	.18	.39
50	.0033	.018	.072	.19	.41
60	.0035	.019	.077	.21	.44
70	.0038	.021	.084	.23	.48
80	.0043	.023	.095	.25	.54
90	.0050	.027	.11	.30	.63
100	.0062	.033	.14	.36	.76
110	.0079	.043	.17	.47	.98
120	.011	.059	.24	.64	1.4

$$\lambda_b = .00224\left[\left(\frac{S}{.17}\right)^3 + 1\right]\exp\left(1.59\left(\frac{T+273}{398}\right)^{10.1}\right)$$

T = Ambient Temperature (°C)
S = Ratio of Operating to Rated Voltage

Operating voltage is the sum of applied D.C. voltage and peak A.C. voltage.

Figure 9.1. Specification tables (from MIL-HDBK-217).

Solution

Following down the left column under (T_A(°C)) to 50, and moving right to the point where 0.3 is intersected, a failure rate to be 0.018 per 10^6 hours is found. (NOTE: The failure rate has decreased by 400% by lowering the stress level by 20%. Lowering the temperature by 10°C (17%) resulted in only 7% decrease in failure rate.)

9.4.1 Application method

Parts application criteria should be developed as part of the equipment designer standards and should be adhered to because of fact that dramatically increasing failure rates can decrease reliability with exposure to increased stress levels. MIL-P-11268 and MIL-E-5400 may be referenced for establishing the parts application criteria.

Deviations to the parts application criteria should be granted only after evaluating the actual part stress conditions. This practice is to assure that the rated value of the part used is equal to, or greater than, the expected stress. Design alternatives, trade-offs, impact on circuit and overall system reliability may then be assessed.

9.4.2 Technique based on part failure rates

Reliability predictions should be accomplished at the lowest equipment level that the preliminary design and configuration analyses permit. While electronic part failure rate prediction techniques are readily available, techniques for predicting failure rates for mechanical and electromechanical devices, are not so readily available. Therefore dialogue between reliability and design engineers is important to ensure that mission and environmental impacts on device performance are accounted. In either the electronic or the non-electronic cases, the part or component failure rates are the basic building blocks for accomplishing higher level predictions.

Equipment level predictions using part failure rates:

1. Provide a basis for identifying design, part selection, application and environmental problem areas.
2. Provide early indication of capability to meet the reliability test requirements.
3. Are essential inputs to system and subsystem level predictions.

Reliability predictions at any equipment level become inputs to higher level predictions. The better the predictions, in terms of reduced uncertainty, the more justifiable are the resulting reliability and design decisions, whether the decision is to maintain the status quo or to take action to improve hardware reliability.

9.5 Summary

The fundamental reason for early predictions based on parts failure rates is to precipitate appropriate action during development, when it is most tolerable from a program standpoint and most effective from the basic reliability and mission reliability viewpoints. Early review of reliability predictions at lowest equipment levels can identify parts or components that have inadequate margins between parts strength and expected applied stresses. In addition, the earlier the review is performed, generally the greater is the range of acceptable options for improving the predictions and the equipment through derating techniques.

Whenever predictions fall short of allocated reliability requirements, alternatives such as the following should be considered:

- identify suitable substitutes,
- re-allocate reliability,
- improve selected parts and components, and
- modify to decrease severity of environments (apply derating techniques).

Some alternatives are more feasible and acceptable than others at given points in development, but all are easier and less expensive to accomplish earlier that later.

9.6 Exercises

1. Describe the benefits of part standardization.

2. Explain some principal functions of a parts control, selection, and integration program.

3. Using the table below, what stress level would cause the greatest failure rate increase if the temperature was raised from 70° to 100°?

   a. .1

   b. .3

c. .5

d. .7

e. .9

T_A(°C)	Stress				
	.1	.3	.5	.7	.9
0	.0028	.015	.061	.16	.35
10	.0028	.015	.061	.16	.35
20	.0029	.016	.064	.17	.35
30	.0030	.016	.066	.18	.37
40	.0031	.017	.068	.18	.39
50	.0033	.018	.072	.19	.141
60	.0035	.019	.077	.21	.44
70	.0038	.021	.084	.23	.48
80	.0043	.023	.095	.25	.54
90	.0050	.027	.11	.30	.63
100	.0062	.033	.14	.36	.76
110	.0079	.043	.17	.47	.98
120	.011	.059	.24	.64	1.4

Extract from MIL-HDBK-217

4. Using the table from Exercise 3, which of the following would have the most positive effect on the item's failure rates?
   a. At 100°C, 50% stress; lower temp to 50°
   b. At 120°C, 50% stress; lower temp to 50°
   c. At 100°C, 50% stress; lower stress to 30%
   d. At 120°C, 50% stress; lower stress to 30%

5. The basic objective of a parts program is to
   a. control the selection and use of standard and nonstandard parts
   b. assure that parts are available when needed
   c. control and maintain cost effective inventories
   d. eliminate unreliable parts

6 A reliability test program should
   a. disclose deficiencies in item design, materiel and workmanship
   b. provide measured reliability data as input for estimates of operational readiness, mission success, maintenance manpower cost, and logistics support cost
   c. determine compliance with quantitative reliability requirements

d. all of the above

7. The Reliability Qualification Test (RQT) is designed primarily to
   a. disclose weak parts
   b. simulate in-service evaluations of the delivered item
   c. assure that minimum acceptable reliability requirements have been met
   d. disclose design deficiencies under simulated conditions

8. Which of the following are considered reliability engineering tests?
   a. Reliability Qualification Test
   b. Production Reliability Acceptance Test
   c. Reliability Development/Growth Test
   d. all of the above
   e. a and c above

9. Applying stresses to an item below the designed strength will
   a. increase reliability
   b. increase the failure rate
   c. decrease the failure rate
   d. a and c above

10. The effects of failure rates on increased temperatures is called the
   a. Thermal Model
   b. Full Scale Model
   c. Arrhenius Model
   d. Derating effects

11. Which of the following has the greatest impact on failure rates?
   a. stress/strength correlation
   b. stress/temperature correlation
   c. actual/simulated correlation
   d. standard/nonstandard parts

12. A test that is primarily done to disclose weak parts is the
   a. Reliability Development/Growth Test
   b. Reliability Qualification Test
   c. Production Reliability Acceptance Test

d. Environmental Stress Screening Test

9.7 References

Rome Air Development Center (RADC). 1985. ***RADC-TR-85-194: RADC Non-electronic Reliability Notebook.*** Document #AD A163900. Springfield, VA: National Technical Information Service.

US Department of Defense. ***MIL-E-5400: General Specification for Electronic Equipment, Airborne.*** Philadelphia: Naval Publications and Forms Center.

US Department of Defense. ***MIL-HDBK-217: Reliability Prediction of Electronic Equipment.*** Philadelphia: Naval Publications and Forms Center.

US Department of Defense. ***MIL-P-11268: Parts, Materials, and Processes Used in Electronic Equipment.*** Philadelphia: Naval Publications and Forms Center.

US Department of Defense. ***MIL-STD-781: Reliability Testing for Engineering Development, Qualification, and Production.*** Philadelphia: Naval Publications and Forms Center.

US Department of Defense. ***MIL-STD-785: Reliability Program for Systems and Equipment Development and Production.*** Philadelphia: Naval Publications and Forms Center.

US Department of Defense. ***MIL-STD-810: Environmental Test Methods and Engineering Guidelines.*** Philadelphia: Naval Publications and Forms Center.

US Department of Defense. ***MIL-STD-965: Parts Control Program.*** Philadelphia: Naval Publications and Forms Center.

♦♦♦

CHAPTER

10

Reliability Design and Management Control

A major problem that confronts all organizations responsible for a reliability program is the selection of tasks that can materially aid in attaining program reliability requirements. Today's schedule and funding constraints mandate a cost-effective selection based on identified program needs. The considerations presented herein are intended to provide the rationale for this selection. They are also intended to jog the memory for "lessons learned" to provoke questions that must be answered and to encourage dialogue with other engineers, operations and support personnel so that answers to questions and solutions to problems may be found.

10.1 Reliability program management principles

Table 10.1 provides general guidance in summary form, of "when and what" to include in a proposal to establish an acceptable and cost effective reliability program. This table can be used to initially identify the elements that are typically included in an effective reliability program for the particular acquisition phase involved. The problem of prioritizing or establishing a base line group from all the tasks in this table cannot be solved unless variables like system complexity, program phase, availability of funds, schedule, etc., are known. The reliability program plan (Task 101) should always be considered for determining the need for this task. However, individual tasks may be cited without requiring a reliability program plan. Sections 10.1.1 through 10.1.10

contain summaries for each task listed in Table 10.1. These summaries may be used to determine which tasks are appropriate to include in the program.

Table 10.1 Application matrix.

TASK	TITLE	TASK TYPE	PROGRAM PHASE CONCEPT	VALID	FSED	PROD
101	RELIABILITY PROGRAM PLAN	MGT	S	S	G	G
102	MONITOR/CONTROL OF SUB-CONTRACTORS AND SUPPLIERS	MGT	S	S	G	G
103	PROGRAM REVIEWS	MGT	S	S	G	G
104	FAILURE REPORTING, ANALYSIS, AND CORRECTIVE ACTION SYSTEM (FRACAS)	ENG	NA	S	G	G
105	FAILURE REVIEW BOARD (FRB)	MGT	NA	S	G	G
201	RELIABILITY MODELING	ENG	S	S	G	GC
202	RELIABILITY ALLOCATIONS	ACC	S	S	G	GC
203	RELIABILITY PREDICTIONS	ACC	S	S	G	GC
204	FAILURE MODES, EFFECTS, AND CRITICALITY ANALYSIS (FMECA)	ENG	S	S	G	GC
205	SNEAK CIRCUIT ANALYSIS (SCA)	ENG	NA	NA	G	GC
206	ELECTRONIC PARTS/CIRCUITS TOLERANCE ANALYSIS	ENG	NA	NA	G	GC
207	PARTS PROGRAM	ENG	S	S	G	G
208	RELIABILITY CRITICAL ITEMS	MGT	S	S	G	G
209	EFFECTS OF FUNCTIONAL TESTING, STORAGE, HANDLING, PACKAGING, TRANSPORTATION, AND MAINTENANCE	ENG	NA	S	G	GC
301	ENVIRONMENTAL STRESS SCREENING (ESS)	ENG	NA	S	G	G

Continued on next page...

Table 10.1—*continued...*

302	RELIABILITY DEVELOPMENT /GROWTH TESTING	ENG	NA	S	G	NA
303	RELIABILITY QUALIFICATION TEST PROGRAM (RGT)	ACC	NA	S	G	G
304	PRODUCTION RELIABILITY ACCEPTANCE TEST PROGRAM (PRAT)	ACC	NA	NA	S	G

CODE DEFINITIONS

TASK TYPE:	PROGRAM PHASE:
ACC—RELIABILITY ACCOUNTING	S—SELECTIVELY APPLICABLE
ENG—RELIABILITY ENGINEERING	G—GENERALLY APPLICABLE
MGT—MANAGEMENT	GC—GENERALLY APPLICABLE TO DESIGN CHANGES ONLY

10.1.1 Task 101

The reliability program plan is a tool to assist in managing an effective reliability program and to evaluate the approach to, understanding of, and execution of reliability tasks. It ensures that procedures for implementing and controlling reliability tasks are adequate, and structured to ensure that appropriate attention will be focused on reliability activities.

The plan should include, but not be limited to, the following:

1. A description of how the program will be conducted, how each task will be performed and procedures to evaluate status and control of each task (Task 101).
2. The description of interrelationships of reliability tasks and how they will interface with other system-oriented tasks. This description includes schedules with estimated start and completion points for each task (Task 104, 204, 208, etc.).
3. The identification of known reliability problems and proposed solutions or the proposed plan to solve the problems (Task 105).
4. The designation of reliability milestones (design and test—Task 201, 202, 203, etc.).
5. Identification of key personnel for managing the reliability program. Include a description of the management structure, the interrelationship between line, service, staff etc. (Task 101).
6. A statement of what source of design guidelines or design review checklist will be utilized. Include a description of how reliability contributes to the total design,

and the level of authority and constraints on this engineering discipline (Task 103).

7. Identification of inputs that the contractor needs. Inputs should include measured basic reliability and mission reliability values, measured environmental stresses, typical failure modes, and critical failure modes (Task 102).

10.1.2 Task 102

Continual visibility of subcontractor activities is essential so that timely and appropriate management action can be taken as the need arises. It is prudent to include contractual provisions that permit the procuring activity to participate, at its discretion, in appropriate formal prime or subcontractor meetings. Information gained at these meetings can provide a basis for follow-up action necessary to maintain adequate visibility of subcontractors' progress.

Active participation in the closed-loop Failure Analysis, Reporting and Corrective Action System (FRACAS) should be required of all equipment contractors. The information about unplanned events which this system can provide is a major factor in assessing and maintaining reliability program effectiveness. It is reasonable to assume that equipment failures will occur during service evaluation testing. During this testing it is very important to determine as rapidly as possible the cause of such failures, the need for corrective action, and the specific action to be taken. For this reason support of these tests is advisable, and should be considered for inclusion in program requirements.

10.1.3 Task 103

Design reviews are an important management and technical tool for the reliability organization. Typically, reviews are held to evaluate the progress, consistency and technical adequacy of a selected design and test approach. Formal review and assessment of reliability requirements should be conducted at major program points, identified as system program reviews. The reviews should be held at least quarterly, although the time can be extended as the program progresses. Representative discussion items include all reliability analyses, failure analysis, test schedules and progress, problems related to vendor and contractor reliability programs, parts and design problems. As the program develops, reliability progress is assessed by the use of additional program reviews as necessary.

These reviews identify and discuss all pertinent aspects of the reliability program. Design reviews can be classified into the following four categories:

Preliminary Design Review—Conducted prior to initial design. According to Dhillon (1985), the purpose of this type of design review is to examine each requirement for such things as completeness, validity, and accuracy. An integral part of this review is the updating of the reliability status, which includes:

a. Reliability modeling
b. Reliability apportionment
c. Reliability predictions
d. FMECA
e. Reliability content of specification
f. Design guideline criteria
g. Other tasks as identified

Test Readiness Review—Conducted primarily to assure appropriate test methods are in place for establishing reliability prediction. Discussions here include, but are not limited to, the following:

a. Reliability analyses status, primarily prediction
b. Test schedules, profiles, plans, failure definitions, and reporting format
c. Failure reporting and corrective action implementation

Critical Design Review—Conducted during the final design stages. During this review a considerable amount of data should be available for the design team (i.e. test results, reports from earlier reviews and other predictive information). Topics for discussion during this review should include, but are not limited to:

a. Reliability content of specifications
b. Reliability prediction and analyses
c. Parts program status
d. Reliability critical items program
e. Other problems affecting reliability
f. Failure Mode, Effects and Criticality Analysis
g. Identification of circuit reference designators whose stress levels exceed the recommended parts application criteria

Production Readiness Review—Conducted to determine if the product is production ready. Discussions will include:

a. Results of the applicable reliability qualification tests
b. Results of the applicable reliability growth testing

In addition to more detailed coverage of those items discussed at the preliminary and critical reviews, formal program reviews should address progress on all reliability tasks specified. All reviews provide an opportunity to review and assign action items and to explore other areas of concern. A mutually acceptable agenda should be generated to ensure that all reliability open items are covered and that all participants are prepared for meaningful discussions.

10.1.4 Tasks 104–105

The purpose of Task 104 is to establish a closed loop failure reporting system, procedures for analysis of failures to determine cause, and documentation for recording corrective action taken. The purpose of Task 105 is to review failure trends and corrective action status, and to assure that adequate corrective actions are taken in a timely manner and recorded during the development and production phases of the program.

10.1.5 Tasks 201–202–203

The purpose of Task 201 is to develop a reliability model for making numerical apportionments and estimates to evaluate system equipment reliability. The purpose of Task 202 is to assure that once quantitative system requirements have been determined, they are allocated or apportioned to lower levels. The purpose of Task 203 is to estimate the basic reliability and mission reliability of the system equipment and to make a determination of whether these reliability requirements can be achieved with the proposed design.

10.1.6 Tasks 204–205–206

The purpose of Task 204 is to identify potential design weaknesses through systematic, documented consideration of all likely ways in which a component or equipment can fail, and causes for each failure mode and the effects of each failure. The purpose of Task 205 is to identify latent paths which cause occurrence of unwanted functions or inhibit desired functions, assuming all components are functioning properly. The purpose of Task 206 is to examine the effects of parts electrical tolerances and parasitic parameters over the range of specified operating temperatures.

10.1.7 Tasks 207–208–209

The purpose of Task 207 is to control the selection and use of standard and nonstandard parts. The purpose of Task 208 is to identify and control those items which require "special attention" because of complexity, application of advanced state-of-the-art techniques, and the impact of potential failure on safety, readiness, mission success, and demand for maintenance support. The purpose of Task 209 is to determine the effects of storage, handling, packaging, transportation, maintenance, and repeated exposure to functional testing on hardware reliability. Details of these tasks are found in Chapter 8.

10.1.8 Tasks 301

The purpose of Task 301 is to establish and implement environmental stress screening procedures so that early failures due to weak parts, workmanship defects, and other non-conformance anomalies can be identified and removed from the equipment.

10.1.9 Task 302

The purpose of Task 302 is to conduct pre-qualification testing to provide a basis for resolving the majority of reliability problems early in the development phase, and for incorporating corrective action to preclude recurrence, prior to the start of production.

10.1.10 Tasks 303–304

The purpose of Task 303 is to determine that the specified reliability requirements have been achieved. The purpose of Task 304 is to assure that the reliability of the hardware is not degraded as the result of changes in tooling, processes, work flow, design, parts quality, or other characteristics identified by procurement.

10.2 Design review teams

There are two key people on a typical design review team: the design engineer, responsible for the preparation of the design; and the reliability engineer, responsible for assuring the specified reliability requirements. Other members, dependent upon the complexity of the design, should include representatives from procurement, manufacturing, production, sales, etc. The purpose of the design review team is to increase reliability and improve design characteristics with minimum costs. There are two criteria that must be met for a successful design review team to be effective: it must be multi-disciplined and authoritative. If a review team is wanted only for advisory purposes, then hiring consultants would be the best alternative. The main thing to remember is that the design review should not be attempted by only one or two people and must be multi-disciplined as well as authoritative to be effective.

10.3 Follow-up and corrective-action

From time to time during the program, transitions from one phase to another have to be made. In addition, there may be other occasions within phases when changes in the program are required. As transition or other change points approach, those responsible for monitoring and achieving system reliability must evaluate the needs for the reliability program and determine its structure if it is needed. A checklist is sometimes used for monitoring the system to assure that all bases are covered throughout the reliability program. Table 10.2 provides a general checklist example that may be used as a guide for evaluating a reliability program. This checklist includes categories for reliability modeling and allocation, which are sometimes delivered as part of a prediction report. O'Connor (1989), offers a word of caution when using checklists: "Relying upon the use of checklists may result in little preparatory work and can become a substitute for thinking." According to O'Connor, there are three golden rules for the use of checklists:

1. Don't use them during the formal review meetings, only in the design office.
2. Avoid vague questions, such as, "Are the grease points accessible?" "What access is provided for lubrication?" is a better question. Try to ask for a more detailed response—avoid "Yes" or "No" answers.
3. Make sure they are pertinent to the design and up to date.

While most tasks can be performed at varying levels of detail, it is important to ensure that only essential tasks are specified, to avoid wasting resources. What has, or what will have been, achieved should be critically appraised at given milestones. For example, as the transition between Full Scale Engineering Development Phase (FSED[4]) and Production Phase (PROD[5]) approaches, judgments regarding reliability tasks during production must be made. In some instances, only minimal testing will be required, while in other instances, a substantial number of FSED tasks will need to be continued. Other cases may call for a

[4] Full-scale engineering development phase: The period when the system and the principal items necessary for its support are designed, fabricated, tested and evaluated.

[5] Production phase: The period from production approval until the last system is delivered and accepted.

Table 10.2 Reliability Analysis Checklist

Major Concerns	Comments
MODELS	
Are all functional elements included in the reliability block model?	System design drawings must be reviewed to be sure that the reliability model agrees with the hardware.
Are all modes of operation considered in the math model?	Duty cycles, alternate paths, degraded conditions and redundant units must be defined and modeled.
Do the math model results show that the design achieves the reliability requirement?	Unit failure rates and redundancy equations are used from the detailed part predictions in the system math model (See MIL-STD-756, Reliability Prediction and Modeling).
ALLOCATION	
Are system reliability requirements allocated (subdivided) to useful levels?	Useful levels are defined as: equipment for subcontractors, assemblies for sub-subcontractors, circuit boards for designers.
Does the allocation process consider complexity, design flexibility, and safety margins?	Conservative values are needed to prevent reallocation at every design change.
PREDICTION	
Does the sum of the parts equal the value of the module or unit?	Many predictions neglect to include all the parts producing optimistic results (check for solder connections, connectors, circuit boards).
Are environmental conditions and part quality representative of the requirements?	Optimistic quality levels and favorable environmental conditions are often assumed causing optimistic results.
Are the circuit and part temperatures defined and do they represent the design?	Temperature is the biggest driver of part failure rates; low temperature assumptions will cause optimistic results.
Are equipment, assembly, subassembly and part reliability drivers identified?	Identification is needed so that corrective actions for reliability Improvement can be considered.
Are alternate (Non-MIL-HDBK-217) failure rates highlighted along with the rationale for their use?	Use of alternate failure rates, if deemed necessary, require submission of backup data to provide credence in the values.
Is the level of detail for the part failure rate models sufficient to reconstruct the result?	Each component type should be sampled and failure rates completely reconstructed for accuracy.
Are critical components such as VHSIC, Monolithic Microwave Integrated Circuits (MMIC), Application Specific Integrated Circuits (ASIC) or Hybrids highlighted?	Prediction methods for advanced technology parts should be carefully evaluated for impact on the module and system.

reliability growth program or a phase-unique task such as Production Reliability Acceptance Test (PRAT[6]). The FSED–PROD transition point was chosen for illustrative purposes; similar reasoning applies whenever program change points occur or are anticipated. The purpose here is to emphasize that the reliability manager should assess and project accomplishments, determine what still needs to be accomplished to achieve reliability requirements, and then tailor a program to meet those requirements. "Tailoring" should not be interpreted as a license to specify a zero reliability program. Necessity and sufficiency are the key criteria to be set in determining whether tasks are tailored into, or excluded from, a reliability program.

10.4 Reliability needs

The elements of a reliability program must be selected to meet reliability needs. Identifying and quantifying these needs should be accomplished prior to release of a proposal for the appropriate acquisition phase so that tasks and requirements commensurate with the needs may be included. The tasks and requirements which are included establish the framework for the continuing reliability dialogue between procurement and those who will ultimately be selected to develop the hardware. It is essential to make appropriate analyses and exercise mature judgment in determining reliability needs.

In making this determination, it is necessary to assemble program data concerning mission and performance requirements (preferably at the sub-system level), anticipated environments, and mission reliability and basic reliability requirements. This information is initially gathered in the conceptual phase and refined throughout development. It is the base upon which the reliability needs are determined and adjusted. The initial life profile definition defines, as a minimum, the boundaries of the performance envelope and provides the timeline, both environmental conditions and induced stresses versus time, typical of operations within that envelope. The quantitative requirements, basic reliability and mission reliability, are then determined for the defined life profile.

Using the information on equipment contemplated to provide the required performance, a separate apportionment or allocation of basic reliability and mission reliability can be made to the equipment level. This apportionment is usually based on available reliability data modified to reflect changes in performance requirements, duty

[6] Production Reliability Acceptance Test: A task assuring that the reliability of the hardware is not degraded as the result of changes in tooling, processes, design, parts quality, or other characteristics.

cycles, and anticipated environments. If the hardware to be procured is either a subsystem or equipment, the allocations would apply down to the lowest assembly level in terms of Mean-Time-Between-Maintenance-Action (MTBMA), or Mean-Time-Between-Failure, (MTBF) or failure rate. The required modifications are largely a matter of judgment, particularly when a new or considerably modified equipment concept must be synthesized to provide a specified function.

A reliability estimate should be made for each item of equipment independent of, and reasonably soon after, completing the initial apportionment. The equipment estimates should be combined to provide an initial estimate of basic reliability and mission reliability. During the conceptual phases[7] and validation phases[8], design details will probably not be available. Therefore, estimates made during these phases and early in FSED will provide "ball park" numbers, which are nevertheless adequate for initial comparisons with, and for establishing the reasonableness of, the initial apportionment. Reapportionment based on a comparison with details of the estimate may be advisable at this time. The apportionment and the estimate procedures should be repeated until reasonable apportioned values are obtained. The apportionment should be frozen prior to awarding subcontracts that have firm reliability requirements.

Some reliability tasks should be accomplished for an entire system, e.g., development and use of a failure reporting system, periodic estimates of basic reliability and mission reliability. In most cases, needs are self evident while others, which must be selected, may apply only to subsystems. While experience plays a key role in the task selection, it should be supplemented by analysis and investigation.

A useful initial analytical procedure is to compare reliability estimates at the subsystem and equipment level, with the corresponding apportionments. If the estimate is less than the apportionment, the need for improvement is indicated. Where "considerable" improvement is required, the subsystem or equipment should be identified as "reliability critical." This identification should be done as early as possible in the program so as to impact the equipment through the proper selection of tasks.

Reasons for the disparity between the apportioned and the estimated values of the reliability critical items should be investigated. Discussions of these reasons and tentative ways to attain the apportioned values, (e.g., relaxed performance requirements, either more or less design redundancy, additional environmental protection), should be held with appropriate project personnel. The object of the investigations and discussions

---

[7] Conceptual phase: The identification and exploration of alternative solutions or solution concepts to satisfy a validated need.

[8] Validation phase: The period when selected candidate solutions are refined through extensive study and analyses; hardware development, testing and evaluations.

is viable recommendations for action to overcome the deficiencies. A significant benefit that can be gained from this process is a consensus on the specific equipment that is considered reliability critical. When systems or equipment performance requirements create a wide and irreconcilable disparity between apportioned and estimated values of required, reliability should be challenged. Elimination of less essential equipment functions can reduce equipment complexity and significantly enhance reliability.

Tasks and requirements can be prioritized and a "rough order of magnitude" estimate made of the time and effort required to complete each task once recommendations for task applications have been determined. This information will be of considerable value in selecting the tasks that can be accomplished within schedule and funding constraints.

10.5 Managing the program

When the technical tasks required to achieve the reliability requirements have been defined, the resources required must be identified and committed to meet objectives efficiently. The task elements should be staffed and time-phased to ensure that reliability objectives are not arbitrarily compromised to meet other program cost or schedule constraints.

The reliability program's success requires that top management be continually informed of program status and unresolved problems that could impact the achievement of program milestones, so that direction and resources can be reoriented as required. In general, the reliability organization should have:

1. A shared responsibility with other disciplines in its engineering department
2. Technical control of reliability disciplines
3. Fiscal control of reliability resources

Management can properly influence the reliability program by placing on the statement of work or contract:

1. Numerical reliability requirements and testing requirements to ensure contractual compliance during development and production
2. The requirement to implement specific reliability tasks during conduct of the program

Working arrangements between reliability and other activities (e.g., maintainability, safety, survivability, vulnerability, testing, quality assurance) should be established to identify mutual interests, maximize benefits of mutually supporting tasks, and minimize effort overlap. Such organizational working relationships can also promote more system-oriented decisions if the work is properly integrated at higher levels.

When all necessary planning for a reliability program has been accomplished, it should be documented as a reliability program plan. A reliability program plan is normally submitted as part of the response to the procuring activity's request for proposals. After mutual agreement has been reached and procurement approval has been granted, the reliability program plan must be made a part of the contract. Since the plan is a contractual tool used to evaluate the reliability program, it should be kept current, subject to procurement approvals.

10.6 General considerations

Both quantitative and qualitative analyses are useful in determining where reliability resources should be applied. For example, modeling, apportionment and estimates of basic reliability and mission reliability can scope the improvement which may be required. A Failure Modes, Effects and Criticality Analysis (FMECA) based on available mission rules and system definition can be used to determine compliance with failure tolerance criteria and to identify single failure points which are critical to either mission success or safety, or both. These kinds of analyses identify improvements which must be made if requirements are to be met.

Reliability analyses are efficient work direction tools, because they can confirm system adequacy or identify the need for design change, providing they are accomplished in conjunction with, or reviewed by, other disciplines. The use of reliability analyses is not limited to the phase traditionally thought of as the design phase. Some of the analyses mentioned above are useful during the early acquisition phases when design criteria, mission requirements, and preliminary designs are being developed. Since the situation is generally fluid during these phases, and firm commitments for full-scale development have not yet been made, a comparison of the reliability benefits of competing configuration concepts may be more readily accepted for use in the decision making process. The ultimate reliability that can be attained by any system or item is limited by the configuration chosen. Therefore, the analyses should be selected to aid configuration definition in light of the existing design criteria and mission requirements. Preliminary reliability estimates and FMECA are generally the most appropriate for this purpose. The depth of the estimates should be sufficient for comparison of the configurations.

The cost of the selected analyses is obviously a function of both the level and breadth requested. For example, an FMECA at the part level for all equipment in a complex system is time consuming, and action taken to reduce reliability risk as a result of such an analysis will probably not be cost effective. However, when the analysis is conducted at a sub-system level, single point failures can be identified and the need for a more detailed analysis or design action can be determined. Similar considerations, which are largely subjective, should be used in tailoring other analyses to fit program needs. The cardinal principles are:

1. For basic reliability, do not analyze below the level at which a failure will cause a demand for maintenance, repair, or logistics support.
2. For mission reliability, do not analyze below the level necessary to identify mission critical failures.

10.7 Summary

Program funding and schedule constraints demand that the limited reliability resources available be used where they are most cost effective. It is also self-evident that major program level requirements and criteria have a more far reaching impact than those developed for lower levels. It is appropriate, therefore, to examine as early as possible the numerical reliability requirements, both basic reliability and mission reliability, mission rules, failure tolerance criteria, etc., so that analyses can be selected to show design compliance or to identify shortcomings. During this examination the numerical requirements and criteria should also be evaluated, and if slight changes to them can improve program cost effectiveness, such information should be presented to program management for appropriate action.

10.8 Exercises

1. Explain the purpose of design reviews.
2. When should formal reviews be conducted?
   a. When all tests have been completed
   b. After failure analysis has been completed
   c. At major program points
   d. Prior to initial design or the Concept phase
3. Explain the function of a failure review board.
4. In which phase of development do most elements of a reliability program plan become "generally applicable?"
   a. CONCEPT
   b. VALID
   c. FSED
   d. PROD
5. List at least six questions which should be asked during the design review.
6. Explain the purpose of the design review team.
7. The design review team may function effectively with two key people.
   a. True
   b. False
8. Explain the "downside" of using a checklist during formal reviews.

10.9 References

[Anonymous]. 1976. ***Engineering Design Handbook, Development Guide for Reliability; Part Two: Design for Reliability.*** AMCP 706–196. Chambersburg, PA: Letter County Army Depot.

Dhillon, B. S. 1985. ***Quality Control, Reliability, and Engineering Design.*** New York: Marcel Dekker.

O'Conner, P. D. T. 1993. ***Practical Reliability Engineering.*** New York: John Wiley & Sons.

US Department of Defense. ***MIL-HDBK-217: Reliability Prediction of Electronic Equipment.*** Philadelphia: Naval Publications and Forms Center.

US Department of Defense. MIL-STD-785: Reliability Program for Systems and Equipment Development and Production. Philadelphia: Naval Publications and Forms Center.

♦♦♦

CHAPTER

11

Product Safety

A system should be in place to assure that risks are eliminated or controlled within established parameters. Such a system requires that internal procedures be in place for the disposition of product related safety incidents that include potentially hazardous conditions not yet involved in an incident.

11.1 Contingency analysis

The first step in any contingency analysis is defining the program. A systematic approach should be used for the product safety program to assure that:

1. Safety is designed into the product in a timely, cost-effective manner
2. Historical safety data, including lessons learned from other products, are considered
3. Minimum risk is sought in accepting and using new technology, materials or new production and operational techniques
4. Retrofit actions required to improve safety is minimized through the timely inclusion of safety features during research and development
5. Changes in design, configuration or requirements are accomplished in a manner that maintains an acceptable level of risk
6. Consideration is given early in the life cycle to both safety and significant safety data documented as "lessons learned"
7. Hazards associated with each system are identified and evaluated then eliminated or reduced to an acceptable level

11.1.1 The product safety program

Military Standard-882, Task 102, describes the System Safety Program Plan in detail and is a recommended reference if development is based on this regulation. Task 102 states that the plan shall describe in detail tasks and activities of safety management and safety engineering required to identify, evaluate, and eliminate or control hazards, or reduce the associated risk to a level acceptable to management throughout the system life cycle. A product safety plan should provide a formal basis of understanding on how the safety program will be executed to meet safety requirements, including general and specific provisions. Sections 11.1.1.1 through 11.1.1.10 are summaries of MIL-STD-882. The System Safety Program Plan (SSPP), in Task 102, will be referred to as the Product Safety Program (PSP).

11.1.1.1 Scope and objectives

The PSP shall describe, as a minimum, four elements of an effective product safety program: a planned approach for task accomplishment, qualified people to accomplish tasks, authority to implement tasks through all levels of management, and appropriate commitment of resources (both personnel and funding) to assure tasks are completed. The PSP shall also define any safety requirements that may be imposed by a contract.

The scope and objectives will describe the interrelationships between system safety and other functional elements of the program and account for all contractually required safety tasks and responsibilities.

11.1.1.2 Organization

The PSP describes the organization of the total program including the organizational and functional relationships, and lines of communication. The organizational relationship between other functional elements having responsibility for safety impacts and safety management all are included in the program plan. The following are probably worth considering for inclusion:

1. Methods are described by which safety personnel may raise issues of concern
2. The organizational unit responsible for executing action needed to eradicate hazards must be identified along with who has authority for the resolution of all identified hazards
3. The qualifications of key safety personnel must be established
4. The process by which management decisions will be made including timely notification of unacceptable risk, necessary action, waivers, deviations, etc.

5. Details of how actions relative to product safety will be effected at the management level possessing resolution authority

11.1.1.3 Milestones

Define product safety program milestones that:

1. Relate to program responsibility and the required inputs and outputs
2. Provide a schedule of tasks including completion dates, reports and reviews
3. Identify subsystem, component, software safety activities as well as integrated system level activities
4. Provide estimates for people-power required to complete each task

11.1.1.4 Requirements and criteria

Establish the risk assessment procedures:

1. Prescribe hazard severity categories, hazard probability levels and system safety precedence
2. Identify any qualitative or quantitative measures to be used, including a description of the acceptable risk levels
3. Describe closed-loop procedures for actions to resolve unacceptable risk

11.1.1.5 Hazard analysis

Describe the analysis techniques and formats to be used in qualitative or quantitative analysis to identify hazards, causes and effects, hazard elimination or risk reduction. The depth to which each technique is used including control of hazards associated with materials must be specified.

11.1.1.6 Safety data

Describe how historical hazard, mishap, and safety lessons learned, data is to be processed. Identify deliverable data and describe procedures for accessibility.

11.1.1.7 Safety verification

Describe verification (test, analysis, inspection, etc.) requirements:

1. Make sure that safety is adequately demonstrated. Identify certification requirements for safety devices, software, etc.
2. Make sure that safety-related verification information gets to the proper authority for review and analysis

11.1.1.8 Audit program

Describe the intended procedures to assure the objectives and requirements of the safety program are being accomplished.

11.1.1.9 Training

Describe the safety training for engineering, technician, operating and maintenance personnel.

11.1.1.10 Interfaces

Describe and identify, in detail, the interface between product safety, product engineering, and all other support disciplines such as: maintainability, quality control, reliability, software development, human factors engineering, medical support (health hazard assessments, and any others).

11.1.2 Safety design requirements

Safety design requirements should be established according to their applicability to the overall design of the product system. Some general system safety design requirements are:

1. Eliminate hazards through design or by material substitution. When elimination is impossible, only those with the least risk should be used.
2. Isolate hazardous substances from other activities or incompatible materials.
3. Locate equipment so that access during operation minimizes personnel exposure to hazards (e.g., hazardous chemicals, high voltage, radiation, cutting edges or sharp points).
4. Minimize risk resulting from excessive environmental conditions (e.g., temperature, pressure, noise or toxicity).
5. Minimize risk created by human error in the operation of the system.
6. Minimize risk from hazards that cannot be eliminated by including interlocks, redundancy, protective clothing, devices or procedures.
7. Protect power sources and critical components by physical separation or shielding.
8. When alternate design approaches cannot eliminate the hazard, place warning devices and caution notes in operation and repair instructions, or provide distinctive markings on hazardous components, materials or facilities to ensure personnel and equipment protection.

11.1.3 Risk assessment

Steps should be taken to eliminate identified hazards or reduce the associated risk to an acceptable level. Decisions regarding the resolve of identified hazards are based on the risk assessment. To aid in achieving the product or system safety objectives, hazards are characterized as to hazard severity categories and hazard probability levels. Since the priority for product or system safety is eliminating hazards by design, a risk assessment procedure considering only hazard severity will generally suffice during the early design phase to minimize the risk. For hazards not eliminated during the early design phase, a risk assessment procedure based upon the hazard probability, hazard severity, as well as risk impact is used to establish priorities for corrective action and resolution of identified hazards.

11.1.4 Hazard severity

Hazard severity categories are defined to provide a qualitative measure of the worst credible mishap resulting from the following: personnel error; environmental conditions; design inadequacies; procedural deficiencies; or system, subsystem or component failure or malfunction as shown in Table 11.1

Table 11.1. Hazard severity categories.

Description	Category	Definition
CATASTROPHIC	I	Death, system loss or severe environmental damage.
CRITICAL	II	Severe injury, severe occupational illness, major system or environmental damage.
MARGINAL	III	Minor injury, minor occupational illness, minor system or environmental damage.
NEGLIGIBLE	IV	Less than minor injury, occupational illness, or less than minor system or environmental damage.

NOTE: These hazard severity categories provide guidance to a wide variety of programs. Adaptation to a particular program requires a mutual understanding as to the meaning of the terms used and what constitutes a failure, major or minor system or environmental damage, severe or minor injury and illness. Other criteria may be used provided it meets the overall program objectives.

11.1.5 Hazard probability

The probability that a hazard will be created during the planned life expectancy of the product or system can be described in potential occurrences per unit of time, events, population, items, or activity. Assigning a quantitative hazard probability to a potential design or procedural hazard is generally not possible early in the design process. A qualitative hazard probability may be derived from research, analysis, and evaluation of historical safety data from similar products or systems. An example of a qualitative hazard probability ranking is shown in Table 11.2.

Table 11.2. Hazard probability categories.

Description	Level	Specific Individual Item
FREQUENT	A	Likely to occur frequently
PROBABLE	B	Will occur several times in the life of an item*
OCCASIONAL	C	Likely to occur some time in the life of an item*
REMOTE	D	Unlikely but possible to occur in the life of an item*
IMPROBABLE	E	So unlikely, it can be assumed occurrence may not be experienced*

* Definitions of descriptive words may have to be modified based on complexity and size of the product or system.

11.1.6 Risk impact

The effects and costs of identified risk having the same hazard risk index are assessed for the risk impact. For example, a periodic release of carbon gases may not cause direct physical harm or equipment damage, but can cause extreme damage to the goodwill of the organization.

11.1.7 Management responsibility

The Product Safety Program Manager is responsible for establishing, planning, organizing, implementing and maintaining an effective safety program, as defined below:

1. Establish definitive product safety program requirements (11.1.1.4) for the procurement or development of the system or product. These requirements identify the specific risk levels that are considered acceptable for the system. Acceptable risk levels are defined in terms of a hazard risk index (11.2.1) that is developed through a hazard severity probability matrix (Tables 11.4 or 11.5).
2. Develop a plan that reflects in detail how the total program is to be conducted.
3. Organize key system safety personnel that meet minimum qualifications. Key system safety personnel are usually limited to the person who has authority for the supervisory or technical responsibilities of the system's safety. Some programs require that key system safety personnel possess special qualifications. These

special qualifications must be specified in the program safety requirements. A guide for appointing key system safety personnel is provided in Table 11.3.

Table 11.3. Minimum qualifications for key system safety personnel.

Program Complexity	Education	Experience	Certification
High	BS in Engineering, Physical Science or other*	Four years in system safety or related discipline	Desired: CSP** or Professional Engineer
Moderate	Bachelor's Degree plus training in system safety	Two years in system safety or related discipline	Enhancement: CSP or Professional Engineer
Low	High School Diploma plus training in system safety	Four years in system safety	None

* Management may specify other degrees or certification

** CSP—Certified Safety Professional

4. Implement the product safety program in a timely and efficient manner.
5. The program is maintained by monitoring the system safety activities to:
    a. Ensure adequate compliance with the safety requirements
    b. Evaluate new design criteria for inclusion into system standards
    c. Establish system safety groups as required to assist in developing and implementing the program
    d. Ensure that technical and historical safety data is available

Management assuring that an acceptably safe product or system will be developed may be summarized into a three-step process:

1. Prevent the initial creation of unnecessary hazards.
2. Tailor product or system safety activities to meet specific program needs*.
3. Manage residual hazards. This is done by understanding their nature and impact, and assuring they are properly dispositioned. For hazards that are to be "accepted," care should be taken to assure that this acceptance of risk occurs at the proper level of authority. Generally, the greater the risk, the higher the approval level needed for acceptance. The higher level of risks must be justified to the decision makers, not the Safety community.

* It is obvious that if step 1 is omitted, a more complex safety effort will be needed to address the greater number and variety of hazards that will populate the design.

11.2 Malfunction and multiple failure analysis

The overall goal of any product or system safety program is to design systems that do not contain hazards that can result in an unacceptable level of risk. The nature of most complex systems often makes it impossible or impractical to design them completely hazard-free. As various hazard analyses are performed, hazards will be identified that will require a resolution. The alternatives for eliminating the specific hazards or controlling the associated risks are evaluated so an acceptable method for risk reduction can be agreed upon. A common method for defining the hazards of a product is much like the FMEA process (Chapter 7). With this process applied to product safety, the hazards, effects, severalties and probabilities associated with the system are identified; and present control measures and recommendations for improvement are recorded. Table 11.4 is an example of a multiple hazard analysis worksheet.

11.2.1 Hazard risk assessment models

To determine what actions to take to eliminate or control identified hazards, a system of determining the level of risk involved must be developed. A good hazard risk assessment model will enable decision makers to properly understand the amount of risk involved relative to what it will cost in schedules and dollars to reduce the risk to an acceptable level. To eliminate or otherwise control as many hazards as possible, hazards must be prioritized for corrective actions. A categorization of hazards may be conducted according to risk level criteria. Categorization may be based on severity since not all hazards are of equal magnitude or criticality to personnel safety system success. In some cases, the anticipated consequences of hazardous events may be minimal, while in others, catastrophic. Hazard categorization may also involve the determination of the likelihood of the hazardous event actually occurring. This may be reported in non-numeric (qualitative) terms, such as frequent, occasional, or improbable; or in numeric (quantitative) terms such as once in 10,000 hours, 10^{-4}/hour. Prioritization may be accomplished either subjectively by qualitative analyses resulting in a comparative hazard risk assessment, or through quantification of the probability of occurrence resulting in a numeric priority factor for that hazardous condition. Tables 11.5 and 11.6 show matrices for hazard risk assessment that can be applied to provide qualitative priority factors for assigning corrective action. In the first matrix an identified hazard assigned a hazard risk index of 1A, 1B, 1C, 2A, 2B, or 3A might require immediate corrective action. A hazard risk index of 1D, 2C, 2D, 3B, or 3C would be tracked for possible corrective action. A hazard risk index of 1E, 2E, 3D, or 3E might have a lower priority for corrective action and may not warrant any tracking actions. In the second matrix, risk indices of 1 through 20 are assigned (1A=1, 1B=2, 1C=4, 2A=3, 3A=7, 4A=13, …, etc., with 1 being the highest risk). This matrix design assigns a different index to each frequency-category pair. This eliminates the situation caused by creating indices as products of numbers assigned to frequency and category which causes common results such as 2 x 6 = 3 x 4 = 4 x 3. This situation could hide information pertinent to prioritization. These are only examples of risk assessment methods and do not fit all programs.

Table 11.4. Hazard analysis worksheet.

System: Caustic Wash Station **Sub-system:** Caustic Tank			**Project Engineer:** Willie B. Safe **Date:** 4-22-93		
Hazard and Number	Hazard Effect	Hazard Severity	Hazard Probability	Present Control	Recommended Control
#1.—Platform slippery when wet	Personnel could fall into tank	I	E	Surface painted with anti-skid coating	Expanded grating plus anti-skid coating
#2.—Walkway on same level with caustic solution	Caustic solution could splash on personnel	II	B	Emergency shower on plat-form; protective clothing	Transparent guard attached to handrail around platform
#3.—Tank leaking	Environment concerns	II	D	NONE	Double wall container around tank
#4.—					

Table 11.5. First hazard risk assessment matrix.

	HAZARD CATEGORY			
FREQUENCY	(1) Catastrophic	(2) Critical	(3) Marginal	(4) Negligible
(A) Frequent ($x>10^{-1}$)*	1A	2A	3A	4A
(B) Probable ($10^{-1}>x>10^{-2}$)*	1B	2B	3B	4B
(C) Occasional ($10^{-2}>x>10^{-3}$)*	1C	2C	3C	4C
(D) Remote ($10^{-3}>x>10^{-6}$)*	1D	2D	3D	4D
(E) Improbable ($10^{-6}>x$)*	1E	2E	3E	4E

* Example of quantitative criteria:

Hazard Risk Index	Suggested Criteria
1A, 1B, 1C, 2A, 2B, 3A	Unacceptable
1D, 2C, 2D, 3B, 3C	Undesirable (management decision required)
1E, 2E, 3D, 3E, 4A, 4B	Acceptable (with management review)
4C, 4D, 4E	Acceptable (without review)

Table 11.6. Second hazard risk assessment matrix.

FREQUENCY	HAZARD CATEGORY			
	Catastrophic	Critical	Marginal	Negligible
(A) Frequent	1	3	7	13
(B) Probable	2	5	9	16
(C) Occasional	4	6	11	18
(D) Remote	8	10	14	19
(E) Improbable	12	15	17	20

Hazard Risk Index	Suggested Criteria
1–5	Unacceptable
6–9	Undesirable (management decision required)
10–17	Acceptable (with management review)
18–20	Acceptable (without review)

11.2.2 Hazard risk impact analysis

This is a means of further prioritizing hazards that may have the same risk hazard index, as well as other factors such as the effects on system operation and economics. An example is the use of a hazardous material that could contaminate the environment, or cause adverse health effects to the general public as well as the user. This material may deserve higher consideration for resolution than a hardware design that could cause a loss of a system through a mishap. Management should identify any hazard risk impacts to be considered when tailoring the system.

11.2.3 Residual risk

Risk associated with significant hazards for which there are no known control measures, no plans to control them, or incomplete control measures, are referred to as residual risk. Any residual risks should be documented as such, along with reasons for non-resolution.

11.2.4 Action on identified hazards

Catastrophic and critical hazards should be resolved through design changes or incorporation of safety devices. These specific hazards should not rely solely on warnings, cautions or procedures for control of the risk. If this is impossible or impractical, alternatives should be recommended. If management desires other hazards to be handled in this way, it must be considered early in the design stages for appropriate action. Actions should be based on process alternative recommendations that are part of the hazard analysis or hazard tracking techniques.

11.2.5 Software hazard risk analysis

The initial assessment of risk for software, and consequently software controlled or software intensive systems, cannot rely solely on the hazard severity and probability. Determination of the probability of failure of a single software function is difficult at best and cannot be based on historical data. Software is generally application specific, and reliability parameters associated with it cannot be estimated in the same manner as hardware. Therefore, another approach is recommended for the initial software risk assessment that considers the potential hazard severity and the degree of control that software exercises. The degree of control is defined using the following software control categories:

I. Software exercises autonomous control over potentially hazardous hardware systems, subsystems or components without the possibility of intervention to preclude the occurrence of a hazard. Failure of the software or a failure to prevent an event leads directly to a hazard's occurrence.

IIa. Software exercises control over potentially hazardous hardware systems, subsystems, or components allowing time for intervention by independent safety systems to mitigate the hazard. However, these systems by themselves are not considered adequate.

IIb. Software item displays information requiring immediate operator action to mitigate a hazard. Software failures will allow, or fail to prevent, the hazard's occurrence.

IIIa. Software item issues commands over potentially hazardous hardware systems, subsystems or components requiring human action to complete the control function. There are several, redundant, independent safety measures for each hazardous event.

IIIb. Software generates information of a safety critical nature used to make safety critical decisions. There are several, redundant, independent safety measures for each hazardous event.

IV. Software does not control safety critical hardware systems, subsystems or components and does not provide safety critical information.

11.2.6 Software hazard criticality matrix

The Software Hazard Criticality Matrix (Table 11.7) is similar to the Hazard Risk Assessment Matrix (Tables 11.5 and 11.6). This matrix is established using the hazard categories for the rows and the Software Control Categories for the columns. The matrix is completed by assigning Software Hazard Risk Index numbers to each element just as Hazard Risk Index (HRI) numbers are assigned in the Hazard Risk Assessment Matrix. A Software Hazard Risk Index (SHRI) of 1 from the matrix implies that the risk may be unacceptable. A SHRI from 2 to 4 is undesirable or requires acceptance from the managing activity. Unlike the hardware related HRI, a low index number does not mean that a design is unacceptable. Rather, it indicates that greater resources need to be applied to the analysis and testing of the software and its interaction with the system.

Table 11.7. Software hazard criticality matrix.

CONTROL CATEGORY	HAZARD CATEGORY			
	Catastrophic	Critical	Marginal	Negligible
I	1	1	3	5
II	1	2	4	5
III	2	3	5	5
IV	3	4	5	5

Hazard Risk Index Suggested Criteria

1 High Risk—Significant analysis and testing resources required

2 Medium Risk—Requirements and design analysis and in-depth testing required

3–4 Moderate Risk—High level analysis and testing acceptable with management review

5 Low Risk—Acceptable

11.3 Exercises

1. A major element for a product safety program is
   a. Identifying hazards
   b. Evaluating hazards
   c. Controlling hazards
   d. Eliminating hazards
   e. All the above are elements of a product safety program
2. Two major elements of a hazard analysis are
   a. Severity and probability
   b. Risk impact and control
   c. Qualitative and quantitative criteria
   d. Criticality and fault tree
3. A contingency analysis should be done to assure that
   a. Changes in design, configuration or requirements are accomplished in a manner that maintains an acceptable level of risk
   b. Consideration is given early in the life cycle to safety and significant safety data documented as "lessons learned"

c. Hazards associated with each system are identified and evaluated then eliminated or reduced to an acceptable level
d. All the above should be considered in the contingency analysis

4. One way to minimize the risk from a hazard that cannot be eliminated is
a. Hazards that cannot be eliminated should always be removed from the design
b. Initiate restrictions on use of the product
c. Include protective devices or procedures in the design
d. Write engineering product safety standards that reclassify the hazard

5. Management is chiefly responsible for the design of
a. The safety qualification system
b. The overall product safety program requirements
c. Hazard identification matrix
d. Hazard risk assessment models

6. Hazards identified as having the same risk index are assessed for
a. Their effect on the environment
b. Their cost
c. Their overall effects
d. Their acceptable risk levels

7. The highest number in a hazard severity category normally represents
a. Something negligible
b. Something of marginal concern
c. Something serious
d. Something catastrophic

8. Hazard probability level "A" assigned to a hazard normally means
a. The hazard is possible
b. The hazard is unlikely to occur
c. The hazard is very likely to occur
d. The hazard will occur sometime
e. The hazard will occur several times

9. The main objective of the hazard analysis during design is to
a. Meet the expectations of safety management
b. Prevent the initial creation of unnecessary hazards
c. Assure that adequate warning or protective devices are in place
d. Demonstrate the ability to design a safe product

10. In the hazard risk assessment matrix below, which category-frequency combinations would be the most undesirable?

FREQUENCY	HAZARD CATEGORY (1) Catastrophic	(2) Critical	(3) Marginal	(4) Negligible
(A) Frequent ($x>10^{-1}$)*	1A	2A	3A	4A
(B) Probable ($10^{-1}>x>10^{-2}$)*	1B	2B	3B	4B
(C) Occasional ($10^{-2}>x>10^{-3}$)*	1C	2C	3C	4C
(D) Remote ($10^{-3}>x>10^{-6}$)*	1D	2D	3D	4D
(E) Improbable ($10^{-6}>x$)*	1E	2E	3E	4E

a. 1A, 3B, 1E

b. 1B, 1C, 4D

c. 3A, 3C, 2E

d. 2C, 3A, 1D

11. From the hazard analysis worksheet below, the most likely order of priorities are:

System: Caustic Wash Station **Project Engineer:** Willie B. Safe
Sub-system: Caustic Tank **Date:** 4-22-93

Hazard and Number	Hazard Effect	Hazard Severity	Hazard Probability	Present Control	Recommended Control
#1.—Platform slippery when wet	Personnel could fall into tank	I	E	Surface painted with anti-skid coating	Expanded grating plus anti-skid coating
#2.—Walkway on same level with caustic solution	Caustic solution could splash on personnel	II	B	Emergency shower on platform; protective clothing	Transparent guard attached to handrail around platform
#3.—Tank leaking	Environment concerns	II	D	NONE	Double wall container around tank
#4.—					

a. 3, 2, 1

b. 2, 3, 1

c. 1, 2, 3

d. 3, 1, 2

12. What is meant by the term "residual risk?"

a. Risks associated with hazards left inherent in design

b. The small risks remaining as a result of removing larger risks

c. Risks associated with hazards for which there are no known controls

d. All of the above

11.4 References

Hammer, Willie. 1980. ***Product Safety Management and Engineering.*** Englewood Cliffs, NJ: Prentice-Hall.

Rome Air Development Center (RADC). 1983. ***The Evaluation and Practical Applications of Failure Modes and Effects Analysis.*** Document #ADA131-358. Springfield, VA: National Technical Information Service.

US Department of Defense. 1977. ***MIL-STD-882: System Safety Program Requirements.*** Philadelphia: Naval Publications and Forms Center.

US Department of Defense. 1980. MIL-STD-1629: Procedures for Performing a Failure Mode, Effects and Criticality Analysis. Philadelphia: Naval Publications and Forms Center.

♦♦♦

CHAPTER

12

Human Factors In Reliability

Human factors engineering should be included in every design that requires human interaction, to achieve the greatest potential for reliability. To increase the human reliability factor, steps should be taken early in the design phase to minimize the probability of error during human interventions. While error in product design may be the "root cause" of a product failure, error in human factors engineering may be the "root cause" of human failure. Human reliability may also be affected by, design characteristics, operational procedures, environments, communications, and standard operational procedures.

This chapter presents human engineering design criteria, principles and practices to be applied in the design of systems, equipment and facilities so as to:

1. Achieve required performance by operator, control and maintenance personnel.
2. Minimize skill and personnel requirements and training time.
3. Achieve required reliability of personnel-equipment combinations.
4. Foster design standardization within and among systems.

Another important aspect of human factors and reliability is training. This chapter presents a comprehensive training program.

12.1 Human factors in design and design principles

Human factors in the reliability engineering process may be defined as: analyzing the desired performance, configuring the physical and mental tasks, and assessing the likelihood for errors that could effect system reliability. Including human factors in design will increase reliability by improving the intractability of humans and components. The objectives of **human factors engineering**, as described by Huchingson (1981), are to improve comfort and performance and to reduce lost time and training cost. McCormick (1957) defines **human engineering** as the adaptation of human tasks to work environments and the working environment to human attributes such as mental, perceptual, sensory, and physical. Edel (1967), states that this adaptation for human use is applicable to various functions, for example, work method development, equipment design, and consumer products.

Human engineering design criteria is defined as, "the summation of available knowledge which defines the nature and limits of human capabilities as they relate to the checkout, operation, maintenance or control of systems or equipment, and which may be applied during engineering design to achieve optimum compatibility between equipment and human performance." It is essential for the designer to understand capabilities such as; sight, touch, audio, thermal tolerance and vibration if human factors are being considered. The remainder of this section provides brief outlines of these human capabilities.

12.1.1 Sight capabilities

The human eye viewing angle and color perception is limited to the position of the head. The ability to see decreases as the viewing angle increases. For example, when a person looks through the normal line of sight (see Figure 12.1), the viewing angle as well as color perception is optimized. For critical functions, indicators should be located within (15°) of the operator's normal line of sight without requiring the operator to assume an uncomfortable, awkward or unsafe position.

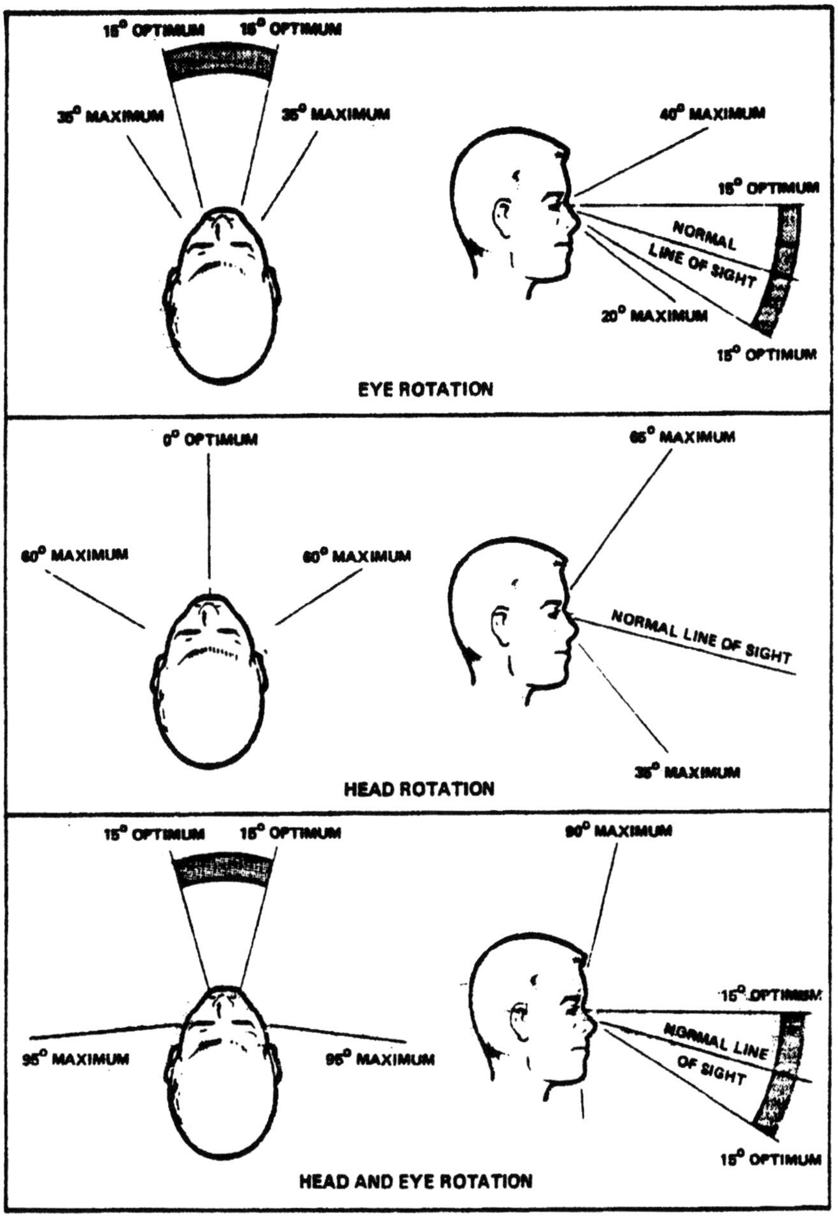

FIGURE 2. VERTICAL AND HORIZONTAL VISUAL FIELD

Figure 12.1. Vertical and horizontal visual field

12.1.2. Touch capabilities

The human capability to interpret visual stimuli by the sense of touch is a natural to be utilized by the designer. For example, control knob shapes can be distinguished by touching, therefore the designer is able to take advantage of this sense to relieve the load on the eyes, thereby reducing eyestrain and fatigue.

The use of coding for a particular application is governed by the relative advantages and disadvantages of each type of coding. Where coding is used to differentiate among controls, application of the codes are uniform throughout the system (see Table 12.1 for advantages and disadvantages). Coding guidelines are:

Location-coding. Controls associated with similar functions should be in the same location relative to the operator from workstation to workstation and from panel to panel.

Size-coding. No more than three different sizes of controls shall be used in coding controls for discrimination by absolute size. Controls used for performing the same function on different items of equipment shall be the same size. When knob diameter is used as the coding parameter, differences between diameters shall not be less than

13 mm (0.5"). When knob thickness is the coding parameter, differences between thickness shall not be less than 10 mm (0.4").

Shape-coding. Primary use of shape coding for controls is for identification of control knobs or handles by "feel"; however, shapes shall be identifiable both visually and tactually. When shape coding is used:

a. The coded feature shall not interfere with ease of control manipulation.
b. Shapes shall be identifiable by the hand regardless of the position and orientation of the control knob or handle.
c. Shapes shall be tactually identifiable when gloves are worn, where applicable.
d. A sufficient number of identifiable shapes shall be provided to cover the expected number of controls that require tactual identification.
e. Shape coded knobs and handles shall be positively and non-reversibly attached to their shafts to preclude incorrect attachment when replacement is required.
f. Shapes should be associated with, or resemble, control function, and not alternate functions.

Table 12.1. Advantages and disadvantages of types of coding.

Advantages	Location	Shape	Size	Mode of Operation	Labeling	Color
Improves visual identification	X	X	X		X	X
Improves non-visual identification (tactual and kinesthetic)	X	X	X	X		
Helps standardization	X	X	X	X	X	X
Aids identification under low levels of illumination and colored lighting	X	X	X	X	(When trans-illuminate d)	(When trans-illuminate d)
May aid in identifying control position (settings)		X		X	X	
Requires little (if any) training; is not subject to forgetting					X	
DISADVANTAGES						
May require extra space	X	X	X	X	X	
Affects manipulation of the control (ease of use)	X	X	X	X		
Limited in number of available coding categories	X	X	X	X		X
May be less effective if operator wears gloves		X	X	X		
Controls must be viewed (i.e., must be within visual areas and with adequate illumination present)					X	X

Color-coding. Strict rules should be applied here, where numbers relate to the different contrasts of the colors. (i.e., controls shall be black; 17038, 27038, or 37038; or gray; 26231 or 36231). If color coding is required, only the following colors identified should be selected for control coding:

a. Red, 11105, 21105, 31105
b. Green, 14187
c. Orange-Yellow, 13538, 23538, 33538
d. White, 17875, 27875, 37875
e. Blue, 15123 shall be used if an additional color is absolutely necessary

Labeling. Labels, legends, placards, signs or markings, or a combination of these shall be provided whenever it is necessary for personnel to identify, interpret, follow procedures or avoid hazards, except where it is obvious to the observer what an item is and what he or she is to do with it.

Label characteristics. Label characteristics shall be consistent with such factors as:

a. Accuracy of identification required
b. Time available for recognition or other responses
c. Distance at which the labels must be read
d. Illuminant level and color
e. Criticality of the function label
f. Consistency of label design within and between systems

The orientation, standardization, content, visibility and color are also covered in this section.

An example of coding specifications for the toggle switch control is shown in Figure 12.2.

These controls should be coded by "and/either/or" shape, location, labeling, color. For visibility the design shall permit viewing of in-line digital read-out from all operator positions. The dimensions, resistance, and separation (distance between adjacent edges) shall all conform to the criteria in Table 12.2.

12.1.3. Audio capabilities

At some point during the design stage of most equipment the design engineer must decide whether to provide audio communications between the operator and the equipment. Audio signals should be provided when:

1. The information to be processed is short, simple, and transitory, requiring immediate or time-based response
2. The common mode of visual display is restricted by: over-burdening; ambient light variability or limitation; operator mobility; degradation of vision by vibration, high g-forces, hypoxia, or other environmental considerations; or anticipated operator inattention
3. The criticality of transmission response makes supplementary or redundant transmission desirable
4. It is desirable to warn, alert, or cue the operator to subsequent additional response
5. Custom or usage has created anticipation of an audio display
6. Voice communication is necessary or desirable

Table 12.2. Minimum separator distances for controls.

	TOGGLE SWITCHES	PUSH-BUTTONS	CONTINUOUS ROTARY CONTROLS	ROTARY SELECTOR \SWITCHES	DISCRETE THUMBWHEEL CONTROLS
TOGGLE SWITCHES	SEE FIG 12.2	13 mm (0.5 in.)	19 mm (0.75 in.)	19 mm (0.75 in.)	13 mm (0.5 in.)
PUSH-BUTTONS	13 mm (0.5 in.)	SEE FIG 12.3	13 mm (0.5 in.)	13 mm (0.5 in.)	13 mm (0.5 in.)
CONTINUOUS ROTARY CONTROLS	19 mm (0.75 in.)	13 mm (0.5 in.)	SEE FIG 12.4	25 mm (1.0 in.)	19 mm (0.75 in.)
ROTARY SELECTOR SWITCHES	19 mm (0.75 in.)	13 mm (0.5 in.)	25 mm (1.0 in.)	SEE FIG 12.5	19 mm (0.75 in.)
DISCRETE THUMBWHEEL CONTROLS	13 mm (0.5 in.)	13 mm (0.5 in.)	19 mm (0.75 in.)	19 mm (0.75 in.)	SEE FIG 12.6

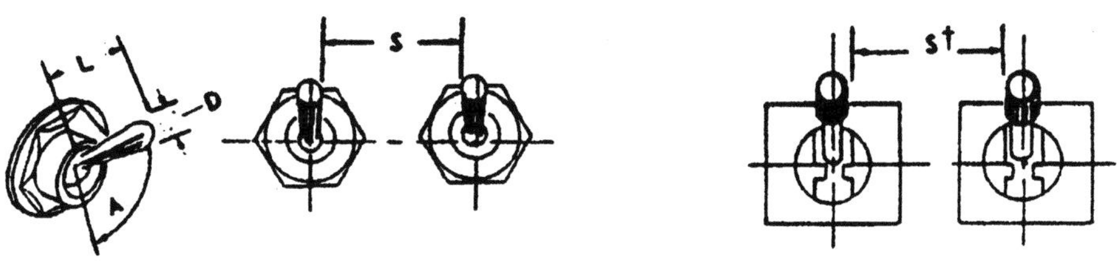

	DIMENSIONS			RESISTANCE	
	L Arm Length *	L Arm Length **	D Control Tip	Small Switch	Large Switch
Maximum	13 mm (1/2 in.)	38 mm (1-1/2 in.)	3 mm (1/8 in.)	2.8 N (10 oz.)	2.8 N (10 oz.)
Minimum	50 mm (2 in.)	50 mm (2 in.)	25 mm (1 in.)	4.5 N (16 oz.)	11 N (40 oz.)

	DISPLACEMENT BETWEEN POSITIONS A	
	2 Position	3 Position
Minimum	525 mrad (30°)	525 mrad (30°)
Maximum	1400 mrad (80°)	700 mrad (40°)
Desired	—	435 mrad (25°)

	SEPARATION S Single Finger Operation	†	Single Finger Sequential Operation	Simultaneous Operation by Different Fingers
Minimum	19 mm (3/4 in.)	25 mm (1 in.)	13 mm (1/2 in.)	16 mm (5/8 in.)
Optimum	50 mm (2 in.)	50 mm (2 in.)	25 mm (1 in.)	19 mm (3/4 in.)

* Use by bare finger ** Use with heavy hand wear † Using a lever lock toggle switch

Figure 12.2. Toggle switches

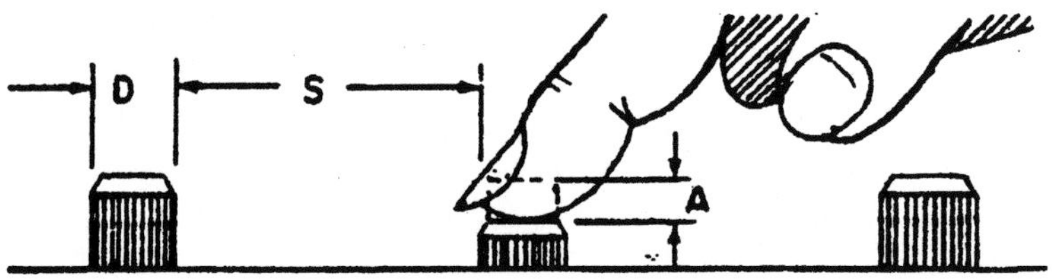

	DIMENSIONS		RESISTANCE		
	DIAMETER D				
	Fingertip	Thumb or Palm	Single Finger	Different Fingers	Thumb or Palm
Minimum	9.5 mm (3/8 in.)	19 mm (3/4 in.)	2.8 N (10 oz.)	1.4 N (5 oz.)	2.8 N (10 oz.)
Maximum	25 mm (1 in.)		11 N (40 oz.)	5.6 N (20 oz.)	23 N (80 oz.)
	DISPLACEMENT A				
	Fingertip		Thumb or Palm		
Minimum	2 mm (5/64 in.)		3 mm (1/8 in.)		
Maximum	6 mm (1/4 in.)		38 mm (1-1/2 in.)		
	SEPARATION S				
	Single Finger	Single Finger Sequential	Different Fingers	Thumb or Palm	
Minimum	13 mm (1/2 in.)	6 mm (1/4 in.)	6 mm (1/4 in.)	25 mm (1 in.)	
Preferred	50 mm (2 in.)	13 mm (1/2 in.)	13 mm (1/2 in.)	150 mm (6 in.)	

Note: Above data for bare hand application. For gloved hand operation, minima should be suitably adjusted.

Figure 12.3. Push-buttons (finger or hand operated)

Audio warning signals should be provided, as necessary, to warn personnel of impending danger, to alert an operator to a critical change in system or equipment status, and to remind the operator of a critical action, or actions, that must be taken. An audio warning system or signal should be provided for the purpose of detecting a condition that the operator's normal observation may have overlooked. If audio is used in conjunction with visual displays, audio warning devices should be supplementary or supportive. The audio signal should be used to alert and direct operator attention to the appropriate visual display. When an audio presentation is required, the optimum type of signal should be presented in accordance with Table 12.3. If visual displays are accepted, then the human sight capacities should be considered.

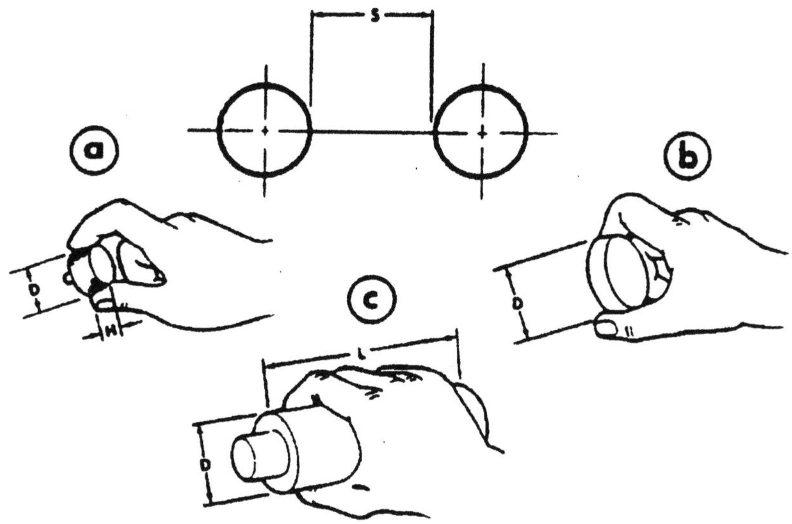

	DIMENSIONS				
	(a) Fingertip Grasp		(b) Thumb and Finger Encircled	(c) Palm Grasp	
	H Height	D Diameter	D Diameter	D Diameter	L Length
Minimum	13 mm (1/2 in.)	10 mm (3/8 in.)	25 mm (1 in.)	38 mm (1-1/2 in.)	75 mm (3 in.)
Maximum	25 mm (1 in.)	100 mm (4 in.)	75 mm (3 in.)	75 mm (3 in.)	—

	TORQUE		SEPARATION	
	*	**	S One Hand Individually	S Two Hands Simultaneously
Minimum	—	—	25 mm (1 in.)	50 mm (2 in.)
Optimum	—	—	50 mm (2 in.)	125 mm (5 in.)
Maximum	32 mN-m (4-1/2 in.-oz.)	42 mN-m (6 in.-oz.)	—	—

* To and including 25 mm (1.0 in.) diameter knobs

** Greater than 25 mm (1.0 in.) diameter knobs

Figure 12.4. Knobs.

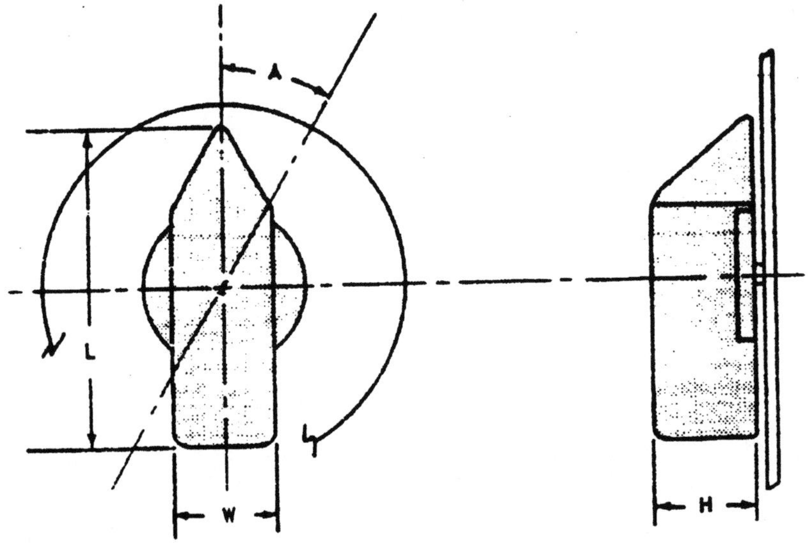

	DIMENSIONS			RESISTANCE
	L Length	W Width	H Depth	
Minimum	25 mm (1 in.)		16 mm (5.8 in.)	115 mN (1 in.-lb.)
Maximum	100 mm (4 in.)	25 mm (1 in.)	75 mm (3 in.)	680 mN (6 in.-lb.)
	DISPLACEMENT		SEPARATION	
	A *	**	One-Hand Random	Two-Hand Operation
Minimum	262 mrad (15°)	262 mrad (15°)	25 mm (1 in.)	25 mm (1 in.)
Maximum	262 mrad (15°)	262 mrad (15°)	—	—
Preferred	—	—	25 mm (1 in.)	25 mm (1 in.)

* For facilitating performance.

** When special engineering requirements demand large separation or when tactually ("blind") positioned controls are required.

Figure 12.5. Rotary selector switch.

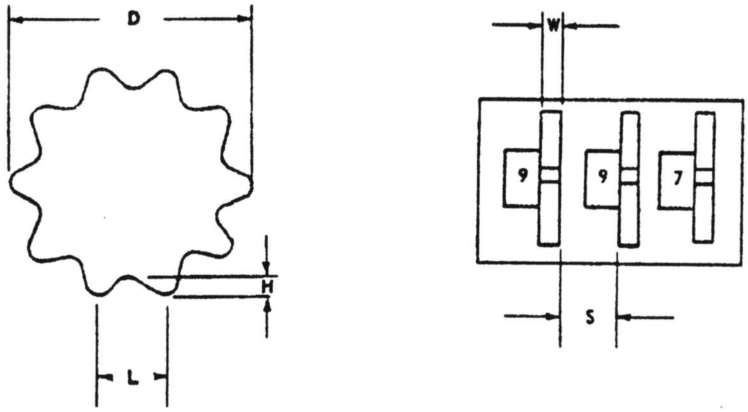

	D DIAMETER	L TROUGH DISTANCE	W WIDTH	H DEPTH	S SEPARATION	RESISTANCE
MINIMUM	30 mm (1-1/8 in.)	11 mm (7/16 in.)	3 mm (1/8 in.)	3 mm (1/8 in.)	10 mm (13/32 in.)	1.7 N (6 oz.)
MAXIMUM	75 mm (3 in.)	19 mm (3/4 in.)		13 mm (1/2 in.)		5.6 N (20 oz.)

Figure 12.6. Discrete thumbwheel control.

12.1.4. Human thermal tolerance capabilities

Some useful guidelines for human thermal tolerance capacity are given below.

Heating—Heating shall be provided within mobile personnel enclosures utilized for detail work or occupied during extended periods of time to maintain interior dry bulb temperature above 10°C (50°F). Within permanent and semi-permanent facilities, provisions shall be made to maintain an effective temperature[9] (ET) or corrected effective temperature[10] (CET) not less than 18°C (65°F) (see Figure 12.7), unless dictated otherwise by workload or extremely heavy clothing. Heating systems shall be designed such that hot air discharge is not directed on personnel.

Ventilating—Ventilating shall be assured by introducing fresh air into any personnel enclosure. If the enclosure volume is 4.25 m^3 (150 ft^3) or less per person, a minimum of 0.85 m^3 (30 ft^3) of ventilation air per minute shall be introduced into the enclosure; approximately two-thirds should be outdoor air.

[9] An arbitrary index which combines into a single value the effect of temperature, humidity, and air movement on the sensation of warmth or cold felt by the human body. The numerical value is that of the temperature of still, saturated air which would induce an identical sensation.

[10] Refer to Wet Bulb Globe Temperature (WBGT).

For larger enclosures, the air supply per person may be in accordance with the curves in Figure 12.8. Air shall be moved past personnel at a velocity not more than 60 m (200 ft) per minute. Where manuals or loose papers are used, airspeed past these items shall be not more that 30 m (100 ft) per minute – 20 m (65 ft) per minute if possible – to preclude pages in manuals from being turned by the air or papers from being blown off work surfaces. Ventilation or other protective measures shall be provided to keep gases, vapors dust, and fumes within permissible exposure limits. Intakes for ventilation systems shall be located to minimize the introduction of contaminated air from such sources as exhaust pipes.

Air Conditioning—The effective temperature or corrected effective temperature within personnel enclosures utilized for detailed work during extended periods shall be maintained at or below 29.5°C (85°F) (see Figure 12.7). Air conditioning systems shall be designed such that cold-air discharge is not directed on personnel.

Humidity—Approximately 45% relative humidity should be provided at 21°C (70°F). This value should decrease with rising temperatures, but should remain above 15% to prevent irritation and drying of body tissues, e.g., eyes, skin, and respiratory tract (see Figure 12.9).

Thermal Tolerance and Comfort Zones—Temperature and humidity exposure should not exceed the effective temperature limits given in Figure 12.9 when corrected for air velocity (Figure 12.8).

Limited Thermal Tolerance Zones—Where hard physical work is to be required for more than two hours, an environment not exceeding a Wet Dry (WD) or Wet Bulb Globe Temperature[11] (WBGT) index of 25°C (77°F) shall be provided. Where the wearing of protective clothing systems (which reduce evaporation of sweat from the skin) is required, this index shall be decreased 5°C (10°F) for complete protective uniforms, 40°C (70°F) for intermediate clothing.

---

[11] A meteorological measurement which can be used as an index to designate conditions of temperature and humidity at which on-set of heat stress can be expected at a particular energy expenditure level. It is calculated as follows:

$$WBGT = 0.7T_{WRnp} + 0.2T_g + 0.1T_A,$$

where: T_{WRnp} = non-psychometric (*np*) wet-bulb (*WB*) temperature,

T_g = temperature at interior center of a 15.2 cm (6 in) black globe, and

T_A = non-psychometric, but shaded, dry bulb (air) temperature.

In the absence of a radiant heat source (e.g., solar, engine, furnace), a modified Wet-Dry (WD_{85} index should be used where

$$WD_{85} = 0.85T_{WRnp} + 0.15T_A$$

Table 12.3. Functional evaluation of audio signals.

FUNCTION	TYPE OF SIGNAL TONES	COMPLEX SOUNDS	SPEECH
QUANTITATIVE INDICATION	POOR Maximum of 5 or 6 tones absolutely recognizable.	POOR Interpolation between signals inaccurate.	GOOD Minimum time and error in obtaining exact value in terms comparable with response.
QUALITATIVE INDICATION	POOR TO FAIR Difficult to judge approximate value and direction of deviation from null setting unless presented in close temporal sequence.	POOR Difficult to judge approximate deviation from desired value.	GOOD Information concerning displacement, direction, and rate presented in form compatible with required response.
STATUS INDICATION	GOOD Start and stop timing. Continuous information where rate of change of input is low.	GOOD Especially suitable for irregularly occurring signals (e.g., alarm signals).	POOR Inefficient; more easily masked; problem of repeatability.
TRACKING	FAIR Null position easily monitored; problem of signal-response compatibility.	POOR Required qualitative indications difficult to provide.	GOOD Meaning intrinsic in signal.
GENERAL	Good for automatic communication of limited information. Meaning must be learned. Easily generated.	Some sounds available with common meaning (e.g., fire bell). Easily generated.	Most effective for rapid (but not automatic) communication of complex, multidimensional information. Meaning intrinsic in signal and context when standardized. Minimum of new learning required.

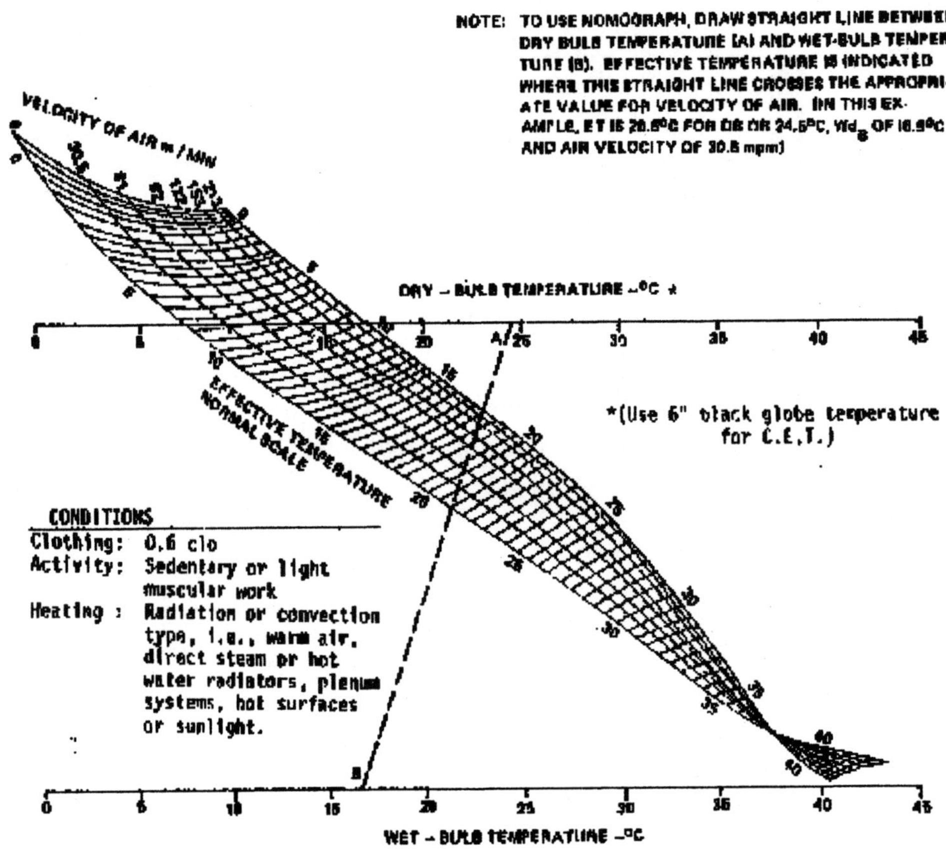

Figure 12.7. Effective temperature (ET) or corrected effective temperature (CET).

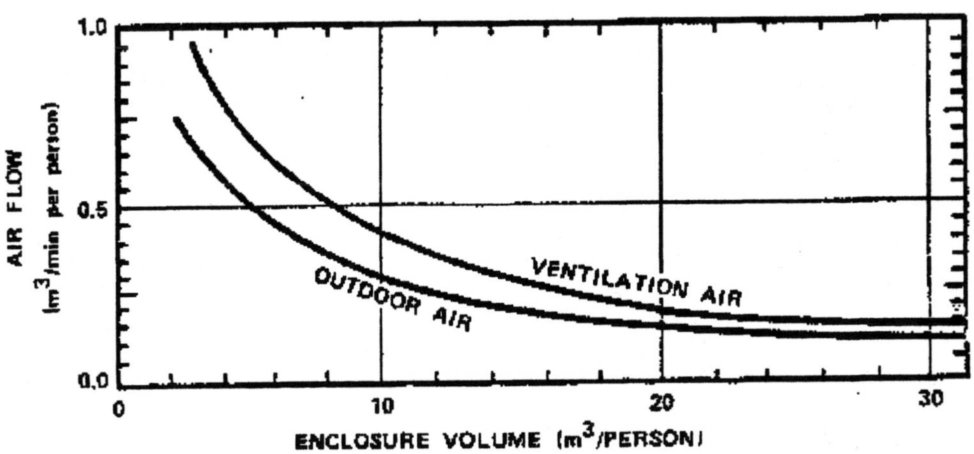

Figure 12.8. Ventilation requirements.

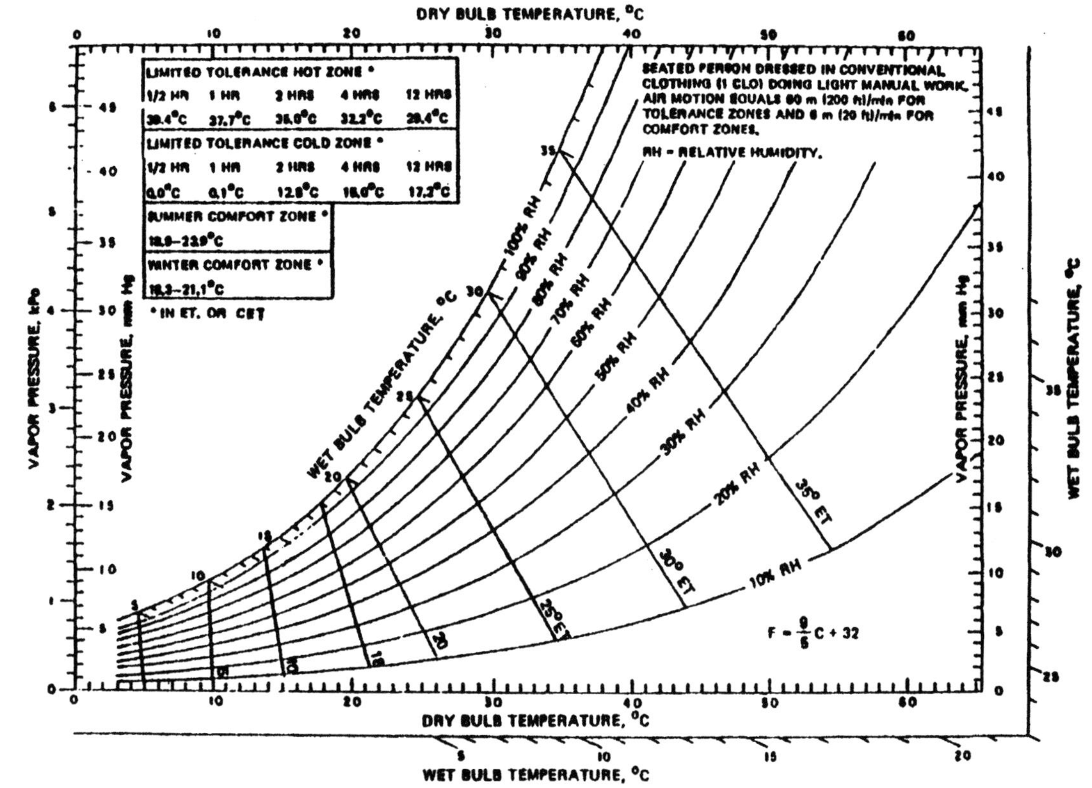

Figure 12.9. Summer and winter comfort zones and thermal tolerance for inhabited compartments.

12.1.5. Human vibration capabilities

Poor performance of mental and physical tasks may be contributed to vibration. For example, when the whole body is exposed to low-frequency, large-amplitude vibrations over a period of time—such things as headaches, motion sickness, eye strain and fatigue can result—all of which affects performance. The human vibration capabilities must be considered to reduce these vibrations. Guidelines are given below.

Safety Level—In order to protect human health, whole body vibration should not exceed twice the acceleration values shown on Figure 12.10 for the time and frequencies indicated.

Proficiency Level—Where proficiency is required for operational and maintenance tasks, whole body vibration should not exceed the acceleration values shown on Figure 12.10 for the time and frequencies indicated.

Comfort Level—Where comfort is to be maintained, the acceleration values shown on Figure 12.10 should be divided by 3.15.

Motion Sickness—In order to prevent motion sickness, very low frequency vibration should not exceed the limits of Figure 12.11.

Building Vibration—Buildings intended for occupation by humans should be designed to control the transmission of whole body vibration levels that are acceptable to the occupants as specified by ISO 2631.

Equipment Vibration—Where whole body vibrations of the human operator, or parts of the body, are not a factor, equipment oscillations should not impair human performance with respect to control manipulations or the readability of numerals or letters. Equipment vibrations in the shaded area of the upper curve of Figure 12.10 should be avoided.

12.2 Man-machine function allocation

It is often determined that human error is the "root cause" of a failure. Further investigation would determine that this is generally accepted by management as a way to merely place "blame" for an unwanted event. This often leads to an assumption that the individual either has a careless attitude or is incompetent for the desired performance. When management communicates such negative connotations, the victims of such accusations usually become alienated from the improvement process. These assumptions often result in less productivity, a lower overall reliability and sometimes lead to a catastrophe.

The amount of human error is proportional to the human ability to interact with the demands of machine and environment. Thus the combined reliability of both human and machine is increased as the ability to interact is increased. The allocation, or apportionment, of functions between humans and machine is dependent upon the desired output with consideration given to: (1) the environment, (2) the thought process, (3) human-machine communication, (4) specific performance, and (5) desired reliability.

A qualitative allocation may be conducted based on Table 12.4. While this table clearly indicates human superiority, it illustrates that the greatest human inefficiencies are the rate of data processing and memory retention (Ireson, 1988).

Table 12.4. Human-machine capabilities.

Human superiority	Machine superiority
Originality (ability to think)	Repetitiveness
Very little reprogramming need for new problems	Quick response to known problems
Recognizing new problems	Recalling data
Detecting new signals	Detecting signals beyond human sensitivity
Performing un-programmed tasks	Functions well under high stress conditions
Interpret data generated from events	Generate large amounts of data
Reasoning inductively (to diagnose general problems)	Reasoning deductively (to identify specific problems)
Evaluating alternate risks	
Utilizing equipment to and beyond known limits	
Ability to step outside boundaries	

Source: NASA SP-6506, An Introduction to the Assurance of Human Performance in Space Systems, NASA, 1968.

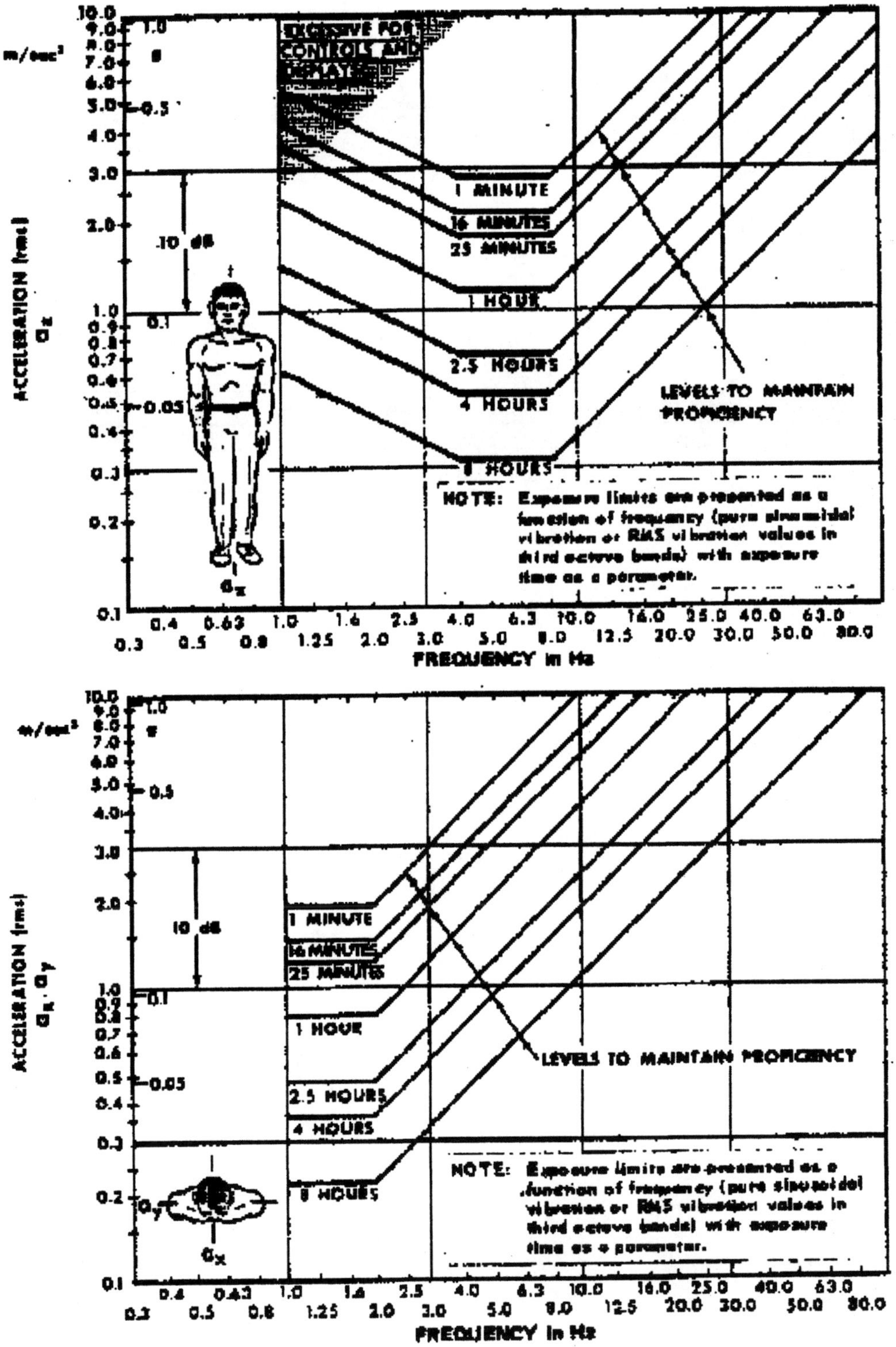

Figure 12.10. Vibration exposure criteria.

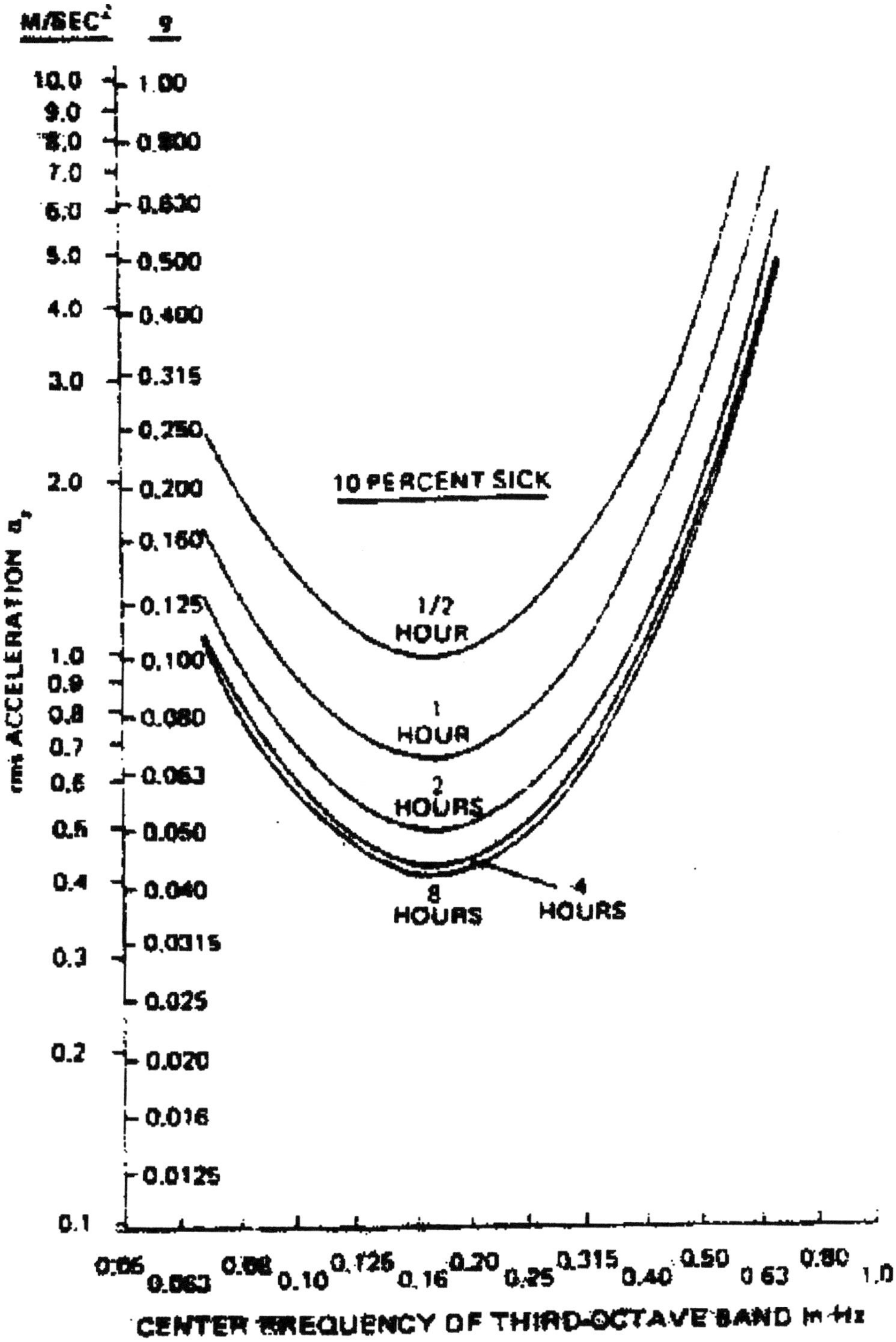

Figure 12.11. The 90% motion sickness protection limits for exposure to vibration.

12.2.1 A case study—crash of a Beech King Air

The following is a failure investigation report that was extracted from a paper published by ASME, presented by Smith (1985). It was included to illustrate a very common mistake in data mis-interpretation caused by improper communications between man and machine. The results show the fatal consequences of such interaction.

"On January 25, 1975, a Beech A90 King Air approached Washington National Airport for a landing. During the instrument approach, the King Air struck a radio antenna, the top of which was shrouded by clouds. The National Transportation Safety Board investigation determined that an unauthorized descent to unsafe altitudes was made for undetermined reasons and inferred incompetence in the crew.

"A second investigation by Galipault (1979), independent of government agencies, arose from a lawsuit filed on behalf of the pilots who not only had lost their lives, but had also been branded as unprofessional.

"Minutes before the crash, air traffic control on two occasions issued incorrect atmospheric pressure information. An electromechanical recorder used to transmit data to the approach control room reproduced handwriting of poor quality. The correct value of 29.41 in Hg (99.50 kPa) was transmitted, but the output was interpreted as 30.41 in Hg (102.89 kPa). When transmitted to the pilots, a barometric setting of 30.41 would have led them to believe they were about 930 ft (283 m) higher than was actually the case.

"The approach of the King Air was later duplicated with a similarly equipped plane and experienced pilot. In six flights the pilot was asked to comply with simulated air traffic control instructions. In each case the pilot accepted the incorrect barometric pressure reading and descended to an altitude about 950 ft (290 m) lower than desired. The large change in barometric pressure produced very little apparent motion of the altimeter needles because of the high rate of descent. The descent profiles were nearly identical to that of the fatal flight.

"Transcripts of the exchanges between the pilots and air traffic control reveal that the King Air's low altitude was recognized, yet no altitude advisory or warning was given. The Federal Aviation Administration (FAA) had issued specific procedures and phraseology for the issuance of approach clearances; however, Washington National was the only facility in the U.S. which had not yet complied with the FAA directive.

> "The second investigation by Galipault suggested a bias in the original investigation by federal agencies. The finding of pilot error, subsequently shown to be incorrect, diverted attention from the inaction of air traffic control. Furthermore, because certain speakers could not be identified, the transcript of air traffic control communications was ruled inadmissible as evidence. The transcript clearly shows both knowledge and inaction regarding the altitude of the King Air."

This problem, assumed to be human unreliability, was found to be in the system not in the pilots themselves. The application of the human engineering design criteria might have prevented this fatal accident. It states that feedback on control response adequacy shall be provided as rapidly as possible. Critical control functions, such as those entered by keyboard, shall provide adequate feedback to the operator to ensure that the data entry is, in fact, errorless and the one that the operator desires to enter. In this case the barometric pressure set by the pilots should have been checked again by ground control and an error message sent immediately to the pilots. Attention to the *general requirements* for human factor engineering related to reliability, outlined later in 12.3.2, could have aided in the prevention of this catastrophe.

12.3 Human factors in production

Manufacturing deals with the production phase of design. Problems encountered here affects the reliability of the finished product. Workstations should be designed to accommodate the individuals involved in the assembly of the product. The problems become obvious when a person is expected to install a part on a machine above the average head level while standing at floor level, or while installing parts below floor level while standing at floor level. Tasks similar to these may be accomplished by humans, under certain conditions, although these tasks might be nearly impossible to perform.

Some human factors for human interaction in the workstation design are needed for the tasks to be performed proficiently. The following section relates to failures due to poor workstation design.

12.3.1 A case study—steam turbine failure

Following is a portion of a failure investigation report, extracted from a paper published by ASME, (Smith, 1985). It was included to demonstrate the mental limitations of humans when required to interact with machines under "high stress" conditions.

"On New Year's Eve, December 31, 1983, a large single train petrochemical plant in the Gulf Coast area experienced severe operational problems. Several days of unusual subfreezing temperatures were wreaking havoc on the plant, which was not designed for the extreme temperatures. At about 8:00 p.m. operators, operating supervision and management were trying to keep the plant on-stream and were attending to numerous process difficulties when a 22,000 hp (16,400 kW) steam turbine experienced a catastrophic thrust bearing failure. The failure precipitated a two-week plant shutdown. This example considers how the failure was allowed to advance to a catastrophic level. Early detection would have limited the damage and the plant outage to about 12 hours.

"The plant operators enjoyed a reputation for operational excellence; plant availability was over 97% in 1983. The plant turbo-machinery was monitored by a computer managed microprocessor based monitoring system. Sensors specifically for monitoring the turbine thrust bearing were two eddy current proximeters for rotor axial position and two thermocouples for thrust bearing temperature.

"Table 12.5 is the computer alarm listing from the early stages of this failure. Procedures would require operators to trip the turbo-machinery train at 20:07:00 or 136 seconds after the failure began. However, the equipment operated several additional minutes resulting in massive damage.

"The computer printout, monitor alert and danger lights, monitor level displays, the audible enunciator and an alarm display light on an adjacent process control board all signaled a problem situation and in some cases the severity of the problem. These signals were either not observed or not investigated by the operators. The operators were faced with a myriad of alarms due to the plant upset. As in the Three Mile Island incident, the operators were flooded with more data than they could absorb and respond to.

"The computer print out [Table 12.5] illustrates that human intervention was needed for interpretation. The conversion of data to decision information is quite difficult during a crisis situation. Alarm light emitting diodes located on the individual monitors are small and monitor liquid crystal displays are difficult to read. It was this inability to quickly distinguish between important and unimportant data, along with information overload, that led to the massive secondary damage to the turbine.

"The principal corrective measure for the previously described event was to automate the shutdown of all turbo-machinery when appropriate sensors indicated thrust bearing failure in progress. A sensor voting scheme was incorporated to preclude false trips."

Inappropriate workstation design and the human inability to react appropriately to large amounts of data in a crisis situation may have been closer to the "root cause" of this catastrophic failure. An assumption based on human error would have failed to identify and correct the true cause of this failure. Also, any inappropriate assumptions about human error by management could have led to human performance problems.

The incident just described is a classic example of what is commonly referred to as **high-stress error**. When humans are faced with large amounts of data to process quickly, it sometimes forces the wrong choice or wrong decision. According to a study done by Swain and Guttman (1980), on the potential for error when humans are subjected to high-stress conditions: with 1 minute to react under high stress, there is a 99.9% chance of making the wrong decision; with 5 minutes to react, a 90% chance; with 30 minutes, 10%; and with 2 hours, still a 1% chance of doing the wrong thing. Even 1% is usually unacceptable – especially when human life is at risk.

High-stress error (Latino, 1987) can be avoided or minimized by:

1. Hazard risk analysis to identify potential hazard characteristics
2. Automation in design using computer logic for short time decision making
3. Designing non-confusing displays to ease interaction of humans under stress
4. Drills or rehearsals on how to react when high-stress conditions occur

12.3.2 General requirements

The general requirements below could have been used in the design phase of the steam turbine support equipment, referred to in 12.3.1., and might have prevented such a catastrophic failure. These requirements are outlined as follows:

The objectives—equipment and facilities shall provide work environments which foster effective procedures, work patterns, and personnel safety and health, and which minimize factors which degrade human performance or increase error. Design shall be such that operator workload, accuracy, time constraint, mental processing and communication requirements do not exceed operator capabilities. Design shall also minimize personnel and training requirements within the limits of time, cost, and performance trade-offs."

Table 12.5. Computer printout showing failure progression.

<table>
<tr><td>ALARMS
3 - 13 - 2 GB501T1A1 Failure onset
T + ALRT/20:04:44 31 Dec Temperature alert</td></tr>
<tr><td>ALARMS
3 - 13 - 2 GB501T1A1 Temperature danger
T +DNGR/20:05:46 31 Dec #2 Thermocouple</td></tr>
<tr><td>ALARMS
3 - 3 - 2 GB501T1A1
3 - 3 - 2 GB501T1A2 Axial thrust position
NOT OK/20:06:17 31 Dec alert position 1 and 2
TP + ALERT/20:06:17 31 Dec</td></tr>
<tr><td>ALARMS
3 - 5 - 2 GB501T1A1
RV ALRT/20:06:17 31 Dec Radial vibration alert
3 - 1 - 2 GB501T1Y inboard and outboard
RV ALRT/20:06:17 31 Dec turbine</td></tr>
<tr><td>ALARMS
3 - 3 - 1 GB501T1A1
TP + DNGR/20:06:19 31 Dec Axial thrust position
3 - 3 - 2 GB501T1A2 danger position 1 and 2</td></tr>
<tr><td>ALARMS
3 - 5 - 2 GB501T1Y Radial vibration danger
RV DNGR/20:06:33 31 Dec</td></tr>
<tr><td>ALARMS
3 - 1- 2 GB501T1Y Radial vibration danger
RV DNGR/20:06:42 31 Dec</td></tr>
<tr><td>ALARMS
3 - 13 - 1 GB501T1TT1 Thrust bearing temperature
T + ALRT/20:07:00 31 Dec alert and danger
T + DNGR/20:07:00 31 Dec #1 Thermocouple shutdown</td></tr>
</table>

Source Smith, Jan B., "Human Reliability Considerations in the Analysis of Mechanical Failures," Presented at the Energy Sources and Technology Conference and Exhibition, February, 1985.

Function allocation—the design shall reflect allocation of functions to personnel, equipment and personnel-equipment combinations to achieve the required sensitivity, and precision with the required reliability of system performance. Also the level of skills of personnel required to operate and maintain the system is considered.

Human engineering design—human engineering factors that affect human performance such as:

a. Adequate physical, visual, auditory, and other communication links between personnel, and between personnel and their equipment, under both normal and emergency conditions.

b. Efficient arrangement of operation and maintenance workplaces, equipment, controls, and displays.

c. Provisions for minimizing psycho-physiological stress effects of mission duration and fatigue.

d. Compatibility of the design, location and layout of controls, displays, workspaces, task allocation and control movements shall be compatible with restrictions imposed on human performance

Fail safe design—one in which a failure will not adversely affect the safe operation of the system, equipment, or facility. A fail safe design should be provided in those areas where failure can cause catastrophe through damage to equipment, injury to personnel or inadvertent operation of critical equipment."

Control/display integration—requirements are covered in part as follows:

a.—The relationships of a control to its associated display and the display to the control shall be immediately apparent and unambiguous to the operator. Controls should be located adjacent to their associated displays and positioned so that neither the control nor the hand normally used for setting the control will obscure the display.

b.—The complexity and precision required of control manipulation and display monitoring shall be consistent with the precision required of the system. Control/Display complexity and precision shall not exceed the capability of the operator (in terms of discrimination of display detail) or exceed the operator's manipulative capability under the dynamic conditions and environment (in terms of manual dexterity, coordination or reaction time) in which human performance is expected to occur.

c.—If more than one crew member must have simultaneous access to a particular group of controls or displays in order to insure proper functioning of a system or subsystem, the operator assigned to control and monitor a particular function or group of related functions shall have physical and visual access to all controls, displays and communication capability necessary to adequately perform assigned tasks.

12.4 Human factors in field testing

A product's useful life is sometimes determined by field testing and this is where all problems, both human and machine, unite to determine the reliability of a product. The following are some interactions to consider during field testing (Ireson, 1988):

1. **Environmental conditions**

   The ability to perform under actual conditions: weather, temperature, vibration, pressures (positive or negative), and the duration, time or speed requirements. Are these requirements appropriate and being addressed with a plan that meets the demands for this interaction?

2. **Application of human thought**

   The ability to make decisions during operational procedures, and the ability to evaluate the consequences of those decisions. The number of decisions may be too great, or difficult, or the sequence of an operation may be too difficult to remember. Artificial intelligence is often built into the system to relieve the probability of human error in both the decision making and sequencing processes. The most simple form of applied artificial intelligence is often referred to as "idiot-proofing" where connectors are designed to prevent swapping during a maintenance operation. Are these requirements appropriate and being addressed with a plan that meets the demands for this interaction?

3. **Communications**

   The ability for the human to communicate with other humans and/or the machine. Are these requirements appropriate and being addressed with a plan that meets the demands for this interaction?

4. **Performance levels**

   The ability for the human to function with the machine in a predetermined manner. Are these requirements appropriate and being addressed with a plan that meets the demands for this interaction?

5. **Human reliability**

How much does the machine depend on human intervention or interaction. Are these requirements appropriate and being addressed with a plan that meets the demands for this interaction?

12.5 Training and education

Although the terms are loosely synonymous, a distinction can be made between *training* and *education*. The dictionary definitions of the two terms are:

Educate

To develop the innate capacities of, especially by schooling or instruction.

To stimulate or develop the mental or moral growth of.

Train

To coach in or accustom to a mode of behavior or performance.

To make proficient with specialized instruction and practice.

Education is seen to be more general and more focused on mental processes. In other words, education teaches people how to *think*. It enhances the mind's ability to deal with reality. Education is more theoretical and conceptual in nature. The *essentials* of past knowledge are taught for the purpose of teaching students how to appraise new events and situations. Education equips students to acquire new knowledge. Generally, people bring a certain *educational background* to their job with them. Employers typically fund ongoing education of employees, but they do not usually provide it for the employees. Education is more likely the responsibility of the individual employee. This may be because education is more generally applicable and more "portable" than is training. Thus, the employee's education is readily transferable to new employers.

Training focuses more on *doing*. Training is more job-focused than education. The emphasis is on maintaining or improving skills needed to perform the current job, or acquiring new job skills. Training is usually paid for, and often provided by the employer. Often, training is employer-specific and the skills acquired may not always transfer readily to another employer.

Quality improvement requires change, and change starts with people. People change when they understand why they must change and how they must change to meet their own personal goals. People join organizations because doing so helps them meet certain of their own goals. Conversely, organizations hire people to help achieve the organization's goals. When organization's set new goals they are, in effect, asking their employees to think differently, perform new tasks, and to engage in new behaviors. Organizations must be prepared to help employees acquire the knowledge, skills, and abilities (KSAs) required by these new expectations. Training and education are the means by which these new KSAs are acquired.

Like all work, training and education is a process. The sequence of steps in this process is (Johnson 1993d, 239-240):

- Consider the training objectives
- Develop a training plan
    - The training plan must dovetail with other company plans such as strategic quality, business, and market plans, and it must be supported by the budget.
    - It must match the organization's training objectives.
    - It must be planned around available assets.
    - It must meet organizational goals.
    - Individual lesson plans must be developed as required.
- Select the training team. The trainer must:
    - Be interested in training.
    - Be committed to quality.
    - Enjoy people.
    - Have patience.
    - Want the training assignment.
    - Be empathetic.
    - Be willing to train to the audience's level.
    - Be respected by colleagues.
    - Have time for training and training preparation.
- Prepare the trainers.
    - Provide trainer training:
        - In-house
        - Outside sources—seminars, toastmasters
    - Provide training methods:
        - Lecture or presentation
        - On-the-job
        - Demonstration

        - Self-study
        - Role playing
        - Case studies
        - Discussion
    - Practice actual training exercises
    - Critique trainer and training material
- Prepare trainees
    - Emphasize the benefit to the employee.
    - Make clear how it will improve the organization and its quality.
    - Make clear any requirements that will transpire as result of the training.
- Train
- Evaluate the training
    - Did training meet the objectives?
    - Trainees can be surveyed about training.
    - Work processes and quality can be reviewed for improvement.
    - Trainees can be tested on training.
    - Results can be tabulated to improve on future training.

12.5.1 Importance of top-down support and strategic planning for quality training

In the traditional organization the training department is another "silo," with its own budget and special interests to protect. In this system the training plans are often not tied to strategic plans. Instead, these plans tend to serve the needs of the trainers and others in the training department, rather than serving the needs of the organization as a whole. The effect on the organization's performance is the same as when any organizational unit pursues its own best interest rather than the organization's interests: negative return on investment, damaged morale, etc.

Modern organizations that adhere to modern quality principles take a different approach entirely. The training plans of these organizations are tied directly to the current and future needs of external customers. These needs are, in turn, the driver behind the organization's strategic plan. The people who develop the strategic plan also develop the strategic training plan. The strategic training plan provides the means of developing the knowledge, skills and abilities that will be needed by employees in the organization in the future. Of course, training is also provided for the maintenance of skills used on a daily basis. Training executives in the modern organizations report directly to senior leadership, reflecting the greater degree of importance placed on human resource development in modern organizations.

The chief deliverable of the planning process is the annual training plan. The annual training plan is a step in the implementation of the longer-term strategic training plan. The plan's focus is neither internal nor external, rather it spans the entire process. The plan makes provision for training of suppliers, employees and customers. New product training is also included.

The annual training plan includes a budget. The annual training budget lists those resources that are required to provide the training. The training budget traditionally includes a brief cost/benefit analysis. Cost/benefit analysis for training, as for all expenditures for intangibles, is challenging. Estimating cost is usually simple enough. Examples of training costs include:

- trainer salaries
- consulting fees
- classroom space and materials
- lost time from the job
- staff salaries
- office space of training staff

Estimating *benefits* with the same degree of precision is problematic. It is usually counterproductive to attempt to get high precision in such estimates. Instead, most organizations settle for rough figures on the value of the trainee to the company. Some examples of training benefit include:

- improved efficiency
- improved quality
- increased customer satisfaction
- improved employee morale
- lower employee turnover
- increased supplier loyalty

Training budgets are tangible evidence of management support for the goals expressed in the training plan. In addition, management support is demonstrated by participating in the development of the strategic training plan. In most cases, senior management delegates the development of annual training plans and budgets to the training department staff.

12.5.2 Leadership training

Management training should include training in leadership principles. Deming's 14 points provide an overview of leadership's role in modern organizations.

Leaders should also receive guidance in the art of "visioning." Visioning involves the ability to develop a mental image of the organization at a future time. The future organization will more closely approximate the ideal organization, where "ideal" is defined as that organization which completely achieves the organization's values. How will such an organization "look?" What will its employees do? Who will be its customers? How will it behave towards its customers, employees, and suppliers? Developing a lucid image of this organization will help the leader see how she should proceed with her primary duty of transforming the present organization. Without such an image in her mind, the executive will lead the organization through a maze with a thousand dead-ends. Conversely, with her vision to guide her, the transformation process will proceed on course. This is not to say that the transformation is ever "easy." But when there is a leader with a vision, it's as if the organization is following an expert scout through hostile territory. The path is clear, but the journey is still difficult.

Leaders need to be masters of communication. Fortunately, most leaders already possess outstanding communication skills; few rise to the top without them. However, training in effective communication is still wise, even if it is only refresher training for some. Also, when large organizations are involved, communications training should include mass communication media, such as video, radio broadcasts, print media, etc. Communicating with customers, investors, and suppliers differs from communicating with employees and colleagues, and special training is often required.

When an individual has a vision of where he wants to go himself, he can pursue this vision directly. However, when dealing with an organization, simply having a clear vision is not enough. The leader must *communicate* the vision to the other members of the organization. Communicating a vision is a much different task than communicating instructions or concrete ideas. Visions of organizations that embody abstract values are necessarily abstract in nature. To effectively convey the vision to others, the leader must convert the abstractions to concretes. One way to do this is by living the vision. The leader demonstrates her values in every action she takes, every decision she makes, which meetings she attends or ignores, when she pays rapt attention and when she doodles absentmindedly on her notepad. Employees who are trying to understand the leader's vision will pay close attention to the behavior of the leader.

Another way to communicate abstract ideas is to tell stories. In organizations there is a constant flow of events. Customers encounter the organization through employees and systems, suppliers meet with engineers, literally thousands of events take place every day. From time-to-time an event occurs that captures the essence of the leader's vision. A clerk provides exceptional customer service, an engineer takes a risk and makes a mistake, a supplier keeps the line running through a mighty effort. These are concrete examples of what the leader wants the future organization to become. She should repeat these stories to others and publicly recognize the people who made the stories. She should also create stories of her own, even if it requires staging an event. There is nothing dishonest about creating a situation with powerful symbolic meaning and using it to communicate a vision. For example, Nordstrom has a story about a sales clerk who accepted a customer return of a defective tire. This story has tremendous symbolic meaning because Nordstrom doesn't sell tires! The story illustrates Nordstrom's policy of allowing employees to use their own best judgment in all situations, even if they make "mistakes," and of going the extra mile to satisfy customers. However, it is doubtful that the event ever occurred. This is irrelevant. When employees hear this story during their orientation training, the message is clear. The story serves its purpose of clearly communicating an otherwise confusing abstraction.

Leaders need training in conflict resolution. In their role as process owner in a traditional organization, leaders preside over a report-based hierarchy trying to deliver value through processes that cut across several functional areas. The inevitable result is competition for limited resources, which creates conflict. Of course, the ideal solution is to resolve the conflict by designing organizations where there is no such destructive competition. Until then, the leader can expect to find a brisk demand for his conflict resolution services.

Finally, leaders should demonstrate strict adherence to the ethical principles. Leadership involves trust, and trust isn't granted to one who violates a moral code that allows people to live and work together. Honesty, integrity and other moral virtues should be second-nature to the leader.

12.5.3 Management training

Leadership involves taking the organization to a future where the organization's members behave differently than they do today. Management involves operating existing systems to deliver value to customers, and improving these systems to provide better value. The skills involved in management are different than those needed to lead the organization. This is why, at times, effective managers make poor leaders and vice-versa.

Managers design processes. Processes combine material and human resources in a particular way to add value. Effectively performing this task requires a background in organization theory and development, theory of motivation, job design, and performance evaluation. While background in these areas is typically part of the requirement for employment as a manager, managers should receive continuous training in all of these areas to maintain and enhance their skills.

In addition to the traditional skills, managers in modern quality-focused organizations also need excellent interpersonal skills. These skills are employed in a variety of roles, such as:

Coaching. A coach doesn't simply tell the players what to do, they help them do it. A baseball pitching coach studies the theory of pitching and the individual pitcher and provides guidance on how to hold the ball, the windup, the delivery, etc. The coach is constantly trying to find ways to help the pitcher do a better job.

Mentoring. The manager-mentor understands the organization to such a degree that he has acquired wisdom regarding the way it works. This allows him to see relationships that are not apparent to the employee. The mentor helps the employee avoid organizational obstacles and to negotiate the barriers.

Negotiation. Managers must negotiate with others in the organization, as well as with suppliers and customers, to acquire the resources necessary to accomplish her department's goals. Obtaining these resources without engendering ill-will requires negotiating skill.

Conflict resolution. Like the leader, the manager is responsible for the work of many people. When these people cannot resolve their own differences, the manager must provide guidance.

Managers should also receive training in the fundamentals of accounting and finance. Such information is essential to such activities as cost-benefit analysis, budgeting and quality costs.

Finally, managers should possess certain technical skills that are crucial to their ability to carry out quality improvement projects. This includes the quality control and quality management tools. Managers must understand enough about measurement issues in the social sciences to be able to monitor the effectiveness of their employee and customer initiatives. Deming lists an understanding of theory of variation as one of the cornerstones of his system of profound knowledge. This requires rudimentary statistical skills. Managers without this training will misdiagnose problems, see trends where none exist, overreact to random variation, and in general make poor decisions.

12.5.4 Employee training and implementation of quality plans

Training should be driven by the organization's needs, which are well understood by the senior leadership. The best way to do this, from a training perspective, is to have the senior leaders conduct the training personally. The primary purpose of training by senior leaders is to communicate the relationship of the training to the organization's mission. Effective leaders will have developed a clear vision of the way the organization will look at some future date, a plan to get there, and a commitment to the plan. This should all be communicated through training conducted by the leaders themselves.

Train-the-trainer programs are quite popular as a way of "leveraging" limited training talent. With more trainers, more people can be trained in a shorter time. However, as Deming pointed out, something is lost when the master trains the trainer who trains the student, etc. Deming's funnel experiments show that the result is to move further and further away from the correct message (off to the "milky way"). Training should be conducted by training masters. Technology such as videotape, CD ROM, cable systems, etc. can provide a means for reaching large numbers of people. It is possible to have interactive training conducted electronically.

Many companies, for example, Ford motor company, believe that technology is easily transferred from one organization to the next, while the knowledge and experience of employees is much more difficult to transfer. This gives the organization an edge over its competitors that cannot be quickly overcome.

When quality improvement plans are implemented, the nature of the work being done changes. People involved in, or impacted by the new approach must receive two different types of training: conceptual and task-based.

Conceptual training involves explanation of the TQM principles driving the change and a shift from an internal, product-based perspective to a customer and process-based focus. Rather than viewing their jobs in isolation, employees must be taught to see all work as a process, connected to other processes in a system. Rather than a goal of "control," where activities are done the same way indefinitely, employees learn that continuous improvement is to be the norm, with processes constantly being changed for the better. Such ideas are radically different and difficult to assimilate. Patient repetition and "walking the talk" are essential elements of such training.

Conceptual training also involves teaching employees the basics of problem-solving. Data driven process improvement demands an understanding of the fundamentals of data collection and analysis.

When jobs are reintegrated, the duties expected of each employee change, often radically. Task based training is necessary to help employees acquire and maintain new skills and proficiencies. Employees are given new responsibilities for self-control of process quality. To effectively handle these new responsibilities employees must learn to use information in ways they never did before. Often employees are asked to help design new information systems, enter data, use computer terminals to access information, read computer output, make management presentations, etc. These skills must be acquired through training and experience. Quality plans that do not include adequate employee training are commonplace, and a primary cause of the high rate of failure of quality plans.

12.5.5 Facilitator training

The facilitator must possess certain unique skills. It is unlikely that an individual who is not already a facilitator will already possess the needed skills. Thus, it is likely that facilitator training will be needed.

A good part of the facilitator's job involves communicating with people who are working on teams. This role involves the following skills:

Communication skills: Quite simply, the facilitator who cannot communicate well is of little use to the modern organization.

Meeting management skills: Schedule the meeting well ahead of time. Be sure that key people are invited and that they plan to attend. Prepare an agenda and stick to it! Start on time. State the purpose of the meeting clearly at the outset. Take minutes. Summarize from time-to-time. Actively solicit input from those less talkative. Curtail the overly talkative members. Manage conflicts. Make assignments and responsibilities explicit and specific. End on time.

Presentation skills: Know why you are speaking to this audience (inform/educate or convince/persuade); perform the task; solicit the desired audience response.

Presentation preparation: Prepare a list of every topic you want to cover. Cull the list to those select few ideas that are most important. Number your points. Analyze each major point. Summarize.

Use of visual aids: A visual aid in a speech is a pictorial used by a speaker to convey an idea. Well designed visual aids add power to a presentation by showing the idea more clearly and easily than words alone. Whereas only 10% of presented material is retained from a verbal presentation after 3 days, 65% is retained when the verbal presentation is accompanied by a visual aid. However, if the visual aids are not properly designed, they can be distracting and even counterproductive. Poor visuals generate more negative comment from conference attendees than any other item.

Facilitators must be sensitive to non-verbal communication. There is much more to communication than mere words. Facilitators should carefully observe posture and body movements, facial expressions, tone of voice, fidgeting, etc. If the facilitator sees these non-verbal signals he should use them to determine whether or not to intervene. For example, a participant who shakes his head when hearing a particular message should be asked to verbalize the reasons why he disagrees with the speaker. A person whose voice tone indicates sarcasm should be asked to explain his attitude. A wallflower who is squirming during a presentation should be asked to tell the group her thoughts.

Facilitators should be active listeners. Active listening involves certain key behaviors:

Look at the speaker
Concentrate on what the speaker is saying, not on how you will respond to it
Wait until the speaker is finished before responding
Focus on the main idea, rather than on insignificant details
Keep emotional reactions under control

Because all of the work of facilitators involves groups, facilitators should have an in-depth understanding of group dynamics and the team process. Also, because the groups and teams involved are usually working on quality improvement projects, the facilitator should be well versed in project management principles and techniques.

12.5.6 Training needs analysis

The first step in the development of the strategic training plan is a training needs assessment. The training needs assessment provides the background necessary for designing the training program and preparing the training plan. The assessment proceeds by performing a task-by-task audit to determine what the organization is doing, and comparing it to what the organization should be doing. The assessment process focuses on three major areas:

Process audit. As stated numerous times, all work is a process. Processes are designed to add values to inputs and deliver values to customers as outputs. Are they operating as designed? Are they operated consistently? Are they measured at key control points? If so, do the measurements show statistical control? The answers to these questions, along with detailed observations of how the process is operated, are input to the development of the training plan.

Assessment of knowledge, skills and abilities. In all probability, there will be deficiencies (opportunities for improvement) observed during the process audits. Some of these deficiencies will involve employee knowledge, skills, or abilities (KSA's). The first principle of self-control is that employees must know what they are doing. Management's job doesn't end by simply giving an employee responsibility for a particular process or task, they must also provide the employee with the opportunity to acquire the KSA's necessary to successfully perform their new duties. This means that if the employee is asked to assume new duties as a member of a quality improvement team, they are given training in team skills, if they are to keep a control chart, they receive training in the maintenance and interpretation of the charts, etc. Since employees are expected to contribute to the implementation of the organization's strategic plan, they should be told what the plan is, and how their job contributes to the plan.

Assessment of employee attitudes. Attitudes are emotions that reflect a response to something taking place within an organization. A person's attitude is, in essence, a judgment about the wisdom of a particular course of events. If an employee's attitude is not positive, they will not use their KSA's to help the organization as effectively as they could. Negative employee attitudes about the direction being taken by the organization indicate that the employee either questions the wisdom of the proposed changes, or doubts the sincerity of the leadership. Regardless, it represents a problem that must be addressed by the training plan.

The assessments above can be conducted using audit techniques or the survey techniques. Assessments can be conducted by either internal or external personnel. In general, employees are more likely to be open and honest when confidentiality is assured, which is more likely when assessments are conducted by outside parties. However, internal assessments can reveal valuable information if proper safeguards are observed to assure the employee's privacy.

It is important that follow up assessments be made to determine if the training conducted closed the gap between the "is" and the "should be." The follow up will also provide a basis for additional training. Reassessment should be conducted first to assure that the desired KSA's were acquired by the target group of employees, then the process should be reassessed to determine if the new KSA's improved the process as predicted. It's common to discover that we made a mistake in assuming that the root cause of the process "is/should-be" gap is a KSA deficiency. If the reassessments indicate that the KSA gap was closed but the process gap persists, another approach must be taken to close the process gap.

12.5.7 Post-training evaluation and reinforcement

Training is said to have "worked" if it accomplishes its objectives. Since the training objectives are (or should be) derived from the strategic plan, the ultimate test is whether or not the organization has accomplished it strategic objectives. However, training is only one of dozens of factors that determine if an organization accomplishes its strategic objectives, and one that is often far removed in time from the final result. To assess training effectiveness we need more direct measures of success, and we need to measure near the time the training has been completed.

Except in academic settings, imparting knowledge or wisdom is seldom the ultimate goal of training. Instead, it is assumed that the knowledge or wisdom will result in improved judgments, lower costs, better quality, higher levels of customer satisfaction, etc. In other words, the training will produce *observable* results. These results were the focus of the training plan development and training needs analysis described earlier in this chapter. Training evaluation requires that they be converted to training metrics or objectives.

Regardless of the format of the presentation, the basic unit of training is the *lesson*. A lesson is a discrete "chunk" of information to be conveyed to a learner. The training objectives form the basis of each lesson, and the lessons provide guidance for development of measurements of success.

Lesson plans provide the basis for measurement at the lowest level. The objectives in the lesson plan are specific and the lesson is designed to accomplish these specific objectives. The assumption is that by accomplishing the set of objectives for each lesson, the objectives of the seminar or other training activity will be met. A further assumption is that by meeting the objectives of all of the training activities, the objectives of the training plan will be met. Finally, it is assumed that by meeting the objectives of the training plan, the objectives of the strategic plan (or strategic quality plan) will be met, or at least will not be compromised due to training inadequacies. All of these assumptions should be subjected to evaluation.

Evaluation

The evaluation process involves four elements (Kirkpatrick 1996):

Reaction. How well did the conferees like the program? This is essentially customer satisfaction measurement. Reaction is usually measured using comment sheets, surveys, focus groups and other customer communication techniques.

Learning. What principles, facts, and techniques were learned? What attitudes were changed? It is entirely possible that conferees react favorably to training, even if learning does not occur. The learning of each conferee should be quantified using pre- and post-tests to identify learning imparted by the training. Results should be analyzed using proper statistical methods. In exceptional cases, e.g., evaluating a consulting company for a large training contract, a formal designed experiment may be justified.

Behavior. What changes in behavior on-the-job occurred? If the conferee leaves the SPC seminar and immediately begins to effectively apply control charts where none were used before, then the training had the desired effect on behavior. However, if the conferee's tests indicate that she gained competence in the subject matter from the training, but no change in behavior took place, the training investment was wasted. Note that behavior change is dependent on a great number of factors besides the training, e.g., management must create systems where the newly learned behaviors are encouraged.

Results. What were the tangible results of the program in terms of reduced cost, improved quality, improved quantity, etc.? This is the real payback on the training investment. The metrics used for measuring results are typically built into the action plans, project plans, budgets, etc. Again, as with behavior change, there are many factors other than training that produce the desired results.

Phillips adds a fifth item to the above list Phillips (1996 316):

Return on Investment. Did the monetary value of the results exceed the cost for the program?

Due to the difficulty and cost involved, it is impractical and uneconomical to insist that every program be evaluated at all five levels. Sampling can be used to obtain evaluations at the higher levels. As an example, one large electric utility set the following sampling targets.

Table 12.6. Targets for percentages of programs to be evaluated.

Level	Percentage
Participant's satisfaction	100
Learning	70
On-the-job-applications (behavior)	50
Results	10
ROI	5

From Phillips, J.J. (1996). "Measuring the Results of Training," in Craig, R.L. editor-in-chief. *The ASTD Training & Development Handbook: A Guide to Human Resources Development.* New York: McGraw-Hill, 317.

Where sampling is used, programs should be selected using a randomization procedure such as random numbers tables. However, to make the results credible finance and accounting personnel should be involved in calculating financial ratios of this type.

Reinforcement

When the subject of reinforcement is raised, monetary remuneration usually comes to mind first. Skill-based pay is gaining favor in some quarters for a variety of reasons:

encourage employees to acquire additional skills

to reward people for training and education

as a reaction to the negative aspects of performance-based pay

While skill-based pay may have merit, cash awards and other such "rewards" are of dubious value and should probably not be used.

Rather than assuming that employees will only engage in training if they receive an immediate tangible reward, research and experience indicate that most employees find value in training that helps them better achieve their personal, job, and career goals. Thus, reinforcement is accomplished by providing the trainee with the opportunity to use the skills they learned about. Proficiency is gained with practice soon after the learning has taken place. Management should provide an environment where the new skills can be honed without pressure and distraction. The "just-in-time"(JIT) principle applies here. Don't provide training for skills that won't be used in the near future.

People who have just learned something new, be it a job skill or a new philosophy such as quality-focus, often have questions arise as they attempt to integrate their new knowledge into their daily thoughts and routine. User groups are very helpful. A user group consists of a number of people who have received similar training. User groups meet from time to time to discuss their understanding of the material with others. The focus of the group varies from "how are you using this?" to "I don't understand this" to "Here's what I am doing!" At times well-versed speakers will be invited to clarify particular aspects of the subject. Presentations of successful applications may be made. Management can encourage user groups by providing facilities, helping with administrative details, and, especially, by attending their meetings on occasion.

Electronic forums are gaining in popularity. Trainers will often make themselves available to answer questions via email. Forum subscribes will send their questions or comments to a "list server." The list server then automatically broadcasts the question or comment to other subscribers to the list. Every subscriber receives every user's message, along with the responses to the message. This often produces "threads." A thread is an exchange of information on a particular topic. E.g., subscriber A has a question about using control charts on financial data, subscriber B responds, then C responds to B and so on. These threads look remarkably like a free-wheeling face-to-face discussion. The result is that a great deal of learning takes place in a format that everyone finds to be more natural (and more interesting) than a traditional classroom environment.

Refreshers

Learning that isn't used right away tends to fade. Even when just-in-time-training (JITT) is used it is unlikely that every skill will be put to immediate and routine use. It is wise to plan for periodic refresher courses to hone the skills acquired during prior training. A refresher course is usually shorter, faster, and more intense than new learning. Instructors need not have the same subject matter mastery as those teaching new material. In fact, it may not even be necessary to have an instructor available. Media such as video and audio tape programs, CD ROM, slide presentations, etc. may be sufficient. These self-study media offer a number of cost and scheduling advantages.

If in-house instructors are used, they may be available to answer occasional questions from previous trainees. Of course, when the instructor is not a full-time trainer, this must be strictly limited. There are a number of ways to reduce the demands on the trainer's time while still answering most questions from participants. If the need is not urgent, the question may be asked using mail or an online forum. House newsletters or bulletin can be used to provide answers to frequently asked questions. More companies now have "intranets" where such information is made available. The trainee may be able to find an Internet news groups devoted to the subject of their concern. There are thousands of news groups covering a huge variety of subjects, many relating to quality and training.

12.5.8 Training tools

This section covers some of the technical aspects of effective training and education. However, the reader should understand that training and education are, in essence, *communication*. Communication within the organization. Pay particular attention to the following sections of that chapter:

- Principles of effective communication
- Criteria for effective communication
- Nonverbal communication
- Communication and presentation skills
- Preparing the presentation
- Visual aids
- Lectures, workbooks, case studies, on-the-job training

Types of Instruction

There are three basic types of instructional formats:

Learner-centered instruction. Learner-centered instruction begins with the needs of the learner and tailors the course accordingly. This approach creates challenges for instructors and administrators when learners bring a wide variety of abilities, backgrounds and needs to the class. Barring strong reasons to the contrary, this approach should be the first choice for most training programs.

Instructor-centered. The instructor determines the direction and pace of the learning. Instructor-centered training takes a "one size fits all" approach to learning. Since student-to-student variability exists, this approach fails a high percentage of students.

Material-centered. This includes most self-paced instruction such as programmed learning, interactive CD ROM, videotape, etc. Students trained using this approach have no instructor with whom they interact.

Lesson Plans

Lesson plans are "roadmaps" of the training material to be presented during a particular training session. Johnson (1993d) provides the following guidelines for developing lesson plans:

- Lesson plans are crucial for training success.
- Every subject must be adequately researched.
- Develop twice as much material as you think you will need.
- Ensure trainees know "what's in it for me" early on so they are motivated to learn.
- Practice prior to presentation.
- Use examples that pertain to the operation where possible.
- Stimulate questions.
- Use training aids.
- Adapt the training to the group's level.
- Never make excuses (e.g., "I've never taught this material before.")
- Remain positive.
- Follow your lesson plan.

There are four main types of lesson plans:

Topical outline. A lesson plan set up in a standard outline format, i.e., major sections are identified using Roman numerals, topics by upper case letters, etc. Most useful when the instructor is intimately familiar with the subject matter and has presented it previously. These lesson plans are easy to construct and can be used as agendas to help keep the class oriented.

Manuscript. Lessons are described in prose in detail. Most useful when teaching complex subjects for the first time. While this format provides a lot of information, the sheer volume of information may make it difficult to locate specific information when required. Careful organization (e.g., tab indexing) is essential. Development time can be considerable.

Sentence. An abbreviated version of the manuscript type of lesson plan. Used for less complex subjects. While providing less information than the manuscript format, this approach is easier to prepare and it is less difficult to find the information.

Key word. A "memory jogger" for the experienced instructor to use to pace the presentation and assure that all subjects are covered. This type of lesson plan normally does not provide information on sequence or time-per-topic. This "quick-and-dirty" approach provides little detail and should not be used unless the instructor is intimately familiar with the material. An advantage to this approach is that it tends to produce less formal presentations.

Lectures and Presentations

In the lecture method an instructor transmits information to students. The lecture method is one of the most used of all presentation methods. It is suitable for conveying information such as procedures, etc. Lectures can be effective providing the speaker keeps in mind that listeners' attention spans are limited to 15-30 minutes, depending on the skill of the speaker, the topic, and the age and education level of the audience. A number of techniques can help increase the retention rate and extend the attention span:

- Cover only one point per lecture; leave out anything not *directly* related to your point
- Vary voice tone
- Have material presented by multiple speakers
- Assure that the room environment is comfortable
- Avoid lecturing after the audience has eaten a meal
- Avoid lecturing late in the evening
- Do not schedule the lecture just prior to another significant event (e.g., 5 p.m. on Christmas eve)

Presentations include lectures, but they can also include oral, visual, and multimedia formats.

Coaching

In modern, quality-focused organizations a supervisor may spend up to 60% of his or her time coaching. On a one-on-one basis coaching refers to the process of helping a single employee improve some aspect of his or her performance. On a group level coaching is a process of developing effective teams and work-groups. The successful coach needs the following skills:

- Communication skills
- Listening skills
- Analysis skills
- Negotiation skills
- Conflict resolution skills

The coach must also possess sufficient subject matter knowledge to assist the employee in achieving results. Expert knowledge of the task enables the coach to describe and demonstrate the desired behavior and to observe the employee performing the task and give feedback. Just as an NBA coach must be an expert in basketball, the business coach must be an expert on the subject they are coaching.

Coaches must also understand how adults learn and grow. Brookfield (1986) identified five factors that influence adult learning:

1. More learning takes place if learning is seen as voluntary, self-initiated activity.
2. More learning takes place in a climate of mutual respect.
3. The best learning takes place in an environment characterized by a spirit of collaboration.
4. Learning involves a balance between action and self-reflection.
5. People learn best when their learning is self-directed. (see the next section for information on self-directed learning)

Coaches should be aware of the essentials of motivation theory, such as Maslow's needs hierarchy, McGregor's theory X and theory Y models of management, and Hertzberg's hygiene theory.

There are many variations on the coaching process, but all involve similar steps (Finnerty 1996):

- ➢ Define performance goals.
- ➢ Identify necessary resources for success.
- ➢ Observe and analyze current performance.
- ➢ Set expectations for performance improvement.
- ➢ Plan a coaching schedule.
- ➢ Meet with the individual or team to get commitment to goals, demonstrate the desired behavior, and establish boundaries.
- ➢ Give feedback on practice and performance.
- ➢ Follow up to maintain goals.

Situations that Require Coaching to Improve Performance

Coaching discussions may be initiated as a result of an administrative situation or because the manager has become aware of an event or incident of concern in relation to a task or project. In their publication *Coaching Skills*, The Center for Management and Organization Effectiveness lists the following situations where coaching may be required:

Administrative Situations

Setting objectives

Performance reviews

Salary discussions

Career planning-development discussions

Job posting and bidding discussions

Project or Task Situations

A specific project or assignment problem: e.g., delays, quality problems, quantity problems, lack of follow-through on commitments.

Absenteeism and/or tardiness

Deficiency in effort or motivation

Behavior which causes problems, e.g., aggressiveness

Training: opportunity or assignment

When someone new joins your group or team

Conflicts between employees or within groups

Communication problems or breakdowns

When your own supervisor makes you aware that one of your employees has a problem

Forms of Coaching

Coaching can take on a variety of forms. The traditional form is, as described above, a supervisor coaching a subordinate or a sponsor coaching a team. However, other forms do occur on occasion and should probably occur more often than they do. A number of the more common alternative forms of coaching are described here (Finnerty 1996):

Mentoring: Mentoring involves a relationship between a senior manager and a less-experienced employee. The mentor is a trusted friend as well as a source of feedback. Mentors provide employees with information and feedback they may not otherwise receive. It was stated earlier that coaches must be subject-matter experts in the area they are coaching. Mentors are experts on the *organization itself.* Because of their superior knowledge of the written and unwritten rules of the organization, the mentor can help the employee traverse the often perplexing maze that must be negotiated to achieve success. Knowledge of the organization's informal leadership, norms, values, and culture can usually only be acquired from either a mentor, or by experience. "*Experience is a stern school, 'tis a fool that will learn in no other.*"

Peer Coaching: Traditional, vertically structured organizations have been the model for mentoring. Typically, the mentor or coach is above the employee in the organizational hierarchy and has formal command authority over her. Peer coaching relationships are those where all parties are approximately the same level within the organization or, at least, where no party in the relationship has command authority over any other party. For all the talk of formal training, policy and procedure manuals, etc., the truth is that in the real world, most "training" takes the form of a coworker telling the employee "the way its *really* done around here." Peer coaching recognizes that this approach has tremendous value. Among the advantages is that there is less hesitancy in asking a peer for help for fear of revealing ignorance or bothering one's superior with a minor problem. Peer coaching can be facilitated by putting people together for the purpose of learning from one another. Peer coaches are provided with training in the skills of coaching.

Executive Coaching: By its nature a command hierarchy discourages upward communication of "bad news." The result of this is that the higher one's position in the hierarchy, the less feedback one obtains on their performance. This is especially true for negative feedback. Employees quickly learn that it is not in their best interest to criticize the boss. While these problems are partially alleviated by such innovations as anonymous 360-degree performance assessments, it may also be useful for the executive to have a coach. The executive coach is usually a consultant hired from outside of the firm. The coaching role is typically to act as coach to the executive team, rather than to a particular executive.

Self-Directed Learning

Self-directed learning (SDL) is defined as a training design in which trainees work at their own pace, without the aid of an instructor, to master predetermined material. SDL is only one of a number of approaches that can be employed to meet the training goal. The choice to use SDL is made after evaluating the various options available. Mastery can be measured via testing, interview, demonstration or some other method.

SDL is the theory that underlies such concepts as self-directed teams and learning organizations. The basis of SDL is "self-directedness." Self-directedness in the SDL context refers to the allowing the trainee to make learning decisions and to follow through on them. The concept of self-directedness has gained in importance with the advent of large *learning centers*. In these facilities large numbers of training programs are available to meet company and individual needs. The advent of *desktop-learning* integrates many of the features of learning centers with modern desktop technology to allow, literally, learning on the trainee's desktop. Technological innovations such as full-screen computer video and network accessible audio-visual presentations continue to expand the availability of SDL materials. Thousands of subjects are available in software and video formats, with more being added every day. These developments increase the importance and effectiveness of SDL.

- There are numerous situations where SDL should be considered (Piskurich 1996):
- Large employee populations with diverse training requirements
- Need for individualized development
- Multiple training sites where the same training must occur concurrently
- Need for a high degree of training consistency
- High turnover rates require continuous training.

Not all people are good candidates for SDL. Not all people are self-directed. Even a particular individual will have varying degrees of self-directedness based on current job circumstances, self-esteem and other factors. Years of experience with an education system that totally controls what is presented, when it is presented, how it is presented, and how the learner is expected to demonstrate their grasp of the material have conditioned many people to a decidedly "other directed" approach to learning. At best, many people will need to be coached in SDL, and some individuals will simply not be able to do it. Fortunately, instruments exist that make it possible to measure the degree of self-directedness of an individual employee. One such instrument is the Gugliamino Self-Directed Learning Readiness Scale (SDLRS). This instrument considers factors such as inner-directedness, achievement motivation, goal-setting skills, and listening ability. It also includes other characteristics of self-directedness such as self-confidence and observational ability which may have a more indirect relationship with SDL readiness.

Self-directedness can be enhanced. There are three components to the process:

Prepare the learner for the SDL experience. Preparation includes providing an organizational culture where SDL is accepted. Supervisors, managers and training professionals must become knowledgeable about SDL and accept its value. Learners can be introduced to the subject through training or less intensive mechanisms, such as brochures.

Develop support systems. Facilitators or counselors should be available to help learners make the transition from other-directed to self-directed learning.

Success breeds success. Work especially hard to assure that early SDL experiences are successful. Start with simple SDL assignments and provide high levels of support to participants.

SDL Assessment

The decision to use the SDL approach should be made only after considering its advantages and disadvantages. Figure 12.12 lists advantages and disadvantages of the SDL approach Piskurich (1996).

	Advantages	Disadvantages
Trainee	Available when trainee is ready Individual choice of material Immediate feedback Work at own pace No surprises Provides review and reference	Not accustomed to SDL Not comfortable setting objectives No instructor No group interaction
Trainer-developer	No constantly repeated classes More time to develop Less time on the road	Difficult to develop properly Must revise more frequently More trainee preparation needed Limited choice of media Must sell concept to others
Corporation	Multiple-site training Fewer trainers Reduced travel expenses Reduced meeting room costs Eliminates trainee travel costs Reduced training downtime Easier to schedule Enables cross-training Just-in-time training Greater training consistency Reduced training time in the aggregate	Initial production costs higher Higher reproduction costs Higher distribution costs Higher revision costs Possible logistical problems

Figure 12.12. Advantages and disadvantages of self-directed learning.

Discussion Format

This learning format is based on a Q&A approach. The instructor develops a set of questions that cover the material of interest, then works to get the class to ask these questions. The approach is based on engaging the students in a discussion of their work, then directing the discussion towards the material the instructor wants to present to the class.

For example, assume that the instructor wants to present material on the theory of variation to a class of machine operators. The discussion might go as follows:

Instructor: Tell me about your work, Joan.

Joan: I run a milling machine. The milling process involves machining a surface of a casting so that it's flat.

Instructor: Are all of the castings that you start with exactly the same?

Joan: Oh, no! There are no two alike!

Instructor: In what way are they different?

Joan: Gosh, in every way you can think of. The hardness varies, so does the thickness. Sometimes the castings are warped, and...

Instructor: Thanks Joan. Has anyone else ever seen variation in the raw materials?

Of course, everyone will say "yes" and the stage is set for the instructor to make her presentation on the theory of variation. Using the discussion format, she'll present a small amount of material, then guide a discussion to create a foundation for the next piece of material.

The discussion format approach makes the class far more interesting to the participants than a simple lecture or presentation. It connects the classroom to the workplace. Discussion classes must be lead by very knowledgeable and experienced instructors. Instructors should plan to have supporting material on hand (examples, etc.) to stimulate the discussion. Facilitation techniques should be used to ensure participation and maintain focus. Care should be take to assure that discussions do not become bogged down in technical details rather than focusing on the training material. The instructor should summarize at the close of the discussion session.

Case Studies

Many different kinds of learning exercises are often referred to as cases (Alden and Kirkhorn 1996):

A predicament facing an organization is explained in great detail and students are expected to pose a solution or strategy for dealing with that complex situation.

A story is told about how some person or some organization dealt with a difficult situation, and students analyze and critique the actions taken, indicating what was appropriate and what might have been done differently.

Students are placed in a dynamic situation in which things change as a result of their actions and they work the situation toward a satisfactory outcome.

Case study materials vary a great deal in content, but most follow a similar structure.

Opening Scenario. This introductory section describes the problem and recaps the major actions.

Questions. This section provides the explicit questions that the students are required to answer. Three kinds of questions may be asked:

Summarize and translate what has been learned.

Assess the situation and critique the actions taken by the characters.

Focus on the decision point and determine what the characters should do.

References. Call attention to the background documents on the case. References might include items such as policies, procedures, grievances, regulations, reports and so on.

Case Information. This section provides the details of the story.

Exhibits. This section includes actual samples of materials that convey important information about the case: promotions materials, memos, letters, etc.

Not all case studies will include all five elements.

When to Use Case Studies

While research on the effectiveness of case studies is skimpy, and the results mixed, there is general agreement on two points: (1) case studies are an effective means for developing problem-solving skills, but (2) they add no value when trying to increase factual knowledge. One survey of training managers produced the following top five reasons for using case studies:

Exercise problem-solving skills

Simulate real life

Sharpen decision-making skills

Generate discussion

Develop judgment

The Instructor's Role

When using case studies the instructor's role is primarily one of orchestration. The instructor sets up the process and assures that the students have the appropriate information and knows what is expected of them. The instructor should orient the students by explaining the purpose of the case, explaining the procedures they should follow, and how they will be evaluated. Students are told how they will receive the case information (e.g., self-study, group study, presentation by instructor, etc.). Then students are told what questions they must answer. Finally, the instructor leads the class discussion of the case—questioning, steering, clarifying, expanding, controlling, protecting, synthesizing, and summarizing—so that the class discussion builds to a successful learning outcome for as many students as possible.

It is important to come to some type of closure before ending the session. The case study approach doesn't permit any one "right" answer (Alden and Kirkhorn 1996, 508). However, the instructor can comment on the discussion, showing how the points related to the course content, where the discussion was especially relevant and where it was superficial.

Availability of Cases

Case studies can be developed in house or purchased from a variety of sources. Many of the organizations that offer case studies also provide assistance in selecting appropriate case studies, teaching notes, and other assistance to instructors. Here is a brief listing of sources:

Harvard Business School. Over 3,000 business cases are included in their catalog. 1-800-545-7685.

The Kennedy School. 1,000 cases. 1-617-495-9523.

Institute for the Study of Diplomacy. 200 cases. 1-202-687-8971.

Hartwick Humanities in Management Institute. Leadership cases. 1-800-94-CASES.

Criticisms of the Case Study Method

Critics of the case study method believe that it is based on a number of false premises and leads to a number of problems (Kirkpatrick 1992-1993):

The case study method is based on a psycho-epistemology that is concrete-bound, anti-conceptual empiricism—the precise opposite of what today's managers need.

It leads to people who are concrete-bound and short-range in their thinking and who act as if they are incapable of grasping principles.

With the case study approach "thinking" does not mean the identification and integration of facts, nor especially does it mean the conceptual grasp and application of principles. Case studies discard at the outset precisely what managers need to learn in order to solve business problems: namely, a body of absolute, abstract principles, which are the essential tools for conceptual thought.

Business cases are contrived problems, not case histories.

Cases are almost always developed from recent situations, thus ignoring the lessons of hundreds of years of business history.

Despite what proponents say, cases are not real-world simulations. Real-world business decisions are made by managers who have been steeped in the concrete detail of their particular company for, at a minimum, usually, years. This familiarity enables managers to separate essentials from details. Students are given only a short time to assimilate a vast number of details, leading to information overload. Yet, decisions are required by the end of the class period. By suggesting that this artificial approach in any way resembles reality, students are being taught to make snap judgments.

There is no fundamental difference between, say, engineering and management sciences (e.g., accounting, advertising). They are all "practical" disciplines. Yet engineering is not taught using the case method. Imagine if it were. The students read a case that provides the detailed description of a highway leading up to a river; the engineer's objective in the case is to design a bridge to span the river. A problem has to be solved, decisions have to be made. No textbooks are provided or lectures given to the students—that would be a "telling" or "dictatorial" method of education. The students engage in "democratic" discussions about the design of the bridge and the materials to be used. There is no "right answer" because a variety of materials and designs would equally solve the problem—and the professor boasts that his answers are no better than the students'. Presumably, thought would be achieved through the social interaction of the discussions and the "currently useful generalizations" of civil engineering would emerge inductively. Would you want to be the first to drive across this bridge?

Workbooks

Workbooks present the trainee with a collection of discussion topics and exercises. Workbooks are usually used as supplements to classroom presentations or self-study materials. Workbooks serve two purposes: (1) as a learning tool, (2) to demonstrate mastery.

A common role of workbooks is as a customized supplement to textbooks. Textbooks are available on most quality subjects and they have the advantage of having been prepared by experts and subjected to peer review. The time invested in creating a textbook is enormous, many textbooks require several thousand man-hours to prepare. In addition, the typesetting, printing, and storage costs runs into tens of thousands of dollars. Most organizations do not have the resources to produce their own textbooks. However, to recoup their costs and make a profit publishers must target their books to a large audience. Thus textbooks tend to be "general" in nature. This makes it more difficult for trainers who have a specific training need. Workbooks can help to bridge this gap.

Workbooks are often designed around the more general textbook. They contain specific examples of how the material in the textbook applies to particular situations encountered by the trainee. Ideally, workbooks will not only be specific to the organization, but to the trainee as well. Workbooks are used to allow the learner to practice the material presented in a classroom. The instructor can review workbooks with the trainee to identify any errors and help the trainee better understand the material.

Job Training

Job training is the vehicle through which the vast majority of training occurs. Job training involves assigning the learners to work with a more experienced employee, either a supervisor, peer, or lead hand, to learn specific tasks in the actual workplace. The learner is usually a new employee who has been either recently hired, transferred, or promoted into the position and who lacks the knowledge and skill to perform some components of her or his job. The experienced employee normally demonstrates and discusses new areas of knowledge and skill and then provides opportunities for practice and feedback. There are three common methods used in job training (Nolan 1996):

Structured on-the-job-training (OJT). Structured OJT allows the learner to acquire skills and knowledge needed to perform the job through a series of structured or planned activities at the work site. All activities are performed under the careful observation and supervision of the OJT instructor. The structured process is based on a thorough analysis of the job and the learner. The OJT instructor introduces the learner systematically to what he or she needs to know to perform competently and meet performance standards and expectations.

Unstructured OJT. Unstructured OJT often means sink or swim. Most activities in unstructured OJT have not been thought through and are done in a haphazard way. A common method of unstructured OJT is to have the learner "sit" with another employee or "follow the employee around" for a few days to see what the employee does and how she does it. This "sit-and-see" technique often leads the learner to pick up as much by trial and error as he does by any instruction given by the more experienced employee. The learner is typically inundated with reading assignments concerning policies, procedures, and other assorted documentation which, when not put into the right context, can cause more confusion than assistance. The learner is often thrust on the experienced employee without notice and is seen as a hindrance, since this training time is interrupting the experienced employee's normal work load and performance outputs. The major drawback of the unstructured approach is that objectives, expectations, and outcomes are not defined in advance and, therefore, results are unpredictable.

Job Instruction Training. Job instruction training was originally developed for use with World War II production workers and is based on a mechanical step procedure requiring the instructor to present the material in an orderly, disciplined manner. It is most frequently used to teach motor skills. Since it is a systematic approach involved, components of it are often found in today's structured OJT.

Developing a Structured OJT Program

Structured OJT has proven to be an efficient and effective means of teaching employees about the skills required to do their job. Developing structured OJT programs is a process that involves the following steps:

Needs analysis.

Job analysis. Job analysis is part of the job design and employee selection processes. Job requirements are matched to the employee's knowledge, skills, and abilities during the selection process. When designing structured OJT programs the characteristics of trainees must be examined in order to target the OJT accurately and develop effective instructional materials, e.g., what works for new hires may not be best for transfers. Trainees should complete employee profile surveys to provide the instructional designer with the information needed to customize the training to the employee's needs.

Course design. This step will produce a course training plan that serves as the blueprint to be used to construct training support materials. The course training plan should include:

- a purpose statement
- performance objectives
- criterion tests

- presentation
- application and feedback methods
- course content outline
- lesson plan
- training schedule.

Material Preparation.

Validation. After training materials have been developed for the OJT program, they must be tested to ensure they fulfill their mandate to train the individual to perform the job. This involves pilot studies with selected individuals to "shake down" the materials. Final validation can only occur by monitoring the program's effectiveness with actual trainees.

Presentation. Effective instructional presentations incorporate a systematic learning process of *presentation, application, and feedback (PAF)* (Nolan 1996, 764). Prior to actual delivery, the OJT instructor needs to review the structured process and methods, collect all materials and tools necessary, and develop a schedule of training.

Presentation	The OJT instructor: States objective Motivates trainee Overviews key steps Presents tasks (tell and show) Tests for understanding
Application	Trainee applies new knowledge and skill through Directed practice Undirected, yet supervised, practice
Feedback	OJT trainer observes and communicates to trainee What was done well What needs improvement How to improve

The final step in a structure OJT program is to evaluate it. The method of evaluation is the same five level process as described above for evaluating training in general: *reaction*, *learning*, *behavior*, *results*, and *return on investment*.

Instructional Games, Simulations, and Role-Plays

These techniques are based on two premises: (1) People learn better through active experience than passive listening; and (2) people learn better through interacting with one another than alone.

Instructional games. An instructional game is an activity that is deliberately designed to produce certain learning outcomes. Instructional games incorporate five characteristics (Thiagarajan 1996):

Conflict. Games specify a goal to be achieved and throw in obstacles to its achievement. A game may involve competition among players, or it may involve player cooperation to achieve a group goal.

Control. Games are governed by rules that specify how to play the game.

Closure. Games have an ending rule, which may be a time limit, completion of a set of tasks, elimination of players from the game, etc. Most effective instructional games use multiple criteria for closure and permit different players or teams to win along different dimensions.

Contrivance. Games contain elements, such as chance, to assure that the game retains a playful character and isn't taken too seriously.

Competency base. The game is designed to help players improve their competencies in specific areas. Learning objectives range from rote recall to complex problem solving and may deal with motor, informational, conceptual, interpersonal, and affective domains.

Simulation games. A simulation game contains the five characteristics of instructional games, but in addition it includes a correspondence between some aspect of the game and reality. Some examples of simulation games that have been used in teaching quality concepts are:

Senge's "Beer Game" in *The Fifth Discipline* (Senge 1990, 26-53). The beer game is designed to teach systems thinking.

Deming's *funnel experiment.* Boardman and Boardman (1990) provide a detailed description of how to set up and conduct the funnel experiment. The funnel experiment illustrates statistical thinking and decision-making.

"The Card Drop Shop." The Card Drop Shop is a small enterprise that has customers, a president, a supervisor, an inspector, a rework operator, several line operators, and an accountant. There is only one process: dropping playing cards onto a target on the floor. The customer ideally wants all cards on the target but will accept the product provided that the total deviation from the target is "not too bad." The customer also specifies that each card is to be held by its center and dropped individually. Like the funnel experiment, the card drop shop illustrates the effects of tampering with a stable process. However, it does so using simpler tools and materials. (Alloway 1994).

Geometric Dimensioning and Tolerancing simulation. (Wearring and Karl, 1995).

Additional applications of simulation to quality are described in Simon and Bruce (1992).

In addition to simulation games, simulation tools such as quincunxes and sampling bead boxes are commonly used in quality training. Simple simulations create a concrete link to abstract concepts, making it easier for people to understand the concept.

Role-plays. In a role-play, players spontaneously act out characters assigned to them in a scenario. Role-plays take a variety of forms, for example:

Media. The scenario for a role-play may be presented through a printed handout, an audiotape, videotape, etc.

Characters. The characters in a role-play may be identified in terms of job functions, personality variables, or attitudes. Some role-plays require people to play their own roles in a different situation (as in a desert survival exercise).

Responses. Most role-plays involve face-to-face communication among the characters, but written or phone communication is also used.

Mode of usage. Groups may be divided into smaller groups or pairs, or a group can watch as others participate. Role-players may be substituted as the role-play progresses. Coaches may be assigned to help the role-players.

Number of players. Most role-plays involve two characters, but there is no fixed limit on the number of players that can be involved.

Replay. The effectiveness of some role-plays can be improved through repetition. Repetition can take place after a presentation of new material by the instructor, or after changing some aspect of the game.

An excellent example of a role-play game is Deming's "Red bead experiment" (Deming 1986, 346ff). The red bead experiment is designed to teach statistical thinking and the use of simple statistical tools.

Correspondence Format

Correspondence schools have been around a long time. This style of learning is self-directed. Traditionally, the student completes a lesson and then mails assignments to an instructor. The instructor reviews the assignment and returns it to the student with comments and suggestions.

The availability of computer networks has greatly expanded the scope of correspondence learning. It is now possible to do most of the work required for college degrees, from the associate degree level to the Ph.D. level, via computer. Students and instructors interact via computer networks. Assignments are posted online, students download the assignments and post their results. Computer communication takes place much faster than mail correspondence, while retaining the advantage of letting students and instructors work at times convenient to them.

Online correspondence is not limited to formal, structured lessons. A great deal of learning takes place on online forums, internet newsgroups, chat sessions, etc. Magazines such as *Quality Digest* publish lists of online resources for quality professionals. On the World-Wide Web most quality sites are linked to other quality-related sites, making it a simple matter to build an extensive list of online quality resources.

Use of Technology in Training

Most training is still conducted using traditional technologies. Human teachers stand in a room before a group of learners and use presentations and lectures to deliver instructional materials. Learners use pens, paper, and books. The most advanced technology is usually the overhead projector.

This situation is changing rapidly. The tremendous advances in computer technology and widespread availability of low cost desktop computers is creating a revolution in learning that is likely to dramatically change the way people will learn in the future. Interactive multimedia training programs represent a synergistic marriage of audiovisual (AV) media and computer-based training (CBT). These programs are designed to integrate text, graphics, animation, audio, and motion video so that each unit of content can be delivered in the medium that's best for it. Because these programs are computer-based, learners can interact with them and go through them in a sequence and depth that meets their particular requirements. And because their delivery systems are designed primarily for micro or personal computers, interactive multimedia training programs can be used at point of need. The interactive multimedia program does the following (Howell and Silvey 1996):

- Meets the individual needs of many learners, accommodating users at all levels of expertise
- Lets each user work at the pace that best suits him or her
- Offers the best presentation for different subjects

- Reverses the traditional student-teacher ratio, bringing many teachers or experts to the individual student
- Provides "creative" learning experiences that give each student a continuing functional reason to learn, beyond merely studying for a test
- Can be designed to encourage learners to explore a topic rather than simply seek a single right answer
- Allows learners at different levels to bring their own expertise and creative capabilities into the learning process
- Can create a "virtual classroom" wherever a microcomputer or workstation can be placed
- Permits small groups of two or three learners to engage in team problem-solving activities
- Can function as an in-class electronic performance-support system, containing the information, guidance, and tools needed to complete complex case activities.

Some interactive multimedia training programs are actually integrated into computer systems used by workers, so that small units of training can be delivered the moment they are needed to support job performance. In fact, the training functions of these systems are so well-integrated that it is often difficult to separate the training functions from the operational aspects of the programs.

Design of Multimedia Training Programs

It is as easy to create a bad multimedia training program as it is with traditional media, if not more so. There are five basic guidelines to developing effective multimedia training programs (Howell and Silvey 1996):

Determine the training goals. Interactive multimedia programs are often assigned goals that fall into one of the following categories:

Transfer of knowledge. The learner acquires information and learns what to do with it. E.g., how to plot an x-bar chart and how to interpret it.

Skill building. The learner acquires certain capabilities by learning steps that address performance. Simulation is one vehicle for achieving this goal.

Performance support. The worker acquires the skill or knowledge to solve a specific job problem without leaving the work environment.

Integrating goals. Assignments require that the student perform complex activities that use the material as it is presented. For example, a lesson might place the learner in the role of a department manager and ask that they use quality tools to reduce variation in a process. The learner would be provided with information on how to do this, e.g., a lesson on run charts.

Understand your customers. In addition to the training needs assessment, the customers' willingness to accept this new approach to learning must be understood.

Decide on best learner experience. Which types of learner experiences will best achieve the goals of the training? It isn't necessary, or desirable, to use every aspect of the medium for every lesson.

Key the design to content stability. Match the program design cycle time to the stability of the content. If a program's design takes too long, the content may be obsolete before the program design is completed.

Use a team approach. Teams should include experts in subject-matter, instructional design, graphics design, programming, and, of course, learners. Design strategy should include a plan and schedule.

Other Training Technology

In addition to multimedia training, there are a number of other training technologies that should be considered:

- Satellite broadcasts
- Cable broadcasts
- Video-conferencing and virtual classrooms
- Videotape
- Narrated slide presentations
- Audio tape

12.6 Exercises

1. The usual consequences of management assuming human error as the "root cause" of a failure is
   a. a repeat of the same failure
   b. disciplinary action for the employee
   c. less productivity
   d. All of the above
   e. a and c
2. Human reliability engineering considerations are most often

   a. difficult to deal with
   b. misjudged by management
   c. neglected during design
   d. all of the above

3. Human engineering design criteria, established by MIL-STD-1472, is best defined as
   a. the qualifications and requirements for the design engineer.
   b. the nature and limits of human capabilities.
   c. criteria furnished to contractors in non-military terms.
   d. criteria established only for military personnel.
4. What is meant by the term "Fail safe design," as defined in MIL-STD-1472?
   a. Designed such that a failure that will not adversely affect the safe operation of the system.
   b. Designed such that failure probability is low if the system is operated safely.
   c. Designed such that the system cannot fail if operated safely.
   d. All of the above.
5. The purpose of MIL-STD-1472 is to present human engineering design criteria so as to
   a. achieve required performance by operator, and maintenance personnel.
   b. minimize skill and personnel requirements and training time.
   c. achieve required reliability of personnel-equipment combinations.
   d. foster design standardization within and among systems.
   e. all of the above.
6. Designing non-confusing displays to ease interaction of humans under stress is a method for reducing
   a. job dissatisfaction
   b. performance problems
   c. the root cause of many problems
   d. high-stress error
7. Rehearsals on how to react when high stress conditions occur are methods of reducing
   a. human mistakes
   b. accidents
   c. high stress error
   d. catastrophic failures
   e. all of the above

8. Human engineering design reflects human engineering factors that affect human performance such as:
   a. Communication links between personnel and their equipment.
   b. Compatibility of the design, location and layout of workspaces.
   c. Psycho physiological stress effects of mission duration.
   d. All of the above.
9. Allocation of functions to personnel and equipment in combinations to achieve the required reliability is defined in MIL-STD-1472 as
   a. human factors allocation
   b. design factors allocation
   c. the function allocation
   d. cross functional allocation
10. Which of the following demonstrates human superiority over machines?
    a. Quick response to known problems
    b. Recalling data
    c. Functioning well under high stress conditions
    d. Interpreting data generated from events
11. Heating shall be provided within mobile personnel enclosures utilized for detail work or occupied during extended periods of time to maintain interior dry bulb temperature above
    a. 50°F
    b. 55°F
    c. 60°F
    d. 65°F
12. Within permanent and semi-permanent facilities, provisions shall be made to maintain an effective temperature (ET) or corrected effective temperature (CET) not less than ______, unless dictated otherwise by workload or extremely heavy clothing.
    a. 50°F
    b. 55°F
    c. 60°F
    d. 65°F

12.7 References

Alden, J. and Kirkhorn, J. 1996. *"Case Studies," in Craig, R.L. editor-in-chief. The ASTD Training & Development Handbook: A Guide to Human Resources Development.* New York, New York: McGraw-Hill.

Alloway, J. A. Jr. 1994. "The Card Drop Shop,". Quality Progress, July.

Brookfield, D. 1986. ***Understanding and Facilitating Adult Learning.*** San Francisco, CA: Jossey-Bass.

Deming, W. E. 1986. ***Out of the Crisis.*** Boston, MA: MIT.

Finnerty, M. F.. 1996. ***"Coaching for Growth and Development," in Craig, R.L. editor-in-chief. The ASTD Training & Development Handbook: A Guide to Human Resources Development.*** New York, New York: McGraw-Hill.

Howell, J. J. and Silvey, L. O. 1996. ***"Interactive Multimedia Training Systems," in Craig, R.L. editor-in-chief. The ASTD Training & Development Handbook: A Guide to Human Resources Development.*** New York, New York: McGraw-Hill.

Ireson, G. W., and C. F. Coombs. 1988. ***Handbook of Reliability Engineering and Management.*** New York, New York: McGraw-Hill.

Johnson, R.S. 1993a. ***TQM: Leadership for the Quality Transformation.*** Milwaukee, WI: ASQ Quality Press.

Johnson, R.S. 1993b. ***TQM: Management Process for Quality Operations.*** Milwaukee, WI: ASQ Quality Press.

Johnson, R.S. 1993c. ***TQM: Quality Training Practices.*** Milwaukee, WI: ASQ Quality Press.

Johnson, R.S. 1993d. ***TQM: The Mechanics of Quality Practices.*** Milwaukee, WI: ASQ Quality Press.

Latino, Charles J. 1987. Solving human-caused failure problems. ***Chemical Engineering Progress*** 83(5):42–45.

Kirkpatrick, D., L. 1996. ***"Evaluation," in Craig, R.L. editor-in-chief. The ASTD Training & Development Handbook: A Guide to Human Resources Development.*** New York, New York: McGraw-Hill.

Martin Marietta Corporation, Baltimore Division. 1968. ***NASA SP-6506: An Introduction to the Assurance of Human Performance in Space Systems.*** Washington, DC: NASA Scientific and Technical Information Division.

Nolan, M. 1996. ***"Job Training," in Craig, R.L. editor-in-chief. The ASTD Training & Development Handbook: A Guide to Human Resources Development.*** New York, New York: McGraw-Hill.

Phillips, J., J. 1996. ***"Measuring the Results of Training," in Craig, R.L. editor-in-chief. The ASTD Training & Development Handbook: A Guide to Human Resources Development.*** New York, New York: McGraw-Hill.

Piskurich, G., M. 1996. ***"Self Directed Learning," in Craig, R.L. editor-in-chief. The ASTD Training & Development Handbook: A Guide to Human Resources Development.*** New York, New York: McGraw-Hill.

Simon, J. and Bruce, P. 1992. "Using Computer Simulation in Quality Control,". Quality Progress, July.

Smith, Jan B. 1985. Human reliability considerations in the analysis of mechanical failures. Paper presented at the Eighth Annual Energy Sources and Technology Conference and Exhibition, February 17–21, 1985, Dallas, TX. Paper 85-Pet-14. New York, NY: ASME, Technical Division.

Swain, A.D., and H. E. Guttman. 1980. ***Handbook of Human Reliability Analysis with Emphasis on Nuclear Power Plant Applications.*** Albuquerque, NM: US Nuclear Regulatory Commission.

US Department of Defense. 1989. ***MIL-STD-1472: Human Engineering Design Criteria for Military Systems, Equipment, and Facilities.*** Philadelphia: Naval Publications and Forms Center.

US General Accounting Office. 1981. ***FRCD-82-5: Guidelines for Assessing whether Human Factors were Considered in the Weapon Systems Acquisition Process.*** Washington, DC: US General Accounting Office.

Wearing, C. and Karl, D. P. 1995. "The Importance of Following GD&T Specifications,". Quality Progress, February.

♦♦♦

CHAPTER

13

Statistical Process Control

A process control system is essentially a feedback system that links process outcomes with process inputs. There are four main elements involved, the process itself, information about the process, action taken on the process, and action taken on the output from the process.

By the process, we mean the whole combination of people, equipment, input materials, methods, and environment that work together to produce output. The performance information is obtained, in part, from evaluation of the process output. The output of a process includes more than product, it also includes information about the operating state of the process such as temperature, cycle times, etc. Action taken on a *process* is future-oriented in the sense that it will affect output yet to come. Action on the *output* is past-oriented because it involves detecting out-of-specification output that has already been produced.

There has been a tendency in the past to concentrate attention on the detection-oriented strategy of product inspection. With this approach we wait until an output has been produced, then the output is inspected and either accepted or rejected. SPC takes you in a completely different direction: improvement in the future. A key concept is *the smaller the variation around the target, the better*. Thus, under this school of thought, it is not enough to merely meet the requirements; continuous improvement is called for even if the requirements are already being met. The concept of never-ending, continuous improvement is at the heart of SPC.

13.1 Sources of variation

Shewhart (1931, 1980) defined control as follows:

"A phenomenon will be said to be controlled when, through the use of past experience, we can predict, at least within limits, how the phenomenon may be expected to vary in the future. Here it is understood that prediction within limits means that we can state, at least approximately, the probability that the observed phenomenon will fall within the given limits."

The critical point in this definition is that control is not defined as the complete absence of variation. Control is simply a state where all variation is *predictable* variation. A controlled process isn't necessarily a sign of good management, nor is an out-of-control process necessarily producing non-conforming product.

In all forms of prediction there is an element of chance. For our purposes we will call any unknown random cause of variation a *chance cause* or a *common cause*, the terms are synonymous and will be used as such. If the influence of any particular chance cause is very small, and if the number of chance causes of variation are very large and relatively constant, we have a situation where the variation is predictable within limits. You can see from the definition above that a system such as this qualifies as a controlled system. Where Dr. Shewhart used the term chance cause, Dr. W Edwards Deming coined the term common cause to describe the same phenomenon. Both terms are encountered in practice.

Needless to say, not all phenomena arise from constant systems of common causes. At times the variation is caused by a source of variation that is not part of the constant system. These sources of variation were called *assignable causes* by Shewhart, *special causes* of variation by Dr. Deming. Experience indicates that special causes of variation can usually be found without undue difficulty, leading to a process that is less variable.

Statistical tools are needed to help us effectively identify the effects of special causes of variation. This leads us to another definition:

Statistical process control is defined as the use of valid statistical methods to identify the existence special causes of variation in a process.

The basic rule of Statistical Process Control is:

Variation from common cause systems should be left to chance, but special causes of variation should be identified and eliminated.

This is Shewhart's original rule. However, the rule should not be misinterpreted as meaning that variation from common causes should be *ignored.* Rather, common-cause variation is explored "off-line." That is, we look for long term process improvements to address common-cause variation.

13.2 Types of control charts

There are two broad categories of control charts: those for use with continuous data (e.g., measurements) and those for use with attributes data (e.g., counts).

In SPC the mean, range and standard deviation are the statistics most often used for analyzing measurement data. Control charts are used to monitor these statistics. An out-of-control point for any of these statistics is an indication that a special cause of variation is present and that an immediate investigation should be made to identify the special cause.

13.2.1 Average and range charts

Average charts are statistical tools used to evaluate the central tendency of a process over time. Range charts are statistical tools used to evaluate the dispersion or spread of a process over time.

Usage: Average charts answer the question "Has a special cause of variation caused the central tendency of this process to change over the time period observed?" Range charts answer the question "Has a special cause of variation caused the process distribution to become more or less erratic?" Average and range charts can be applied to any continuous variable like weight, size, etc.

The basis of the control chart is the *rational subgroup*. Rational subgroups are composed of items which were produced under essentially the same conditions. The average and range is computed for each subgroup separately, then plotted on the control chart. Each subgroup's statistics are compared to the control limit, and patterns of variation between subgroups are analyzed.

Control limits for both the average and the range charts are computed such that it is highly unlikely that a subgroup average or range from a stable process would fall outside of the limits. All control limits are set at plus and minus three standard deviations from the center line of the chart. Thus, the control limits for subgroup averages are plus and minus three standard deviations of the mean from the grand average, the control limits for the subgroup ranges are plus and minus three standard deviations of the range from the average range. These control limits are quite robust with respect to non-normality in the process distribution.

To facilitate calculations, constants are used in the control limit equations. Table 13.1 provides control chart constants for subgroups of 10 or less.

Table 13.1. Control chart constants.

n	A_2	A_3	D_3	D_4	d_2	B_3	B_4
2	1.880	2.659	0	3.267	1.128	0	3.267
3	1.023	1.954	0	2.574	1.693	0	2.568
4	0.729	1.628	0	2.282	2.059	0	2.266
5	0.5770	1.427	0	2.115	2.326	0	2.089
6	0.483	1.287	0	2.004	2.534	0.03	1.970
7	0.419	1.182	0.076	1.924	2.704	0.118	1.882
8	0.373	1.099	0.136	1.864	2.847	0.185	1815
9	0.337	1.032	0.184	1.816	2.970	0.239	1.761
10	0.308	0.975	0.223	1.777	3.078	0.284	1.716

The control limits for the averages chart are computed using the following expressions.

$$\bar{\bar{X}} = \frac{\text{sum of subgroup averages}}{\text{number of subgroups}} \tag{13. 1}$$

$$LCL = \bar{\bar{X}} - A_2 R \tag{13. 2}$$

$$UCL = \bar{\bar{X}} + A_2 R \tag{13. 3}$$

The control limits for the range chart are computed using the following expressions.

$$\bar{X} = \frac{\text{sum of subgroup measurements}}{\text{subgroup size}} \tag{13. 4}$$

$$R = \text{Largest in subgroup - Smallest in subgroup} \tag{13.5}$$

$$\bar{R} = \frac{\text{sum of subgroup ranges}}{\text{number of subgroups}} \tag{13.6}$$

$$LCL = D_3 \bar{R} \tag{13.7}$$

$$UCL = D_4 \bar{R} \tag{13.8}$$

Example 13.1

Table 13.2 contains 25 subgroups of 5 observations each.

Table 13.2. Data for Averages and Ranges Control Charts

Sample 1	Sample 2	Sample 3	Sample 4	Sample 5	Average	Range
110	93	99	98	109	101.8	17
103	95	109	95	98	100.0	14
97	110	90	97	100	98.8	20
96	102	105	90	96	97.8	15

105	110	109	93	98	103.0	17
110	91	104	91	101	99.4	19
100	96	104	93	96	97.8	11
93	90	110	109	105	101.4	20
90	105	109	90	108	100.4	19
103	93	93	99	96	96.8	10
97	97	104	103	92	98.6	12
103	100	91	103	105	100.4	14
90	101	96	104	108	99.8	18
97	106	97	105	96	100.2	10
99	94	96	98	90	95.4	9
106	93	104	93	99	99.0	13
90	95	98	109	110	100.4	20
96	96	108	97	103	100.0	12
109	96	91	98	109	100.6	18
90	95	94	107	99	97.0	17
91	101	96	96	109	98.6	18
108	97	101	103	94	100.6	14
96	97	106	96	98	98.6	10
101	107	104	109	104	105.0	8
96	91	96	91	105	95.8	14
					2,487.2	369

The control limits calculated for the average chart are:

$$\overline{\overline{X}} = \frac{\textit{sum of subgroup averages}}{\textit{number of subgroups}} = \frac{2487.2}{25} = 99.48$$

$$LCL = \overline{\overline{X}} - A_2\overline{R} = 99.48 - 0.577(14.76) = 90.975$$
$$UCL = \overline{\overline{X}} + A_2\overline{R} = 99.48 + 0.577(14.76) = 108.001$$

The control limits calculated for the range chart are:

$$\overline{R} = \frac{\text{sum of subgroup ranges}}{\text{number of subgroups}} = \frac{369}{25} = 14.76$$

$$LCL = D_3\overline{R} = 0(14.76) = 0$$
$$UCL = D_4\overline{R} = 2.115(14.76) = 31.21$$

Since it is not possible to have a subgroup range less than zero, the lower control limit is not shown on the control chart for ranges. The completed average and range control charts are shown in Figure 13.1.

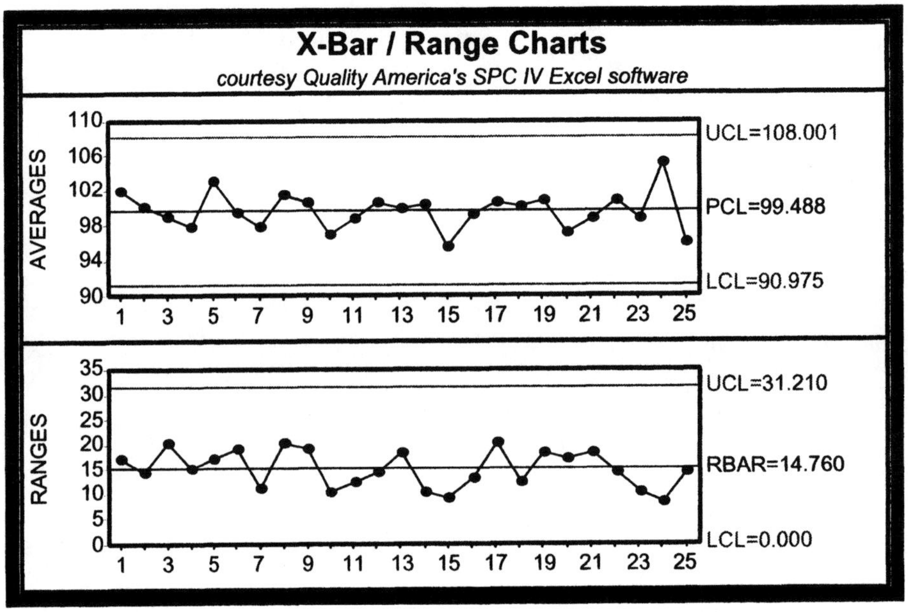

Figure 13.1. Completed average and range control charts.

The charts in Figure 13.1 show a process in statistical control. This merely means that we can predict the limits of variability for this process. To determine the capability of the process with respect to requirements one must use the methods described later in this chapter.

13.2.2 Average and standard deviation control charts

Average and standard deviation control charts are conceptually identical to average and range control charts. The difference is that the subgroup sample standard deviation is used to measure dispersion rather than the subgroup range. The subgroup standard deviation is statistically more efficient than the subgroup range for subgroup sizes greater than 2. This efficiency advantage increases as the subgroup size increases. However, the range is easier to compute and easier for most people to understand. If the analyses are to be interpreted by statistically knowledgeable personnel and calculations are not a problem, the standard deviation chart may be preferred for all subgroup sizes.

Control limits for both the average and the sigma charts are computed such that it is highly unlikely that a subgroup average or sigma from a stable process would fall outside of the limits. All control limits are set at plus and minus three standard deviations from the center line of the chart. Thus, the control limits for subgroup averages are plus and minus three standard deviations of the mean from the grand average. The control limits for the subgroup sigma are plus and minus three standard deviations of sigma from the average sigma. These control limits are quite robust with respect to non-normality in the process distribution.

To facilitate calculations, constants are used in the control limit equations. Table 13.1 provides control chart constants for subgroups of 10 or less. The control limits for the X-bar chart are:

$$\bar{\bar{X}} = \frac{\text{sum of subgroup averages}}{\text{number of subgroups}} \tag{13.9}$$

$$LCL = \bar{\bar{X}} - A_3\bar{s} \tag{13.10}$$

$$UCL = \bar{\bar{X}} + A_3\bar{s} \tag{13.11}$$

The control limits for the s chart are:

$$\bar{s} = \frac{\text{sum of subgroup sigmas}}{\text{number of subgroups}} \tag{13.12}$$

$$LCL = B_3\bar{s} \tag{13.13}$$

$$UCL = B4\bar{s} \tag{13.14}$$

Example 13.2

To illustrate the calculations and to compare the range method to the standard deviation results, the data used in the previous example will be reanalyzed using the subgroup standard deviation rather than the subgroup range.

Table 13.3. Data for Averages and Ranges Control Charts

Sample 1	Sample 2	Sample 3	Sample 4	Sample 5	Average	s
110	93	99	98	109	101.8	7.369
103	95	109	95	98	100.0	6.000
97	110	90	97	100	98.8	7.259
96	102	105	90	96	97.8	5.848
105	110	109	93	98	103.0	7.314
110	91	104	91	101	99.4	8.325
100	96	104	93	96	97.8	4.266
93	90	110	109	105	101.4	9.290
90	105	109	90	108	100.4	9.607
103	93	93	99	96	96.8	4.266

97	97	104	103	92	98.6	4.930
103	100	91	103	105	100.4	5.550
90	101	96	104	108	99.8	7.014
97	106	97	105	96	100.2	4.868
99	94	96	98	90	95.4	3.578
106	93	104	93	99	99.0	6.042
90	95	98	109	110	100.4	8.792
96	96	108	97	103	100.0	5.339
109	96	91	98	109	100.6	8.081
90	95	94	107	99	97.0	6.442
91	101	96	96	109	98.6	6.804
108	97	101	103	94	100.6	5.413
96	97	106	96	98	98.6	4.219
101	107	104	109	104	105.0	3.082
96	91	96	91	105	95.8	5.718
					2,487.2	155.44

Solution

The control limits for the average chart are:

$$\overline{\overline{X}} = \frac{\textit{sum of subgroup averages}}{\textit{number of subgroups}} = \frac{2487.2}{25} = 99.48$$

$$LCL = \overline{\overline{X}} - A_3\bar{s} = 99.48 - 1.427(6.218) = 90.613$$

$$UCL = \overline{\overline{X}} + A_3\bar{s} = 99.48 + 1.427(6.218) = 108.363$$

The control limits for the *s* chart are:

$$\bar{s} = \frac{\text{sum of subgroup sigmas}}{\text{number of subgroups}} = \frac{155.45}{25} = 6.218$$

$$LCL = 0(6.218) = 0$$

$$UCL = 2.089(6.218) = 12.989$$

The completed average and sigma control charts are shown in Figure 13.2.

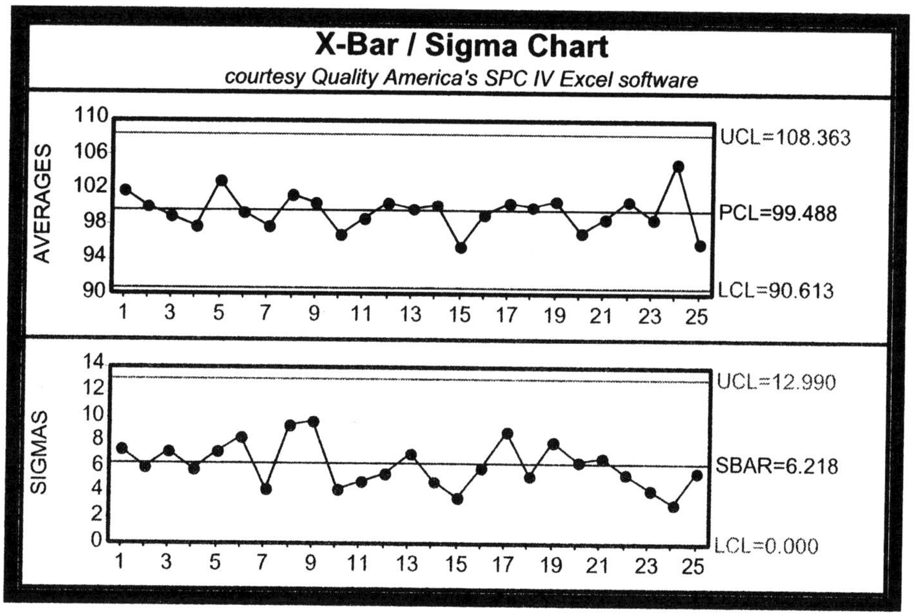

Figure 13.2. Completed Average and Sigma Control Charts.

Note that the control limits for the averages chart are only slightly different than the limits calculated using ranges, and the conclusions reached are the same as when ranges were used.

13.2.3 Control charts for individual measurements

Individuals control charts are statistical tools used to evaluate the central tendency of a process over time. They are also called *X charts* and *moving range charts.* Individuals control charts are used when it is not feasible to use averages for process control. There are many possible reasons why averages control charts may not be desirable: observations may be expensive to get (e.g., destructive testing), output may be too homogeneous over short time intervals (e.g., pH of a solution), the production rate may be slow and the interval between successive observations long, et cetera. Control charts for individuals are often used to monitor batch process, such as chemical processes, where the within-batch variation is so small relative to between-batch variation that the control limits on a standard x bar chart would be too close together. Range charts are used in conjunction with individuals charts to help monitor dispersion.[12]

The calculations for the X and moving range chart are the same as with averages and ranges charts where the subgroup is a consecutive pair of process measurements. The *individuals* chart limits are:

$$LCL = \overline{X} - A_2\overline{R} = \overline{X} - 2.66\overline{R} \tag{13.15}$$

$$UCL = \overline{X} + A_2\overline{R} = \overline{X} + 2.66\overline{R} \tag{13.16}$$

The *Moving Range* chart limits are:

$$LCL = 0 \tag{13.17}$$

$$UCL = 3.267\overline{R} \tag{13.18}$$

Example 13.3

Table 13.4 contains 25 measurements. To facilitate comparison, the measurements are the first observations in each subgroup used in the previous 2 examples.

Table 13.4. Data for Individuals and Moving Ranges Control Charts

Sample	Range
110	None
103	7
97	6
96	1
105	9
110	5

12 There is considerable debate over the value of moving R charts. Academic researchers have failed to show statistical value in them. However, many practitioners (including this author) believe that moving R charts provide valuable additional information that can be used in troubleshooting.

100	10
93	7
90	3
103	13
97	6
103	6
90	13
97	7
99	2
106	7
90	16
96	6
109	13
90	19
91	1
108	17
96	12
101	5
96	5

Solution

The control limits for the individuals chart are:

$$\overline{X} = \frac{sum\ of\ measurements}{number\ of\ measurements} = \frac{2476}{25} = 99.04$$

$$LCL = \overline{X} - A_2\overline{R} = 99.04 - 2.66(8.17) = 77.328$$

$$UCL = \overline{X} + A_2\overline{R} = 99.04 + 2.66(8.17) = 120.752$$

The Moving Range chart limits are:

$$LCL = 0$$

$$UCL = 3.267(8.17) = 26.69$$

The completed individual and moving range control charts are shown in Figure 13.3.

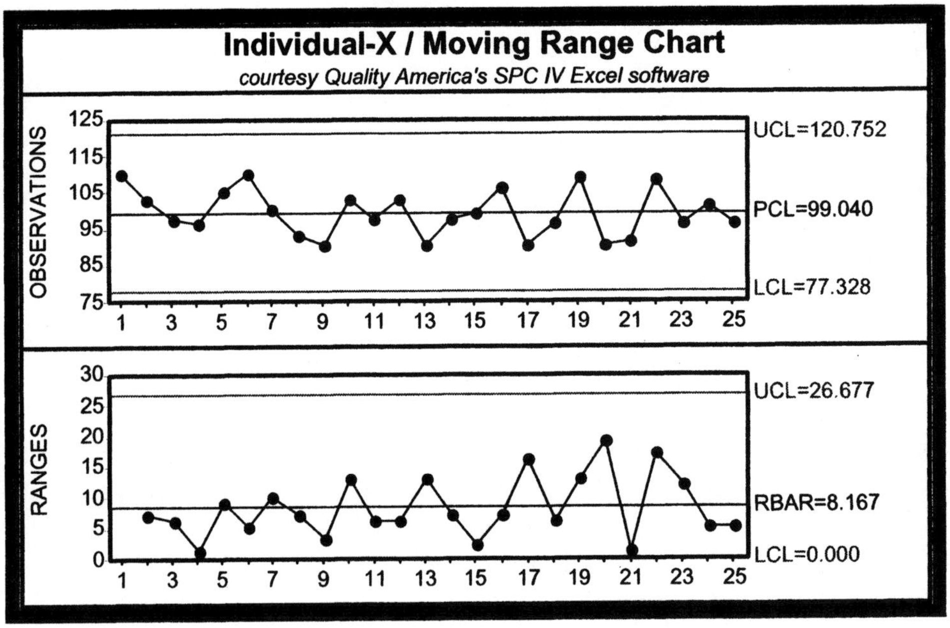

Figure 13.3. Completed Individual and Moving Range Control Charts

In this case the conclusions are the same as with averages charts. However, averages charts always provide tighter control than X charts. In some cases the additional sensitivity provided by averages charts may not be justified on either an economic or an engineering basis. When this happens the use of averages charts will merely lead to wasting money by investigating special causes that are of minor importance.

13.2.4 Proportion defective charts

Proportion defective charts (*P* charts) are statistical tools used to evaluate the proportion defective, or proportion non-conforming, produced by a process. *P* charts can be applied to any variable where the appropriate performance measure is a unit count. *P* charts answer the question: "has a special cause of variation caused the central tendency of this process to produce an abnormally large or small number of defective units over the time period observed?"

The center for a *P* chart line is the average proportion defective and the two control limits are set at plus and minus three standard deviations. If the process is in statistical control then virtually all proportions should be between the control limits and they should fluctuate randomly about the center line. The P chart characteristics are:

$$p = \frac{\text{subgroup defective count}}{\text{subgroup size}} \tag{13.21}$$

$$\bar{p} = \frac{\text{sum of subgroup defective counts}}{\text{sum of subgroup sizes}} \tag{13.22}$$

$$LCL = \bar{p} - 3\sqrt{\frac{\bar{p}(1-\bar{p})}{n}} \tag{13.23}$$

$$UCL = \bar{p} + 3\sqrt{\frac{\bar{p}(1-\bar{p})}{n}} \tag{13.24}$$

In the above equations, n is the subgroup size. If the subgroup sizes varies, the control limits will also vary, becoming closer together as n increases.

As with all control charts, a special cause is probably present if there are any points beyond either the upper or the lower control limit. Analysis of P chart patterns between the control limits is extremely complicated if the sample size varies because the distribution of p varies with the sample size.

Example 13.4

The data in Table 13.5 were obtained by opening randomly selected crates from each shipment and counting the number of bruised peaches. There are 250 peaches per crate. Normally samples consist of one crate per shipment. However, when part-time help is available, samples of 2 crates are taken.

Table 13.5. Raw data for P chart.

Shipment No.	Crates	Peaches	Bruised	p
1	1	250	47	0.188
2	1	250	42	0.168
3	1	250	55	0.220
4	1	250	51	0.204
5	1	250	46	0.184
6	1	250	61	0.244
7	1	250	39	0.156
8	1	250	44	0.176
9	1	250	41	0.164
10	1	250	51	0.204
11	2	500	88	0.176
12	2	500	101	0.202
13	2	500	101	0.202
14	1	250	40	0.160
15	1	250	48	0.192
16	1	250	47	0.188

17	1	250	50	0.200
18	1	250	48	0.192
19	1	250	57	0.228
20	1	250	45	0.180
21	1	250	43	0.172
22	2	500	105	0.210
23	2	500	98	0.196
24	2	500	100	0.200
25	2	500	96	0.192
	TOTALS	8,000	1,544	

Solution

The center-line is the same for subgroups of size 250 and 500. The center-line is

$$\bar{p} = \frac{\text{sum of subgroup defective counts}}{\text{sum of subgroup sizes}} = \frac{1{,}544}{8{,}000} = 0.193$$

The control limits for subgroups of size 250 are:

$$LCL = \bar{p} - 3\sqrt{\frac{\bar{p}(1-\bar{p})}{n}} = 0.193 - 3\sqrt{\frac{0.193(1-0.193)}{250}} = 0.118$$

$$UCL = \bar{p} + 3\sqrt{\frac{\bar{p}(1-\bar{p})}{n}} = 0.193 + 3\sqrt{\frac{0.193(1-0.193)}{250}} = 0.268$$

The control limits for subgroups of size 500 are:

$$LCL = 0.193 - 3\sqrt{\frac{0.193(1-0.193)}{500}} = 0.140$$

$$UCL = 0.193 + 3\sqrt{\frac{0.193(1-0.193)}{500}} = 0.246$$

The control limits and the subgroup proportions are shown in Figure 13.4.

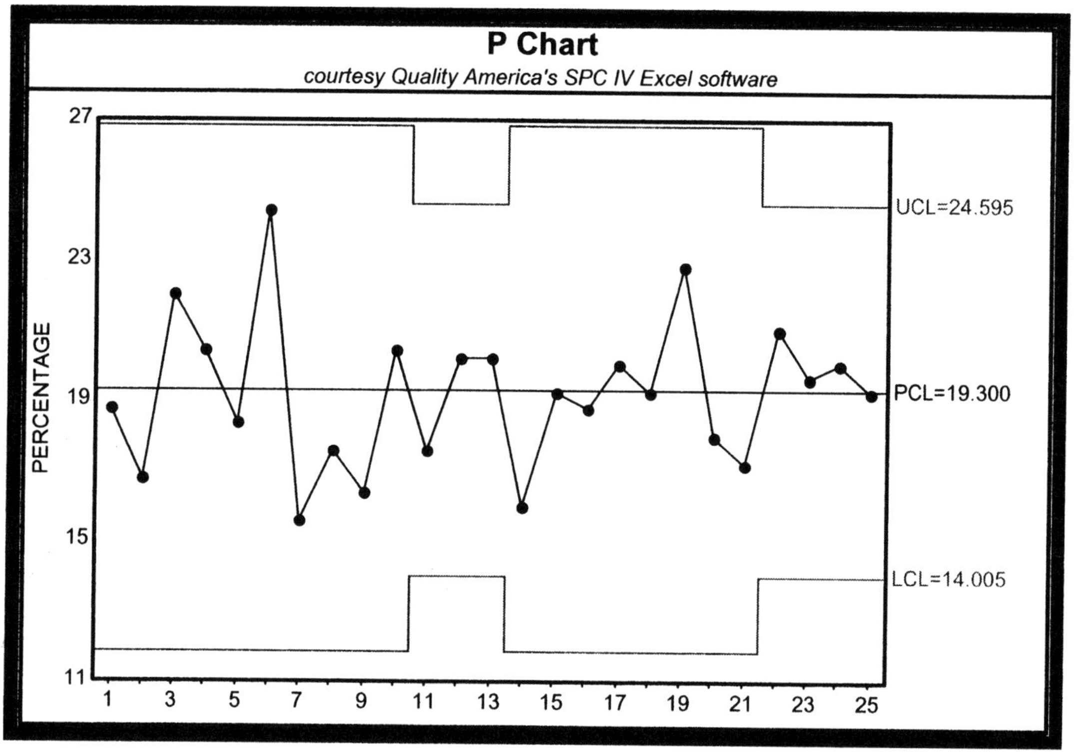

Figure 13.4. Example *P* chart.

13.2.5 Count of defectives charts

Count of defectives charts (*np* charts) are statistical tools used to evaluate the count of defectives, or count of items non-conforming, produced by a process. This type of chart can be applied to any variable where the appropriate performance measure is a unit count and the subgroup size is held constant. Note that wherever an *np* chart can be used, a *p* chart can be used too.

The control limits for an *np* chart are:

$$LCL = n\bar{p} - 3\sqrt{n\bar{p}(1-\bar{p})} \tag{13.25}$$

$$LCL = n\bar{p} + 3\sqrt{n\bar{p}(1-\bar{p})} \tag{13.26}$$

Example 13.5

The data in Table 13.6 were obtained by opening randomly selected crates from each shipment and counting the number of bruised peaches. There are 250 peaches per crate (constant n is required for ***np*** charts.)

Table 13.6. Data for ***np*** chart.

Shipment	Bruised Peaches
1	20
2	28
3	24
4	21
5	32
6	33
7	31
8	29
9	30
10	34
11	32
12	24
13	29
14	27
15	37
16	23
17	27
18	28
19	31
20	27
21	30
22	23
23	23
24	27
25	35
26	29
27	23
28	23
29	30
30	28
TOTAL	838

Solution

The lower control limit for the ***np*** chart is

$$n\bar{p} = \frac{\text{sum of subgroup defective counts}}{\text{number of subgroups}} = \frac{838}{30} = 27.93$$

$$LCL = n\bar{p} - 3\sqrt{n\bar{p}(1-\bar{p})} = 27.93 - 3\sqrt{27.93\left(1 - \frac{27.93}{250}\right)} = 12.99$$

The upper control limit for the ***np*** chart is

$$UCL = n\bar{p} + 3\sqrt{n\bar{p}(1-\bar{p})} = 27.93 + 3\sqrt{27.93\left(1 - \frac{27.93}{250}\right)} = 42.88$$

The control limits and the subgroup defective counts are shown in Figure 13.5.

13.2.6 Average occurrence charts

Average occurrence charts (*U* charts) are statistical tools used to evaluate the average number of occurrences-per-unit produced by a process. *U* charts can be applied to any variable where the appropriate performance measure is a count of how often a particular event occurs. *U* charts answer the question: "has a special cause of variation caused the central tendency of this process to produce an abnormally large or small number of occurrences over the time period observed?" Note that, unlike *P* or *np* charts, *U* charts do not necessarily involve counting physical items. Rather, they involve counting of *events*. For example, when using a *P* chart one would count bruised peaches. When using a *U* chart they would count the *bruises*.

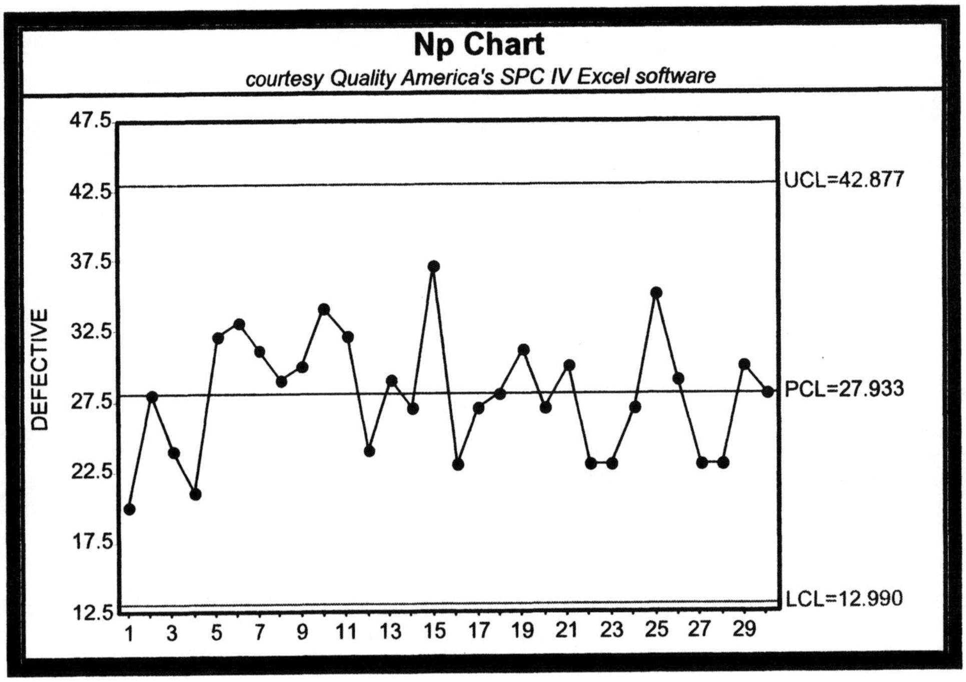

Figure 13.5. Example ***np*** chart.

The center line for a *U* chart is the average number of occurrences-per-unit and the two control limits are set at plus and minus three standard deviations. The lower control limit is

$$LCL = \bar{u} - 3\sqrt{\frac{\bar{u}}{n}} \tag{13.27}$$

where

$$u = \frac{\text{subgroup count of occurrences}}{\text{subgroup size in units}} \tag{13.28}$$

$$\bar{u} = \frac{\text{sum of subgroup occurrences}}{\text{sum of subgroup sizes in units}} \tag{13.29}$$

The upper control limit is

$$UCL = \bar{u} + 3\sqrt{\frac{\bar{u}}{n}} \tag{13.30}$$

In the above equations, *n* is the subgroup size in units. If the subgroup sizes varied, the control limits would also vary. One way of helping determine whether or not a particular set of data is suitable for a *u* chart or a *p* chart is to examine the equation used to compute the center line for the control chart. If the unit of measure is the same in both the numerator and the denominator, then a *p* chart is indicated, otherwise a *u* chart is indicated. For example, if the center-line is the average number of bruises per crate, then the numerator is in terms of bruises while the denominator is in terms of crates, indicating a *u* chart.

The unit size is arbitrary but once determined it cannot be changed without re-computing all subgroup occurrences-per-unit and control limits. For example, if the occurrences were accidents and a unit was 100,000 hours worked, then a month with 250,000 hours worked would be 2.5 units and a month with 50,000 hours worked would be 0.5 units. If the unit size were 200,000 hours then the two months would have 1.25 and 0.2 units respectively. The equations for the center line and control limits would "automatically" take into account the unit size, so the control charts would give identical results regardless of which unit size is used.

Example 13.6

The data in Table 13.7 were obtained by opening randomly selected crates from each shipment and counting the number of bruised peaches. There are 250 peaches per crate. Our unit size will be taken as one full crate, i.e., we will be counting crates rather than the peaches themselves. Normally samples consist of one crate per shipment. However, when part-time help is available, samples of 2 crates are taken.

Table 13.7. Raw Data for u chart

Shipment No.	Units (Crates)	Flaws	Flaws per Unit
1	1	47	47
2	1	42	42
3	1	55	55
4	1	51	51
5	1	46	46
6	1	61	61
7	1	39	39
8	1	44	44
9	1	41	41
10	1	51	51
11	2	88	44
12	2	101	50.5
13	2	101	50.5
14	1	40	40
15	1	48	48
16	1	47	47
17	1	50	50
18	1	48	48
19	1	57	57
20	1	45	45
21	1	43	43
22	2	105	52.5
23	2	98	49
24	2	100	50
25	2	96	48
TOTALS	32	1,544	

Solution

When a single crate is inspected the center line and control limits are

$$\bar{u} = \frac{\text{sum of subgroup count of occurrences}}{\text{sum of subgroup unit sizes}} = \frac{1{,}544}{32} = 48.25$$

$$LCL = \bar{u} - 3\sqrt{\frac{\bar{u}}{n}} = 48.25 - 3\sqrt{\frac{48.25}{1}} = 27.411$$

$$UCL = \bar{u} + 3\sqrt{\frac{\bar{u}}{n}} = 48.25 + 3\sqrt{\frac{48.25}{1}} = 69.089$$

When 2 crates are inspected the control limits are

$$LCL = 48.25 - 3\sqrt{\frac{48.25}{2}} = 33.514$$

$$UCL = 48.25 + 3\sqrt{\frac{48.25}{2}} = 62.986$$

The control limits and the subgroup occurrences-per-unit are shown in Figure 13.6.

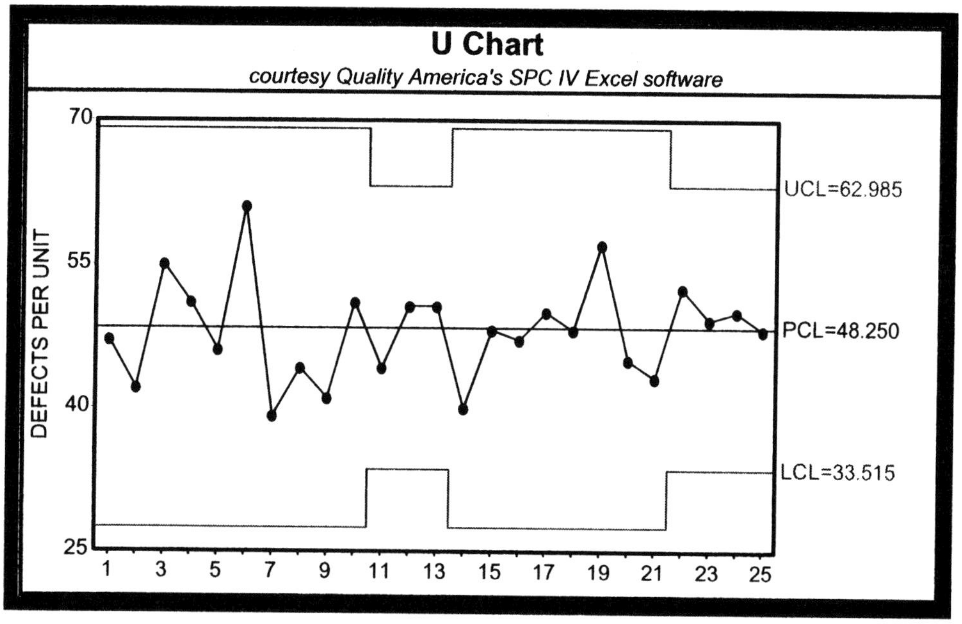

Figure 13.6. Example *U* chart.

Note that the data used to construct the *U* chart were the same as that used for the *P* chart, except the counts are considered occurrences (flaws) instead physical items (bruised peaches). The practical implications of using a *U* chart when a *P* chart should have been used, or vice versa, are usually not serious. The decisions based on the control charts will be quite similar in most cases encountered in quality engineering regardless of whether a *U* or a *P* chart is used.

13.2.7 Counts of occurrence charts

Counts of occurrence charts (*C* charts) are statistical tools used to evaluate the number of occurrences-per-unit produced by a process. *C* charts can be applied to any variable where the appropriate performance measure is a count of how often a particular event occurs and samples of constant size are used. *C* charts answer the question: "has a special cause of variation caused the central tendency of this process to produce an abnormally large or small number of occurrences over the time period observed?" Note that, unlike *P* or *np* charts, *C* charts do not involve counting physical items. Rather, they involve counting of *events*. For example, when using an *np* chart one would count bruised peaches. When using a *C chart* they would count the *bruises*.

The center-line for a *C* chart is the average number of occurrences-per-unit.

$$\overline{c} = \frac{\text{sum of subgroup occurrences}}{\text{number of subgroups}} \tag{13.31}$$

The control limits for a *C* chart are

$$LCL = \overline{c} - 3\sqrt{\overline{c}} \tag{13.32}$$

$$UCL = \overline{c} + 3\sqrt{\overline{c}} \tag{13.33}$$

One way of helping determine whether or not a particular set of data is suitable for a *C chart* or an *np* chart is to examine the equation used to compute the center-line for the control chart. If the unit of measure is the same in both the numerator and the denominator, then a *P* chart is indicated, otherwise a *C chart* is indicated. For example, if the center line is the number of bruises divided by the number of crates, then the numerator is in terms of bruises while the denominator is in terms of crates, indicating a *C chart*. The unit size is arbitrary but, once determined, it cannot be changed without re-computing all subgroup occurrences-per-unit and control limits.

Example 13.7

The data in Table 13.8 were obtained by opening randomly selected crates from each shipment and counting the number of flaws. There are 250 peaches per crate. The unit size will be taken as one full crate. Every subgroup consists of one crate. If the subgroup size varied, a *U* chart would be used.

Table 13.8. *C* chart data.

Shipment	Flaws
1	27
2	32
3	24
4	31
5	42
6	38
7	33
8	35
9	35
10	39
11	41
12	29

13	34
14	34
15	43
16	29
17	33
18	33
19	38
20	32
21	37
22	30
23	31
24	32
25	42
26	40
27	21
28	23
29	39
30	29
TOTAL	1,006

Solution

The lower control limit is:

$$\bar{c} = \frac{\text{sum of subgroup occurrences}}{\text{number of subgroups}} = \frac{1{,}006}{30} = 33.53$$

$$LCL = \bar{c} - 3\sqrt{\bar{c}} = 33.53 - 3(5.791) = 16.161$$

The upper control limit is:

$$UCL = \bar{c} + 3\sqrt{\bar{c}} = 33.53 + 3(5.791) = 50.906$$

The control limits and the occurrence counts are shown in Figure 13.7.

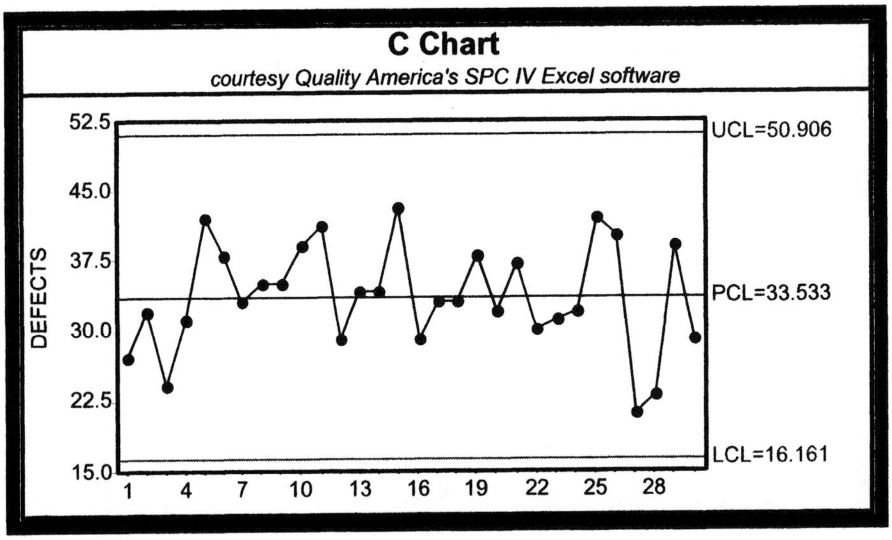

Figure 13.7. Completed *C* chart.

13.3 Advanced control charts

The *X*-bar and *Range* control charts were developed in the days of slide rules. Without computers or even electronic calculators, simple methods were required for estimating standard deviation and controlling variation, thus, the *X*-bar and *Range* charts were developed. Because of their ease of use and simplicity, these charts are still very popular. The simplicity of these charts comes with a price; they are inefficient in comparison to more sophisticated control charts. The cumulative sum (CUSUM) control chart and the exponential weighted moving average chart (EWMA) are much better at detecting out-of-control conditions than the *X*-bar chart.

The average run length (ARL) for a control chart is the average number of sample points that are plotted before an out-of-control condition is found. For an *X*-bar chart with 3-sigma control limits, the average run length is

$$ARL = \frac{1}{p} \tag{13.34}$$

where p is the probability of a single point falling outside the control limits. Table 13.9 shows the average run length as a function of the shift in the mean for an *X*-bar chart.

Table 13.9. ARL for X-bar charts.

Mean Shift (Standard Deviations)	ARL

0.00	370.4
0.25	281.1
0.50	155.2
0.75	81.2
1.00	43.9
1.25	25.0
1.50	15.0
1.75	9.5
2.00	6.3
2.25	4.4
2.50	3.2
2.75	2.5
3.00	2.0

This table is disturbing. When the mean shifts by 1 standard deviation, on average 43.9 points must be plotted on an *X*-bar chart before this shift will be discovered. For a shift in the mean of 1/2 standard deviation, an average of more than 155 points will be plotted on an *X*-bar chart before the out-of-control condition will be discovered.

The CUSUM and EWMA charts were developed to provide more sensitivity to small shifts in the mean. The CUSUM chart is based on the principle of using all the information in a sequence of samples. The value plotted for a CUSUM chart is

$$S_i = \sum_{j=1}^{i} \left(x_j - \mu \right) \tag{13.35}$$

where x_j is the average of the *i*th sample of size *n*, and μ is the process target.

By summing the deviation from the process target, even a small shift in the mean will produce either an upward or downward trend in S_i. For manual calculations, a V-mask can be used to determine out-of-control conditions. A V-mask funnel shaped with the narrow end pointing to the right on the control chart. With computers widely available, the tabular CUSUM is more popular. The upper and lower control limits for the tabular CUSUM chart are set at ±*H*, and 2 values of S_i are plotted; one on the upper side and one on the lower side. These values are

$$S_{ui} = \max\left[0, \bar{x}_i - \left(\mu + K\right) + S_{ui-1}\right] \tag{13.36}$$

$$S_{Li} = \max\left[0, \left(\mu - K\right) - \bar{x}_i + S_{Li-1}\right] \tag{13.37}$$

where *K* is the slope of the sides of the funnel in the corresponding V-mask. Note that the sample for *x*-bar is the set of observations since $S_i = 0$. The values of *H* and *K* determine the sensitivity of the chart. The recommend setting for K is 1/2σ the size of the mean shift the chart is attempting to detect. The table below shows the ARL for a CUSUM chart with $K = 1/2\sigma$.

Table 13.10. CUSUM chart ARL.

Mean Shift (Standard Deviations)	ARL $H = 4\sigma$	ARL $H = 5\sigma$
0	168	465
0.25	74.2	139
0.50	26.6	38.0
0.75	13.3	17.0
1.00	8.38	10.4
1.50	4.75	5.75
2.00	3.34	4.01
2.50	2.62	3.11
3.00	2.19	2.57

Example 13.8

The first 30 data points in the table below were produced by a random number generator with a mean of 50 and a standard deviation of 3. The next 20 points were produced by a random number generator with a mean of 51.5 and a standard deviation of 3. Since the standard deviation is 3, the values of H and K will be set at

$$H = 5\sigma = 5(3) = 15$$

$$K = 0.5\sigma = 0.5(3) = 1.5$$

A sample sub-group size of 1 will be used, so x-bar will be equal to x_i. The data points along with S_{Hi} and S_{Li} are given in Table 13.11.

Table 13.11. Example CUSUM chart computations.

i	Observation	x-Bar	(x-bar) -(μ+K)	(μ -K) -(x-bar)	SH	SL
1	50.99	50.99	-0.51	-2.49	-0.51	-2.49
2	56.36	56.36	4.86	-7.86	4.35	0.00
3	47.51	47.51	-3.99	0.99	0.36	0.99
4	46.81	46.81	-4.69	1.69	0.00	2.68
5	46.34	46.34	-5.16	2.16	0.00	4.84
6	46.38	46.38	-5.12	2.12	0.00	6.96
7	50.11	50.11	-1.39	-1.61	0.00	5.35
8	45.35	45.35	-6.15	3.15	0.00	8.50

9	47.7	47.70	-3.80	0.80	0.00	9.30
10	47.24	47.24	-4.26	1.26	0.00	10.56
11	46.21	46.21	-5.29	2.29	0.00	12.85
12	55.26	55.26	3.76	-6.76	3.76	6.09
13	48.69	48.69	-2.81	-0.19	0.95	5.90
14	50.07	50.07	-1.43	-1.57	0.00	4.33
15	54.69	54.69	3.19	-6.19	3.19	0.00
16	51.33	51.33	-0.17	-2.83	3.02	0.00
17	49.98	49.98	-1.52	-1.48	1.50	0.00
18	46.37	46.37	-5.13	2.13	0.00	2.13
19	49.41	49.41	-2.09	-0.91	0.00	1.22
20	50.16	50.16	-1.34	-1.66	0.00	0.00
21	55.33	55.33	3.83	-6.83	3.83	0.00
22	50.15	50.15	-1.35	-1.65	2.48	0.00
23	56.07	56.07	4.57	-7.57	7.05	0.00
24	47.37	47.37	-4.13	1.13	2.92	1.13
25	48.82	48.82	-2.68	-0.32	0.24	0.81
26	48.95	48.95	-2.55	-0.45	0.00	0.36
27	48.2	48.20	-3.30	0.30	0.00	0.66
28	54.13	54.13	2.63	-5.63	2.63	0.00
29	50.88	50.88	-0.62	-2.38	2.01	0.00
30	50.9	50.90	-0.60	-2.40	1.41	0.00
31	53.02	53.02	1.52	-4.52	2.93	0.00
32	52.04	52.04	0.54	-3.54	3.47	0.00
33	51.14	51.14	-0.36	-2.64	3.11	0.00
34	54.6	54.60	3.10	-6.10	6.21	0.00
35	53.01	53.01	1.51	-4.51	7.72	0.00
36	57.66	57.66	6.16	-9.16	13.88	0.00
37	54.11	54.11	2.61	-5.61	16.49	0.00
38	52.83	52.83	1.33	-4.33	17.82	0.00
39	48.72	48.72	-2.78	-0.22	15.04	0.00
40	47.51	47.51	-3.99	0.99	11.05	0.99
41	49.42	49.42	-2.08	-0.92	8.97	0.07
42	48.94	48.94	-2.56	-0.44	6.41	0.00
43	50.23	50.23	-1.27	-1.73	5.14	0.00
44	52.12	52.12	0.62	-3.62	5.76	0.00
45	50.41	50.41	-1.09	-1.91	4.67	0.00
46	53.96	53.96	2.46	-5.46	7.13	0.00
47	47.93	47.93	-3.57	0.57	3.56	0.57
48	53.63	53.63	2.13	-5.13	5.69	0.00
49	47.13	47.13	-4.37	1.37	1.32	1.37
50	54.77	54.77	3.27	-6.27	4.59	0.00

The CUSUM plot is shown below. Note that S_{Li} is plotted as a negative.

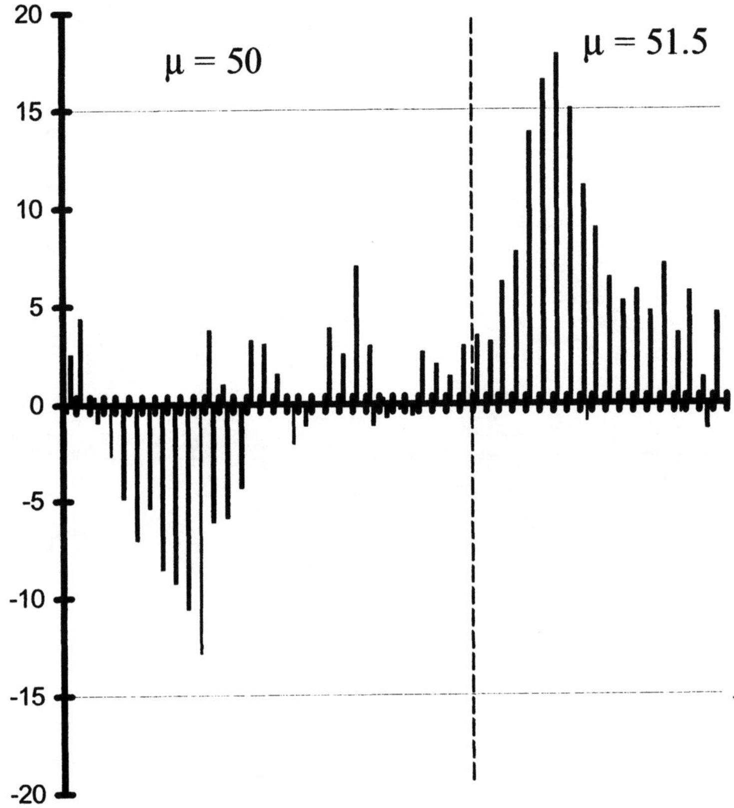

Figure 13.8. Example CUSUM chart.

The EWMA chart has ARL performance similar to the CUSUM chart. The plotted value in an EWMA chart is

$$z_i = \lambda \bar{x}_i + (1-\lambda) z_{i-1} \tag{13.38}$$

$$z_0 = \bar{\bar{x}} \tag{13.39}$$

where λ is a constant that determines the sensitivity of the chart. The value of λ must be $0 < \lambda \leq 1$, but the recommended value is $0 < \lambda \leq 0.1$. In practice, the most common values are λ = 0.08, λ = 0.1, and λ = 0.15. The upper and lower control limits for the EWMA chart are

$$LCL = \bar{\bar{x}} - 3\frac{\sigma}{\sqrt{n}}\sqrt{\left(\frac{\lambda}{2-\lambda}\right)\left[1-(1-\lambda)^{2i}\right]} \tag{13.40}$$

$$UCL = \bar{\bar{x}} + 3\frac{\sigma}{\sqrt{n}}\sqrt{\left(\frac{\lambda}{2-\lambda}\right)\left[1-(1-\lambda)^{2i}\right]} \tag{13.41}$$

As the subgroup size increases (i) to a large value the above equations simplify to

$$LCL = \bar{\bar{x}} - 3\sigma\sqrt{\frac{\lambda}{(2-\lambda)n}} \tag{13.42}$$

$$UCL = \bar{\bar{x}} + 3\sigma\sqrt{\frac{\lambda}{(2-\lambda)n}} \tag{13.43}$$

The term $\bar{\bar{x}}$ in the equations above is often replaced with the target value.

Example 13.9

Using the example data for the CUSUM chart above, the computations for an EWMA chart and the chart itself are shown below. The standard deviation, σ was computed from the first 30 observations, and the target, 50, was used for $\bar{\bar{x}}$.

Table 13.12. EWMA example computations.

Observation Number	Observation Value	z_i	Lower Limit	Upper Limit
1	50.99	50.10	49.10	50.90
2	56.36	50.73	48.79	51.21
3	47.51	50.41	48.59	51.41
4	46.81	50.05	48.44	51.56
5	46.34	49.68	48.33	51.67
6	46.38	49.35	48.25	51.75
7	50.11	49.43	48.19	51.81
8	45.35	49.02	48.14	51.86
9	47.70	48.89	48.10	51.90
10	47.24	48.73	48.06	51.94
11	46.21	48.48	48.04	51.96
12	55.26	49.16	48.02	51.98
13	48.69	49.11	48.00	52.00
14	50.07	49.21	47.99	52.01
15	54.69	49.76	47.98	52.02
16	51.33	49.92	47.97	52.03
17	49.98	49.93	47.96	52.04
18	46.37	49.57	47.96	52.04
19	49.41	49.55	47.95	52.05
20	50.16	49.61	47.95	52.05
21	55.33	50.18	47.95	52.05
22	50.15	50.18	47.95	52.05

23	56.07	50.77	47.94	52.06
24	47.37	50.43	47.94	52.06
25	48.82	50.27	47.94	52.06
26	48.95	50.14	47.94	52.06
27	48.20	49.95	47.94	52.06
28	54.13	50.37	47.94	52.06
29	50.88	50.42	47.94	52.06
30	50.90	50.47	47.94	52.06
31	53.02	50.73	47.94	52.06
32	52.04	50.86	47.94	52.06
33	51.14	50.89	47.94	52.06
34	54.60	51.26	47.94	52.06
35	53.01	51.44	47.94	52.06
36	57.66	52.06	47.94	52.06
37	54.11	52.27	47.94	52.06
38	52.83	52.33	47.94	52.06
39	48.72	51.97	47.94	52.06
40	47.51	51.52	47.94	52.06
41	49.42	51.31	47.94	52.06
42	48.94	51.07	47.94	52.06
43	50.23	50.99	47.94	52.06
44	52.12	51.10	47.94	52.06
45	50.41	51.03	47.94	52.06
46	53.96	51.32	47.94	52.06
47	47.93	50.98	47.94	52.06
48	53.63	51.25	47.94	52.06
49	47.13	50.84	47.94	52.06
50	54.77	51.23	47.94	52.06

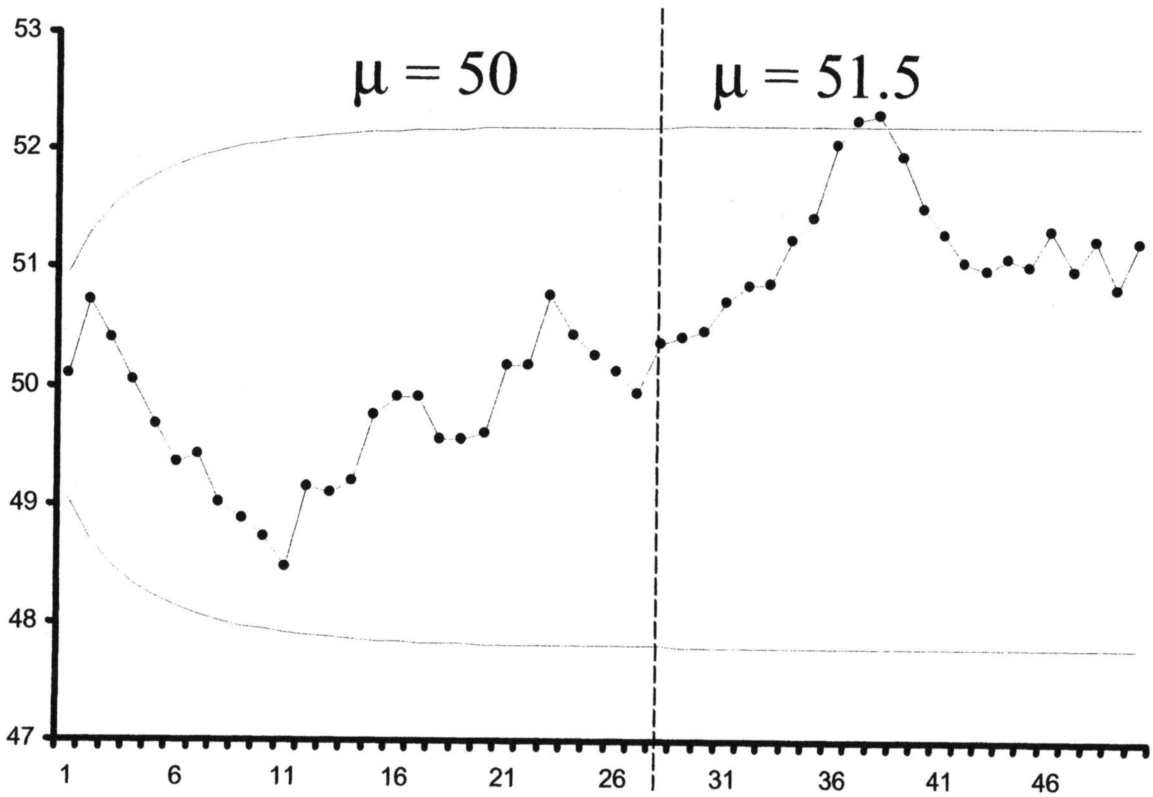

Figure 13.9. Example EWMA chart.

13.4 Implementation

Assuming that the organization's leadership has created an environment where open and honest communication can flourish, SPC implementation becomes a matter of (1) selecting processes for applying the SPC approach and (2) selecting variables within each process. This section describes an approach to this activity.

13.4.1 Control plans

Process control plans should be prepared for each key process. The plans should be prepared by teams of people who understand the process. The team should begin by creating a flowchart of the process using the process elements determined in creating the house of quality. The flowchart will show how the process elements relate to one another and it will help in the selection of control points. It will also show the point of delivery to the customer, which is usually an important control point. Note that the customer may be an internal customer.

For any given process there are a number of different types of process elements. Some process elements are internal to the process, others external. The rotation speed of a drill is an internal process element, while the humidity in the building is external. Some process elements, while important, are easy to hold constant at a given value so that they do not change unless deliberate action is taken. These are *fixed* elements. Other process elements vary of their own accord and must be watched, these are *variable* elements. The drill rotation speed can be set in advance, but the line voltage for the drill press may vary, which causes the drill speed to change in spite of its initial setting. Figure 13.10 provides a planning guide based on the internal/external and fixed/variable classification scheme. Of course, other classification schemes may be more suitable on a given project and the engineer is encouraged to develop the approach that best serves his or her needs. For convenience each class is identified with a roman numeral; I = fixed-internal, II = fixed-external, III = variable-internal and IV = variable-external.

	INTERNAL	EXTERNAL
FIXED	I Setup approval Periodic audits Preventive maintenance	II Audit Certification
VARIABLE	III Control charts Fool-proofing product Fool-proofing process Sort the output	IV Supplier SPC Receiving inspection Supplier sorting Fool-proof product Fool-proofing process

Figure 13.10. Guide to selecting and controlling process variables.

In selecting the appropriate method of control for each process element, pay particular attention to those process elements which received high importance rankings in the house of quality analysis. In some cases an important process element is very expensive to control. When this happens, look at the QFD correlation matrix or the statistical correlation matrix for possible assistance. The process element may be correlated with other process elements that are less costly to control. Either correlation matrix will also help you to minimize the number of control charts. It is usually unnecessary to keep control charts on several variables that are correlated with one another. In these cases it may be possible to select the process element that is least expensive (or most sensitive) to monitor as the control variable.

As Figure 13.10 indicates, control charts are not always the best method of controlling a given process element. In fact, control charts are seldom the method of choice. When process elements are important we would prefer that they *not vary at all*! Only when this cannot be accomplished economically should the engineer resort to the use of control charts to monitor the element's variation. Control charts may be thought of as a control mechanism of last resort. Control charts are useful only when the element being monitored can be expected to exhibit measurable and "random-looking" variation when the process is properly controlled. A process element that always measures 10 if everything is okay is not a good candidate for control charting. Nor is one that measures 10 or 12, but never anything else. Ideally, the measurements being monitored with variables control charts will be capable of taking on any value, i.e., the data will be continuous. Discrete measurement data can be used if it's not too discrete; indeed, all real-world data are somewhat discrete. As a rule of thumb, at least 10 different values should appear in the data set and no one value should comprise more than 20% of the data set. When the measurement data becomes too discrete for SPC, monitor it with checksheets or simple time-ordered plots. Of course, the above discussion applies to measurement data. Attribute control charts can be used to monitor process elements that are discrete counts.

Any process control plan must include instructions on the action to be taken if problems appear. This is particularly important where control charts are being used for process control. Unlike process control procedures such as audits or setup approvals, it is not always apparent just what is wrong when a control chart indicates a problem. The investigation of special causes of variation usually consists of a number of predetermined actions (such as checking the fixture or checking a cutting tool) followed by notifying someone if the items checked don't reveal the source of the problem. Also verify that the arithmetic was done correctly and that the point was plotted in the correct position on the control chart.

Note that Figure 13.10 includes "sort output" as part of the process control plan. Sorting the output implies that the process is not capable of meeting the customer's requirements, as determined by a process capability study and the application of Deming's all-or-none rules. However, even if sorting is taking place, SPC is still advisable. SPC will help assure that things don't get any worse. SPC will also reveal improvements that may otherwise be overlooked. The improvements may result in a process that is good enough to eliminate the need for sorting.

13.4.2 Rational subgroups

The basis of all control charts is the *rational subgroup*. Rational subgroups are composed of items which were produced under essentially the same conditions. The statistics, for example the average and range, are computed for each subgroup separately, then plotted on the control chart. When possible, rational subgroups are formed by using consecutive units. Each subgroup's statistics are compared to the control limit, and patterns of variation between subgroups are analyzed. Note the sharp contrast between this approach and the *random sampling* approach used for enumerative statistical methods.

The idea of rational subgrouping becomes a bit fuzzy when dealing with x charts, or individuals control charts. The reader may well wonder about the meaning of the term subgrouping when the "subgroup" is a single measurement. The basic idea underlying control charts of all types is to identify the capability of the process. The mechanism by which this is accomplished is careful formation of rational subgroups as defined above. When possible, rational subgroups are formed by using consecutive units. The measure of process variability, either the subgroup standard deviation or the subgroup range, is the basis of the control limits for averages. Conceptually, this is akin to basing the control limits on short-term variation. These control limits are used to monitor variation over time.

As far as possible, this approach also forms the basis of establishing control limits for individual measurements. This is done by forming quasi-subgroups using pairs of consecutive measurements. These subgroups of 2 are used to compute ranges. The ranges are used to compute the control limits for the individual measurements.

13.4.3 Interpretation

Control charts provide the operational definition of the term "special cause." A special cause is simply anything which leads to an observation beyond a control limit. However, this simplistic use of control charts does not do justice to their power. Control charts are running records of the performance of the process and, as such, they contain a vast store of information on potential improvements. While some guidelines are presented here, control chart interpretation is an art that can only be developed by looking at many control charts and probing the patterns to identify the underlying system of causes at work.

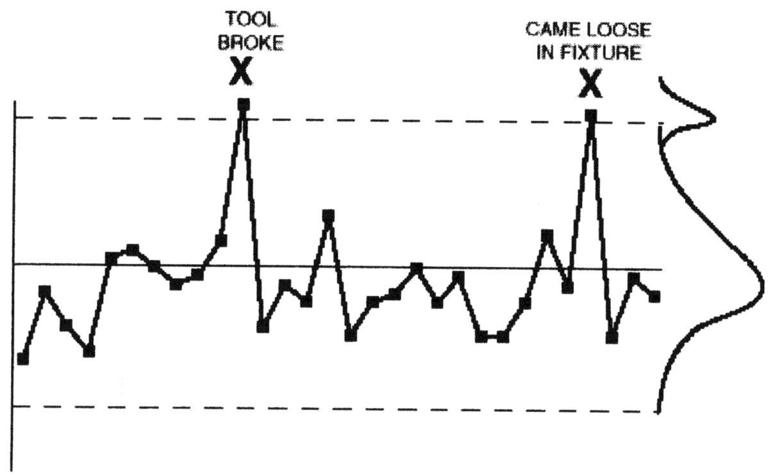

Figure 13.11. Control chart patterns: freaks.

Freak patterns are the classical special cause situation. Freaks result from causes that have a large effect but that occur infrequently. When investigating freak values look at the cause-and-effect diagram for items that meet these criteria. The key to identifying freak causes is timelines in collecting and recording the data. If you have difficulty, try sampling more frequently.

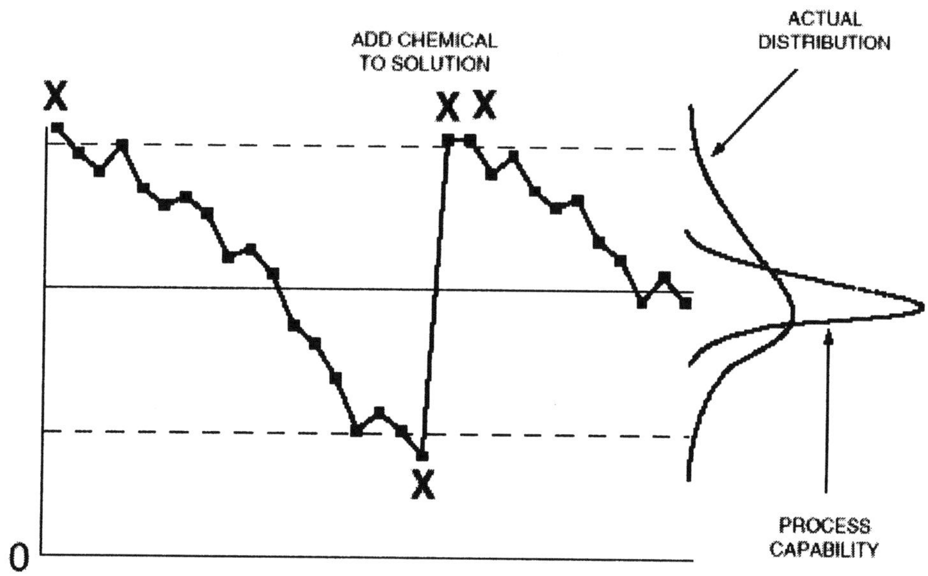

Figure 13.12. Control chart patterns: drift.

Drift is generally seen in processes where the current process value is partly determined by the previous process state. For example, if the process is a plating bath, the content of the tank cannot change instantaneously, instead it will change gradually.

Another common example is tool wear: the size of the tool is related to its previous size. Once the cause of the drift has been determined, the appropriate action can be taken. Whenever economically feasible the drift should be eliminated, e.g., install an automatic chemical dispenser for the plating bath, or make automatic compensating adjustments to correct for tool wear. Note that the total process variability increases when drift is allowed, which adds cost. When this is not possible, the control chart can be modified in one of two ways:

Make the slope of the center line and control limits match the natural process drift. The control chart will then detect departures from the natural drift.

Plot *deviations* from the natural or expected drift.

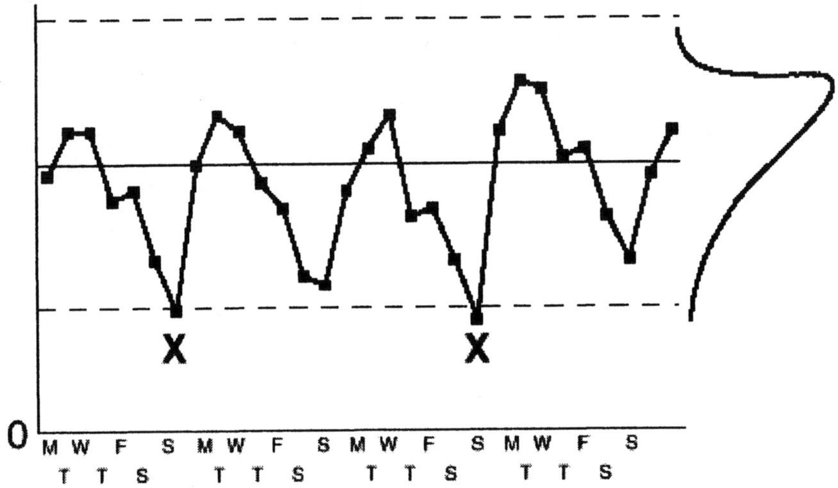

Figure 13.13. Control chart patterns: cycles.

Cycles often occur due to the nature of the process. Common cycles include hour of the day, day of the week, month of the year, quarter of the year, week of the accounting cycle, etc. Cycles are caused by modifying the process inputs or methods according to a regular schedule. The existence of this schedule and its effect on the process may or may not be known in advance. Once the cycle has been discovered, action can be taken. The action might be to adjust the control chart by plotting the control measure against a variable base. For example, if a day-of-the-week cycle exists for shipping errors because of the workload, you might plot shipping errors per 100 orders shipped instead of shipping errors per day. Alternatively, it may be worthwhile to change the system to smooth out the cycle. Most processes operate more efficiently when the inputs are relatively stable and when methods are changed as little as possible.

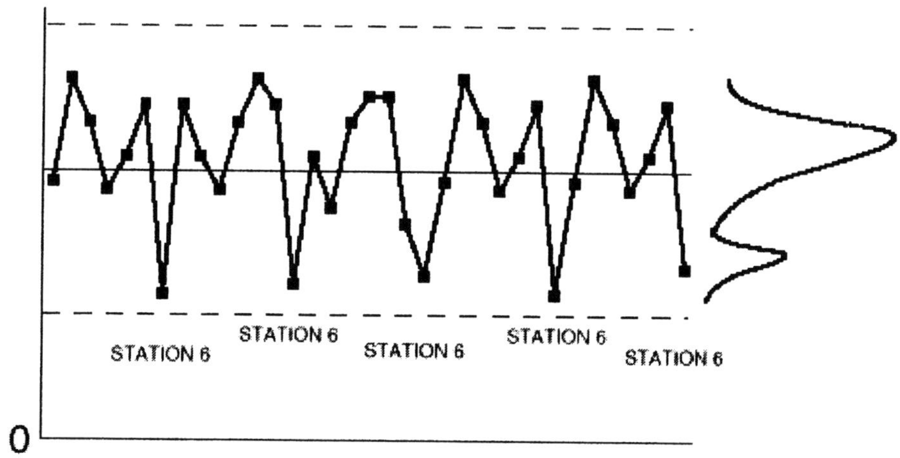

Figure 13.14. Control chart patterns: repeating patterns.

A controlled process will exhibit only "random looking" variation. A pattern where every n^{th} item is different is, obviously, non-random. These patterns are sometimes quite subtle and difficult to identify. It is sometimes helpful to see if the average fraction defective is close to some multiple of a known number of process streams. For example, if the machine is a filler with 40 stations, look for problems that occur 1/40, 2/40, 3/40, etc. of the time.

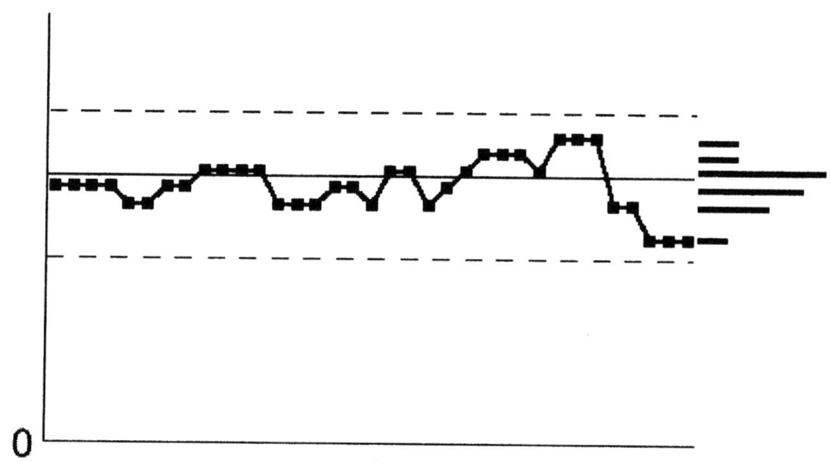

Figure 13.15. Control chart patterns: discrete data.

When plotting measurement data the assumption is that the numbers exist on a continuum, i.e., there will be many different values in the data set. In the real world, the data are never completely continuous. It usually doesn't matter much if there are, say, 10 or more different numbers. However, when there are only a few numbers that appear over-and-over it can cause problems with the analysis. A common problem is that the R chart will underestimate the average range, causing the control limits on both the average and range charts to be too close together. The result will be too many "false alarms" and a general loss of confidence in SPC.

The usual cause of this situation is inadequate gage resolution. The ideal solution is to obtain a gage with greater resolution. Sometimes the problem occurs because operators, inspectors or computers are rounding the numbers. The solution here is to record additional digits.

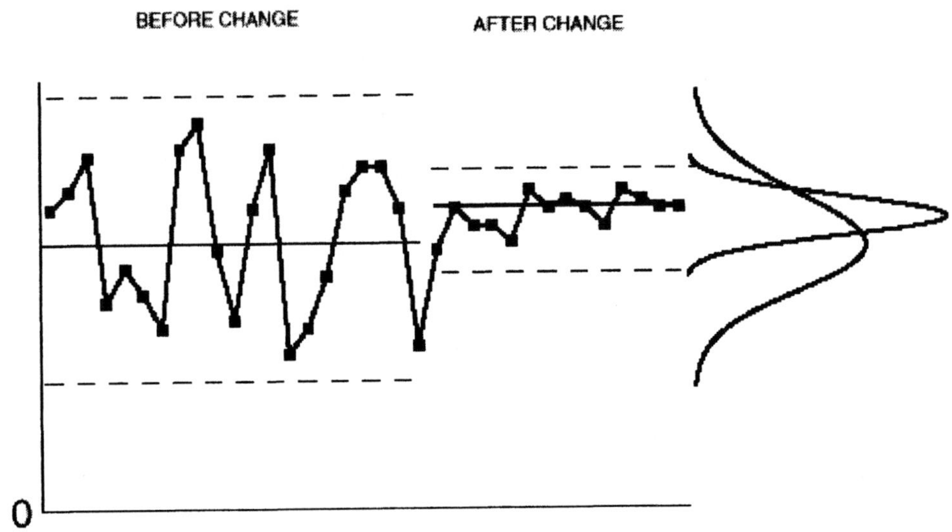

Figure 13.16. Control chart patterns: planned changes.

The reason SPC is done is to accelerate the learning process and to eventually produce an improvement. Control charts serve as historical records of the learning process and they can be used by others to improve other processes. When an improvement is realized the change should be written on the old control chart, its effect will show up as a less variable process. These charts are also useful in communicating the results to leaders, suppliers, customers and others interested in quality improvement.

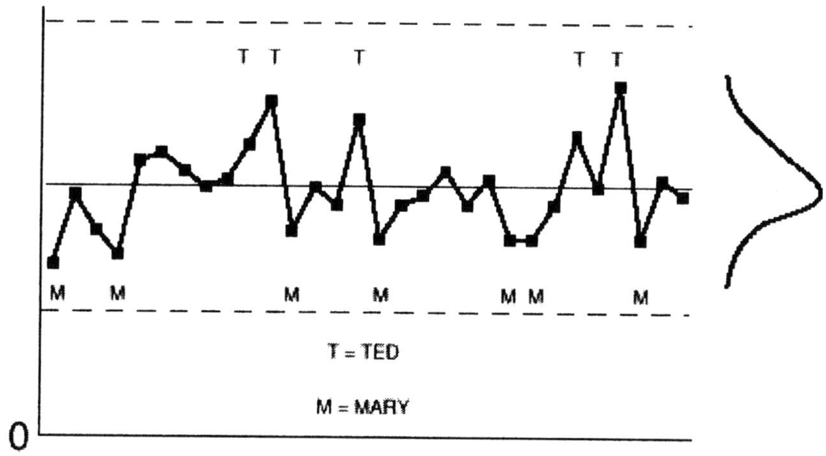

Figure 13.17. Control chart patterns: suspected differences.

Seemingly random patterns on a control chart are evidence of unknown causes of variation, which is not the same as *uncaused* variation. There should be an ongoing effort to reduce the variation from these so-called common causes. Doing so requires that the unknown causes of variation be identified. One way of doing this is a retrospective evaluation of control charts. This involves brainstorming and preparing cause-and-effect diagrams, then relating the control chart patterns to the causes listed on the diagram. For example, if "operator" is a suspected cause of variation, place a label on the control chart points produced by each operator. If the labels exhibit a pattern, there is evidence to suggest a problem. Conduct an investigation into the reasons and set up controlled experiments (prospective studies) to test any theories proposed. If the experiments indicate a true cause-and-effect relationship, make the appropriate process improvements. Keep in mind that a statistical *association* is not the same thing as a causal *correlation*. The observed association must be backed up with solid subject-matter expertise and experimental data.

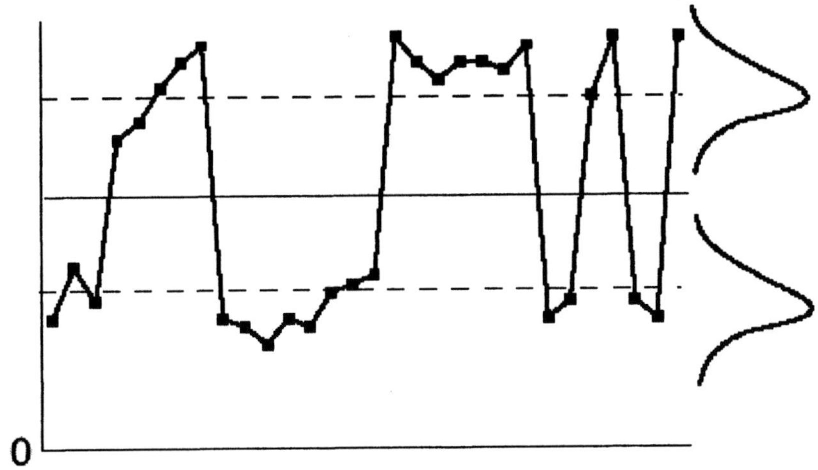

Figure 13.18. Control chart patterns: mixture.

Mixture exists when there data from two different cause-systems are plotted on a single control chart. It indicates a failure in creating rational subgroups. The underlying differences should be identified and corrective action taken. The nature of the corrective action will determine how the control chart should be modified.

Mixture example #1 - The mixture represents two different operators who can be made more consistent. A single control chart can be used to monitor the new, consistent process.

Mixture example #2 - The mixture is in the number of emergency room cases received on Saturday evening, versus the number received during a normal work day. Separate control charts should be used to monitor patient load during the two different time periods.

13.4.4 Process control rules

If the process is stable, then the distribution of subgroup averages will be approximately normal. With this in mind we can also analyze the *patterns* on the control charts to see if they might be attributed to a special cause of variation. To do this we divide a normal distribution into zones, with each zone 1 standard deviation wide. Figure 13.19 shows the approximate percentage we expect to find in each zone from a stable process.

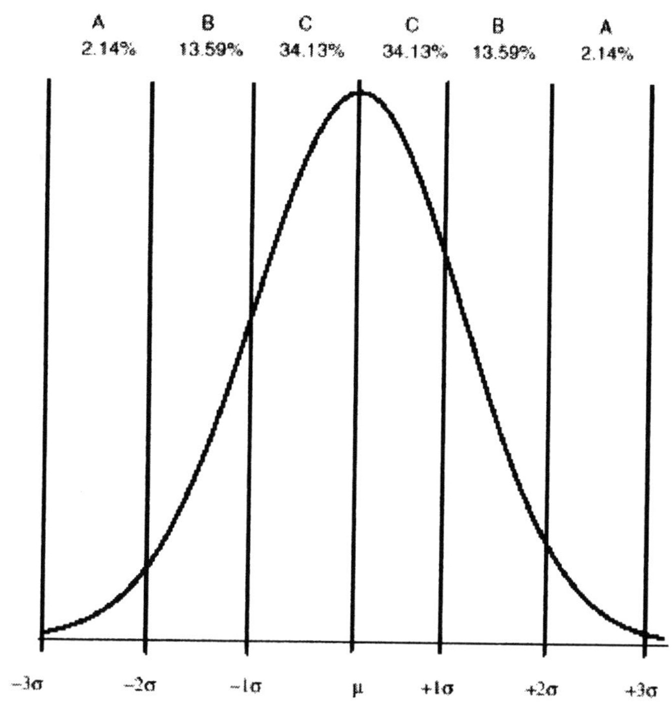

Figure 13.19. Percentiles for a normal distribution.

Zone *C* is the area from the mean to the mean plus or minus one sigma, zone *B* is from plus or minus one to plus or minus two sigma, and zone *A* is from plus or minus two to plus or minus three sigma. Of course, any point beyond three sigma (i.e., outside of the control limit) is an indication of an out-of-control process.

Since the control limits are at plus and minus 3 standard deviations, finding the 1 and 2 sigma lines on a control chart is as simple as dividing the distance between the grand average and either control limit into thirds, which can be done using a ruler. This divides each half of the control chart into 3 zones. The 3 zones are labeled *A*, *B*, and *C* as shown on Figure 13.20.

Figure 13.20. Zones on a control chart.

Based on the expected percentages in each zone, sensitive run tests can be developed for analyzing the patterns of variation in the various zones. Remember, the existence of a non-random pattern means that a special cause of variation was (or is) probably present. The averages control chart run tests are:

1. One point beyond zone A.
2. Nine consecutive points on one side of the center-line.
3. Six consecutive points either increasing or decreasing.
4. Fourteen consecutive points alternating up and down.
5. Two of 3 consecutive points in zone A.
6. Four of 5 consecutive points in zone B or beyond.
7. Fifteen consecutive points in zone C.
8. Eight consecutive points with no points in zone C.

Note that, when a point responds to an out-of-control test it is marked with an "X" to make the interpretation of the chart easier. Using this convention, the patterns on the control charts can be used as an aid in troubleshooting.

13.5 Tampering effects and diagnostics

Tampering occurs when adjustments are made to a process that is already in statistical control. Adjusting a controlled process will always increase process variability, an obviously undesirable result. The best means of diagnosing tampering is to conduct a process capability study and to use a control chart to provide guidelines for adjusting the process.

Perhaps the best analysis of the effects of tampering is from Deming (1986). Deming describes four common types of tampering by drawing an analogy with moving a funnel to hit a desired target. These "funnel rules" are:

- Leave the funnel fixed, aimed at the target, no adjustment.
- At drop k (k = 1, 2, 3,...) the marble will come to rest at point z_k, measured from the target. (In other words, z_k is the error at drop k.) Move the funnel the distance - z_k from the last position - memory 1.
- Set the funnel at each drop right over the spot z_k, measured from the target - no memory.
- Set the funnel at each drop right over the spot (z_k) where it last came to rest - no memory.

Rule #1 is the best rule for stable processes. By following this rule the process average will remain stable and the variance will be minimized. Rule #2 produces a stable output but one with twice the variance of rule #1. Rule #3 results in a system that "explodes", i.e., a symmetrical pattern will appear with a variance that increases without bound. Rule #4 creates a pattern that steadily moves away from the target, without limit.

At first glance one might wonder about the relevance of such apparently abstract rules. However, upon more careful consideration, one finds many practical situations where these rules apply.

Rule #1 is the ideal situation and it can be approximated by using control charts to guide decision making. If process adjustments are made only when special causes are indicated and identified, a pattern similar to that produced by rule #1 will result.

Rule #2 has intuitive appeal for many people. It is commonly encountered in such activities as gage calibration (check the standard once and adjust the gage accordingly) or in some automated equipment (using an automatic gage, check the size of the last feature produced and make a compensating adjustment). Since the system produces a stable result this situation can go unnoticed indefinitely. However, as shown by Taguchi, increased variance translates to poorer quality and higher cost.

The rationale that leads to rule #3 goes something like this: "A measurement was taken and it was found to be 10 units above the desired target. This happened because the process was set 10 units too high. I want the average to equal the target. To accomplish this I must try to get the next measurement to be 10 units too low." This might be used, for example, in preparing a chemical solution. While reasonable on its face, the result of this approach is a wildly oscillating system.

A common example of rule #4 is the "train-the-trainer" method. A master educator spends a short time training a group of "experts," who then train others, who train others, et cetera. A similar example is on-the-job training. Another is creating a setup by using a piece from the last job. Yet another is a gage calibration system where standards are used to create other standards, which are used to create still others, and so on. Just how far the final result will be from the ideal depends on how many levels deep the scheme has progressed.

13.6 Process capability studies and indices

Process capability analysis is a two-stage process that involves:

bringing a process into a state of statistical control for a reasonable period of time, and

comparing the long-term process performance to management or engineering requirements.

Process capability analysis can be done with either attribute data or continuous data *if and only if the process is in statistical control*, and has been for a reasonable period of time. Application of Process capability methods to processes that are not in statistical control results in unreliable estimates of process capability and should never be done.

13.6.1 How to perform a process capability study

Following is a step-by-step approach to process capability analysis (Pyzdek, 1985.)

Select a Candidate for the Study - This step should be institutionalized. A goal of any organization should be ongoing process improvement. However, because a company has only a limited resource base and can't solve all problems simultaneously, it must set priorities for its efforts. The tools for this include Pareto analysis and fishbone diagrams.

Define the Process - It is all too easy to slip into the trap of solving the wrong problem. Once the candidate area has been selected in step 1, define the scope of the study. A process is a unique combination of machines, tools, methods, and personnel engaged in adding value by providing a product or service. Each element of the process should be identified at this stage. This is not a trivial exercise. The input of many people may be required. There are likely to be a number of conflicting opinions about what the process actually involves.

Procure Resources for the Study - Process capability studies disrupt normal operations and require significant expenditures of both material and human resources. Since it is a project of major importance, it should be managed as such. All of the usual project management techniques should be brought to bear. This includes planning, scheduling and management status reporting.

Evaluate the Measurement System - Evaluate the measurement system's ability to do the job. Again, be prepared to spend the time necessary to get a valid means of measuring the process before going ahead.

Prepare a Control Plan - The purpose of the control plan is twofold: (a) isolate and control as many important variables as possible and, (b) provide a mechanism for tracking variables that cannot be completely controlled. The object of the capability analysis is to determine what the process can do if it is operated the way it is designed to be operated. This means that such obvious sources of potential variation as operators and vendors will be controlled while the study is conducted. In other words, a single well trained operator will be used and the material will be from a single vendor. There are usually some variables that are important, but that are not controllable. One example is the ambient environment, such as temperature, barometric pressure or humidity. Certain process variables may degrade as part of the normal operation; for example tools wear and chemicals are used. These variables should still be tracked using logsheets and similar tools.

Select a Method for the Analysis - The SPC method will depend on the decisions made up to this point. If the performance measure is an attribute, one of the attribute charts will be used. Variables charts will be used for process performance measures assessed on a continuous scale. Also considered will be the skill level of the personnel involved, need for sensitivity, and other resources required to collect, record, and analyze the data.

Gather and Analyze the Data - Use one of the control charts described in this chapter, plus common sense. It is usually advisable to have at least two people go over the data analysis to catch inadvertent errors in transcribing data or performing the analysis.

Track Down and Remove Special Causes - A special cause of variation may be obvious, or it may take months of investigation to find it. The effect of the special cause may be good or bad. Removing a special cause that has a bad effect usually involves eliminating the cause itself. For example, if poorly trained operators are causing variability the special cause is the training system (not the operator) and it is eliminated by developing an improved training system or a process that requires less training. However, the removal of a beneficial special cause may actually involve incorporating the special cause into the normal operating procedure. For example, if it is discovered that materials with a particular chemistry produce better product the special cause is the newly discovered material and it can be made a common cause simply by changing the specification to assure that the new chemistry is always used.

Estimate the Process Capability - One point cannot be overemphasized: the process capability cannot be estimated until a state of statistical control has been achieved! After this stage has been reached, the methods described later in this chapter may be used. After the numerical estimate of process capability has been arrived at it must be compared to management's goals for the process, or it can be used as an input into economic models. Deming's all-or-none rules provide a simple model that can be used to determine if the output from a process should be sorted 100% or shipped as is.

Establish a Plan for Continuous Process Improvement - Once a stable process state has been attained steps should be taken to maintain it and improve upon it. SPC is just one means of doing this. Far more important than the particular approach taken is a company environment that makes continuous improvement a normal part of the daily routine of everyone.

13.6.2 Analysis of process capability data

There are several process capability indices that measure the ability of a process to create parts that meet specifications. Perhaps the biggest drawback of using process capability indexes is that they take the analysis a step away from the data. The danger is that the analyst will lose sight of the purpose of the capability analysis, which is to improve quality. To the extent that capability indexes help accomplish this goal, they are worthwhile. To the extent that they distract from the goal, they are harmful. The quality engineer should continually refer to this principle when interpreting capability indexes.

One of the first capability indexes used was Cp. The *natural tolerance* of the process is computed as 6σ. The index simply makes a direct comparison of the process natural tolerance to the engineering requirements. Assuming the process distribution is normal and the process average is exactly centered between the engineering requirements, a Cp index of 1.0 gives a minimally capable process. However, to allow a bit of room for process drift, the generally accepted minimum value for Cp is 1.33.

The Cp index has two major shortcomings. First, it can't be used unless there are both upper and lower specifications. Second, it does not account for process centering. If the process average is not exactly centered relative to the engineering requirements the Cp index will give misleading results. In recent years the Cp index has largely been replaced by the Cpk index. The Cp index is computed from the following expression.

$$Cp = \frac{\text{Engineering Tolerance}}{6\sigma} \tag{13.44}$$

where σ is the process standard deviation.

The Cpk index compensates for un-centered processes. The difference between C_{PK} and C_P represents the potential gain to be had from centering the process, and gives a much better indication of the ability of a process to produce parts to specification. The Cpk index assumes a normally distributed process and overstates process capability as the skewness of the process increases. A Cpk of at least 1.33 is preferred. During evaluation of new process with small sizes a Cpk of 1.67 is usually required. Since this index is usually computed prior to actual production it is commonly referred to as Ppk. The Cpk index is computed from the following expression.

$$Cpk = \min\left(\frac{\text{Upper Specification - }\mu}{3\sigma}, \frac{\mu - \text{Lower Specification}}{3\sigma}\right) \tag{13.45}$$

where μ is the process average.

The C_R index is equivalent to the Cp index. The index simply makes a direct comparison of the process to the engineering requirements. Assuming the process distribution is normal and the process average is exactly centered between the engineering requirements, a C_R index of 100% would give a minimally acceptable process. However, to allow a bit of room for process drift, the generally accepted maximum value for C_R is 75%. In general, the smaller C_R is, the better. The C_R index suffers from the same shortcomings as the C_P index. The C_R index is computed from the following expression.

$$C_R = \frac{6\sigma}{\text{Engineering Tolerance}} \tag{13.46}$$

The C_M index is generally used to evaluate machine capability studies, rather than full-blown process capability studies. Since variation will increase when normal sources of process variation are added (e.g., tooling, fixtures, materials, etc.), C_M uses a four sigma spread rather than a three sigma spread. The C_M index is computed from the following expression.

$$C_M = \frac{\text{Engineering Tolerance}}{8\sigma} \tag{13.47}$$

Example 13.10

A process is normally distributed with a mean of 210 and a standard deviation of 4.3. The lower specification limit is 180 and the upper specification is 230. Compute the Cp, and Cpk indexes.

Solution

The Cp index is

$$Cp = \frac{230\text{-}180}{6(4.3)} = 1.94$$

The Cpk index is

$$Cpk = \min\left(\frac{230\text{-}210}{3(4.3)}, \frac{210-180}{3(4.3)}\right) = 1.74$$

Example 13.11

A process has a Cpk of 0.9. Assuming a normal distribution what is the scrap rate of the process?

Solution

A Cpk of 0.9 occurs in a process that is centered 2.7 standard deviations from the nearest specification.

$$Cpk = \frac{\text{Upper Specification} - \mu}{3\sigma}$$

$$0.9 = \frac{\Delta}{3\sigma}$$

$$\Delta = 2.7\sigma$$

The area under the standard normal curve beyond 2.7 standard deviations is 0.003467 or 3,467 parts per million.

Example 13.12

A process has a mean of 13.51 and is exponentially distributed. The upper specification is 50, and there is no lower specification. What is the Cpk index and what is the scrap rate of the process?

Solution

Recall that the standard deviation of the exponential distribution is equal to the mean. The Cpk index is

$$Cpk = \frac{50 - 13.51}{3(13.51)} = 0.90$$

The scrap rate is

$$Scrap = e^{-\frac{50}{13.51}} = 0.0247$$

This is 24,700 parts per million. Recall from Example 13.11 that a normally distributed process with a Cpk of 0.9 has a scrap rate of 3,467 parts per million. The scrap rate for the exponentially distributed process is more than 7 times the scrap rate of the normally distributed process even though the two processes have identical Cpk indexes. Capability indexes are **meaningless** unless the process is normally distributed. As the skewness of a process increases capability indices increases their overestimation of process capability.

13.7 Exercises

1. Based on subgroups of n=5 the grand average is 105 and the average range is 10. Compute the upper and lower control limits for the ***X***-bar and R charts.

2. Based on subgroups of n=15 the grand average is 10 and the average subgroup sigma is 0.5. Compute the upper and lower control limits for the ***X***-bar and s charts.

3. The average hospital census over the past 24 months has been stable at 68 patients. The average variation month-to-month is 6 patients and it is also stable. The most recent census is 80 patients, is the process still in statistical control?

4. Construct a p chart using the following data:

Month	IV Lines	Infections	%
Nov-92	29	1	3.4%
Dec-92	28	1	3.6%
Jan-93	37	4	10.8%
Feb-93	27	0	0.0%
Mar-93	26	1	3.8%
Apr-93	29	2	6.9%
May-93	22	1	4.5%
Jun-93	25	1	4.0%
Jul-93	32	0	0.0%
Aug-93	41	1	2.4%
Sep-93	38	2	5.3%
Oct-93	33	0	0.0%

5. A process has a mean of 40.8, a standard deviation of 0.25 and is normally distributed. The lower specification is 40, and there is no upper specification. What is the Cpk?

6. A process has a mean of 40.8, a standard deviation of 0.25 and is normally distributed. The upper specification is 42, and there is no lower specification. What is the Cpk?

7. A process is normally distributed with a mean of 110 and a standard deviation of 6.2. The lower specification limit is 95 and the upper specification is 130. Compute the Cp, and Cpk indexes.

8. A process has a Cpk of 0.95. Assuming a normal distribution what is the scrap rate of the process?

9. Determine the correct chart for the following: (1) waiting time for customers in a bank, each individual customer's waiting time will be plotted; (2) the salt content of potato chips using samples of three every 15 minutes; (3) the thickness of ceramic substrates using samples of 10 from each batch produced; (4) the number of customers who complain per 100 customers served; (5) the proportion of customers served each day who complain; (6) the number of falls in the hospital reported each week; (7) the number of medication order errors per day.

13.8 References

Abraham, B. and Whitney, J.B. (1990). "Applications of EWMA Charts to Data from Continuous Processes," Annual Quality Congress Transactions, ASQ, Milwaukee, WI.

Alloway, J.A., Jr. (1994). "The Card Drop Shop," Quality Progress, July.

Alwan, L.C. and Roberts, H.V. (1989). "Time Series Modeling for Statistical Process Control," in Statistical Process Control in Automated Manufacturing, Keats, J.B. and Hubel, N.F. editors, Marcel Dekker, New York.

Boardman, T.J. and Boardman, E.C. (1990). "Don't Touch That Funnel!" Quality Progress, December.

Deming, W.E. (1975). "On Probability as a Basis for Action," The American Statistician, 29(4), 146–152.

Deming, W.E. (1986). Out of the Crisis, MIT Center for Advanced Engineering Study, Cambridge, MA.

Edwards, C.D. (1968). "The Meaning of Quality," Quality Progress, October.

Efron, B. 1982. The Jackknife, the Bootstrap, and Other Resampling Plans, Philadelphia, PA: Society for Industrial and Applied Mathematics.

Fivars, G. (1980). "The Critical Incident Technique: a Bibliography," Research and Publication Service, American Institutes for Research ED 195 681.

Flanagan, J.C. (1954). "The Critical Incident Technique." Psychological Bulletin, 51(4), July, pp. 327–358.

Harrington, H.J. (1992). "Probability and Statistics," in Pyzdek, T. and R.W. Berger, Editors. 1992. Quality Engineering Handbook, 513–577. Milwaukee, WI: ASQ Quality Press.

Hillier, F.S. (199). "X-bar and R-chart Control Limits Based on a Small Number of Subgroups", Journal of Quality Technology, Vol. 1, No. 1, January, pp. 17–26.

Holland, John H. (1996). Hidden Order: How Adaptation Builds Complexity Reading, MA: Perseus Books, p.56.

Hunter, J.S. (1986). "The Exponentially Weighted Moving Average," Journal of Quality Technology 18, pp. 203–210.

Hunter, J.S. (1989). "A One Point Plot Equivalent to the Shewhart Chart with Western Electric Rules," Quality Engineering 2, pp. 13–19.

Hurley, H. and Loew, C. (1996). "A Quality Change for New Product Development," Quality Observer, January, pp. 10–13.

Hutton, D.W. (1994). The Change Agent's Handbook: A Survival Guide for Quality Improvement Champions. Milwaukee, WI: ASQ Quality Press.

Imai, M. (1986). Kaizen, New York: Random House.

Johnson, R.S. (1993b). TQM: Management Processes for Quality Operations. Milwaukee, WI: ASQ Quality Press.

Johnson, R.S. (1993d). TQM: The Mechanics of Quality Processes. Milwaukee, WI: ASQ Quality Press.

Joiner, B.L. (1994). Fourth Generation Management: The New Business Consciousness. New York: McGraw-Hill, Inc.

Juran, J.M. (1994). "The Upcoming Century of Quality," address delivered at the ASQ Annual Quality Congress, May 24, 1994.

Juran, J.M. and Gryna, F.M. (1993). Quality Planning and Analysis, 3rd Edition. New York: McGraw-Hill.

Juran, J.M., (1994). "The Upcoming Century of Quality," Wilton, CT: Juran Institute, Inc.

King, J.R. (1971). "Probability Charts for Decision Making," TEAM, Tamworth, NH 03886, (603) 323–8843.

Koch, G.G. and D.B. Gillings. (1983). "Statistical Inference, Part I," in Encyclopedia of Statistical Sciences, vol. 4, 84–88, Samuel Kotz and Normal L. Johnson, editors-in-chief, New York: John Wiley & Sons.

Mizuno, S., Editor. (1988). Management for Quality Improvement: The 7 New QC Tools, Cambridge, MA: Productivity Press.

Palm, A.C. (1990). "SPC Versus Automatic Process Control," Annual Quality Congress Transactions, ASQ, Milwaukee, WI.

Proschan, F. and Savage, I.R. (1960). "Starting a Control Chart," Industrial Quality Control, Vol. 17, No. 3, September, pp. 12–13.

Provost, L.P. (1988). "Interpretation of Results of Analytic Studies," paper presented at 1989 NYU Deming Seminar for Statisticians, March 13. New York: New York University.

Pyzdek, T. (1976). "The Impact of Quality Cost Reduction on Profits," Quality Progress, May.

Pyzdek, T. (1985). "A Ten-step Plan for Statistical Process Control Studies," Quality Progress, April, 77–81.

Pyzdek, T. (1985). "A Ten-step Plan for Statistical Process Control Studies," Quality Progress, July, 99. 18–20.

Pyzdek, T. (1989), What every engineer should know about quality control, New York: Marcel-Dekker, Inc.

Pyzdek, T. (1990). SPC-PC: Statistical Process Control for Personal Computers, Quality America, Tucson, Arizona.

Pyzdek, T. (1991), What Every Manager Should Know About Quality, New York: Marcel-Dekker, Inc.

Pyzdek, T. (1991–92). "A Simulation of Receiving Inspection," Quality Engineering, 4(1), 9–19.

Pyzdek, T. (1992). Pyzdek's Guide to SPC Volume Two—Applications and Special Topics, Tucson, AZ: Quality Publishing, Inc.

Pyzdek, T. (1992). Turn-key SPC Instructor's Manual. Tucson, AZ: Quality Publishing.

Pyzdek, T. (1994). Pocket Guide to Quality Tools. Tucson, AZ: Quality Publishing.

Pyzdek, T. (1996). The Quality Engineering bible. Tucson, AZ: Quality Publishing.

Pyzdek, T. (1996b). "The Pursuit of Happiness: Use Market-based Management to Improve Performance and Go Beyond TQM." Quality Progress, May, pp. 92–96

Pyzdek, T. and R.W. Berger, Editors. (1992). Quality Engineering Handbook. Milwaukee, WI: ASQ Quality Press.

Rand, A. (1964). The Virtue of Selfishness, New York: Penguin Books.

Reichheld, F.F. (1996), The Loyalty Effect: The Hidden Force Behind Growth, Profits, and Lasting Value. New York: McGraw-Hill.

Roberts, L. (1994). Process Reengineering: The Key to Achieving Breakthrough Success. Milwaukee, WI: ASQ Quality Press.

Rose, K.H. (1995). "A Performance Measurement Model." Quality Progress, February, pp. 63–66.

Saaty, T.L. (1988). Decision Making for Leaders: The Analytic Hierarchy Process for Decisions in a Complex World. Pittsburgh, PA: RWS Publications.

Shainin, Dorian (1984). "Better than Good Old X-bar and R-charts Asked by Vendees." ASQ Quality Congress Transactions, Milwaukee, pp. 302–307.

Sheridan, B.M. (1993). Policy Deployment: The TQM Approach to Long-Range Planning. Milwaukee, WI: ASQ Quality Press.

Shewhart (1939, 1986). Statistical Method from the Viewpoint of Quality Control. New York: Dover Publications.

Shewhart, W.A. (1930, 1980). Economic Control of Quality of Manufactured Product, ASQ Quality Press, Milwaukee, WI.

Shewhart, W.A. (1931, 1980). Economic Control of Quality of Manufacturing. Milwaukee, WI: ASQ Quality Press.

Simon, J. and Bruce, P. (1992). "Using Computer Simulation in Quality Control," Quality Progress, Statistics Corner Column.

Simon, J.L. 1992. Resampling: the New Statistics, Arlington, VA: Resampling Stats, Inc.

Thiagarajan, S. (1996). "Instructional Games, Simulations, and Role-plays," The ASTD Training and Development Handbook: A Guide to Human Resources Development in Craig, R.L. editor-in-chief, New York: McGraw-Hill, pp. 517–533.

Wearring, C. and Karl, D.P. (1995). "The Importance of Following GD&T Specifications," Quality Progress, February.

Wheeler, D.J. (1991). "Shewhart's Charts: Myths, Facts, and Competitors," Annual Quality Congress Transactions, ASQ, Milwaukee, WI.

♦♦♦

CHAPTER

14

Experimentation

In engineering, many concepts are well defined using mathematics. However, there are many factors not included in mathematical models. For example consider a chemical processing plant. The mathematical model describing the process may assume that all processing chemicals are pure when in reality they are not, or the model may not include environmental factors such as humidity. An experiment could be used to further enhance the textbook model to include these factors.

In its most fundamental form, experimentation is used to reveal a model of a phenomenon. Experimental design is used to plan the experiment so a maximum amount of information can be gained from a minimum of resources. After an experiment has been designed and conducted, an engineer analyzes the data and makes conclusions and recommendations, one of which may be that more experimentation is required.

14.1 Experimental design

An experiment is designed to determine the effects of inputs or **factors** on outputs or **responses**. For example, an engineer wants a plastic composite to have tensile strength of at least 20 ksi, and to be flexible enough to be bent at a 75° angle without breaking. This engineer, along with other experts, has determined that the following factors may have an influence on the outputs tensile strength and flexibility:

- Ingredient A as a percent of the total
- Ingredient B as a percent of the total

- Ingredient C as a percent of the total
- Injection pressure
- Mold dwell time
- Mixing time

This experiment is described graphically in Figure 14.1.

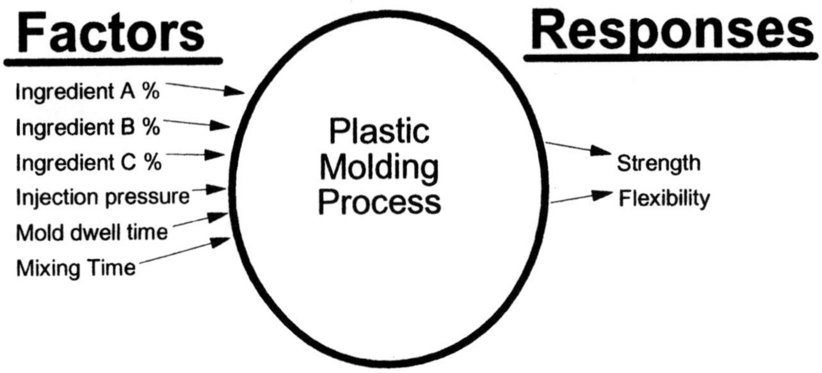

Figure 14.1. Plastic processing experiment.

Factors are referred to as **independent variables**, because the experimenter has control over these variables and can change the values of these variables (except for noise factors). Responses are referred to as **dependent variables**; the experimenter can only influence the value of a response variable adjusting the values of the factors. In many cases, there are independent variables that the experimenter cannot control. For example, the relative humidity may effect the output of a process. In this case, the experimenter cannot determine the humidity levels for the purpose of experimentation or to optimize the process. This type of variable is called a **noise factor**. Ideally, levels of the other factors can be determined and set so that the effect of the noise factor is minimized.

The experimental design strategy is dependent on the knowledge of the product or process. Early in the experimental process, the goal is usually to determine which factors have a significant effect on the response variable, and which do not. This type of experimental design is called a **screening design**. An intermediate design step is to determine if there are any factor interactions, and sometimes to determine non-linear effects. The final stage is usually the optimization of the product or process.

14.1.1 Screening designs

Screening designs are used to determine which factors have a statistically significant effect on the response variable. At the screening stage the goal is not to define a mathematical model, but to determine which factors should be included in the mathematical model. The model will be refined at later stages of experimentation. Interactions are only included in the model if there is a high belief that the interactions exist. Because interactions are assumed not to exist, the factor effects are confounded with interactions effects. This allows many factors to be considered with a minimum of experimental trials. A typical screening model with one response variable and n factors, assuming no interactions exist, is

$$y = c_0 + c_1x_1 + c_2x_2 + c_3x_3 + \ldots + c_nx_n + e \tag{14.1}$$

where c_1, c_2, ... c_n are the coefficients to be determined, and e is the error in the prediction model.

A typical screening model with one response variable and n factors, assuming no interactions exist other than the interaction between factors x_1 and x_3 , is

$$y = c_0 + c_1x_1 + c_2x_2 + c_3x_3 + \ldots + c_nx_n + c_{n+1}x_1x_3 + e \tag{14.2}$$

If an experimental design contains n factors, n+1 coefficients must be determined, which requires a minimum of n+1 experimental trials. After the values of the coefficients have been determined, it is necessary to determine if the estimated values are significantly different from zero. Only the factors with coefficients that are statistically different from zero are considered in further experimentation. To determine if the coefficients are statistically different from zero requires experimental trials for degrees-of-freedom. For example, if an experimental design contains n factors, and there are n+5 experimental trials, since n+1 trials are required to determine the values of the coefficients, the remaining 4 trials are available to determine the experimental error; these extra trials are referred to as **degrees-of-freedom**. At least one degree-of-freedom is required for experimental error.

Screening designs almost always use 2 levels for each factor. Two points define a straight line, thus, a linear effect for each factor is assumed. This assumption is not important, because at the screening stage of experimentation, the goal is not to determine the exact mathematical effect of each factor, but to determine which factors have an effect in the region of experimentation. Limiting the design to 2 levels maximizes the number of factors that can be studied in a given number of trials.

When designing an experiment, it is best to use a balanced or **orthogonal** design if possible. The factor effects in orthogonal designs are independent, which maximizes the discriminatory power of the experiment. Using 1 to designate the high level of a factor and -1 to designate the low level of a factor, a design is orthogonal if the sum of every column is zero and the dot product for all possible pairs of factors is zero (the tables that define orthogonal designs typically use 1 to designate the low level and 2 to designate the high level). Plackett and Burman (1946) introduced several orthogonal designs, and others such as Taguchi, have added additional designs. The L_4, L_8, and L_{12} designs are shown below with their corresponding linear graphs and interaction tables. For additional orthogonal designs, consult Phadke (1989).

Table 14.1. L_4 (2^3).

	Factor		
Trial	A	B	C
1	1	1	1
2	1	2	2
3	2	1	2
4	2	2	1

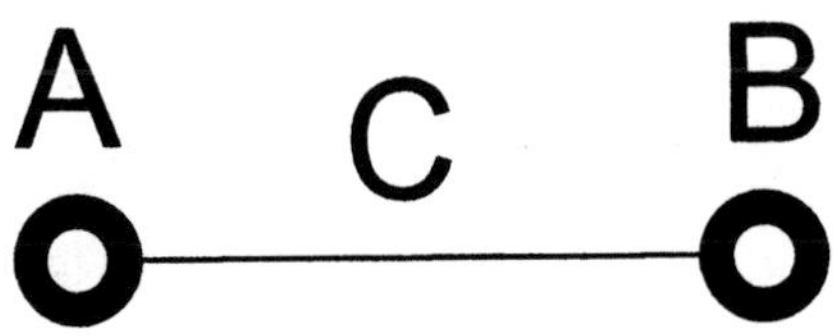

Figure 14.2. L_4 linear graph.

Table 14.2. L_4 interaction table.

Factor	A	B	C
A		C	B
B			A
C			

Table 14.3. L_8 (2^7).

Trial	Factor						
	A	B	C	D	E	F	G
1	1	1	1	1	1	1	1
2	1	1	1	2	2	2	2
3	1	2	2	1	1	2	2
4	1	2	2	2	2	1	1
5	2	1	2	1	2	1	2
6	2	1	2	2	1	2	1
7	2	2	1	1	2	2	1
8	2	2	1	2	1	1	2

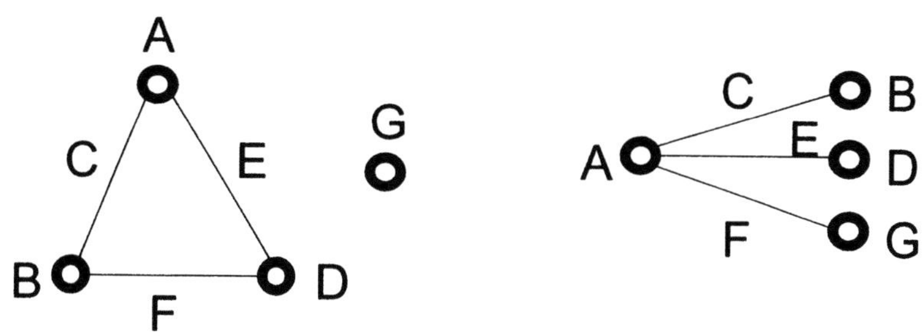

Figure 14.3. L_8 linear graphs.

Table 14.4. L_8 interaction table.

	Factor						
Factor	A	B	C	D	E	F	G
A		C	B	E	D	G	F
B			A	F	G	D	E
C				G	F	E	D
D					A	B	C
E						C	B
F							A

Table 14.5. L_{12} (2^{11})*.

	Factor										
Trial	A	B	C	D	E	F	G	H	I	J	K
1	1	1	1	1	1	1	1	1	1	1	1
2	1	1	1	1	1	2	2	2	2	2	2
3	1	1	2	2	2	1	1	1	2	2	2
4	1	2	1	2	2	1	2	2	1	1	2
5	1	2	2	1	2	2	1	2	1	2	1
6	1	2	2	2	1	2	2	1	2	1	1
7	2	1	2	2	1	1	2	2	1	2	1
8	2	1	2	1	2	2	2	1	1	1	2
9	2	1	1	2	2	2	1	2	2	1	1
10	2	2	2	1	1	1	1	2	2	1	2
11	2	2	1	2	1	2	1	1	1	2	2
12	2	2	1	1	2	1	2	1	2	2	1

* There is no linear graph for this orthogonal design; all 2-way interactions are partially confounded with the factors assigned to the remaining 9 columns.

Linear graphs and interaction tables are used to determine how factors are confounded. Referring to the left part of Figure 14.3, the linear graph for the L_8 design, it can be seen that the interaction between the factors assigned to columns B and D is confounded with the factor assigned to column F. The right part of Figure 14.3 shows another linear graph for the L_8 design, and the effect of the factor assigned to column F is confounded with the interaction between the factors assigned to columns A and G. Which of these linear graphs is correct? Both linear graphs are partially correct. Linear graphs graphically show the confounding of interactions with main effects, but they do not show all of the confounding. The graphical layout is useful, but to be sure all confounding is accounted for, the interaction table must be consulted. In Table 14.4, the interaction table for the L_8 design, column F appears 3 times; it can be seen that the factor assigned to this column is confounded with:

- the interaction between the factors assigned to columns B and D,
- the interaction between the factors assigned to columns A and G, and
- the interaction between the factors assigned to columns C and E.

The subscript of a design's name represents the number of experimental trials; for example in the L_8 design there are 8 trials. In parentheses following the design name is a designation for the number of levels and factors; for example in the L_8 design 2^7 represents 2 levels for 7 factors. Thus, an L_8 design can accommodate a maximum of 7 factors, however, this is not recommended. To estimate the overall mean and the coefficients for 7 factors requires 8 trials; this leaves no trials for estimation of the experimental error. This means that it is not possible to determine if any of the factors have a statistically significant effect on the response variable. In practice many experimenters eliminate one or two of the factors that appear to have the least effect on the response variable from the experiment after all trials have been completed, and then use the degrees of freedom for the eliminated factors to estimate the experimental error. This practice violates a fundamental principle of data analysis, but seems to work in practice; however, this author does not recommend the practice.

Example 14.1

A porosity in a casting has historically averaged 1.1%. Changes in specifications require a porosity rate below 0.5% to be profitable. A team of engineers, operators and maintenance personnel identified the following factors as having a potential impact on the scrap rate:

1) Mold pre-heat temperature
2) Casting material pouring temperature
3) Mold post-processing holding area temperature
4) Pouring speed
5) Funnel type (Type 1 or Type 2)

It is also suspected that there is an interaction between the mold pre-heat temperature and the casting material pouring temperature.

Solution

To identify the main effects, all 2-way interactions, all 3-way interactions, all 4-way interactions, and the 5-way interaction would require 32 trials (2 to the 5th), and still leaves no degrees of freedom to determine experimental error. This is the first stage of experimentation, and the purpose is to determine which of the 5 factors have an impact on porosity. By ignoring the interactions, with the exception of the interaction between the mold pre-heat temperature and the casting material pouring temperature, only 8 trials are required. Eight trials gives 1 degree-of-freedom for estimating the overall mean, 1 degree-of-freedom for estimating each of the 5 factor effects, 1 degree-of-freedom for estimating the effect of the interaction between the mold pre-heat temperature and the casting material pouring temperature, and 1 degree-of-freedom for estimating the experimental error.

Since a minimum of 8 trials are required, an L_8 design is a logical choice. When assigning the factors to columns, the L_8 linear graphs can be used to ensure that the interaction between the mold pre-heat temperature and the casting material pouring temperature is not confounded with any other factors. Referring to the first linear graph in Figure 14.2, the interaction between Factor A and Factor B is confounded with Factor C. Thus, by assigning mold pre-heat temperature to column A and the casting material pouring temperature to column B, and not assigning any factor to column C, the effect of the interaction between the mold pre-heat temperature and the casting material pouring temperature can be estimated. The complete experimental design is shown in Table 14.6.

Table 14.6. Example experimental design.

Trial	Factor						
	Mold Pre-heat Temp.	Casting Material Pouring Temp.	Not Assigned	Mold Post Processing Holding Area Temp.	Not Assigned	Pouring Speed	Funnel Type
1	1	1	1	1	1	1	1
2	1	1	1	2	2	2	2
3	1	2	2	1	1	2	2
4	1	2	2	2	2	1	1
5	2	1	2	1	2	1	2
6	2	1	2	2	1	2	1
7	2	2	1	1	2	2	1
8	2	2	1	2	1	1	2

14.1.2 Other designs

A **factorial** design investigates all possible combinations of the factors and levels in an experiment. For example, if the factors in an experiment are temperature, pressure and time, with the levels for these factors as follows:

- temperature 0°C, 50°C and 100°C
- pressure 10 bar, 20 bar
- time 1hr, 2 hr

then the experimental trials required are shown in Table 14.7.

Table 14.7. Full factorial experiment.

Trial	Temperature	Pressure	Time
1	0	10	1
2	0	10	2
3	0	20	1
4	0	20	2
5	50	10	1
6	50	10	2
7	50	20	1
8	50	20	2
9	100	10	1
10	100	10	2
11	100	20	1
12	100	20	2

If the number of levels is the same for all factors in a factorial design, the number of experimental trials is L^F; where L is the number of levels and F is the number of factors.

For example, the number of experimental trials required for a full factorial experiment with 6 factors each at 2 levels is

$$\text{Trials} = 2^6 = 64$$

The primary disadvantage of a factorial is its inefficiency. As seen from the above example, 64 trials are needed to investigate 6 factors. Using a screening design, and ignoring interactions, 6 factors can be investigated with only 8 trials. The advantage of a factorial is that there are enough degrees-of-freedom to avoid any confounding of main effects with interactions.

The design shown in Table 14.7 is commonly referred to as a **full factorial** design to differentiate between fractional or partial factorial designs. To reduce the number of trials required, full factorial designs are often reduced to fractional factorial designs. The screening designs discussed earlier are fractional factorial designs. Fractional factorial designs allow fewer experimental trials to be run, by allowing the confounding of main effects with some interactions. The most common approach is to disregard interactions of more than 2 factors.

Consider the situation of 6 factors each at 2 levels. The experimental model contains:

- the overall mean
- 6 main effects
- 15 2-way interactions
- 20 3-way interactions
- 15 4-way interactions
- 6 5-way interactions, and
- 1 6-way interaction

The overall mean, the main effects and the interactions of this design consume 64 trials or degrees-of-freedom to estimate, but interactions involving 3 or more factors consume 42 degree-of-freedom. High-level interactions (those involving more than 2 factors) have been shown to be negligible in most cases, thus, resources can be saved by ignoring them.

A 1/2 factorial requires 1/2 the trials of a full factorial, for designs with 2 levels for each factor, the 1/2 factorial with k factors and is typically referred to as a 2^{k-1} design because 2^{k-1} trials are required. Again considering the experiment with 6 factors at 2 levels, a 1/2 factorial requires 32 trials, which provides enough degrees-of-freedom to estimate the overall mean, all main effects, all 2-way interactions, and 9 of the 20 3-way interactions.

A 1/4 factorial requires 1/4 the trials of a full factorial, for designs with 2 levels for each factor, the 1/4 factorial with k factors and is typically referred to as a 2^{k-2} design because 2^{k-2} trials are required. Using a 1/4 factorial, the experiment with 6 factors at 2 levels requires 16 trials, which provides enough degrees-of-freedom to estimate the overall mean, all 6 main effects, and 9 of the 15 2-way interactions.

Fractional factorials are characterized by their resolution. The resolution designates the degree of confounding between the main effects and interactions. Table 14.8 describes some common design resolutions.

Table 14.8. Design resolutions.

Resolution	Description
III	No factor (main) effects are confounded with each other, but factors effects may be confounded with two-way interactions.
IV	No factor (main) effects are confounded with each other or with any two-way interactions, but two-way interactions are confounded with each other.
V	No factor (main) effects are confounded with each other or with any two-way interactions. Two-way interactions are not confounded with each other, but may be confounded with three-way interactions.

14.2 Analyzing experimental data

The two primary methods for analyzing experimental data are Analysis of Variance (ANOVA) and regression. ANOVA is generally used to identify which variables should be included in the experimental model and regression is used to estimate the coefficients for the variables, but either ANOVA or regression can be used to identify the important variables and determine the coefficients for these variables. Another consideration is that when using ANOVA data must be collected in a systematic manner, usually from a designed experiment, while regression can generally be used on any data set.

Consider the industrial rolling of aluminum sheet metal, and there is an attempt to determine which factors influence the variability of the thickness of the sheet metal.

This may not be possible by analyzing existing production data. The factors of interest are often co-linear when analyzing non-experimental data. In a designed experiment, great care is taken to be able to prevent confounding of the main effects and some if not all of the interactions, but this is not the case for production data. Back to the aluminum sheet example; the factors that are of interest are speed, roll pressure, sheet temperature, and the rate of injection for coolant. After analyzing the data, it is found the rate of coolant injection is statistically significant, but the speed is not. However, if the rate of coolant injection is not included in the model, the speed is the most significant variable. How can this happen? This is common occurrence when 2 variables are co-linear. Unlike a designed experiment where the settings of the rate of coolant injection and speed would be carefully set, in the historical data, when the speed is reduced, the rate of coolant injection was also reduced, thus, it is impossible to separate the effects of these 2 variables.

14.2.1 Analysis of variance

Analysis of Variance (ANOVA) is used to determine which factors have a statistically significant effect on the response variable. Any ANOVA has the following assumptions:

- the treatment data is normally distributed,
- the variance is the same for all treatments, and
- all samples are randomly selected and are independent.

One-Way ANOVA

A one-way ANOVA is used to determine if the mean is equal for different levels of a **SINGLE** variable. If there are only 2 levels for the variable, then a t-test is used, but when there are more than 2 levels, ANOVA is required. For example, is the mean hardness of a rubber compound equal if a component of the rubber is supplied by 5 different vendors.

The null hypothesis being tested in a one-way ANOVA is

H_0: All treatment means are equal

The alternate hypothesis is

H_1: At least one of the treatment means is different

The one-way ANOVA table is shown in Table 14.9.

Table 14.9. One-way ANOVA table.

Source of Variation	Sum of Squares	Degrees of Freedom	Mean Square	F Statistic
Between Treatments	SST	k-1	$MST = \frac{SST}{k-1}$	$F = \frac{MST}{MSE}$
Error	SSE	N-k	$MSE = \frac{SSE}{N-k}$	
Total	TSS	N-1		

SST = sum of squares between treatments

$$SST = \sum_{j=1}^{k} \frac{(Y_{\bullet j})^2}{n} - \frac{Y_{\bullet\bullet}^2}{N} \qquad (14.3)$$

$Y_{.j}$ = the sum of observations for treatment j.

$Y_{..}$ = the sum of all observations.

SSE = sum of squares due to error

$SSE = TSS - SST$

TSS = total sum of squares

$$TSS = \sum_{i=1}^{k}\sum_{j=1}^{n} y_{ij}^2 - \frac{Y_{\bullet\bullet}^2}{N} \qquad (14.4)$$

MST = mean square for treatments

MSE = mean square for error

k = number of treatment levels

n = number of runs at a particular level

N = total number of runs

F = the calculated F statistic with k-1 and N-k degrees of freedom

If the calculated F is greater than the critical F value then reject the null hypothesis (i.e., all the means are not statistically the same at the α value chosen.)

Example 14.2

An automobile bearing is experiencing a higher than expected failure rate due to low strength. Upon investigation, it is discovered that the supplier produces this bearing on 4 different machines. The strength of randomly selected bearings from the 4 machines is given below; at 5% significance, is there any difference in the bearing strength based on the production machine?

Machine A	Machine B	Machine C	Machine D
47.6	52.8	50.4	52.4
52.5	51.4	55.8	51.6
45.9	49.7	49.8	57.8
51.6	54.2	51.8	53.2
51.7	53.6	51.8	49.1
50.4	52.8	53.2	51.6
48.0	54.0	51.2	44.2
50.3			49.7

Solution

The sum of all 30 observations is 1,540.1. The sum of the square of the 30 observations is 79,281.35. The total sum of squares is

$$TSS = 79,281.35 - \frac{1,540.1^2}{30} = 217.75$$

The sum of the 8 observations for Machine A is 398.0. The sum of the 7 observations for Machine B is 368.5. The sum of the 7 observations for Machine C is 364.0. The sum of the 8 observations for Machine D is 409.6. The sum of squares between treatments is

$$SST = \frac{398.0^2}{8} + \frac{368.5^2}{7} + \frac{364.0^2}{7} + \frac{409.6^2}{8} - \frac{1,540.1^2}{30} = 35.31$$

The sum of squares for the error term is the difference between the total sum of squares and the sum of squares between treatments.

$$SSE = 217.75 - 35.31 = 182.44$$

The degrees of freedom between treatments is equal to the number of treatments minus 1, which equals 3. The degrees of freedom for the error is the total number of observations less the number of treatments which is 30-4 = 26. The critical value for the F-distribution at 5% significance with 3 and 26 degrees of freedom is 2.98. The complete ANOVA table is shown in Table 14.10. Since the computed F-statistic is less than the critical value, the null hypothesis is not rejected - there is not enough evidence to show that the mean strength is not the same for bearings produced on all machines.

Table 14.10. Example One-way ANOVA table.

Source of Variation	Sum of Squares	Degrees of Freedom	Mean Square	F Statistic
Between Treatments	35.31	3	11.77	1.68
Error	182.44	26	7.02	
Total	217.75	29		

Multi-factor ANOVA

When more than one factor is of interest, a more general ANOVA is required for analysis. The 2-way and 3-way ANOVA table are shown below, and the 3-way ANOVA can be expanded to accommodate any desired number of factors. With any ANOVA the mathematics is tedious, but there are several commercial software packages available to perform these computations.

Table 14.11. Two-way ANOVA table.

Source of Variation	Sum of Squares	Degrees of Freedom	Mean Square	F Statistic
A Treatments	SSA	a-1	$MSA = \frac{SSA}{a-1}$	$F = \frac{MSA}{MSE}$
B Treatments	SSB	b-1	$MSB = \frac{SSB}{b-1}$	$F = \frac{MSB}{MSE}$
Interaction	SSAB	(a-1)(b-1)	$MSAB = \frac{SSAB}{(a-1)(b-1)}$	$F = \frac{MSAB}{MSE}$
Error	SSE	ab(n-1)	$MSE = \frac{SSE}{ab(n-1)}$	
Total	TSS	N-1		

$$SSA = \sum_{i=1}^{a} \frac{(Y_{i..})^2}{bn} - \frac{Y_{\bullet\bullet}^2}{N} \tag{14.5}$$

$$SSB = \sum_{j=1}^{b} \frac{(Y_{.i.})^2}{an} - \frac{Y_{\bullet\bullet}^2}{N} \qquad (14.6)$$

$$SSAB = \sum_{i=1}^{a} \sum_{i=1}^{b} \frac{(Y_{ij.})^2}{n} - \frac{Y_{\bullet\bullet}^2}{N} - SSA - SSB \qquad (14.7)$$

$$TSS = \sum_{i=1}^{a} \sum_{j=1}^{b} \sum_{k=1}^{n} Y_{ijk}^2 - \frac{Y_{\bullet\bullet}^2}{N} \qquad (14.8)$$

a = number of treatment levels for factor A

b = number of treatment levels for factor B

n = the number of repetitions for each experimental setting

N = total number of runs (abn)

Table 14.12. Three factor ANOVA table.

Source of Variation	Sum of Squares	Degrees of Freedom	Mean Square	F Statistic
Factor A	SSA	a-1	$MSA = \frac{SSA}{a-1}$	$F = \frac{MSA}{MSE}$
Factor B	SSB	b-1	$MSB = \frac{SSB}{b-1}$	$F = \frac{MSB}{MSE}$
Factor C	SSC	c-1	$MSC = \frac{SSC}{c-1}$	$F = \frac{MSC}{MSE}$
AB Interaction	SSAB	(a-1)(b-1)	$MSAB = \frac{SSAB}{(a-1)(b-1)}$	$F = \frac{MSAB}{MSE}$
AC Interaction	SSAC	(a-1)(c-1)	$MSAC = \frac{SSAC}{(a-1)(c-1)}$	$F = \frac{MSAC}{MSE}$
BC Interaction	SSBC	(b-1)(c-1)	$MSBC = \frac{SSBC}{(b-1)(c-1)}$	$F = \frac{MSBC}{MSE}$
ABC Interaction	SSABC	(a-1)(b-1)(c-1)	$MSABC = \frac{SSABC}{(a-1)(b-1)(c-1)}$	$F = \frac{MSABC}{MSE}$
Error	SSE	abc(n-1)	$MSE = \frac{SSE}{ab(n-1)}$	
Total	TSS	N-1		

$$SSA = \sum_{i=1}^{a} \frac{Y_{i...}^2}{bcn} - \frac{Y_{....}^2}{N} \qquad (14.9)$$

$$SSB = \sum_{j=1}^{b} \frac{Y_{.j..}^2}{acn} - \frac{Y_{....}^2}{N} \tag{14.10}$$

$$SSC = \sum_{k=1}^{c} \frac{Y_{..k.}^2}{abn} - \frac{Y_{....}^2}{N} \tag{14.11}$$

$$SSAB = \sum_{i=1}^{a}\sum_{j=1}^{b} \frac{Y_{ij..}^2}{cn} - \sum_{i=1}^{a} \frac{Y_{i...}^2}{bcn} - \sum_{j=1}^{b} \frac{Y_{.j..}^2}{acn} + \frac{Y_{....}^2}{N} \tag{14.12}$$

$$SSAC = \sum_{i=1}^{a}\sum_{k=1}^{c} \frac{Y_{i.k.}^2}{bn} - \sum_{i=1}^{a} \frac{Y_{i...}^2}{bcn} - \sum_{k=1}^{c} \frac{Y_{..k.}^2}{abn} + \frac{Y_{....}^2}{N} \tag{14.13}$$

$$SSBC = \sum_{j=1}^{b}\sum_{k=1}^{c} \frac{Y_{.jk.}^2}{an} - \sum_{j=1}^{b} \frac{Y_{.j..}^2}{acn} - \sum_{k=1}^{c} \frac{Y_{..k.}^2}{abn} + \frac{Y_{....}^2}{N} \tag{14.14}$$

$$SSABC = \sum_{i=1}^{a}\sum_{j=1}^{b}\sum_{k=1}^{c} \frac{Y_{ijk.}^2}{n} - \sum_{i=1}^{a}\sum_{j=1}^{b} \frac{Y_{ij..}^2}{cn} - \sum_{i=1}^{a}\sum_{k=1}^{c} \frac{Y_{i.k.}^2}{bn} - \sum_{j=1}^{b}\sum_{k=1}^{c} \frac{Y_{.jk.}^2}{an} + \sum_{i=1}^{a} \frac{Y_{i...}^2}{bcn} + \sum_{j=1}^{b} \frac{Y_{.j..}^2}{acn} + \sum_{k=1}^{c} \frac{Y_{..k.}^2}{abn} - \frac{Y_{....}^2}{N} \tag{14.15}$$

$$TSS = \sum_{i=1}^{a}\sum_{j=1}^{b}\sum_{k=1}^{c}\sum_{l=1}^{n} Y_{ijkl}^2 - \frac{Y_{....}^2}{N} \tag{14.16}$$

a = number of treatment levels for factor A

b = number of treatment levels for factor B

c = number of treatment levels for factor C

n = the number of repetitions for each experimental setting

N = total number of runs ($abcn$)

Example 14.3

The porosity of aluminum casting is not satisfactory. To improve the situation, it is suspected that the cooling temperature and the pour rate affect the porosity. An experiment was conducted using 3 cooling temperatures and 4 pour rates. The results are shown in Table 14.13. Determine if either of these factors affect porosity at a significance of 5%.

Table 14.13. Two way ANOVA sample data.

Pour Rate	Cooling Temperature			$\sum y$	$\sum y^2$
	300	400	500		

5	1.51%	1.87%	2.26%	5.64	10.88
10	1.92%	2.42%	2.88%	7.22	17.84
15	2.30%	2.84%	3.46%	8.60	25.33
20	2.75%	3.41%	4.09%	10.25	35.92
$\sum y$	8.48	10.54	12.69		
$\sum y^2$	18.82	29.05	42.10		

Solution

The treatment totals and the sum of the squared treatment values are given in the table above, the total of all 12 observations is 31.71. The sum of the square of all 12 observations is 89.97. The total sum of squares is

$$TSS = 89.97 - \frac{31.71^2}{12} = 6.17$$

The sum of the 3 observations with a pour rate of 5 is 5.64; 5.64 squared is 31.81. The sum of the 3 observations with a pour rate of 10 is 7.22; 7.22 squared is 52.13. The sum of the 3 observations with a pour rate of 15 is 8.60; 8.60 squared is 73.96. The sum of the 3 observations with a pour rate of 20 is 10.25; 10.25 squared is 105.06. The sum of squares for the pour rate is

$$SS_{pour} = \frac{31.81}{3} + \frac{52.13}{3} + \frac{73.96}{3} + \frac{105.06}{3} - \frac{31.71^2}{12} = 3.86$$

The sum of squares for the coolant temperature is

$$SS_{coolant} = \frac{71.91}{4} + \frac{111.09}{4} + \frac{161.04}{4} - \frac{31.71^2}{12} = 2.22$$

Since there are no replications in this experiment, there are no degrees of freedom for experimental error. To overcome this problem, the degrees of freedom for the interaction are used for the experimental error. The sum of squares for error is

$$SSE = 6.17 - 3.86 - 2.22 = 0.09$$

The critical values of the F-distribution are determined by the degrees of freedom for the factor of interest and the error. The critical value of the F-distribution for the pour rate has 3 and 6 degrees of freedom, and at 5% significance is equal to 4.76. The critical value of the F-distribution for the coolant temperature has 2 and 6 degrees of freedom, and at 5% significance is equal to 5.14. The remainder of the ANOVA table, including the computed values from the F-distribution is shown in Table 14.14. Since the computed value from the F-distribution is greater than the critical value for both factors, the null hypothesis is rejected for both factors, and it is concluded that the mean response is not equal for all levels of either factor. Simply stated, both factors significantly affect the porosity.

Table 14.14. Two-way ANOVA table.

Source of Variation	Sum of Squares	Degrees of Freedom	Mean Square	F Statistic
Pour Rate	3.86	3	1.29	86
Coolant Temp.	2.22	2	1.11	74
Error	0.09	6	0.015	
Total	6.17	11		

Example 14.4

The porosity of aluminum casting is not satisfactory. To improve the situation, it is suspected that the cooling temperature and the pour rate affect the porosity. An experiment was conducted using 3 cooling temperatures and 4 pour rates and each experimental condition was replicated. The results are shown in Table 14.15. Determine if either of these factors affect porosity at a significance of 5%.

Table 14.15. Two way ANOVA with replications sample data.

	Cooling Temperature				
Pour Rate	300	400	500	$\sum y$	$\sum y^2$
5	1.51% 1.53%	1.87% 1.81%	2.26% 2.25%	11.24	21.61
10	1.92% 1.87%	2.42% 2.45%	2.88% 2.82%	14.36	35.29
15	2.30% 2.34%	2.84% 2.84%	3.46% 3.43%	17.21	50.63
20	2.75% 2.75%	3.41% 3.43%	4.09% 4.07%	20.50	71.81
$\sum y$	16.97	21.07	25.27	63.31	
$\sum y^2$	37.69	58.19	83.49		179.34

Solution

The treatment totals and the sum of the squared treatment values are given in the table above, the total of all 24 observations is 63.31. The sum of the square of all 24 observations is 179.34. The total sum of squares is

$$TSS = 179.34 - \frac{63.31^2}{24} = 12.334$$

The sum of squares for the pour rate is

$$SS_{pour} = \frac{11.24^2}{(3)(2)} + \frac{14.36^2}{(3)(2)} + \frac{17.21^2}{(3)(2)} + \frac{20.50^2}{(3)(2)} - \frac{63.31^2}{24} = 7.824$$

The sum of squares for the coolant temperature is

$$SS_{coolant} = \frac{16.97^2}{(4)(2)} + \frac{21.07^2}{(4)(2)} + \frac{25.27^2}{(4)(2)} - \frac{63.31^2}{24} = 4.306$$

The sum of squares for the interaction between the pour rate and the coolant temperature is

$$SS_{\text{interaction}} = \frac{358.67}{2} - \frac{1048.98}{(3)(2)} - \frac{1370.50}{(4)(2)} + \frac{63.31^2}{24} = 0.199$$

The sum of squares for error is

$$SSE = 12.334 - 7.824 - 4.306 - 0.199 = 0.0045$$

The critical values of the F-distribution are determined by the degrees of freedom for the factor of interest and the error. The critical value of the F-distribution for the pour rate has 3 and 12 degrees of freedom, and at 5% significance is equal to 3.49. The critical value of the F-distribution for the coolant temperature has 2 and 12 degrees of freedom, and at 5% significance is equal to 3.89. The critical value of the F-distribution for the interaction has 6 and 12 degrees of freedom, and at 5% significance is equal to 3.00. The remainder of the ANOVA table, including the computed values from the F-distribution is shown in Table 14.16. Both factors and the interaction between them have a statistically significant affect on porosity.

Table 14.16. Two-way ANOVA with replications table.

Source of Variation	Sum of Squares	Degrees of Freedom	Mean Square	F Statistic
Pour Rate	7.824	3	2.61	6868
Coolant Temp.	4.306	2	2.15	5684
Interaction	0.199	6	0.033	87
Error	0.0045	12	0.00038	
Total	12.334	23		

Example 14.5

In an effort to understand the variables that impact the brightness of rolled sheet metal, the engineering department conducted an experiment using 6 factors with 2 levels each. The experimental results are shown in Table 14.17. Which of these factors affect brightness at 5% significance?

Table 14.17. Brightness experimental data.

Trial	Temperature	Speed	Angle	Pressure	Water	Kerosene	Brightness
1	400	1000	0	1.5	0%	2%	4.2
2	400	1500	0	1.5	1%	5%	5.1
3	400	1000	15	2.5	0%	5%	3.7
4	400	1500	15	2.5	1%	2%	5.6
5	600	1000	15	1.5	1%	2%	2.9
6	600	1500	15	1.5	0%	5%	3.8
7	600	1000	0	2.5	1%	5%	4.7
8	600	1500	0	2.5	0%	2%	3.6

Solution

Since there are only 8 trials, there is 1 degree of freedom for estimating the overall mean, 6 degrees of freedom for estimating the effect of the 6 factors, and 1 degree of freedom for experimental error. There are no degrees of freedom available for estimating the effect of any interactions. The brightness sum for all 8 trials is 33.6. The total sum-of-squares is

$$TSS = \sum_{i=1}^{8} y_i^2 - \frac{\left(\sum_{i=1}^{n} y_i\right)^2}{n} = 4.2^2 + 5.1^2 + 3.7^2 + 5.6^2 + 2.9^2 + 3.8^2$$

$$+4.7^2 + 3.6^2 - \frac{33.6^2}{8} = 5.48$$

The sum-of-squares due to temperature is computed from the squared differences between the average of all experimental trials, 4.2, and the average of the trials at each setting of temperature. When the temperature is set at 400, the average of the 4 trials was 4.65; when temperature 600, the average of the 4 trials was 3.75. The sum-of-squares for temperature is

$$SS_{Temp} = 4(4.65-4.2)^2 + 4(3.75-4.2)^2 = 1.62$$

The sum-of-squares for the remaining factors are

$$SS_{Speed} = 4(3.875-4.2)^2 + 4(4.525-4.2)^2 = 0.845$$
$$SS_{Angle} = 4(4.4-4.2)^2 + 4(4.0-4.2)^2 = 0.32$$
$$SS_{\text{Pressure}} = 4(4.0-4.2)^2 + 4(4.4-4.2)^2 = 0.32$$
$$SS_{Water} = 4(3.825-4.2)^2 + 4(4.575-4.2)^2 = 1.125$$
$$SS_{Kerosene} = 4(4.075-4.2)^2 + 4(4.325-4.2)^2 = 0.125$$

The sum-of-squares error is the difference between the total sum-of-squares and the sum-of-squares for the factors, and is 1.125. The degrees-of-freedom for the sum-of-squares for each factor is equal to the number of levels for that factor minus 1. The degrees-of-freedom for the total is equal to the number of experimental trials minus 1. The mean square for each factor is equal to the sum-of-squares divided by the degrees-of-freedom. The F-statistic for each factor is equal to the mean square for that factor divided by the mean square error. This is summarized in Table 14.18. The critical value of the F-statistic for each of the factors at 5% significance using 1 and 1 degrees of freedom is 161.4. Since none of the computed values from the F-distribution exceed the critical value, none of the factors are considered significant.

Table 14.18. Multi-factor ANOVA example.

Source of Variation	Sum of Squares	Degrees of Freedom	Mean Square	F Statistic
Temperature	1.62	1	1.62	1.44

Speed	0.845	1	0.845	0.75
Angle	0.32	1	0.32	0.28
Pressure	0.32	1	0.32	0.28
Water	1.125	1	1.125	1.00
Kerosene	0.125	1	0.125	0.11
Error	1.125	1	1.125	
Total	5.48	7		

14.2.2 Interactions

Interaction plots provide a graphical test for the presence of interactions. If the lines are parallel, there is no interaction between the variables, non-parallel lines indicate an interaction; the strength of the interaction is indicated by the degree the lines differ from parallel. Figure 14.4 shows the lack of an interaction between factors A and B, and Figure 14.5 shows a strong interaction between factors A and B.

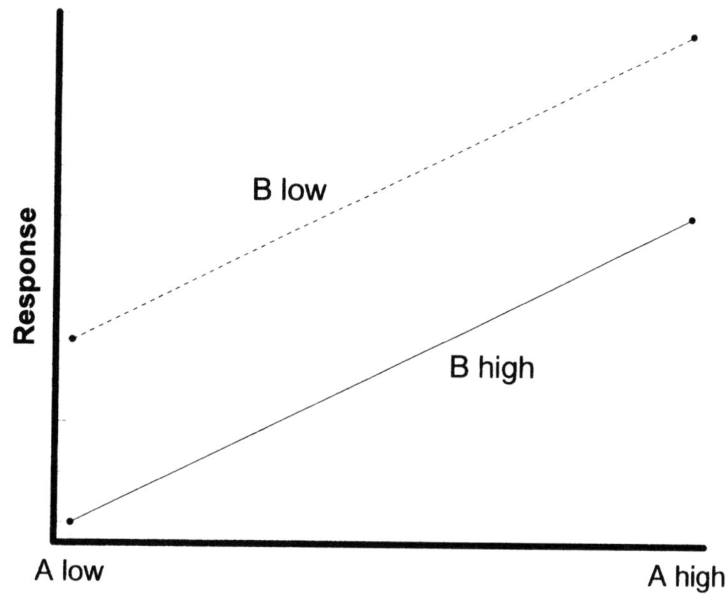

Figure 14.4. Interaction plot with no interaction.

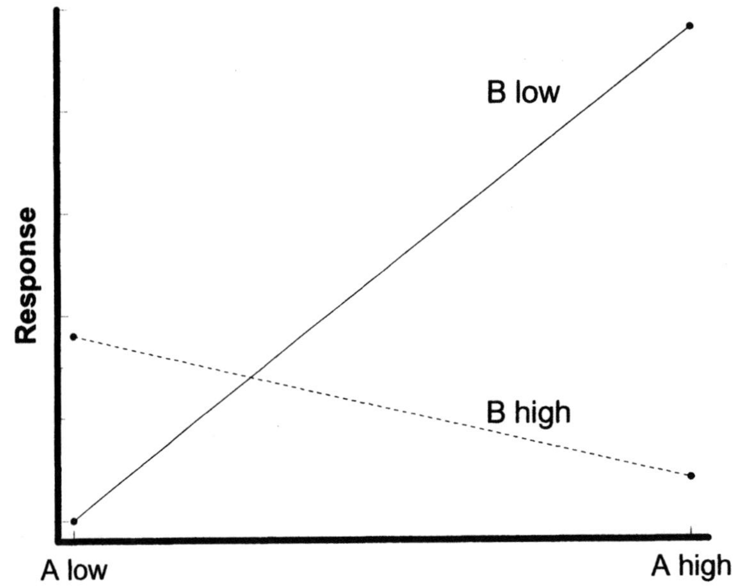

Figure 14.5. Interaction plot with a strong interaction.

Caution should be taken when using interaction plots. Remember that when using screening designs the interaction between 2 variables is often confounded with the main effect of another factor. Consider the simple model

$$Y = 1 + 2A + 3B + 4C$$

In this model, factors A, B and C all have a statistically significant affect on the response, and there is no interaction between any of the factors. Now consider a simple L_4 screening design. The levels for each factor will be set at 0 and 1; these levels, and the corresponding response are shown in the table below.

Table 14.19. Interaction plot demonstration.

Factor A	Factor B	Factor C	Response
0	0	0	1
0	1	1	8
1	0	1	7
1	1	0	6

An interaction plot for factors A and B is shown in the figure below. A strong interaction is indicated, even though there is no interaction in the model. This is because the interaction between the factors assigned to the first 2 columns of a L_4

design is confounded with the main effect of the factor assigned to the third column. Since factor C has a statistically significant affect on the response, an interaction plot for factors A and B falsely indicates an interaction is present.

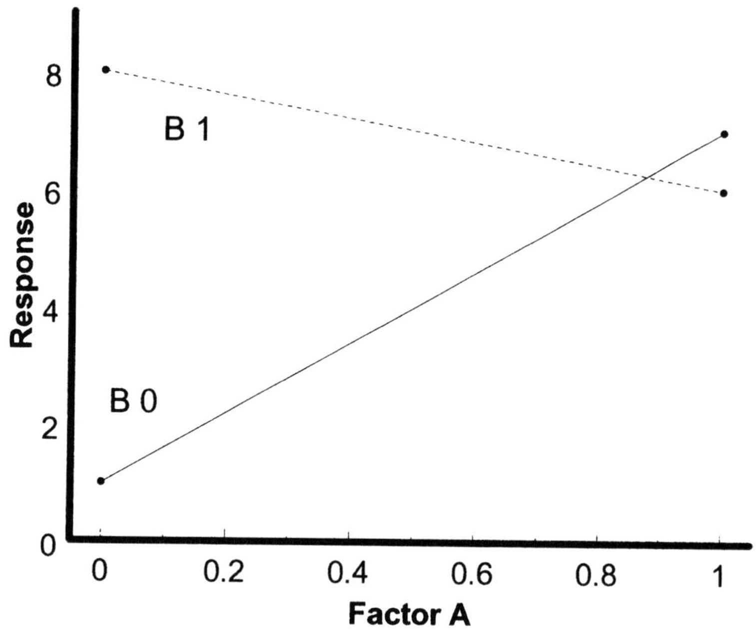

Figure 14.6. Misleading interaction plot.

14.3 Regression

Regression is a method of fitting an equation to a data set. Simple regression involves one independent variable and one dependent variable. Multiple regression allows more than 1 independent variable and 1 dependent variable. Least squares is the most common method of regression. With least squares regression a straight line is fit to the data by minimizing the sum of the squared distance of each point from the line. The arrows in Figure 14.7 display the vertical distance from the best-fit line. This best fit line is determined by minimizing the squared distances represented by the arrow.

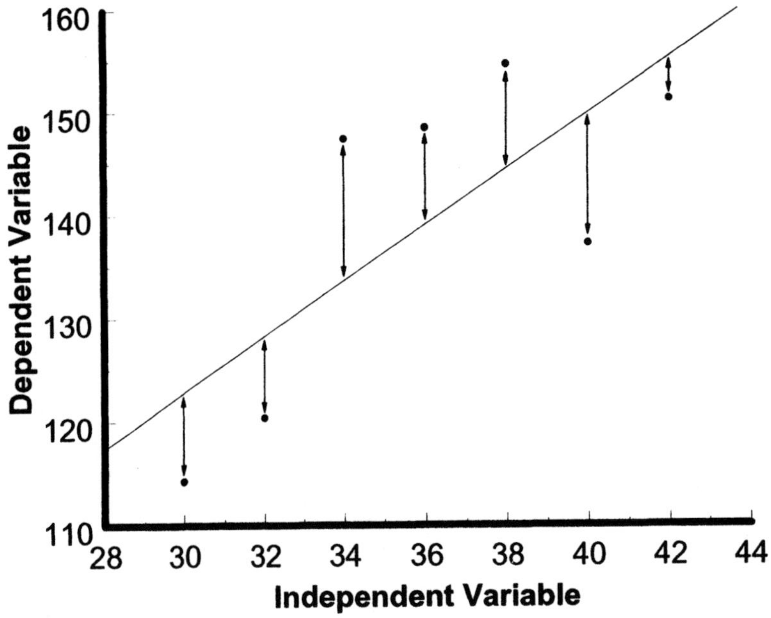

Figure 14.7. Least squares regression.

The coefficient of determination r^2 is a measure of the amount of variability in the data explained by the regression model. For example, if $r^2 = 0.91$, then 91% of the variability in the model is explained by the regression model. The correlation coefficient, r, measures the correlation between the independent variable and the dependent variable. Perfect positive correlation is achieved when $r = 1$. Perfect negative correlation is achieved when $r = -1$. If there is no correlation between the variables (the variables are independent) then $r = 0$. Figure 14.7 displays positive correlation; if the independent variable increases, the dependent variable increases. Figure 14.8 demonstrated negative correlation; if the independent variable increases, the dependent variable decreases.

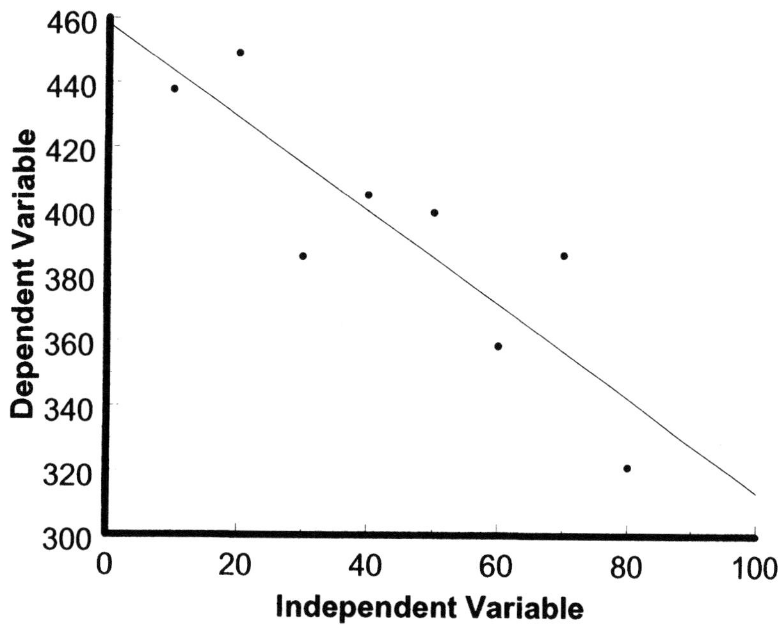

Figure 14.8. Negative correlation.

14.3.1 Simple regression

Simple linear regression is used to estimate the coefficients of the model

$$y_i = a + bx_i + e_i \tag{14.17}$$

where y_i is the dependent variable,

x_i is the independent variable, and

e_i is the residual; the error in the fit of the model.

Using the method of least squares, the coefficients are estimated from the following expressions.

$$b = \frac{n\sum_{i=1}^{n} x_i y_i - \left(\sum_{i=1}^{n} x_i\right)\left(\sum_{i=1}^{n} y_i\right)}{n\sum_{i=1}^{n} x_i^2 - \left(\sum_{i=1}^{n} x_i\right)^2} \tag{14.18}$$

$$a = \frac{\sum_{i=1}^{n} y_i - b\sum_{i=1}^{n} x_i}{n} \tag{14.19}$$

where n is the sample size.

The sum of squares of x is

$$S_{xx} = \sum_{i=1}^{n} x_i^2 - \frac{\left(\sum_{i=1}^{n} x_i\right)^2}{n} \tag{14.20}$$

The sum of squares of y is

$$S_{yy} = \sum_{i=1}^{n} y_i^2 - \frac{\left(\sum_{i=1}^{n} y_i\right)^2}{n} \tag{14.21}$$

The sum of squares of x and y is

$$S_{xy} = \sum_{i=1}^{n} x_i y_i - \frac{\left(\sum_{i=1}^{n} x_i\right)\left(\sum_{i=1}^{n} y_i\right)}{n} \tag{14.22}$$

The sample coefficient of determination is

$$r^2 = \frac{S_{xy}^2}{S_{xx}S_{yy}} \tag{14.23}$$

The residual in the model is a random variable with a mean of zero, and each individual e_i has the same variance, σ^2. An estimate of this variance is

$$s^2 = \frac{S_{yy} - bS_{xy}}{n-2} \tag{14.24}$$

The variance of the estimate of *b* is

$$\operatorname{var}(b) = \frac{s^2}{S_{xx}} \tag{14.25}$$

The variance of the estimate of *a* is

$$\operatorname{var}(a) = \frac{\sum_{i=1}^{n} x_i^2}{nS_{xx}} \tag{14.26}$$

The covariance of the estimates of *a* and *b* is

$$\operatorname{cov}(a,b) = -\bar{x}\operatorname{var}(b) \tag{14.27}$$

A prediction interval that contains a **single** response with a confidence of (1-α) is

$$\hat{y}_0 \pm t_{\alpha/2} s \sqrt{1 + \frac{1}{n} + \frac{(x_0 - \bar{x})^2}{S_{xx}}} \tag{14.28}$$

$t_{\alpha/2}$ has n-2 degrees of freedom.

A confidence interval that contains the **mean** response with a confidence of (1-α) is

$$\hat{y}_0 \pm t_{\alpha/2} s \sqrt{\frac{1}{n} + \frac{(x_0 - \bar{x})^2}{S_{xx}}} \tag{14.29}$$

$t_{\alpha/2}$ has n-2 degrees of freedom.

A confidence interval that contains b response with a confidence of (1-α) is

$$b \pm t_{\alpha/2} \sqrt{\frac{s^2}{S_{xx}}} \tag{14.30}$$

$t_{\alpha/2}$ has n-2 degrees of freedom.

Example 14.6

The time to fail for a bearing and the load placed on the bearing are shown below. Using simple regression, develop a model to predict the time to fail as a function of the load placed on the bearing.

Load	Time to Fail
10	912
20	940
30	953
40	955
50	902
60	867
70	875
80	887

Solution

Table 14.20 shows the computations required for the following equations. The slope of the regression line is

$$b = \frac{8(324{,}040) - (360)(7{,}291)}{8(20{,}400) - 360^2} = -0.965$$

The intercept of the regression line is

$$a = \frac{7{,}291 - (-0.965)(360)}{8} = 954.8$$

The time to fail can be estimated from the expression

$T = 954.8 - 0.965L$

The regression line is shown in Figure 14.9.

Table 14.20. Example regression computations.

Load	Time	Load Squared	Time Squared	Load x Time
10	912	100	831,744	9,120
20	940	400	883,600	18,800
30	953	900	908,209	28,590
40	955	1,600	912,025	38,200
50	902	2,500	813,604	45,100
60	867	3,600	751,689	52,020
70	875	4,900	765,625	61,250
80	887	6,400	786,769	70,960
360	7,291	20,400	6,653,265	324,040

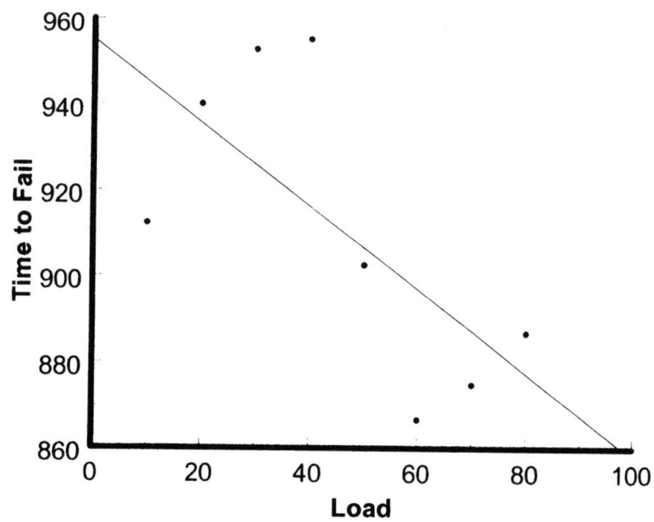

Figure 14.9. Time to fail regression model.

14.3.2 Multiple regression

In many cases estimation models require more than one independent variable. For example, a model to estimate a person's weight may include height, sex, and age. For multiple regression, the model being estimated is

$$y_i = b_0 + b_1 x_{1i} + b_2 x_{2i} + \ldots + b_k x_{ki} + e_i$$

where y_i is the dependent variable,

$x_{1i}, x_{2i}, \ldots, x_{ki}$, are the independent variables, and

e_i is the residual; the error in the fit of the model.

Using least squares regression, the parameters of the model are found by solving the matrix equation

$$Y = XB + e \tag{14. 31}$$

where

$$Y = \begin{bmatrix} y_1 \\ y_2 \\ \vdots \\ y_n \end{bmatrix} \tag{14. 32}$$

$$X = \begin{bmatrix} 1 & (x_{11} - \bar{x}_1) & (x_{21} - \bar{x}_2) & \cdots & (x_{k1} - \bar{x}_k) \\ 1 & (x_{12} - \bar{x}_1) & (x_{22} - \bar{x}_2) & \cdots & (x_{k2} - \bar{x}_k) \\ \vdots & \vdots & \vdots & & \vdots \\ 1 & (x_{1n} - \bar{x}_1) & (x_{2n} - \bar{x}_2) & \cdots & (x_{kn} - \bar{x}_k) \end{bmatrix} \tag{14. 33}$$

$$B = \begin{bmatrix} b_0 \\ b_1 \\ b_2 \\ \vdots \\ b_k \end{bmatrix} \tag{14. 34}$$

and

$$b_0 = b_0' + \sum_{i=1}^{k} b_i \bar{x}_i \tag{14.35}$$

The matrix equation is solved using matrix algebra

$$B = (X'X)^{-1} X'Y \tag{14.36}$$

The transpose of the matrix X is written as X', and the inverse of matrix X is written as X^{1}.

The amount of variability in the data explained by the regression model is measured by the coefficient of multiple correlation

$$R^2 = \frac{\sum_{i=1}^{n}(\hat{y}_i - \bar{y})^2}{\sum_{i=1}^{n}(y_i - \bar{y})^2} \tag{14.37}$$

where $\hat{y}_i$ is the response estimated from the regression model.

For example, if $R^2 = 0.971$, then 97.1% of the variability in the model is explained by the regression model.

Example 14.7

The tensile strength of a seal is dependent on the casting temperature, the casting pressure, and the volume of a chemical agent. Given the experimental data below, develop a model to predict the tensile strength.

Strength	Temperature	Pressure	Chemical Agent
39	192	1.2	12.2
29	199	1.4	13.8
39	196	1.5	12.8
39	208	1.3	12.6
35	202	1.2	13.1
32	186	1.4	13.6
33	183	1.2	13.1
30	180	1.3	13.5

Solution

The tensile strength prediction model is

$$S = b_0 + b_1 T + b_2 P + b_3 C$$

where S is the tensile strength, T is the casting temperature, P is the casting pressure, and C is the volume of the chemical agent. The average of the 8 tensile strength readings is 34.5. The average of the 8 casting temperature readings is 193.25. The average of the 8 casting pressure readings is 1.3125. The average of the chemical agent volumes is 13.0875. The matrix equation used to determine the constants of the prediction model is

$$\begin{bmatrix} 39 \\ 29 \\ 39 \\ 39 \\ 35 \\ 32 \\ 33 \\ 30 \end{bmatrix} = \begin{bmatrix} 1 & -1.25 & -0.1125 & -0.8875 \\ 1 & 5.75 & 0.0875 & 0.7125 \\ 1 & 2.75 & 0.1875 & -0.2875 \\ 1 & 14.75 & -0.0125 & -0.4875 \\ 1 & 8.75 & -0.1125 & 0.0125 \\ 1 & -7.25 & 0.0875 & 0.5125 \\ 1 & -10.25 & -0.1125 & 0.0125 \\ 1 & -13.25 & -0.0125 & 0.4125 \end{bmatrix} \begin{bmatrix} b_0' \\ b_1 \\ b_2 \\ b_3 \end{bmatrix}$$

After completing the appropriate matrix algebra, the regression model is

$$S = 106.47 - 0.658T + 9.3P - 7.4P + e$$

The computations required for the example are extremely tedious if done by hand. Fortunately, there are several computer routines that can be used. Microsoft® Excel™ and Lotus® 123™ are both capable of performing multiple linear regression. Excel™ includes an add-in and 123™ has functions to perform regression. In Excel™, select "Data Analysis" from the "Tools" menu and select the regression option. In 123™, the function "@Regression(*X-range, Y-range, Attribute*). The *X-range* specifies the independent variable(s) and includes more than 1 column for multiple regression. The *X-range* specifies the dependent variable. Setting the *attribute* to 1 returns, b_0, setting the *attribute* to 3 returns the coefficient of determination, setting the *attribute* to 101 returns the coefficient of the first independent variable, setting the *attribute* to 102 returns the coefficient of the second independent variable, setting the *attribute* to 103 returns the coefficient of the third independent variable, etc.

14.3.3 Non-linear regression

Assuming a linearity is a restrictive, but this assumption can be overcome by making the appropriate transformations before performing regression. Consider the linear regression line fit to the data in the figure below.

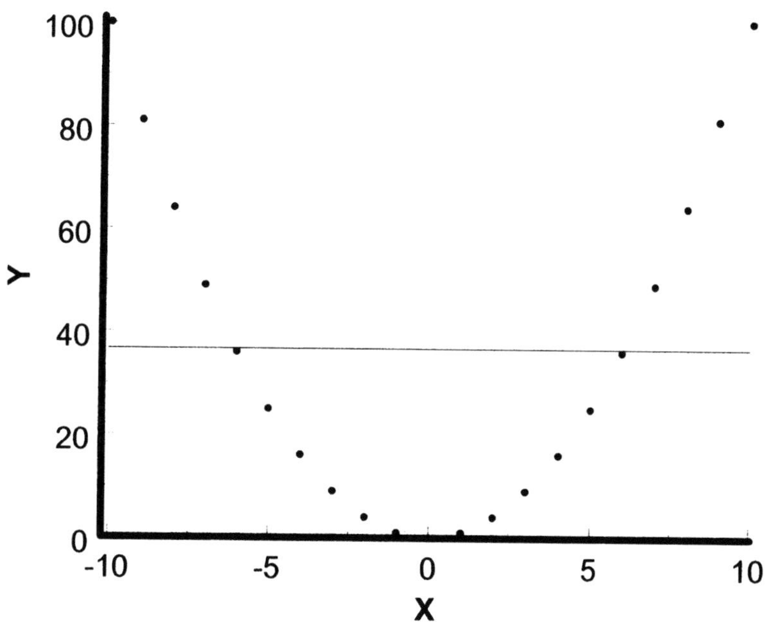

Figure 14.10. Non-linear regression.

The regression model displayed in Figure 14.10 does not fit the data; a non-linear regression model is required. Instead of using the linear model of

$$y = b_0 + b_1 x$$

the non-linear model of

$$y = b_0 + b_1 x^2$$

can be used to model this data. This is accomplished by transforming the data prior to regression. In this case, the square of x is regressed against y. The resulting regression model is shown in the figure below.

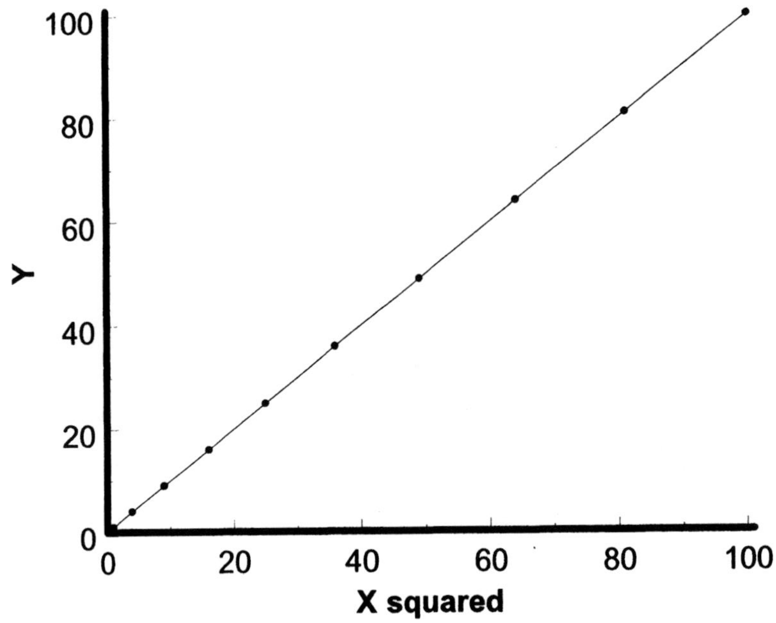

Figure 14.11. Non-linear regression example.

Example 14.8

The following data was collected, and it is believed that a logarithmic model describes the data. Fit a model of the form

$$y = a + b \ln x$$

to the following data.

x	y
45	245.8
53	247.6
78	251.1
97	256.5
199	265.6

Solution

First, the model must be transformed to a linear form. By regressing the natural logarithm of x instead of x, a linear form is achieved. The value of y, x and $\ln x$ are given below. The independent variable in the regression is $\ln x$, and the dependent variable is y.

y	x	$\ln x$
245.8	453.807	
247.6	533.970	
251.1	784.357	
256.5	974.575	
265.6	199	5.293

The regression coefficients are

$a = 193.64$, and
$b = 13.56$

Thus, the regression model is

$$y = 193.64 + 13.56 \ln x$$

The regression in linear and non-linear form is shown in the figures below.

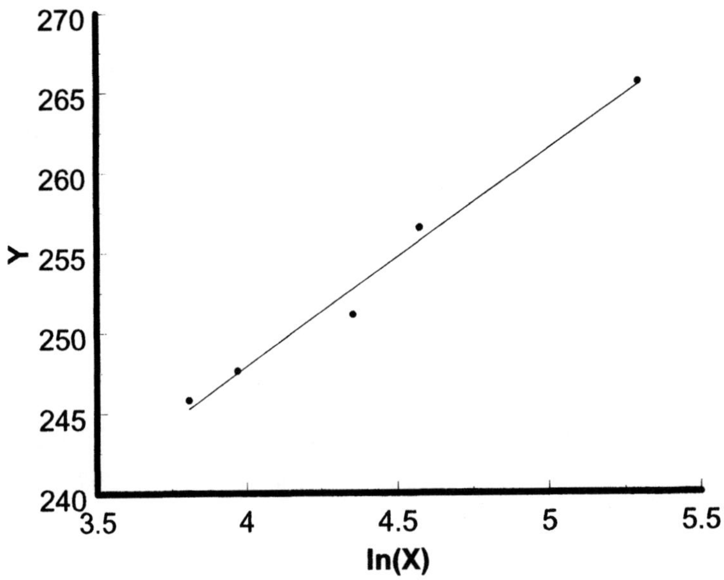

Figure 14.12. Logarithmic regression in non-linear format.

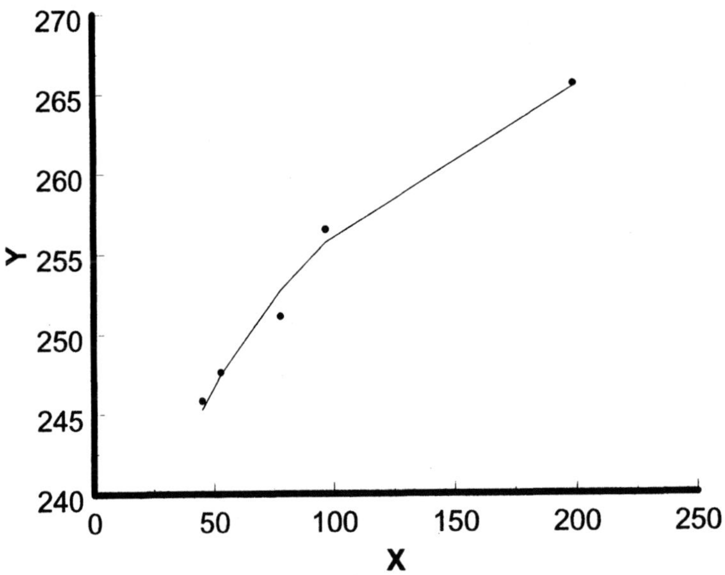

Figure 14.13. Logarithmic regression in linear format.

Example 14.9

The following data was collected, and it is believed that a second order polynomial fits the data. Fit a second order polynomial to the data.

x	y
1	6.3
2	8.9
3	11.1
4	16.2
5	18.2

Solution

A second order polynomial is

$$y = b_0 + b_1 x + b_2 x^2$$

Multiple regression is required to fit this model. The independent variables are x and x squared. The variable x squared is treated like it is completely independent of x for the purposes of the regression. The dependent variable is y.

y	x	x^2
6.3	11	
8.9	24	
11.1	3	9
16.2	4	16
18.2	5	25

The regression coefficients are

$b_0 = 3.66$, and

$b_1 = 2.381$

$b_2 = 0.121$

Thus, the regression model is

$$y = 3.66 + 2.381x + 0.121x^2$$

14.4 Surface response methodology

Additional topics of interest to experimenters are evolutionary operation (EVOP), response surface methodology, and sequential simplex optimization.

After experimentation has been completed, and a process or design has been optimized, it is not uncommon for the experimentation team to consider the problem to be solved. In many cases, the optimum solution does not stay optimum; equipment changes as it ages, the composition of components change over time, and many other factors make small changes that affect the response variable. EVOP is a method of addressing the constant small changes in a process in attempt to maintain an optimum level. Experimentation is carried out on a near continuous basis by introducing small changes to the system. The experiments usually are factorial designs with 2 levels.

Surface response methodology consists of several statistical techniques designed to fit non-linear equations with multiple independent variables to a single response variable. If a suitable model is found the response variable can be optimized. The optimization may be to

- minimize the response variable,
- maximize the response variable, or to
- minimize the variability in the response variable.

Multiple regression, factorial experiments with more than 2 levels, and methods for steepest assent are used to fit response surfaces.

One of the most popular methods of steepest accent is sequential simplex optimization. The simplex is a geometric figure with the number of vertexes equal to the number of factors plus one. A simplex with one factor is a line, a simplex with two factors is a triangle, etc. The simplex is a simplistic model of the surface. a new simplex is formed by eliminating the vertex with the worst response and replacing it by projecting it through the average coordinates of the remaining vertexes. A new experiment is performed with the factor levels determined by the coordinates of the new vertex, and the process is repeated until an optimum response is found. Using the sequential simplex method optimization has two advantages over using factorial designed experiments for EVOP.

The number of trials in the initial simplex is k+1 (k is the number of experimental factors), while a factorial approach has at least 2^k and possibly 3^k or 4^k.

Only one new trial is required to move to a new area in the space defined by the factors; while a factorial design requires at least 2^{k-1} trials.

An example of a sequential simplex optimization is shown in the figure below.

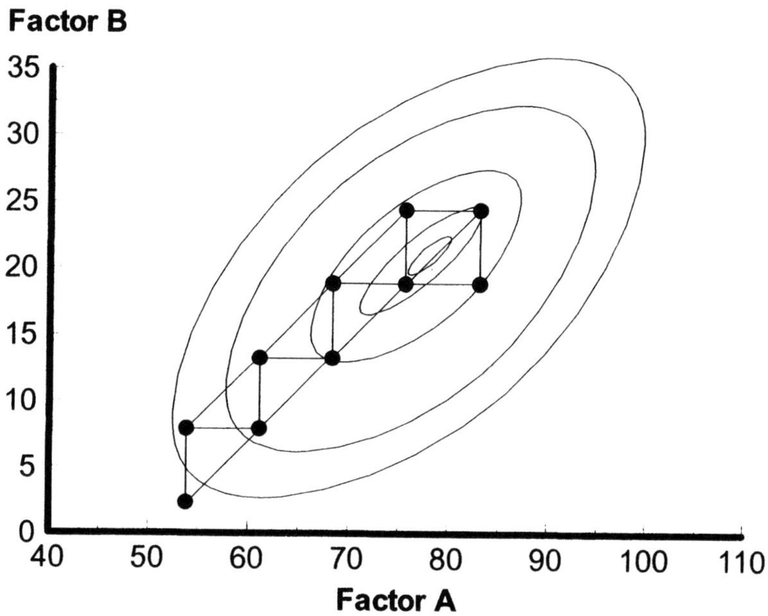

Figure 14.14. Sequential simplex optimization with 2 factors.

To compute the coordinates of the new vertex (R), refer to the following table, where *B* is the vertex with the best response, *N* is the vertex with the next best response, and *W* is the vertex with the worst response. This table shows an example for a 3 factor optimization ($k = 3$). The logic is the same for any number of factors.

Table 14.21. Determining the simplex vertex.

	Factor A	Factor B	Factor C	Response	Vertex Number
Best response	10	15	75	424.3	3
Next best response	20	45	75	398.7	1
Sum	30	60	150		
Sum/k (P)	10	20	50		
Worst response (W)	15	15	85	212.4	2
P-W	-5	2	-35		
New vertex P+(P-w)	5	25	15		4

The optimum point is never reached in Figure 14.14 because the size of the individual simplexes is too large. If the simplex size is too small, an excessive number of steps is required to reach the optimum point. This problem has been solved by using a variable size simplex. Instead of a simple reflection (*R*), there are other options:

- double the length of the reflection (*E*),
- reduce the length of the expansion by 50% (*Cr*),
- produce the reflection in the opposite direction (*W*), and

- produce the reflection in the opposite direction but at 50% of the normal length (*Cw*).

The additional rules for determining the location of the new vertex are:

If the response for the new vertex (*R*) is better than the response for the next best vertex (N), but worse than the response of the best (*B*) vertex, then use *R* as the new vertex.

If the response for the new vertex (*R*) is better than the response of the best vertex (*B*), then compute $E = P + 2(P\text{-}W)$. If the response of *E* is better than the response of *B*, use *E* as the new vertex; if not, use *R* as the new vertex.

If the response for the new vertex (*R*) is worse than the response of the next best vertex (*N*) but better than the response of the worst vertex (*W*), then compute $Cr = P + 0.5(P\text{-}W)$ as the new vertex.

If the response for the new vertex (*R*) is worse than the response of the worst vertex (*W*), then compute $Cw = P - 0.5(P\text{-}W)$ as the new vertex.

Consider the parameter *Y* determined from the function

$$y = 40A + 35B + 25AB - 15A^2 - 15B^2$$

Starting with the vertices

Vertex	A	B	Response
1	100	100	-42,500
2	100	120	-57,800
3	120	120	-63,000

The response *Y* is maximized using a variable-step sequential simplex as shown below.

Coordinates	A	B	Response	Rank	Vertex #	Rank of New Vertex
Best (B)	100	100	-42500	B	1	
Next Best (N)	100	120	-57800	N	2	
Sum	200	220				
P bar	100	110				
Worst (W)	120	120	-63000	W	3	
P bar - W	-20	-10				

Next Vertex (R)	80	100	-39300	R		Compute E
E =	60	90	-34950			Use E

Step 2						Rank of
Coordinates	A	B	Response	Rank	Vertex #	New Vertex
Best (B)	60	90	-34950	B	4	
Next Best (N)	100	100	-42500	N	1	
Sum	160	190				
P bar	80	95				
Worst (W)	100	120	-57800	W	2	
P bar - W	-20	-25				
Next Vertex (R)	60	70	-17650	R		Compute E
E =	40	45	-6200			Use E

Step 3						Rank of
Coordinates	A	B	Response	Rank	Vertex #	New Vertex
Best (B)	40	45	-6200	B	5	
Next Best (N)	60	90	-34950	N	4	
Sum	100	135				
P bar	50	67.5				
Worst (W)	100	100	-42500	W	1	
P bar - W	-50	-32.5				
Next Vertex (R)	0	35	-17150	R		Use R

Step 4						Rank of
Coordinates	A	B	Response	Rank	Vertex #	New Vertex
Best (B)	40	45	-6200	B	5	
Next Best (N)	0	35	-17150	N	6	
Sum	40	80				
P bar	20	40				
Worst (W)	60	90	-34950	W	4	
P bar - W	-40	-50				
Next Vertex (R)	-20	-10	-3650	R		Compute E
E =	-60	-60	-22500			Use R

Step 5						Rank of
Coordinates	A	B	Response	Rank	Vertex #	New Vertex
Best (B)	-20	-10	-3650	B	7	
Next Best (N)	40	45	-6200	N	5	
Sum	20	35				
P bar	10	17.5				
Worst (W)	0	35	-17150	W	6	
P bar - W	10	-17.5				
Next Vertex	20	0	-5200	R		Use R

(R)

Step 6

Coordinates	A	B	Response	Rank	Vertex #	Rank of New Vertex
Best (B)	-20	-10	-3650	B	7	
Next Best (N)	20	0	-5200	N	8	
Sum	0	-10				
P bar	0	-5				
Worst (W)	40	45	-6200	W	5	
P bar - W	-40	-50				
Next Vertex (R)	-40	-55	-17900	R		Compute Cw
Cw =	-20	-30	-6350			Use Cw

Step 7

Coordinates	A	B	Response	Rank	Vertex #	Rank of New Vertex
Best (B)	-20	-10	-3650	B	7	
Next Best (N)	20	0	-5200	N	8	
Sum	0	-10				
P bar	0	-5				
Worst (W)	-20	-30	-6350	W	9	
P bar - W	20	25				
Next Vertex (R)	20	20	-500	R		Compute E
E =	40	45	-6200			Use R

Step 8

Coordinates	A	B	Response	Rank	Vertex #	Rank of New Vertex
Best (B)	20	20	-500	B	10	
Next Best (N)	-20	-10	-3650	N	7	
Sum	0	10				
P bar	0	5				
Worst (W)	20	0	-5200	W	8	
P bar - W	-20	5				
Next Vertex (R)	-20	10	-12950	R		Compute Cw
Cw =	-10	7.5	-4356			Use Cw

Step 9

Coordinates	A	B	Response	Rank	Vertex #	Rank of New Vertex
Best (B)	20	20	-500	B	10	
Next Best (N)	-20	-10	-3650	N	7	
Sum	0	10				
P bar	0	5				
Worst (W)	-10	7.5	-4356	W	11	
P bar - W	10	-2.5				
Next Vertex (R)	10	2.5	-481	R		Compute E

E =	20	0	-5200			Use R

Step 10

Coordinates	A	B	Response	Rank	Vertex #	Rank of New Vertex
Best (B)	10	2.5	-481	B	12	
Next Best (N)	20	20	-500	N	10	
Sum	30	22.5				
P bar	15	11.25				
Worst (W)	-20	-10	-3650	W	7	
P bar - W	35	21.25				
Next Vertex (R)	50	32.5	-9581	R		Compute Cw
Cw =	32.5	21.875	-3182			Use Cw

Step 11

Coordinates	A	B	Response	Rank	Vertex #	Rank of New Vertex
Best (B)	10	2.5	-481	B	12	
Next Best (N)	20	20	-500	N	10	
Sum	30	22.5				
P bar	15	11.25				
Worst (W)	32.5	21.875	-3182	W	13	
P bar - W	-17.5	-10.625				
Next Vertex (R)	-2.5	0.625	-217	R		Compute E
E =	-20	-10	-3650			Use R

Step 12

Coordinates	A	B	Response	Rank	Vertex #	Rank of New Vertex
Best (B)	-2.5	0.625	-217	B	14	
Next Best (N)	10	2.5	-481	N	12	
Sum	7.5	3.125				
P bar	3.75	1.5625				
Worst (W)	20	20	-500	W	10	
P bar - W	-16.25	-18.4375				
Next Vertex (R)	-12.5	-16.875	-2432	R		Compute Cw
Cw =	-4.375	-7.65625	-772			Use Cw

Step 13

Coordinates	A	B	Response	Rank	Vertex #	Rank of New Vertex
Best (B)	-2.5	0.625	-217	B	14	
Next Best (N)	10	2.5	-481	N	12	
Sum	7.5	3.125				
P bar	3.75	1.5625				
Worst (W)	-4.375	-7.65625	-772	W	15	
P bar - W	8.125	9.21875				
Next Vertex (R)	11.875	10.78125	194	R		Compute E

E =	20	20	-500			Use R

Step 14 Coordinates	A	B	Response	Rank	Vertex #	Rank of New Vertex
Best (B)	11.875	10.78125	194	B	16	
Next Best (N)	-2.5	0.625	-217	N	14	
Sum	9.375	11.40625				
P bar	4.6875	5.703125				
Worst (W)	10	2.5	-481	W	12	
P bar - W	-5.3125	3.203125				
Next Vertex (R)	-0.625	8.90625	-1048	R		Compute Cw
Cw =	2.03125	7.304688	-154			Use Cw

Step 15 Coordinates	A	B	Response	Rank	Vertex #	Rank of New Vertex
Best (B)	11.875	10.78125	194	B	16	
Next Best (N)	2.03125	7.304688	-154	N	17	
Sum	13.90625	18.08594				
P bar	6.953125	9.042969				
Worst (W)	-2.5	0.625	-217	W	14	
P bar - W	9.453125	8.417969				
Next Vertex (R)	16.40625	17.46094	-182	R		Compute Cr
Cr =	21.13281	21.66992	-690			Use Cr

Step 16 Coordinates	A	B	Response	Rank	Vertex #	Rank of New Vertex
Best (B)	11.875	10.78125	194	B	16	
Next Best (N)	2.03125	7.304688	-154	N	17	
Sum	13.90625	18.08594				
P bar	6.953125	9.042969				
Worst (W)	21.13281	21.66992	-690	W	18	
P bar - W	-14.1797	-12.627				
Next Vertex (R)	-7.22656	-3.58398	-743	R		Compute Cw
Cw =	-0.13672	2.729492	-31			Use Cw

Step 17 Coordinates	A	B	Response	Rank	Vertex #	Rank of New Vertex
Best (B)	11.875	10.78125	194	B	16	
Next Best (N)	-0.13672	2.729492	-31	N	19	
Sum	11.73828	13.51074				
P bar	5.869141	6.755371				
Worst (W)	2.03125	7.304688	-154	W	17	
P bar - W	3.837891	-0.54932				
Next Vertex (R)	9.707031	6.206055	120	R		Use R

Step 18						Rank of
Coordinates	A	B	Response	Rank	Vertex #	New Vertex
Best (B)	11.875	10.78125	194	B	16	
Next Best (N)	9.707031	6.206055	120	N	20	
Sum	21.58203	16.9873				
P bar	10.79102	8.493652				
Worst (W)	-0.13672	2.729492	-31	W	19	
P bar - W	10.92773	5.76416				
Next Vertex (R)	21.71875	14.25781	-1016	R		Compute Cw
Cw =	16.25488	11.37573	-233			Use Cw

Step 19						Rank of
Coordinates	A	B	Response	Rank	Vertex #	New Vertex
Best (B)	11.875	10.78125	194	B	16	
Next Best (N)	9.707031	6.206055	120	N	20	
Sum	21.58203	16.9873				
P bar	10.79102	8.493652				
Worst (W)	16.25488	11.37573	-233	W	21	
P bar - W	-5.46387	-2.88208				
Next Vertex (R)	5.327148	5.611572	259	R		Compute E
E =	-0.13672	2.729492	-31			Use R

Step 20						Rank of
Coordinates	A	B	Response	Rank	Vertex #	New Vertex
Best (B)	5.327148	5.611572	259	B	22	
Next Best (N)	11.875	10.78125	194	N	16	
Sum	17.20215	16.39282				
P bar	8.601074	8.196411				
Worst (W)	9.707031	6.206055	120	W	20	
P bar - W	-1.10596	1.990356				
Next Vertex (R)	7.495117	10.18677	166	R		Compute Cr
Cr =	6.942139	11.18195	11			Use Cr

Step 21						Rank of
Coordinates	A	B	Response	Rank	Vertex #	New Vertex
Best (B)	5.327148	5.611572	259	B	22	
Next Best (N)	11.875	10.78125	194	N	16	
Sum	17.20215	16.39282				
P bar	8.601074	8.196411				
Worst (W)	6.942139	11.18195	11	W	23	
P bar - W	1.658936	-2.98553				
Next Vertex (R)	10.26001	5.210876	-57	R		Compute Cw
Cw =	9.430542	6.703644	184			Use Cw

Step 22 Coordinates	A	B	Response	Rank	Vertex #	Rank of New Vertex
Best (B)	5.327148	5.611572	259	B	22	
Next Best (N)	11.875	10.78125	194	N	16	
Sum	17.20215	16.39282				
P bar	8.601074	8.196411				
Worst (W)	9.430542	6.703644	184	W	24	
P bar - W	-0.82947	1.492767				
Next Vertex (R)	7.771606	9.689178	218	R		Use R

Step 23 Coordinates	A	B	Response	Rank	Vertex #	Rank of New Vertex
Best (B)	5.327148	5.611572	259	B	22	
Next Best (N)	7.771606	9.689178	218	N	24	
Sum	13.09875	15.30075				
P bar	6.549377	7.650375				
Worst (W)	11.875	10.78125	194	W	25	
P bar - W	-5.32562	-3.13087				
Next Vertex (R)	1.223755	4.519501	17	R		Compute Cw
Cw =	3.886566	6.084938	178			Use Cw

Step 24 Coordinates	A	B	Response	Rank	Vertex #	Rank of New Vertex
Best (B)	5.327148	5.611572	259	B	22	
Next Best (N)	7.771606	9.689178	218	N	24	
Sum	13.09875	15.30075				
P bar	6.549377	7.650375				
Worst (W)	3.886566	6.084938	178	W	26	
P bar - W	2.662811	1.565437				
Next Vertex (R)	9.212189	9.215813	267	R		Compute E
E =	11.875	10.78125	194			Use R

After 24 step the best response is 267. Using traditional optimization techniques the optimum is 281 when A = 7.54 and 7.46. The surface response for this example is shown in Figure 14.15.

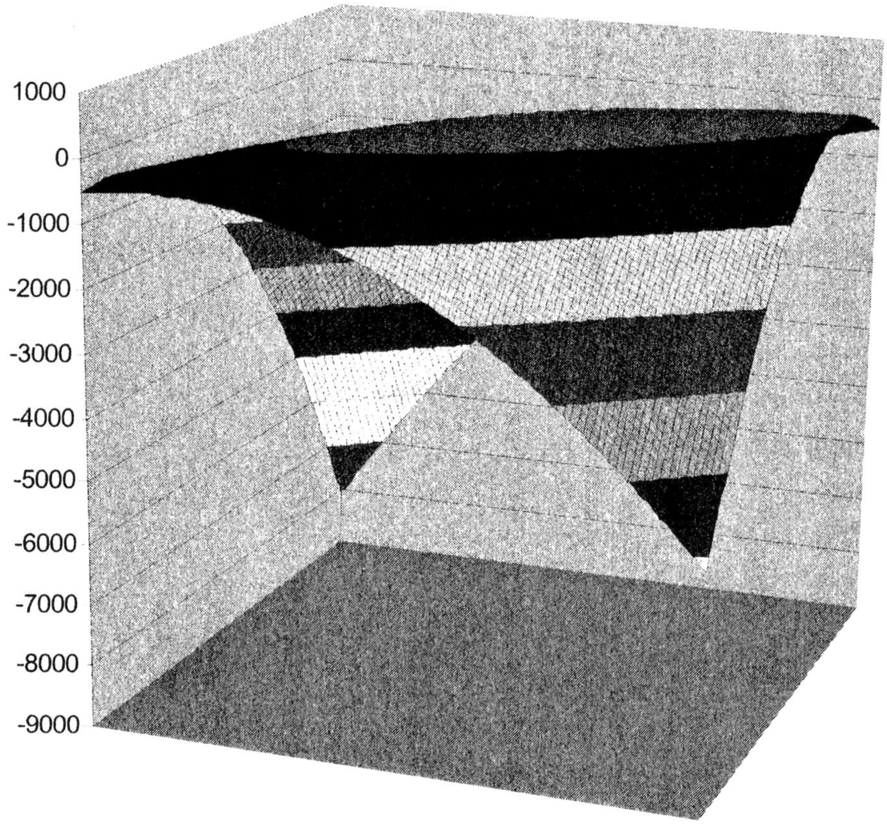

Figure 14.15. Response surface for example simplex problem.

Remember that surface response methods are usually the last step of a designed experiment. In a real situation the response for each of the 24 steps above would be the average of many trials.

14.5 Exercises

1. Taguchi designs are most appropriate when
   a. during the early stages of experimentation
   b. when all factors are orthogonal
   c. when the experimental model is non-linear
   d. when large amounts of data are available

2. A screening experiment was conducted to narrow the list of contributing factors from 8 to 3. Now the task is to determine the setting of these 3 factors that minimizes the process variability while producing at the specified target. The most appropriate experimental design technique would be

a. a Taguchi design with 2 levels

b. a Taguchi design with 3 levels

c. a full factorial design

d. sequential simplex optimization

3. Temperature, humidity and pressure affect the hardness of an alloy. Which of the following statements are true?

I. Hardness is the response variable.

II. Temperature is considered a factor.

III. Humidity is considered a noise variable.

a. Only I is true

b. Only II is true

c. I and II are true

d. I, II and III are true

4. Which of the following is not a true statement?

a. Screening designs never account for interactions.

b. Screening designs always set 2 levels for the factors.

c. Screening designs are always orthogonal.

d. None of the above are true.

5. A full factorial experiment is going to be conducted with 4 factors with 2 levels and 2 additional factors with 3 levels. How many trials are required?

a. 14

b. 11

c. 144

d. 6

6. Which of the following is least likely to be a dependent variable in a designed experiment?

a. processing pressure

b. tensile strength

c. machine speed

d. production shift

7. A fractional factorial experiment was conducted with 6 factors at 2 levels each. If 12 trials are conducted, how many two-way interactions can be considered if 1 degree of freedom is used for experimental error?

   a. 2

   b. 4

   c. 6

   d. 0

8. The following are the results of a designed experiment. Which of the following statements best describes the 2-way interaction between Factor A and Factor B.

Factor A	Factor B	Factor C	Response
-1	-1	-1	14.3
-1	1	1	17.8
1	-1	1	9.1
1	1	-1	8.6

   a. There is no interaction.

   b. The interaction is weak.

   c. The interaction is strong

   d. There are not enough experimental trials to make any inferences about the interaction.

9. In an L_8 design, which of the following statements are true?

   I. The interaction between Factors A and D is confounded with the main effect of Factor E.

   II. The interaction between Factors C and F is confounded with the main effect of Factor E.

   III. The interaction between Factors B and G is confounded with the main effect of Factor E.

   a. I only

   b. I and II only

   c. III only

   d. I, II and III

10. Given the data in below, determine the total sum of squares.

Factor A	Factor B	Factor C
1.3	2.1	1.2
1.6	1.3	1.3
1.8	1.9	1.3
1.2	1.2	1.2
1.3	2.2	1.1
1.4		1.4
1.1		

a. 1.943

b. 2.874

c. 14.398

d. 17.547

11. Given the data in Exercise 10, determine the sum of squares between treatments.

a. 3.657

b. 1.256

c. 0.0057

d. 0.687

12. Given the data in Exercise 10, determine the sum of squares for error.

a. 3.657

b. 7.101

c. 0.0057

d. 0.687

13. Given the data in Exercise 10, determine the mean square between treatments.

a. 0.3435

b. 1.5298

c. 0.0057

d. none of the above

14. Given the data in Exercise 10, determine the mean square error.

a. 0.3435

b. 1.5298

c. 0.0057

d. none of the above

15. Given the data in Exercise 10, determine the value of the F-statistic.

 a. 0.3435

 b. 1.5298

 c. 4.1039

 d. none of the above

16. Given the data in Exercise 10, determine the critical value of the F-statistic at 5% significance.

 a. 1.239

 b. 1.698

 c. 3.682

 d. None of the above

17. Sequential simplex optimization is most appropriate when

 a. There are a large number of factors to be explored.

 b. No interactions are present.

 c. The important factors have been identified.

 d. A factorial experiment has failed.

18. Given the figure below, the value of the correlation coefficient is

 a. positive

 b. greater than the coefficient of determination

 c. both a and b above

 d. none of the above

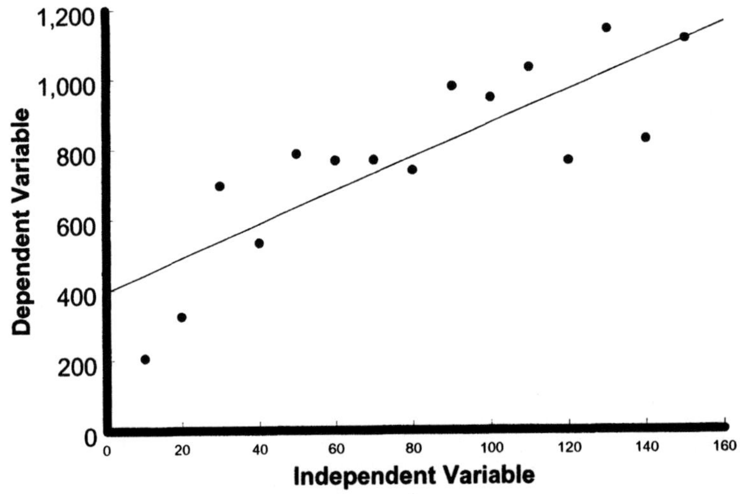

19. Given the figure below, the value of the correlation coefficient is

    a. negative

    b. greater than the coefficient of determination

    c. both a and b above

    d. none of the above

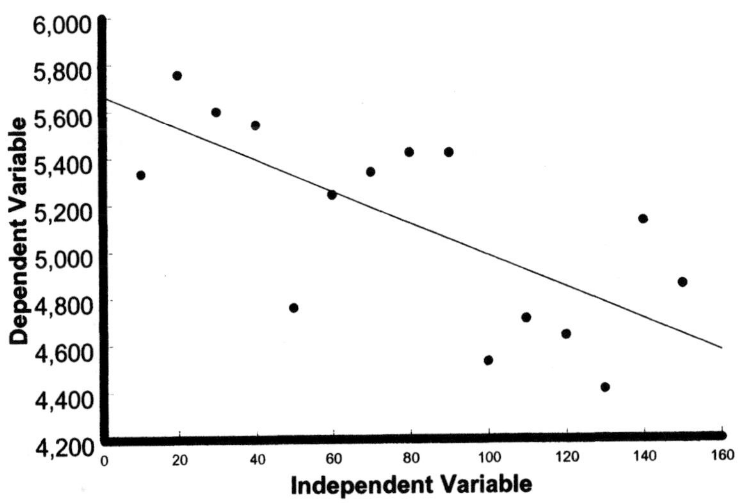

20. Comparing "Situation A" and "Situation B" below, which of the following is NOT a true statement?

    a. the correlation coefficient for Situation A is greater than correlation coefficient for situation B

b. the coefficient of determination for Situation A is greater than the coefficient of determination for situation B

c. the independent variable is a better predictor for the dependent variable in Situation A than Situation B.

d. All of the above are true statements.

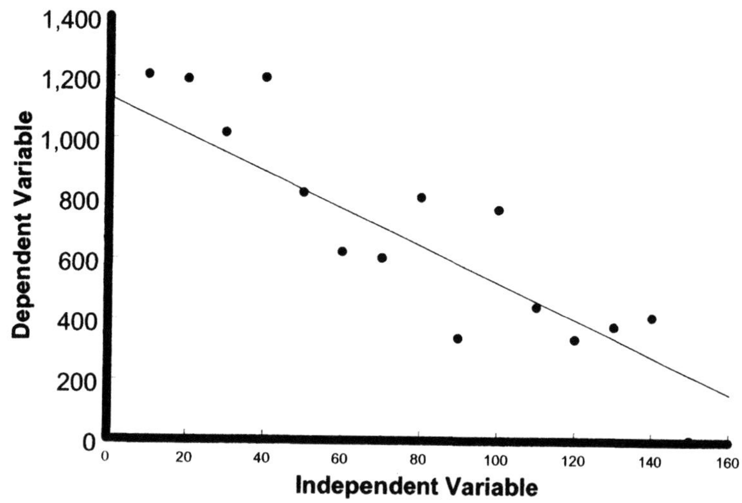

Situation A

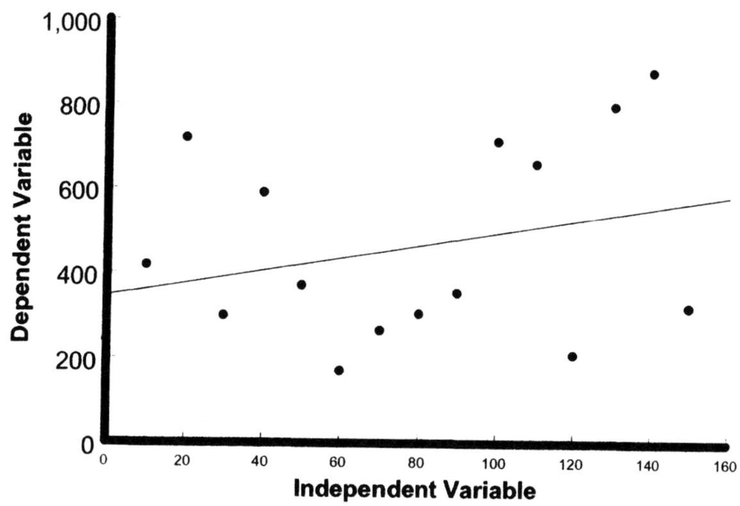

Situation B

21. Which of the following is NOT true?

a. the coefficient of determination is always greater than the correlation coefficient

b. the coefficient of determination always increases when an independent variable is added to the regression model

c. linear regression can be used to model non-linear systems

d. all of the above statements are true

22. Which of the following statement is true regarding the figure below?

a. the coefficient of determination is positive

b. the correlation coefficient is positive

c. linear regression cannot be applied to the this system

d. all of the above statements are true.

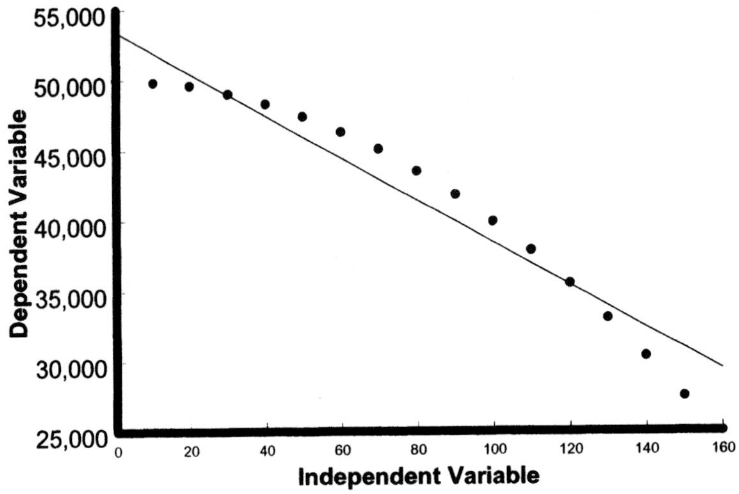

14.6 References

Blueprints for Product Reliability, Reliability Analysis Center (RAC), Rome, New York

O'Conner, P.D.T., 1993, ***Practical Reliability Engineering***, New York: John Wiley & Sons.

Alwan, L.C. and Roberts, H.V. (1989). "Time Series Modeling for Statistical Process Control," in Statistical Process Control in Automated Manufacturing, Keats, J.B. and Hubel, N.F. editors, Marcel Dekker, New York.

Berry, Michael J.A. and Linoff, Gordon. (1997). Data Mining Techniques: for Marketing, Sales, and Customer Support, New York: John Wiley & Sons.

Box, G.E.P. and Draper, N.R. (1969). Evolutionary Operation: A statistical method for process improvement, New York: John Wiley & Sons.

Box, G.E.P. and Draper, N.R. (1987). Empirical Model-Building and Response Surfaces, New York: John Wiley & Sons.

Box, Hunter and Hunter. (1978). Statistics for Experimenters, New York: John Wiley & Sons.

Draper, N. and H. Smith. (1981). Applied Regression Analysis, 2nd edition, New York: John Wiley & Sons.

Duncan, A.J. (1974). Quality Control and Industrial Statistics, 4th Edition, Homewood, IL, Irwin.

Hicks, Charles R. (1973). Fundamental Concepts in the Design of Experiments, 2nd Edition, New York: Holt, Rinehart, Winston, pp. 31–38.

Hicks, Charles R. (1993). Fundamental Concepts in the Design of Experiments, New York: Holt, Rinehart, Winston.

Holland, John H. (1996). Hidden Order: How Adaptation Builds Complexity Reading, MA: Perseus Books, p.56.

Koch, G.G. and D.B. Gillings. (1983). "Statistical Inference, Part I," in Encyclopedia of Statistical Sciences, vol. 4, 84–88, Samuel Kotz and Normal L. Johnson, editors-in-chief, New York: John Wiley & Sons.

Meyers, Raymond H. and Montgomery, Douglas C. (1995). Response Surface Methodology: Process and Product Optimization Using Designed Experiments, New York: John Wiley & Sons.

Michael J.A. Berry and Gordon Linoff. (1997). Data Mining Techniques, New York: John Wiley & Sons, pp. 369–370.

Montgomery, D.C. (1984). Design and Analysis of Experiments, 2nd Edition. New York: John Wiley & Sons.

Montgomery, Douglas C. (1996) Design and Analysis of Experiments, New York: John Wiley & Sons.

Natrella, M.G. (1963). Experimental Statistics: NBS Handbook 91. Washington, DC: US Government Printing Office.

Phadke, M. S. (1989). ***Quality Engineering Using Robust Design***, Prentice Hall, Englewood Cliffs, New Jersey.

Plackett, R.L. and Burman, J.P. (1946). "The Design of Optimum Multifactorial Experiments", ***Biometrika***, Vol: 33, pp. 305-325.

Taguchi, G. (1983). Introduction to Quality Engineering: Designing Quality into Products and Processes, White Plains, NY: Quality Resources.

Taguchi, G. 1987. System of Experimental Design. White Plains, NY: Quality Resources.

Tukey, J.W. (1977). Exploratory Data Analysis, Reading, MA: Addison-Wesley.

Walters, F. H, Parker, L. R., Morgan, S. L., and Deming, S. N. (1991). ***Sequential Simplex Optimization***, CRC Press, Boca Raton, Florida.

♦♦♦

APPENDIX

A

Area Under the Standard Normal Curve

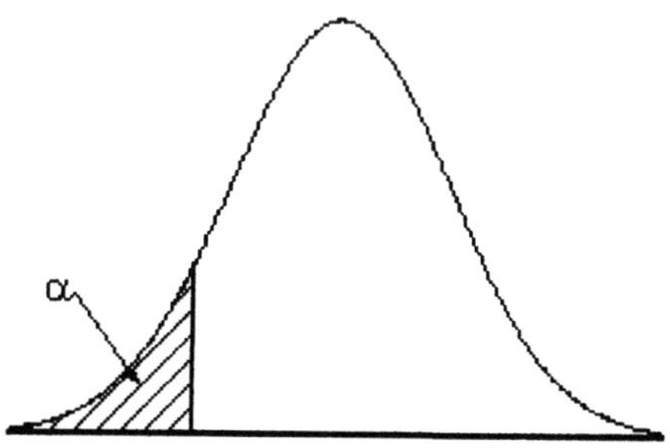

Table A.1. Area under the standard normal curve.

z	0.00	0.01	0.02	0.03	0.04	0.05	0.06	0.07	0.08	0.09
−3.4	0.0003	0.0003	0.0003	0.0003	0.0003	0.0003	0.0003	0.0003	0.0003	0.0002
−3.3	0.0005	0.0005	0.0005	0.0004	0.0004	0.0004	0.0004	0.0004	0.0004	0.0003
−3.2	0.0007	0.0007	0.0006	0.0006	0.0006	0.0006	0.0006	0.0005	0.0005	0.0005
−3.1	0.0010	0.0009	0.0009	0.0009	0.0008	0.0008	0.0008	0.0008	0.0007	0.0007

Continued on next page…

Table A.1—continued...

z	0.00	0.01	0.02	0.03	0.04	0.05	0.06	0.07	0.08	0.09
–3.0	0.0013	0.0013	0.0013	0.0012	0.0012	0.0011	0.0011	0.0011	0.0010	0.0010
–2.9	0.0019	0.0018	0.0018	0.0017	0.0016	0.0016	0.0015	0.0015	0.0014	0.0014
–2.8	0.0026	0.0025	0.0024	0.0023	0.0023	0.0022	0.0021	0.0021	0.0020	0.0019
–2.7	0.0035	0.0034	0.0033	0.0032	0.0031	0.0030	0.0029	0.0028	0.0027	0.0026
–2.6	0.0047	0.0045	0.0044	0.0043	0.0041	0.0040	0.0039	0.0038	0.0037	0.0036
–2.5	0.0062	0.0060	0.0059	0.0057	0.0055	0.0054	0.0052	0.0051	0.0049	0.0048
–2.4	0.0082	0.0080	0.0078	0.0075	0.0073	0.0071	0.0069	0.0068	0.0066	0.0064
–2.3	0.0107	0.0104	0.0102	0.0099	0.0096	0.0094	0.0091	0.0089	0.0087	0.0084
–2.2	0.0139	0.0136	0.0132	0.0129	0.0125	0.0122	0.0119	0.0116	0.0113	0.0110
–2.1	0.0179	0.0174	0.0170	0.0166	0.0162	0.0158	0.0154	0.0150	0.0146	0.0143
–2.0	0.0228	0.0222	0.0217	0.0212	0.0207	0.0202	0.0197	0.0192	0.0188	0.0183
–1.9	0.0287	0.0281	0.0274	0.0268	0.0262	0.0256	0.0250	0.0244	0.0239	0.0233
–1.8	0.0359	0.0351	0.0344	0.0336	0.0329	0.0322	0.0314	0.0307	0.0301	0.0294
–1.7	0.0446	0.0436	0.0427	0.0418	0.0409	0.0401	0.0392	0.0384	0.0375	0.0367
–1.6	0.0548	0.0537	0.0526	0.0516	0.0505	0.0495	0.0485	0.0475	0.0465	0.0455
–1.5	0.0668	0.0655	0.0643	0.0630	0.0618	0.0606	0.0594	0.0582	0.0571	0.0559
–1.4	0.0808	0.0793	0.0778	0.0764	0.0749	0.0735	0.0721	0.0708	0.0694	0.0681
–1.3	0.0968	0.0951	0.0934	0.0918	0.0901	0.0885	0.0869	0.0853	0.0838	0.0823
–1.2	0.1151	0.1131	0.1112	0.1093	0.1075	0.1056	0.1038	0.1020	0.1003	0.0985
–1.1	0.1357	0.1335	0.1314	0.1292	0.1271	0.1251	0.1230	0.1210	0.1190	0.1170
–1.0	0.1587	0.1562	0.1539	0.1515	0.1492	0.1469	0.1446	0.1423	0.1401	0.1379
–0.9	0.1841	0.1814	0.1788	0.1762	0.1736	0.1711	0.1685	0.1660	0.1635	0.1611
–0.8	0.2119	0.2090	0.2061	0.2033	0.2005	0.1977	0.1949	0.1922	0.1894	0.1867
–0.7	0.2420	0.2389	0.2358	0.2327	0.2296	0.2266	0.2236	0.2206	0.2177	0.2148
–0.6	0.2743	0.2709	0.2676	0.2643	0.2611	0.2578	0.2546	0.2514	0.2483	0.2451
–0.5	0.3085	0.3050	0.3015	0.2981	0.2946	0.2912	0.2877	0.2843	0.2810	0.2776
–0.4	0.3446	0.3409	0.3372	0.3336	0.3300	0.3264	0.3228	0.3192	0.3156	0.3121
–0.3	0.3821	0.3783	0.3745	0.3707	0.3669	0.3632	0.3594	0.3557	0.3520	0.3483
–0.2	0.4207	0.4168	0.4129	0.4090	0.4052	0.4013	0.3974	0.3936	0.3897	0.3859
–0.1	0.4602	0.4562	0.4522	0.4483	0.4443	0.4404	0.4364	0.4325	0.4286	0.4247
–0.0	0.5000	0.4960	0.4920	0.4880	0.4840	0.4801	0.4761	0.4721	0.4681	0.4641
0.0	0.5000	0.5040	0.5080	0.5120	0.5160	0.5199	0.5239	0.5279	0.5319	0.5359
0.1	0.5398	0.5438	0.5478	0.5517	0.5557	0.5596	0.5636	0.5675	0.5714	0.5753
0.2	0.5793	0.5832	0.5871	0.5910	0.5948	0.5987	0.6026	0.6064	0.6103	0.6141
0.3	0.6179	0.6217	0.6255	0.6293	0.6331	0.6368	0.6406	0.6443	0.6480	0.6517
0.4	0.6554	0.6591	0.6628	0.6664	0.6700	0.6736	0.6772	0.6808	0.6844	0.6879

Continued on next page...

Table A.1—*continued...*

z	0.00	0.01	0.02	0.03	0.04	0.05	0.06	0.07	0.08	0.09
0.5	0.6915	0.6950	0.6985	0.7019	0.7054	0.7088	0.7123	0.7157	0.7190	0.7224
0.6	0.7257	0.7291	0.7324	0.7357	0.7389	0.7422	0.7454	0.7486	0.7517	0.7549
0.7	0.7580	0.7611	0.7642	0.7673	0.7704	0.7734	0.7764	0.7794	0.7823	0.7852
0.8	0.7881	0.7910	0.7939	0.7967	0.7995	0.8023	0.8051	0.8078	0.8106	0.8133
0.9	0.8159	0.8186	0.8212	0.8238	0.8264	0.8289	0.8315	0.8340	0.8365	0.8389
1.0	0.8413	0.8438	0.8461	0.8485	0.8508	0.8531	0.8554	0.8577	0.8599	0.8621
1.1	0.8643	0.8665	0.8686	0.8708	0.8729	0.8749	0.8770	0.8790	0.8810	0.8830
1.2	0.8849	0.8869	0.8888	0.8907	0.8925	0.8944	0.8962	0.8980	0.8997	0.9015
1.3	0.9032	0.9049	0.9066	0.9082	0.9099	0.9115	0.9131	0.9147	0.9162	0.9177
1.4	0.9192	0.9207	0.9222	0.9236	0.9251	0.9265	0.9279	0.9292	0.9306	0.9319
1.5	0.9332	0.9345	0.9357	0.9370	0.9382	0.9394	0.9406	0.9418	0.9429	0.9441
1.6	0.9452	0.9463	0.9474	0.9484	0.9495	0.9505	0.9515	0.9525	0.9535	0.9545
1.7	0.9554	0.9564	0.9573	0.9582	0.9591	0.9599	0.9608	0.9616	0.9625	0.9633
1.8	0.9641	0.9649	0.9656	0.9664	0.9671	0.9678	0.9686	0.9693	0.9699	0.9706
1.9	0.9713	0.9719	0.9726	0.9732	0.9738	0.9744	0.9750	0.9756	0.9761	0.9767
2.0	0.9772	0.9778	0.9783	0.9788	0.9793	0.9798	0.9803	0.9808	0.9812	0.9817
2.1	0.9821	0.9826	0.9830	0.9834	0.9838	0.9842	0.9846	0.9850	0.9854	0.9857
2.2	0.9861	0.9864	0.9868	0.9871	0.9875	0.9878	0.9881	0.9884	0.9887	0.9890
2.3	0.9893	0.9896	0.9898	0.9901	0.9904	0.9906	0.9909	0.9911	0.9913	0.9916
2.4	0.9918	0.9920	0.9922	0.9925	0.9927	0.9929	0.9931	0.9932	0.9934	0.9936
2.5	0.9938	0.9940	0.9941	0.9943	0.9945	0.9946	0.9948	0.9949	0.9951	0.9952
2.6	0.9953	0.9955	0.9956	0.9957	0.9959	0.9960	0.9961	0.9962	0.9963	0.9964
2.7	0.9965	0.9966	0.9967	0.9968	0.9969	0.9970	0.9971	0.9972	0.9973	0.9974
2.8	0.9974	0.9975	0.9976	0.9977	0.9977	0.9978	0.9979	0.9979	0.9980	0.9981
2.9	0.9981	0.9982	0.9982	0.9983	0.9984	0.9984	0.9985	0.9985	0.9986	0.9986
3.0	0.9987	0.9987	0.9987	0.9988	0.9988	0.9989	0.9989	0.9989	0.9990	0.9990
3.1	0.9990	0.9991	0.9991	0.9991	0.9992	0.9992	0.9992	0.9992	0.9993	0.9993
3.2	0.9993	0.9993	0.9994	0.9994	0.9994	0.9994	0.9994	0.9995	0.9995	0.9995
3.3	0.9995	0.9995	0.9995	0.9996	0.9996	0.9996	0.9996	0.9996	0.9996	0.9997
3.4	0.9997	0.9997	0.9997	0.9997	0.9997	0.9997	0.9997	0.9997	0.9997	0.9998

APPENDIX

B

The Gamma Function

Table B.1. The gamma function

x	$\Gamma(x)$	x	$\Gamma(x)$	x	$\Gamma(x)$	x	$\Gamma(x)$
1.00	1.00000	1.25	0.90640	1.50	0.88623	1.75	0.91906
1.01	0.99433	1.26	0.90440	1.51	0.88659	1.76	0.92137
1.02	0.98884	1.27	0.90250	1.52	0.88704	1.77	0.92376
1.03	0.98355	1.28	0.90072	1.53	0.88757	1.78	0.92623
1.04	0.97844	1.29	0.89904	1.54	0.88818	1.79	0.92877
1.05	0.97350	1.30	0.89747	1.55	0.88887	1.80	0.93138
1.06	0.96874	1.31	0.89600	1.56	0.88964	1.81	0.93408
1.07	0.96415	1.32	0.89464	1.57	0.89049	1.82	0.93685
1.08	0.95973	1.33	0.89338	1.58	0.89142	1.83	0.93969
1.09	0.95546	1.34	0.89222	1.59	0.89243	1.84	0.94261
1.10	0.95135	1.35	0.89115	1.60	0.89352	1.85	0.94561
1.11	0.94740	1.36	0.89018	1.61	0.89468	1.86	0.94869
1.12	0.94359	1.37	0.88931	1.62	0.89592	1.87	0.95184
1.13	0.93993	1.38	0.88854	1.63	0.89724	1.88	0.95507
1.14	0.93642	1.39	0.88785	1.64	0.89864	1.89	0.95838

Continued on next page...

Table B.1—*continued...*

x	$\Gamma(x)$	x	$\Gamma(x)$	x	$\Gamma(x)$	x	$\Gamma(x)$
1.15	0.93304	1.40	0.88726	1.65	0.90012	1.90	0.96177
1.16	0.92980	1.41	0.88676	1.66	0.90167	1.91	0.96523
1.17	0.92670	1.42	0.88636	1.67	0.90330	1.92	0.96877
1.18	0.92373	1.43	0.88604	1.68	0.90500	1.93	0.97240
1.19	0.92089	1.44	0.88581	1.69	0.90678	1.94	0.97610
1.20	0.91817	1.45	0.88566	1.70	0.90864	1.95	0.97988
1.21	0.91558	1.46	0.88560	1.71	0.91057	1.96	0.98374
1.22	0.91311	1.47	0.88563	1.72	0.91258	1.97	0.98768
1.23	0.91075	1.48	0.88575	1.73	0.91467	1.99	0.99581
1.24	0.90852	1.49	0.88595	1.74	0.91683	2.00	1.00000

$$\Gamma(x) = \int_0^\infty e^{-\tau}\tau^{(x-1)}d\tau$$

$$\Gamma(x+1) = x\Gamma(x)$$

APPENDIX

C

Rank Tables

Table C.1. Median ranks.

Order	Sample Size									
	1	2	3	4	5	6	7	8	9	10
1	0.5000	0.2929	0.2063	0.1591	0.1294	0.1091	0.0943	0.0830	0.0741	0.0670
2		0.7071	0.5000	0.3857	0.3138	0.2644	0.2285	0.2011	0.1796	0.1623
3			0.7937	0.6143	0.5000	0.4214	0.3641	0.3205	0.2862	0.2586
4				0.8409	0.6862	0.5786	0.5000	0.4402	0.3931	0.3551
5					0.8706	0.7356	0.6359	0.5598	0.5000	0.4517
6						0.8909	0.7715	0.6795	0.6069	0.5483
7							0.9057	0.7989	0.7138	0.6449
8								0.9170	0.8204	0.7414
9									0.9259	0.8377
10										0.9330

Continued on next page...

Table C.1—*continued...*

Order	Sample Size									
	11	12	13	14	15	16	17	18	19	20
1	0.0611	0.0561	0.0519	0.0483	0.0452	0.0424	0.0400	0.0378	0.0358	0.0341
2	0.1480	0.1360	0.1258	0.1170	0.1094	0.1027	0.0968	0.0915	0.0868	0.0825
3	0.2358	0.2167	0.2004	0.1865	0.1743	0.1637	0.1542	0.1458	0.1383	0.1315
4	0.3238	0.2976	0.2753	0.2561	0.2394	0.2247	0.2118	0.2002	0.1899	0.1805
5	0.4119	0.3785	0.3502	0.3258	0.3045	0.2859	0.2694	0.2547	0.2415	0.2297
6	0.5000	0.4595	0.4251	0.3954	0.3697	0.3471	0.3270	0.3092	0.2932	0.2788
7	0.5881	0.5405	0.5000	0.4651	0.4348	0.4082	0.3847	0.3637	0.3449	0.3280
8	0.6762	0.6215	0.5749	0.5349	0.5000	0.4694	0.4423	0.4182	0.3966	0.3771
9	0.7642	0.7024	0.6498	0.6046	0.5652	0.5306	0.5000	0.4727	0.4483	0.4263
10	0.8520	0.7833	0.7247	0.6742	0.6303	0.5918	0.5577	0.5273	0.5000	0.4754
11	0.9389	0.8640	0.7996	0.7439	0.6955	0.6529	0.6153	0.5818	0.5517	0.5246
12		0.9439	0.8742	0.8135	0.7606	0.7141	0.6730	0.6363	0.6034	0.5737
13			0.9481	0.8830	0.8257	0.7753	0.7306	0.6908	0.6551	0.6229
14				0.9517	0.8906	0.8363	0.7882	0.7453	0.7068	0.6720
15					0.9548	0.8973	0.8458	0.7998	0.7585	0.7212
16						0.9576	0.9032	0.8542	0.8101	0.7703
17							0.9600	0.9085	0.8617	0.8195
18								0.9622	0.9132	0.8685
19									0.9642	0.9175
20										0.9659

Order	Sample Size									
	21	22	23	24	25	26	27	28	29	30
1	0.0325	0.0310	0.0297	0.0285	0.0273	0.0263	0.0253	0.0245	0.0236	0.0228
2	0.0786	0.0751	0.0719	0.0690	0.0662	0.0637	0.0614	0.0592	0.0572	0.0553
3	0.1253	0.1197	0.1146	0.1099	0.1055	0.1015	0.0978	0.0944	0.0911	0.0881
4	0.1721	0.1644	0.1573	0.1509	0.1449	0.1394	0.1343	0.1296	0.1252	0.1210
5	0.2189	0.2091	0.2001	0.1919	0.1843	0.1774	0.1709	0.1648	0.1592	0.1540
6	0.2657	0.2538	0.2430	0.2330	0.2238	0.2153	0.2074	0.2001	0.1933	0.1869
7	0.3126	0.2986	0.2858	0.2741	0.2632	0.2532	0.2440	0.2354	0.2274	0.2199
8	0.3594	0.3433	0.3286	0.3151	0.3027	0.2912	0.2806	0.2707	0.2614	0.2528
9	0.4063	0.3881	0.3715	0.3562	0.3422	0.3292	0.3171	0.3059	0.2955	0.2858
10	0.4531	0.4329	0.4143	0.3973	0.3816	0.3671	0.3537	0.3412	0.3296	0.3187

Continued on next page...

Table C.1—*continued...*

Order	Sample Size 21	22	23	24	25	26	27	28	29	30
11	0.5000	0.4776	0.4572	0.4384	0.4211	0.4051	0.3903	0.3765	0.3637	0.3517
12	0.5469	0.5224	0.5000	0.4795	0.4605	0.4431	0.4268	0.4118	0.3977	0.3846
13	0.5937	0.5671	0.5428	0.5205	0.5000	0.4810	0.4634	0.4471	0.4318	0.4176
14	0.6406	0.6119	0.5857	0.5616	0.5395	0.5190	0.5000	0.4824	0.4659	0.4506
15	0.6874	0.6567	0.6285	0.6027	0.5789	0.5569	0.5366	0.5176	0.5000	0.4835
16	0.7343	0.7014	0.6714	0.6438	0.6184	0.5949	0.5732	0.5529	0.5341	0.5165
17	0.7811	0.7462	0.7142	0.6849	0.6578	0.6329	0.6097	0.5882	0.5682	0.5494
18	0.8279	0.7909	0.7570	0.7259	0.6973	0.6708	0.6463	0.6235	0.6023	0.5824
19	0.8747	0.8356	0.7999	0.7670	0.7368	0.7088	0.6829	0.6588	0.6363	0.6154
20	0.9214	0.8803	0.8427	0.8081	0.7762	0.7468	0.7194	0.6941	0.6704	0.6483
21	0.9675	0.9249	0.8854	0.8491	0.8157	0.7847	0.7560	0.7293	0.7045	0.6813
22		0.9690	0.9281	0.8901	0.8551	0.8226	0.7926	0.7646	0.7386	0.7142
23			0.9703	0.9310	0.8945	0.8606	0.8291	0.7999	0.7726	0.7472
24				0.9715	0.9338	0.8985	0.8657	0.8352	0.8067	0.7801
25					0.9727	0.9363	0.9022	0.8704	0.8408	0.8131
26						0.9737	0.9386	0.9056	0.8748	0.8460
27							0.9747	0.9408	0.9089	0.8790
28								0.9755	0.9428	0.9119
29									0.9764	0.9447
30										0.9772

Order	Sample Size 31	32	33	34	35	36	37	38	39	40
1	0.0221	0.0214	0.0208	0.0202	0.0196	0.0191	0.0186	0.0181	0.0176	0.0172
2	0.0536	0.0519	0.0503	0.0489	0.0475	0.0462	0.0449	0.0438	0.0427	0.0416
3	0.0853	0.0827	0.0802	0.0779	0.0757	0.0736	0.0716	0.0697	0.0680	0.0663
4	0.1172	0.1136	0.1101	0.1069	0.1039	0.1011	0.0983	0.0958	0.0933	0.0910
5	0.1491	0.1444	0.1401	0.1360	0.1322	0.1285	0.1251	0.1218	0.1187	0.1158
6	0.1809	0.1753	0.1701	0.1651	0.1605	0.1560	0.1519	0.1479	0.1441	0.1406
7	0.2128	0.2063	0.2001	0.1942	0.1887	0.1836	0.1786	0.1740	0.1696	0.1654
8	0.2447	0.2372	0.2301	0.2234	0.2170	0.2111	0.2054	0.2001	0.1950	0.1901
9	0.2766	0.2681	0.2600	0.2525	0.2453	0.2386	0.2322	0.2261	0.2204	0.2149
10	0.3086	0.2990	0.2900	0.2816	0.2736	0.2661	0.2590	0.2522	0.2458	0.2397

Continued on next page...

Table C.1—*continued...*

Order	Sample Size									
	31	32	33	34	35	36	37	38	39	40
11	0.3405	0.3299	0.3200	0.3107	0.3019	0.2936	0.2857	0.2783	0.2712	0.2645
12	0.3724	0.3609	0.3500	0.3398	0.3302	0.3211	0.3125	0.3044	0.2966	0.2893
13	0.4043	0.3918	0.3800	0.3690	0.3585	0.3486	0.3393	0.3305	0.3221	0.3141
14	0.4362	0.4227	0.4100	0.3981	0.3868	0.3762	0.3661	0.3565	0.3475	0.3389
15	0.4681	0.4536	0.4400	0.4272	0.4151	0.4037	0.3929	0.3826	0.3729	0.3637
16	0.5000	0.4845	0.4700	0.4563	0.4434	0.4312	0.4197	0.4087	0.3983	0.3884
17	0.5319	0.5155	0.5000	0.4854	0.4717	0.4587	0.4464	0.4348	0.4237	0.4132
18	0.5638	0.5464	0.5300	0.5146	0.5000	0.4862	0.4732	0.4609	0.4492	0.4380
19	0.5957	0.5773	0.5600	0.5437	0.5283	0.5138	0.5000	0.4870	0.4746	0.4628
20	0.6276	0.6082	0.5900	0.5728	0.5566	0.5413	0.5268	0.5130	0.5000	0.4876
21	0.6595	0.6391	0.6200	0.6019	0.5849	0.5688	0.5536	0.5391	0.5254	0.5124
22	0.6914	0.6701	0.6500	0.6310	0.6132	0.5963	0.5803	0.5652	0.5508	0.5372
23	0.7234	0.7010	0.6800	0.6602	0.6415	0.6238	0.6071	0.5913	0.5763	0.5620
24	0.7553	0.7319	0.7100	0.6893	0.6698	0.6514	0.6339	0.6174	0.6017	0.5868
25	0.7872	0.7628	0.7400	0.7184	0.6981	0.6789	0.6607	0.6435	0.6271	0.6116
26	0.8191	0.7937	0.7699	0.7475	0.7264	0.7064	0.6875	0.6695	0.6525	0.6363
27	0.8509	0.8247	0.7999	0.7766	0.7547	0.7339	0.7143	0.6956	0.6779	0.6611
28	0.8828	0.8556	0.8299	0.8058	0.7830	0.7614	0.7410	0.7217	0.7034	0.6859
29	0.9147	0.8864	0.8599	0.8349	0.8113	0.7889	0.7678	0.7478	0.7288	0.7107
30	0.9464	0.9173	0.8899	0.8640	0.8395	0.8164	0.7946	0.7739	0.7542	0.7355
31	0.9779	0.9481	0.9198	0.8931	0.8678	0.8440	0.8214	0.7999	0.7796	0.7603
32		0.9786	0.9497	0.9221	0.8961	0.8715	0.8481	0.8260	0.8050	0.7851
33			0.9792	0.9511	0.9243	0.8989	0.8749	0.8521	0.8304	0.8099
34				0.9798	0.9525	0.9264	0.9017	0.8782	0.8559	0.8346
35					0.9804	0.9538	0.9284	0.9042	0.8813	0.8594
36						0.9809	0.9551	0.9303	0.9067	0.8842
37							0.9814	0.9562	0.9320	0.9090
38								0.9819	0.9573	0.9337
39									0.9824	0.9584
40										0.9828

Continued on next page...

Table C.1—continued...

Order	Sample Size									
	41	42	43	44	45	46	47	48	49	50
1	0.0168	0.0164	0.0160	0.0156	0.0153	0.0150	0.0146	0.0143	0.0140	0.0138
2	0.0406	0.0396	0.0387	0.0379	0.0370	0.0362	0.0355	0.0347	0.0340	0.0333
3	0.0647	0.0632	0.0617	0.0603	0.0590	0.0577	0.0565	0.0553	0.0542	0.0531
4	0.0888	0.0867	0.0847	0.0828	0.0810	0.0792	0.0776	0.0760	0.0744	0.0729
5	0.1130	0.1103	0.1078	0.1054	0.1030	0.1008	0.0987	0.0966	0.0947	0.0928
6	0.1372	0.1339	0.1308	0.1279	0.1251	0.1224	0.1198	0.1173	0.1149	0.1126
7	0.1614	0.1575	0.1539	0.1504	0.1471	0.1439	0.1409	0.1380	0.1352	0.1325
8	0.1855	0.1812	0.1770	0.1730	0.1692	0.1655	0.1620	0.1587	0.1555	0.1524
9	0.2097	0.2048	0.2000	0.1955	0.1912	0.1871	0.1831	0.1794	0.1757	0.1722
10	0.2339	0.2284	0.2231	0.2181	0.2133	0.2087	0.2043	0.2000	0.1960	0.1921
11	0.2581	0.2520	0.2462	0.2406	0.2353	0.2302	0.2254	0.2207	0.2162	0.2119
12	0.2823	0.2756	0.2693	0.2632	0.2574	0.2518	0.2465	0.2414	0.2365	0.2318
13	0.3065	0.2992	0.2923	0.2857	0.2794	0.2734	0.2676	0.2621	0.2568	0.2517
14	0.3307	0.3229	0.3154	0.3083	0.3015	0.2950	0.2888	0.2828	0.2770	0.2715
15	0.3549	0.3465	0.3385	0.3308	0.3235	0.3166	0.3099	0.3035	0.2973	0.2914
16	0.3790	0.3701	0.3616	0.3534	0.3456	0.3381	0.3310	0.3242	0.3176	0.3113
17	0.4032	0.3937	0.3846	0.3760	0.3677	0.3597	0.3521	0.3448	0.3379	0.3311
18	0.4274	0.4173	0.4077	0.3985	0.3897	0.3813	0.3733	0.3655	0.3581	0.3510
19	0.4516	0.4410	0.4308	0.4211	0.4118	0.4029	0.3944	0.3862	0.3784	0.3709
20	0.4758	0.4646	0.4539	0.4436	0.4338	0.4245	0.4155	0.4069	0.3987	0.3907
21	0.5000	0.4882	0.4769	0.4662	0.4559	0.4460	0.4366	0.4276	0.4189	0.4106
22	0.5242	0.5118	0.5000	0.4887	0.4779	0.4676	0.4577	0.4483	0.4392	0.4305
23	0.5484	0.5354	0.5231	0.5113	0.5000	0.4892	0.4789	0.4690	0.4595	0.4503
24	0.5726	0.5590	0.5461	0.5338	0.5221	0.5108	0.5000	0.4897	0.4797	0.4702
25	0.5968	0.5827	0.5692	0.5564	0.5441	0.5324	0.5211	0.5103	0.5000	0.4901
26	0.6210	0.6063	0.5923	0.5789	0.5662	0.5540	0.5423	0.5310	0.5203	0.5099
27	0.6451	0.6299	0.6154	0.6015	0.5882	0.5755	0.5634	0.5517	0.5405	0.5298
28	0.6693	0.6535	0.6384	0.6240	0.6103	0.5971	0.5845	0.5724	0.5608	0.5497
29	0.6935	0.6771	0.6615	0.6466	0.6323	0.6187	0.6056	0.5931	0.5811	0.5695
30	0.7177	0.7008	0.6846	0.6692	0.6544	0.6403	0.6267	0.6138	0.6013	0.5894
31	0.7419	0.7244	0.7077	0.6917	0.6765	0.6619	0.6479	0.6345	0.6216	0.6093
32	0.7661	0.7480	0.7307	0.7143	0.6985	0.6834	0.6690	0.6552	0.6419	0.6291
33	0.7903	0.7716	0.7538	0.7368	0.7206	0.7050	0.6901	0.6758	0.6621	0.6490
34	0.8145	0.7952	0.7769	0.7594	0.7426	0.7266	0.7112	0.6965	0.6824	0.6689
35	0.8386	0.8188	0.8000	0.7819	0.7647	0.7482	0.7324	0.7172	0.7027	0.6887

Continued on next page...

Table C.1—*continued...*

Order	Sample Size 41	42	43	44	45	46	47	48	49	50
36	0.8628	0.8425	0.8230	0.8045	0.7867	0.7698	0.7535	0.7379	0.7230	0.7086
37	0.8870	0.8661	0.8461	0.8270	0.8088	0.7913	0.7746	0.7586	0.7432	0.7285
38	0.9112	0.8897	0.8692	0.8496	0.8308	0.8129	0.7957	0.7793	0.7635	0.7483
39	0.9353	0.9133	0.8922	0.8721	0.8529	0.8345	0.8169	0.8000	0.7838	0.7682
40	0.9594	0.9368	0.9153	0.8946	0.8749	0.8561	0.8380	0.8206	0.8040	0.7881
41	0.9832	0.9604	0.9383	0.9172	0.8970	0.8776	0.8591	0.8413	0.8243	0.8079
42		0.9836	0.9613	0.9397	0.9190	0.8992	0.8802	0.8620	0.8445	0.8278
43			0.9840	0.9621	0.9410	0.9208	0.9013	0.8827	0.8648	0.8476
44				0.9844	0.9630	0.9423	0.9224	0.9034	0.8851	0.8675
45					0.9847	0.9638	0.9435	0.9240	0.9053	0.8874
46						0.9850	0.9645	0.9447	0.9256	0.9072
47							0.9854	0.9653	0.9458	0.9271
48								0.9857	0.9660	0.9469
49									0.9860	0.9667
50										0.9862

Table C.2. 95% ranks.

Order	Sample Size 1	2	3	4	5	6	7	8	9	10
1	0.9500	0.7764	0.6316	0.5271	0.4507	0.3930	0.3482	0.3123	0.2831	0.2589
2		0.9747	0.8646	0.7514	0.6574	0.5818	0.5207	0.4707	0.4291	0.3942
3			0.9830	0.9024	0.8107	0.7287	0.6587	0.5997	0.5496	0.5069
4				0.9873	0.9236	0.8468	0.7747	0.7108	0.6551	0.6066
5					0.9898	0.9372	0.8712	0.8071	0.7486	0.6965
6						0.9915	0.9466	0.8889	0.8312	0.7776
7							0.9927	0.9536	0.9023	0.8500
8								0.9936	0.9590	0.9127
9									0.9943	0.9632
10										0.9949

Continued on next page...

Table C.2—continued...

Order	Sample Size									
	11	12	13	14	15	16	17	18	19	20
1	0.2384	0.2209	0.2058	0.1926	0.1810	0.1707	0.1616	0.1533	0.1459	0.1391
2	0.3644	0.3387	0.3163	0.2967	0.2794	0.2640	0.2501	0.2377	0.2264	0.2161
3	0.4701	0.4381	0.4101	0.3854	0.3634	0.3438	0.3262	0.3103	0.2958	0.2826
4	0.5644	0.5273	0.4946	0.4657	0.4398	0.4166	0.3956	0.3767	0.3594	0.3437
5	0.6502	0.6091	0.5726	0.5400	0.5108	0.4844	0.4605	0.4389	0.4191	0.4010
6	0.7288	0.6848	0.6452	0.6096	0.5774	0.5483	0.5219	0.4978	0.4758	0.4556
7	0.8004	0.7547	0.7130	0.6750	0.6404	0.6090	0.5803	0.5540	0.5300	0.5078
8	0.8649	0.8190	0.7760	0.7364	0.7000	0.6666	0.6360	0.6078	0.5819	0.5580
9	0.9212	0.8771	0.8343	0.7939	0.7563	0.7214	0.6892	0.6594	0.6319	0.6064
10	0.9667	0.9281	0.8873	0.8473	0.8091	0.7733	0.7399	0.7088	0.6799	0.6531
11	0.9953	0.9695	0.9340	0.8960	0.8583	0.8222	0.7881	0.7560	0.7261	0.6980
12		0.9957	0.9719	0.9389	0.9033	0.8679	0.8336	0.8010	0.7703	0.7413
13			0.9961	0.9740	0.9432	0.9097	0.8762	0.8437	0.8125	0.7829
14				0.9963	0.9758	0.9469	0.9154	0.8836	0.8525	0.8227
15					0.9966	0.9773	0.9501	0.9203	0.8901	0.8604
16						0.9968	0.9787	0.9530	0.9247	0.8959
17							0.9970	0.9799	0.9555	0.9286
18								0.9972	0.9810	0.9578
19									0.9973	0.9819
20										0.9974

Order	Sample Size									
	21	22	23	24	25	26	27	28	29	30
1	0.1329	0.1273	0.1221	0.1173	0.1129	0.1088	0.1050	0.1015	0.0981	0.0950
2	0.2067	0.1981	0.1902	0.1829	0.1761	0.1698	0.1640	0.1585	0.1534	0.1486
3	0.2706	0.2595	0.2492	0.2398	0.2310	0.2229	0.2153	0.2082	0.2016	0.1953
4	0.3292	0.3159	0.3036	0.2923	0.2817	0.2719	0.2627	0.2542	0.2461	0.2386
5	0.3844	0.3691	0.3549	0.3418	0.3296	0.3182	0.3076	0.2977	0.2884	0.2796
6	0.4370	0.4198	0.4039	0.3891	0.3754	0.3626	0.3506	0.3394	0.3289	0.3190
7	0.4874	0.4685	0.4510	0.4347	0.4195	0.4054	0.3921	0.3797	0.3680	0.3570
8	0.5359	0.5155	0.4964	0.4787	0.4622	0.4468	0.4323	0.4187	0.4060	0.3939
9	0.5828	0.5609	0.5405	0.5214	0.5036	0.4870	0.4714	0.4567	0.4429	0.4299
10	0.6281	0.6048	0.5832	0.5629	0.5439	0.5262	0.5095	0.4938	0.4790	0.4651

Continued on next page...

Table C.2—continued...

Order	Sample Size									
	21	22	23	24	25	26	27	28	29	30
11	0.6719	0.6475	0.6246	0.6032	0.5832	0.5643	0.5466	0.5300	0.5143	0.4994
12	0.7142	0.6887	0.6649	0.6424	0.6214	0.6016	0.5829	0.5654	0.5488	0.5331
13	0.7550	0.7287	0.7039	0.6806	0.6586	0.6379	0.6184	0.6000	0.5825	0.5661
14	0.7943	0.7673	0.7418	0.7176	0.6949	0.6734	0.6530	0.6338	0.6156	0.5984
15	0.8318	0.8044	0.7784	0.7536	0.7301	0.7079	0.6869	0.6669	0.6480	0.6301
16	0.8676	0.8401	0.8137	0.7884	0.7644	0.7416	0.7199	0.6993	0.6797	0.6611
17	0.9012	0.8740	0.8475	0.8220	0.7976	0.7743	0.7521	0.7309	0.7107	0.6915
18	0.9322	0.9059	0.8798	0.8543	0.8297	0.8060	0.7834	0.7617	0.7411	0.7213
19	0.9599	0.9354	0.9102	0.8851	0.8605	0.8367	0.8138	0.7918	0.7707	0.7505
20	0.9828	0.9618	0.9383	0.9141	0.8899	0.8662	0.8432	0.8209	0.7995	0.7789
21	0.9976	0.9836	0.9635	0.9410	0.9177	0.8944	0.8715	0.8491	0.8275	0.8067
22		0.9977	0.9843	0.9650	0.9434	0.9210	0.8985	0.8763	0.8547	0.8337
23			0.9978	0.9850	0.9665	0.9457	0.9241	0.9023	0.8808	0.8598
24				0.9979	0.9856	0.9678	0.9478	0.9269	0.9058	0.8850
25					0.9980	0.9862	0.9690	0.9497	0.9295	0.9091
26						0.9980	0.9867	0.9702	0.9515	0.9319
27							0.9981	0.9872	0.9712	0.9531
28								0.9982	0.9876	0.9722
29									0.9982	0.9880
30										0.9983

Order	Sample Size									
	31	32	33	34	35	36	37	38	39	40
1	0.0921	0.0894	0.0868	0.0843	0.0820	0.0798	0.0778	0.0758	0.0739	0.0722
2	0.1441	0.1398	0.1359	0.1321	0.1285	0.1251	0.1219	0.1189	0.1160	0.1132
3	0.1895	0.1839	0.1787	0.1738	0.1692	0.1647	0.1605	0.1566	0.1528	0.1492
4	0.2315	0.2248	0.2185	0.2125	0.2069	0.2015	0.1964	0.1916	0.1870	0.1826
5	0.2714	0.2636	0.2563	0.2493	0.2427	0.2365	0.2305	0.2249	0.2195	0.2144
6	0.3096	0.3008	0.2925	0.2846	0.2772	0.2701	0.2634	0.2570	0.2509	0.2450
7	0.3467	0.3369	0.3276	0.3189	0.3106	0.3027	0.2952	0.2880	0.2812	0.2747
8	0.3826	0.3719	0.3618	0.3522	0.3430	0.3344	0.3262	0.3183	0.3108	0.3037
9	0.4177	0.4061	0.3951	0.3847	0.3748	0.3654	0.3564	0.3479	0.3398	0.3320
10	0.4519	0.4394	0.4276	0.4165	0.4058	0.3957	0.3861	0.3769	0.3682	0.3598

Continued on next page...

Table C.2—continued...

Order	Sample Size									
	31	32	33	34	35	36	37	38	39	40
11	0.4854	0.4721	0.4596	0.4476	0.4363	0.4255	0.4152	0.4054	0.3960	0.3871
12	0.5182	0.5042	0.4909	0.4782	0.4661	0.4547	0.4438	0.4333	0.4234	0.4139
13	0.5504	0.5356	0.5216	0.5082	0.4955	0.4834	0.4719	0.4609	0.4503	0.4403
14	0.5820	0.5665	0.5518	0.5378	0.5244	0.5117	0.4995	0.4880	0.4769	0.4663
15	0.6130	0.5968	0.5814	0.5668	0.5528	0.5395	0.5268	0.5147	0.5031	0.4919
16	0.6434	0.6266	0.6106	0.5953	0.5808	0.5669	0.5537	0.5410	0.5289	0.5172
17	0.6733	0.6559	0.6393	0.6234	0.6083	0.5939	0.5801	0.5669	0.5543	0.5422
18	0.7025	0.6846	0.6674	0.6511	0.6354	0.6205	0.6062	0.5925	0.5794	0.5669
19	0.7312	0.7127	0.6951	0.6782	0.6621	0.6467	0.6319	0.6178	0.6042	0.5912
20	0.7592	0.7403	0.7223	0.7049	0.6883	0.6725	0.6572	0.6426	0.6286	0.6152
21	0.7866	0.7674	0.7489	0.7312	0.7142	0.6978	0.6822	0.6672	0.6528	0.6389
22	0.8134	0.7938	0.7750	0.7569	0.7395	0.7228	0.7068	0.6913	0.6765	0.6623
23	0.8394	0.8196	0.8005	0.7821	0.7644	0.7473	0.7309	0.7152	0.7000	0.6854
24	0.8646	0.8447	0.8254	0.8068	0.7888	0.7715	0.7547	0.7386	0.7231	0.7082
25	0.8889	0.8691	0.8497	0.8309	0.8127	0.7951	0.7781	0.7617	0.7459	0.7306
26	0.9122	0.8926	0.8732	0.8544	0.8360	0.8182	0.8010	0.7844	0.7683	0.7527
27	0.9342	0.9150	0.8960	0.8772	0.8588	0.8409	0.8235	0.8066	0.7903	0.7745
28	0.9547	0.9363	0.9177	0.8992	0.8809	0.8629	0.8454	0.8284	0.8119	0.7959
29	0.9731	0.9562	0.9383	0.9202	0.9022	0.8843	0.8668	0.8497	0.8331	0.8169
30	0.9884	0.9740	0.9575	0.9402	0.9226	0.9050	0.8876	0.8705	0.8538	0.8375
31	0.9983	0.9888	0.9748	0.9588	0.9420	0.9248	0.9077	0.8907	0.8740	0.8576
32		0.9984	0.9891	0.9755	0.9600	0.9436	0.9269	0.9102	0.8936	0.8773
33			0.9984	0.9894	0.9762	0.9611	0.9452	0.9289	0.9126	0.8964
34				0.9985	0.9898	0.9769	0.9622	0.9467	0.9308	0.9149
35					0.9985	0.9900	0.9775	0.9632	0.9481	0.9326
36						0.9986	0.9903	0.9781	0.9642	0.9494
37							0.9986	0.9906	0.9787	0.9651
38								0.9987	0.9908	0.9792
39									0.9987	0.9910
40										0.9987

Continued on next page...

Table C.2—*continued...*

Order	Sample Size									
	41	42	43	44	45	46	47	48	49	50
1	0.0705	0.0688	0.0673	0.0658	0.0644	0.0630	0.0618	0.0605	0.0593	0.0582
2	0.1106	0.1080	0.1056	0.1033	0.1011	0.0990	0.0970	0.0951	0.0932	0.0914
3	0.1457	0.1424	0.1393	0.1363	0.1334	0.1306	0.1280	0.1254	0.1230	0.1206
4	0.1784	0.1744	0.1706	0.1669	0.1634	0.1600	0.1568	0.1537	0.1507	0.1478
5	0.2095	0.2048	0.2004	0.1961	0.1920	0.1880	0.1843	0.1806	0.1771	0.1738
6	0.2395	0.2342	0.2291	0.2242	0.2195	0.2151	0.2108	0.2066	0.2027	0.1988
7	0.2685	0.2626	0.2569	0.2515	0.2463	0.2413	0.2365	0.2319	0.2274	0.2232
8	0.2969	0.2904	0.2841	0.2781	0.2724	0.2669	0.2616	0.2565	0.2516	0.2469
9	0.3246	0.3175	0.3107	0.3042	0.2980	0.2920	0.2862	0.2807	0.2754	0.2702
10	0.3518	0.3442	0.3368	0.3298	0.3231	0.3166	0.3104	0.3044	0.2986	0.2931
11	0.3785	0.3703	0.3625	0.3549	0.3477	0.3408	0.3341	0.3277	0.3215	0.3156
12	0.4048	0.3961	0.3877	0.3797	0.3720	0.3646	0.3575	0.3507	0.3441	0.3378
13	0.4307	0.4214	0.4126	0.4041	0.3960	0.3881	0.3806	0.3734	0.3664	0.3597
14	0.4562	0.4464	0.4371	0.4282	0.4196	0.4113	0.4034	0.3957	0.3884	0.3813
15	0.4813	0.4711	0.4613	0.4519	0.4429	0.4342	0.4259	0.4178	0.4101	0.4026
16	0.5061	0.4955	0.4852	0.4754	0.4659	0.4568	0.4481	0.4397	0.4316	0.4237
17	0.5306	0.5195	0.5088	0.4986	0.4887	0.4792	0.4701	0.4613	0.4528	0.4446
18	0.5548	0.5433	0.5322	0.5215	0.5112	0.5013	0.4918	0.4827	0.4738	0.4653
19	0.5787	0.5667	0.5552	0.5441	0.5335	0.5232	0.5133	0.5038	0.4946	0.4858
20	0.6023	0.5899	0.5780	0.5665	0.5555	0.5449	0.5346	0.5248	0.5152	0.5060
21	0.6256	0.6128	0.6005	0.5887	0.5773	0.5663	0.5557	0.5455	0.5356	0.5261
22	0.6486	0.6354	0.6228	0.6106	0.5988	0.5875	0.5766	0.5660	0.5558	0.5460
23	0.6713	0.6578	0.6448	0.6322	0.6201	0.6085	0.5972	0.5863	0.5759	0.5657
24	0.6938	0.6799	0.6665	0.6536	0.6412	0.6292	0.6176	0.6065	0.5957	0.5852
25	0.7159	0.7017	0.6880	0.6748	0.6621	0.6498	0.6379	0.6264	0.6153	0.6046
26	0.7377	0.7232	0.7092	0.6957	0.6827	0.6701	0.6579	0.6461	0.6348	0.6238
27	0.7592	0.7444	0.7302	0.7164	0.7030	0.6902	0.6777	0.6657	0.6540	0.6427
28	0.7804	0.7653	0.7508	0.7368	0.7232	0.7100	0.6973	0.6850	0.6731	0.6615
29	0.8012	0.7859	0.7712	0.7569	0.7431	0.7297	0.7167	0.7041	0.6920	0.6802
30	0.8216	0.8062	0.7912	0.7767	0.7627	0.7491	0.7358	0.7230	0.7106	0.6986
31	0.8417	0.8261	0.8110	0.7963	0.7820	0.7682	0.7548	0.7417	0.7291	0.7169
32	0.8613	0.8456	0.8304	0.8155	0.8011	0.7871	0.7735	0.7602	0.7474	0.7349
33	0.8804	0.8647	0.8494	0.8345	0.8199	0.8057	0.7919	0.7785	0.7655	0.7528
34	0.8990	0.8834	0.8680	0.8530	0.8383	0.8240	0.8101	0.7965	0.7833	0.7705
35	0.9170	0.9015	0.8862	0.8712	0.8564	0.8420	0.8280	0.8143	0.8009	0.7879

Continued on next page...

Table C.2—*continued...*

Order	Sample Size 41	42	43	44	45	46	47	48	49	50
36	0.9343	0.9191	0.9039	0.8889	0.8742	0.8597	0.8456	0.8317	0.8183	0.8051
37	0.9507	0.9359	0.9210	0.9062	0.8915	0.8770	0.8628	0.8490	0.8354	0.8221
38	0.9660	0.9519	0.9374	0.9229	0.9083	0.8940	0.8798	0.8658	0.8522	0.8388
39	0.9798	0.9668	0.9530	0.9389	0.9246	0.9104	0.8963	0.8824	0.8687	0.8553
40	0.9913	0.9802	0.9676	0.9541	0.9403	0.9263	0.9124	0.8985	0.8849	0.8714
41	0.9987	0.9915	0.9807	0.9683	0.9552	0.9416	0.9280	0.9143	0.9007	0.8873
42		0.9988	0.9917	0.9812	0.9691	0.9562	0.9429	0.9295	0.9161	0.9028
43			0.9988	0.9919	0.9816	0.9698	0.9571	0.9441	0.9310	0.9178
44				0.9988	0.9920	0.9820	0.9704	0.9580	0.9453	0.9324
45					0.9989	0.9922	0.9824	0.9710	0.9589	0.9464
46						0.9989	0.9924	0.9827	0.9716	0.9598
47							0.9989	0.9925	0.9831	0.9722
48								0.9989	0.9927	0.9834
49									0.9990	0.9928
50										0.9990

Table C.3. 5% ranks.

Order	Sample Size 1	2	3	4	5	6	7	8	9	10
1	0.0500	0.0253	0.0170	0.0127	0.0102	0.0085	0.0073	0.0064	0.0057	0.0051
2		0.2236	0.1354	0.0976	0.0764	0.0628	0.0534	0.0464	0.0410	0.0368
3			0.3684	0.2486	0.1893	0.1532	0.1288	0.1111	0.0977	0.0873
4				0.4729	0.3426	0.2713	0.2253	0.1929	0.1688	0.1500
5					0.5493	0.4182	0.3413	0.2892	0.2514	0.2224
6						0.6070	0.4793	0.4003	0.3449	0.3035
7							0.6518	0.5293	0.4504	0.3934
8								0.6877	0.5709	0.4931
9									0.7169	0.6058
10										0.7411

Continued on next page...

Table C.3—*continued...*

Order	Sample Size									
	11	12	13	14	15	16	17	18	19	20
1	0.0047	0.0043	0.0039	0.0037	0.0034	0.0032	0.0030	0.0028	0.0027	0.0026
2	0.0333	0.0305	0.0281	0.0260	0.0242	0.0227	0.0213	0.0201	0.0190	0.0181
3	0.0788	0.0719	0.0660	0.0611	0.0568	0.0531	0.0499	0.0470	0.0445	0.0422
4	0.1351	0.1229	0.1127	0.1040	0.0967	0.0903	0.0846	0.0797	0.0753	0.0714
5	0.1996	0.1810	0.1657	0.1527	0.1417	0.1321	0.1238	0.1164	0.1099	0.1041
6	0.2712	0.2453	0.2240	0.2061	0.1909	0.1778	0.1664	0.1563	0.1475	0.1396
7	0.3498	0.3152	0.2870	0.2636	0.2437	0.2267	0.2119	0.1990	0.1875	0.1773
8	0.4356	0.3909	0.3548	0.3250	0.3000	0.2786	0.2601	0.2440	0.2297	0.2171
9	0.5299	0.4727	0.4274	0.3904	0.3596	0.3334	0.3108	0.2912	0.2739	0.2587
10	0.6356	0.5619	0.5054	0.4600	0.4226	0.3910	0.3640	0.3406	0.3201	0.3020
11	0.7616	0.6613	0.5899	0.5343	0.4892	0.4517	0.4197	0.3922	0.3681	0.3469
12		0.7791	0.6837	0.6146	0.5602	0.5156	0.4781	0.4460	0.4181	0.3936
13			0.7942	0.7033	0.6366	0.5834	0.5395	0.5022	0.4700	0.4420
14				0.8074	0.7206	0.6562	0.6044	0.5611	0.5242	0.4922
15					0.8190	0.7360	0.6738	0.6233	0.5809	0.5444
16						0.8293	0.7499	0.6897	0.6406	0.5990
17							0.8384	0.7623	0.7042	0.6563
18								0.8467	0.7736	0.7174
19									0.8541	0.7839
20										0.8609

Order	Sample Size									
	21	22	23	24	25	26	27	28	29	30
1	0.0024	0.0023	0.0022	0.0021	0.0020	0.0020	0.0019	0.0018	0.0018	0.0017
2	0.0172	0.0164	0.0157	0.0150	0.0144	0.0138	0.0133	0.0128	0.0124	0.0120
3	0.0401	0.0382	0.0365	0.0350	0.0335	0.0322	0.0310	0.0298	0.0288	0.0278
4	0.0678	0.0646	0.0617	0.0590	0.0566	0.0543	0.0522	0.0503	0.0485	0.0469
5	0.0988	0.0941	0.0898	0.0859	0.0823	0.0790	0.0759	0.0731	0.0705	0.0681
6	0.1324	0.1260	0.1202	0.1149	0.1101	0.1056	0.1015	0.0977	0.0942	0.0909
7	0.1682	0.1599	0.1525	0.1457	0.1395	0.1338	0.1285	0.1237	0.1192	0.1150
8	0.2057	0.1956	0.1863	0.1780	0.1703	0.1633	0.1568	0.1509	0.1453	0.1402
9	0.2450	0.2327	0.2216	0.2116	0.2024	0.1940	0.1862	0.1791	0.1725	0.1663
10	0.2858	0.2713	0.2582	0.2464	0.2356	0.2257	0.2166	0.2082	0.2005	0.1933

Continued on next page...

Table C.3—*continued...*

Order	Sample Size 21	22	23	24	25	26	27	28	29	30
11	0.3281	0.3113	0.2961	0.2824	0.2699	0.2584	0.2479	0.2383	0.2293	0.2211
12	0.3719	0.3525	0.3351	0.3194	0.3051	0.2921	0.2801	0.2691	0.2589	0.2495
13	0.4172	0.3952	0.3754	0.3576	0.3414	0.3266	0.3131	0.3007	0.2893	0.2787
14	0.4641	0.4391	0.4168	0.3968	0.3786	0.3621	0.3470	0.3331	0.3203	0.3085
15	0.5126	0.4845	0.4595	0.4371	0.4168	0.3984	0.3816	0.3662	0.3520	0.3389
16	0.5630	0.5315	0.5036	0.4786	0.4561	0.4357	0.4171	0.4000	0.3844	0.3699
17	0.6156	0.5802	0.5490	0.5213	0.4964	0.4738	0.4534	0.4346	0.4175	0.4016
18	0.6708	0.6309	0.5961	0.5653	0.5378	0.5130	0.4905	0.4700	0.4512	0.4339
19	0.7294	0.6841	0.6451	0.6109	0.5805	0.5532	0.5286	0.5062	0.4857	0.4669
20	0.7933	0.7405	0.6964	0.6582	0.6246	0.5946	0.5677	0.5433	0.5210	0.5006
21	0.8671	0.8019	0.7508	0.7077	0.6704	0.6374	0.6079	0.5813	0.5571	0.5349
22		0.8727	0.8098	0.7602	0.7183	0.6818	0.6494	0.6203	0.5940	0.5701
23			0.8779	0.8171	0.7690	0.7281	0.6924	0.6606	0.6320	0.6061
24				0.8827	0.8239	0.7771	0.7373	0.7023	0.6711	0.6430
25					0.8871	0.8302	0.7847	0.7458	0.7116	0.6810
26						0.8912	0.8360	0.7918	0.7539	0.7204
27							0.8950	0.8415	0.7984	0.7614
28								0.8985	0.8466	0.8047
29									0.9019	0.8514
30										0.9050

Order	Sample Size 31	32	33	34	35	36	37	38	39	40
1	0.0017	0.0016	0.0016	0.0015	0.0015	0.0014	0.0014	0.0013	0.0013	0.0013
2	0.0116	0.0112	0.0109	0.0106	0.0102	0.0100	0.0097	0.0094	0.0092	0.0090
3	0.0269	0.0260	0.0252	0.0245	0.0238	0.0231	0.0225	0.0219	0.0213	0.0208
4	0.0453	0.0438	0.0425	0.0412	0.0400	0.0389	0.0378	0.0368	0.0358	0.0349
5	0.0658	0.0637	0.0617	0.0598	0.0580	0.0564	0.0548	0.0533	0.0519	0.0506
6	0.0878	0.0850	0.0823	0.0798	0.0774	0.0752	0.0731	0.0711	0.0692	0.0674
7	0.1111	0.1074	0.1040	0.1008	0.0978	0.0950	0.0923	0.0898	0.0874	0.0851
8	0.1354	0.1309	0.1268	0.1228	0.1191	0.1157	0.1124	0.1093	0.1064	0.1036
9	0.1606	0.1553	0.1503	0.1456	0.1412	0.1371	0.1332	0.1295	0.1260	0.1227
10	0.1866	0.1804	0.1746	0.1691	0.1640	0.1591	0.1546	0.1503	0.1462	0.1424

Continued on next page...

Table C.3—*continued...*

Order	Sample Size									
	31	32	33	34	35	36	37	38	39	40
11	0.2134	0.2062	0.1995	0.1932	0.1873	0.1818	0.1765	0.1716	0.1669	0.1625
12	0.2408	0.2326	0.2250	0.2179	0.2112	0.2049	0.1990	0.1934	0.1881	0.1831
13	0.2688	0.2597	0.2511	0.2431	0.2356	0.2285	0.2219	0.2156	0.2097	0.2041
14	0.2975	0.2873	0.2777	0.2688	0.2605	0.2527	0.2453	0.2383	0.2317	0.2255
15	0.3267	0.3154	0.3049	0.2951	0.2858	0.2772	0.2691	0.2614	0.2541	0.2473
16	0.3566	0.3441	0.3326	0.3218	0.3117	0.3022	0.2932	0.2848	0.2769	0.2694
17	0.3870	0.3734	0.3607	0.3489	0.3379	0.3275	0.3178	0.3087	0.3000	0.2918
18	0.4180	0.4032	0.3894	0.3766	0.3646	0.3533	0.3428	0.3328	0.3235	0.3146
19	0.4496	0.4335	0.4186	0.4047	0.3917	0.3795	0.3681	0.3574	0.3472	0.3377
20	0.4818	0.4644	0.4482	0.4332	0.4192	0.4061	0.3938	0.3822	0.3714	0.3611
21	0.5146	0.4958	0.4784	0.4622	0.4472	0.4331	0.4199	0.4075	0.3958	0.3848
22	0.5481	0.5279	0.5091	0.4918	0.4756	0.4605	0.4463	0.4331	0.4206	0.4088
23	0.5823	0.5606	0.5404	0.5218	0.5045	0.4883	0.4732	0.4590	0.4457	0.4331
24	0.6174	0.5939	0.5724	0.5524	0.5339	0.5166	0.5005	0.4853	0.4711	0.4578
25	0.6533	0.6281	0.6049	0.5835	0.5637	0.5453	0.5281	0.5120	0.4969	0.4828
26	0.6904	0.6631	0.6382	0.6153	0.5942	0.5745	0.5562	0.5391	0.5231	0.5081
27	0.7286	0.6992	0.6724	0.6478	0.6252	0.6043	0.5848	0.5667	0.5497	0.5337
28	0.7685	0.7364	0.7075	0.6811	0.6570	0.6346	0.6139	0.5946	0.5766	0.5597
29	0.8105	0.7752	0.7437	0.7154	0.6894	0.6656	0.6436	0.6231	0.6040	0.5861
30	0.8559	0.8161	0.7815	0.7507	0.7228	0.6973	0.6738	0.6521	0.6318	0.6129
31	0.9079	0.8602	0.8213	0.7875	0.7573	0.7299	0.7048	0.6817	0.6602	0.6402
32		0.9106	0.8641	0.8262	0.7931	0.7635	0.7366	0.7120	0.6892	0.6680
33			0.9132	0.8679	0.8308	0.7985	0.7695	0.7430	0.7188	0.6963
34				0.9157	0.8715	0.8353	0.8036	0.7751	0.7491	0.7253
35					0.9180	0.8749	0.8395	0.8084	0.7805	0.7550
36						0.9202	0.8781	0.8434	0.8130	0.7856
37							0.9222	0.8811	0.8472	0.8174
38								0.9242	0.8840	0.8508
39									0.9261	0.8868
40										0.9278

Continued on next page...

Table C.3—*continued...*

Order	Sample Size 41	42	43	44	45	46	47	48	49	50
1	0.0013	0.0012	0.0012	0.0012	0.0011	0.0011	0.0011	0.0011	0.0010	0.0010
2	0.0087	0.0085	0.0083	0.0081	0.0080	0.0078	0.0076	0.0075	0.0073	0.0072
3	0.0202	0.0198	0.0193	0.0188	0.0184	0.0180	0.0176	0.0173	0.0169	0.0166
4	0.0340	0.0332	0.0324	0.0317	0.0309	0.0302	0.0296	0.0290	0.0284	0.0278
5	0.0493	0.0481	0.0470	0.0459	0.0448	0.0438	0.0429	0.0420	0.0411	0.0402
6	0.0657	0.0641	0.0626	0.0611	0.0597	0.0584	0.0571	0.0559	0.0547	0.0536
7	0.0830	0.0809	0.0790	0.0771	0.0754	0.0737	0.0720	0.0705	0.0690	0.0676
8	0.1010	0.0985	0.0961	0.0938	0.0917	0.0896	0.0876	0.0857	0.0839	0.0822
9	0.1196	0.1166	0.1138	0.1111	0.1085	0.1060	0.1037	0.1015	0.0993	0.0972
10	0.1387	0.1353	0.1320	0.1288	0.1258	0.1230	0.1202	0.1176	0.1151	0.1127
11	0.1583	0.1544	0.1506	0.1470	0.1436	0.1403	0.1372	0.1342	0.1313	0.1286
12	0.1784	0.1739	0.1696	0.1655	0.1617	0.1580	0.1544	0.1510	0.1478	0.1447
13	0.1988	0.1938	0.1890	0.1845	0.1801	0.1760	0.1720	0.1683	0.1646	0.1612
14	0.2196	0.2141	0.2088	0.2037	0.1989	0.1943	0.1899	0.1857	0.1817	0.1779
15	0.2408	0.2347	0.2288	0.2233	0.2180	0.2129	0.2081	0.2035	0.1991	0.1949
16	0.2623	0.2556	0.2492	0.2431	0.2373	0.2318	0.2265	0.2215	0.2167	0.2121
17	0.2841	0.2768	0.2698	0.2632	0.2569	0.2509	0.2452	0.2398	0.2345	0.2295
18	0.3062	0.2983	0.2908	0.2836	0.2768	0.2703	0.2642	0.2583	0.2526	0.2472
19	0.3287	0.3201	0.3120	0.3043	0.2970	0.2900	0.2833	0.2770	0.2709	0.2651
20	0.3514	0.3422	0.3335	0.3252	0.3173	0.3098	0.3027	0.2959	0.2894	0.2831
21	0.3744	0.3646	0.3552	0.3464	0.3379	0.3299	0.3223	0.3150	0.3080	0.3014
22	0.3977	0.3872	0.3772	0.3678	0.3588	0.3502	0.3421	0.3343	0.3269	0.3198
23	0.4213	0.4101	0.3995	0.3894	0.3799	0.3708	0.3621	0.3539	0.3460	0.3385
24	0.4452	0.4333	0.4220	0.4113	0.4012	0.3915	0.3824	0.3736	0.3652	0.3573
25	0.4694	0.4567	0.4448	0.4335	0.4227	0.4125	0.4028	0.3935	0.3847	0.3762
26	0.4939	0.4805	0.4678	0.4559	0.4445	0.4337	0.4234	0.4137	0.4043	0.3954
27	0.5187	0.5045	0.4912	0.4785	0.4665	0.4551	0.4443	0.4340	0.4241	0.4148
28	0.5438	0.5289	0.5148	0.5014	0.4888	0.4768	0.4654	0.4545	0.4442	0.4343
29	0.5693	0.5536	0.5387	0.5246	0.5113	0.4987	0.4867	0.4752	0.4644	0.4540
30	0.5952	0.5786	0.5629	0.5481	0.5341	0.5208	0.5082	0.4962	0.4848	0.4739

Continued on next page...

Table C.3—*continued...*

Order	Sample Size									
	41	42	43	44	45	46	47	48	49	50
31	0.6215	0.6039	0.5874	0.5718	0.5571	0.5432	0.5299	0.5173	0.5054	0.4940
32	0.6482	0.6297	0.6123	0.5959	0.5804	0.5658	0.5519	0.5387	0.5262	0.5142
33	0.6754	0.6558	0.6375	0.6203	0.6040	0.5887	0.5741	0.5603	0.5472	0.5347
34	0.7031	0.6825	0.6632	0.6451	0.6280	0.6119	0.5966	0.5822	0.5684	0.5554
35	0.7315	0.7096	0.6893	0.6702	0.6523	0.6354	0.6194	0.6043	0.5899	0.5763
36	0.7605	0.7374	0.7159	0.6958	0.6769	0.6592	0.6425	0.6266	0.6116	0.5974
37	0.7905	0.7658	0.7431	0.7219	0.7020	0.6834	0.6659	0.6493	0.6336	0.6187
38	0.8216	0.7952	0.7709	0.7485	0.7276	0.7080	0.6896	0.6723	0.6559	0.6403
39	0.8543	0.8256	0.7996	0.7758	0.7537	0.7331	0.7138	0.6956	0.6785	0.6622
40	0.8894	0.8576	0.8294	0.8039	0.7805	0.7587	0.7384	0.7193	0.7014	0.6844
41	0.9295	0.8920	0.8607	0.8331	0.8080	0.7849	0.7635	0.7435	0.7246	0.7069
42		0.9312	0.8944	0.8637	0.8366	0.8120	0.7892	0.7681	0.7484	0.7298
43			0.9327	0.8967	0.8666	0.8400	0.8157	0.7934	0.7726	0.7531
44				0.9342	0.8989	0.8694	0.8432	0.8194	0.7973	0.7768
45					0.9356	0.9010	0.8720	0.8463	0.8229	0.8012
46						0.9370	0.9030	0.8746	0.8493	0.8262
47							0.9382	0.9049	0.8770	0.8522
48								0.9395	0.9068	0.8794
49									0.9407	0.9086
50										0.9418

APPENDIX

D

Beta Table

Table D.1. Values of reliability.

Sample Size	Successes	Confidence Level					
		0.5	0.75	0.90	0.95	0.99	0.995
1	0	0.293	0.134	0.051	0.025	0.005	0.002
	1	0.707	0.500	0.316	0.224	0.100	0.071
2	0	0.206	0.092	0.034	0.017	0.003	0.002
	1	0.500	0.326	0.196	0.135	0.059	0.041
	2	0.794	0.630	0.464	0.368	0.216	0.171
3	1	0.386	0.243	0.142	0.098	0.042	0.029
	2	0.614	0.456	0.320	0.249	0.141	0.111
	3	0.841	0.707	0.562	0.473	0.316	0.266
4	2	0.500	0.359	0.247	0.189	0.106	0.083
	3	0.686	0.546	0.416	0.343	0.222	0.185
	4	0.871	0.758	0.631	0.549	0.398	0.346
5	3	0.579	0.447	0.794	0.271	0.173	0.144
	4	0.736	0.610	0.333	0.418	0.294	0.254
	5	0.891	0.794	0.681	0.607	0.464	0.414

Continued on next page...

Table D.1—*continued...*

Sample Size	Successes	Confidence Level 0.5	0.75	0.90	0.95	0.99	0.995
6	4	0.636	0.514	0.404	0.341	0.236	0.203
	5	0.772	0.659	0.547	0.479	0.357	0.315
	6	0.906	0.820	0.720	0.652	0.518	0.469
7	5	0.679	0.567	0.462	0.400	0.293	0.258
	6	0.799	0.697	0.594	0.529	0.410	0.368
	7	0.917	0.841	0.750	0.688	0.562	0.516
8	6	0.714	0.609	0.510	0.450	0.344	0.307
	7	0.820	0.728	0.632	0.571	0.456	0.415
	8	0.926	0.857	0.774	0.717	0.600	0.555
9	7	0.742	0.645	0.550	0.493	0.388	0.352
	8	0.838	0.753	0.663	0.606	0.496	0.456
	9	0.933	0.871	0.794	0.741	0.631	0.589
10	7	0.676	0.580	0.489	0.436	0.340	0.307
	8	0.764	0.674	0.585	0.530	0.428	0.392
	9	0.852	0.773	0.690	0.636	0.530	0.492
12	9	0.725	0.638	0.556	0.505	0.412	0.379
	10	0.800	0.720	0.640	0.590	0.494	0.459
	11	0.874	0.806	0.732	0.684	0.587	0.551
	12	0.948	0.899	0.838	0.794	0.702	0.665
15	12	0.775	0.702	0.629	0.583	0.497	0.466
	13	0.836	0.769	0.700	0.656	0.570	0.537
	14	0.897	0.840	0.778	0.736	0.651	0.619
	15	0.958	0.917	0.866	0.829	0.750	0.718

Continued on next page...

Table D.1—*continued...*

Sample Size	Successes	Confidence Level 0.5	0.75	0.90	0.95	0.99	0.995
20	17	0.828	0.769	0.709	0.671	0.596	0.568
	18	0.875	0.822	0.766	0.729	0.656	0.628
	19	0.921	0.877	0.827	0.793	0.723	0.696
	20	0.968	0.936	0.896	0.867	0.803	0.777
25	22	0.861	0.811	0.761	0.728	0.663	0.638
	23	0.898	0.855	0.808	0.777	0.714	0.690
	24	0.936	0.900	0.858	0.830	0.771	0.747
	25	0.974	0.948	0.915	0.891	0.838	0.816
30	27	0.883	0.841	0.797	0.768	0.711	0.689
	28	0.915	0.877	0.837	0.810	0.755	0.734
	29	0.946	0.916	0.880	0.856	0.804	0.784
	30	0.978	0.956	0.928	0.908	0.862	0.843
35	31	0.871	0.831	0.791	0.764	0.710	0.690
	32	0.898	0.862	0.824	0.798	0.747	0.727
	33	0.926	0.894	0.859	0.835	0.786	0.767
	34	0.954	0.927	0.896	0.875	0.829	0.811
	35	0.981	0.962	0.938	0.920	0.880	0.863
40	36	0.887	0.851	0.814	0.790	0.742	0.724
	37	0.911	0.879	0.844	0.822	0.775	0.757
	38	0.935	0.907	0.875	0.854	0.810	0.793
	39	0.959	0.936	0.908	0.889	0.849	0.832
	40	0.983	0.967	0.945	0.930	0.894	0.879
45	41	0.899	0.867	0.834	0.812	0.768	0.751
	42	0.921	0.892	0.860	0.840	0.798	0.781
	43	0.942	0.917	0.888	0.869	0.830	0.814
	44	0.964	0.942	0.918	0.901	0.864	0.849
	45	0.985	0.970	0.951	0.937	0.905	0.891
50	46	0.909	0.880	0.849	0.829	0.789	0.773
	47	0.928	0.902	0.874	0.855	0.816	0.801
	48	0.948	0.925	0.899	0.882	0.845	0.831
	49	0.967	0.948	0.926	0.910	0.877	0.863
	50	0.985	0.973	0.956	0.943	0.914	0.901

Continued on next page...

Table D.1—*continued...*

Sample Size	Successes	Confidence Level					
		0.5	0.75	0.90	0.95	0.99	0.995
75	71	0.939	0.919	0.898	0.884	0.855	0.843
	72	0.952	0.934	0.914	0.901	0.874	0.863
	73	0.965	0.949	0.932	0.920	0.894	0.884
	74	0.978	0.965	0.950	0.939	0.916	0.906
	75	0.991	0.982	0.970	0.961	0.941	0.933
100	95	0.944	0.928	0.910	0.899	0.875	0.866
	96	0.954	0.939	0.922	0.912	0.889	0.880
	97	0.964	0.950	0.935	0.925	0.904	0.896
	98	0.974	0.962	0.948	0.939	0.919	0.911
	99	0.983	0.974	0.962	0.954	0.936	0.929
	100	0.993	0.986	0.978	0.971	0.955	0.949
150	145	0.962	0.951	0.939	0.932	0.916	0.909
	146	0.969	0.959	0.948	0.940	0.925	0.919
	147	0.976	0.966	0.956	0.950	0.935	0.929
	148	0.982	0.974	0.965	0.959	0.946	0.940
	149	0.989	0.982	0.974	0.969	0.957	0.952
	150	0.996	0.991	0.985	0.980	0.970	0.966
200	195	0.972	0.963	0.954	0.948	0.963	0.931
	196	0.977	0.969	0.961	0.955	0.943	0.939
	197	0.982	0.975	0.967	0.962	0.951	0.946
	198	0.987	0.981	0.974	0.969	0.959	0.955
	199	.0992	0.987	0.981	0.977	0.967	0.964
	200	0.996	0.993	0.989	0.985	0.977	0.974

◆◆◆

APPENDIX

E

The *F* Distribution

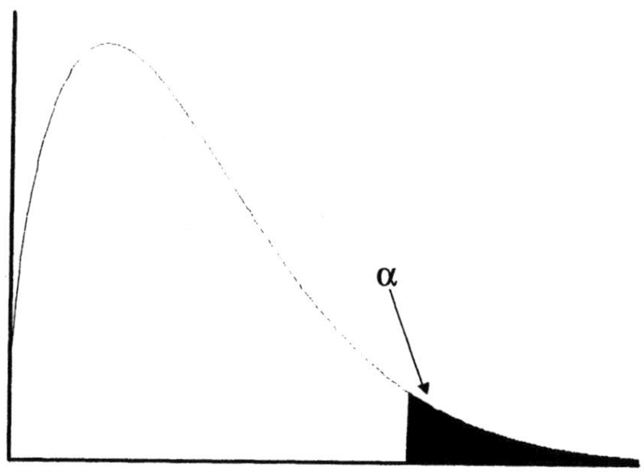

Table E.1. Upper 0.5% points.

$$F_{0.995;\nu 1,\nu 2} = 1/F_{0.005;\nu 2,\nu 1}$$

v1	v2									
	1	2	3	4	5	6	7	8	9	10
1	16212	198.50	55.552	31.332	22.785	18.635	16.235	14.688	13.614	12.827
2	19997	199.01	49.800	26.284	18.314	14.544	12.404	11.043	10.107	9.427
3	21614	199.16	47.468	24.260	16.530	12.917	10.883	9.597	8.717	8.081
4	22501	199.24	46.195	23.154	15.556	12.028	10.050	8.805	7.956	7.343
5	23056	199.30	45.391	22.456	14.939	11.464	9.522	8.302	7.471	6.872

Continued on next page...

Table E.1—*continued...*

	v2									
v1	1	2	3	4	5	6	7	8	9	10
6	23440	199.33	44.838	21.975	14.513	11.073	9.155	7.952	7.134	6.545
7	23715	199.36	44.434	21.622	14.200	10.786	8.885	7.694	6.885	6.303
8	23924	199.38	44.125	21.352	13.961	10.566	8.678	7.496	6.693	6.116
9	24091	199.39	43.881	21.138	13.772	10.391	8.514	7.339	6.541	5.968
10	24222	199.39	43.685	20.967	13.618	10.250	8.380	7.211	6.417	5.847
12	24427	199.42	43.387	20.705	13.385	10.034	8.176	7.015	6.227	5.661
15	24632	199.43	43.085	20.438	13.146	9.814	7.968	6.814	6.032	5.471
20	24837	199.45	42.779	20.167	12.903	9.589	7.754	6.608	5.832	5.274
25	24959	199.45	42.590	20.003	12.756	9.451	7.623	6.482	5.708	5.153
30	25041	199.48	42.466	19.892	12.656	9.358	7.534	6.396	5.625	5.071
40	25146	199.48	42.310	19.751	12.530	9.241	7.422	6.288	5.519	4.966
60	25254	199.48	42.150	19.611	12.402	9.122	7.309	6.177	5.410	4.859
120	25358	199.49	41.990	19.469	12.274	9.001	7.193	6.065	5.300	4.750

	v2							
v1	12	15	20	15	30	40	60	120
1	16212	198.50	55.552	31.332	22.785	18.635	16.235	14.688
2	19997	199.01	49.800	26.284	18.314	14.544	12.404	11.043
3	21614	199.16	47.468	24.260	16.530	12.917	10.883	9.597
4	22501	199.24	46.195	23.154	15.556	12.028	10.050	8.805
5	23056	199.30	45.391	22.456	14.939	11.464	9.522	8.302
6	23440	199.33	44.838	21.975	14.513	11.073	9.155	7.952
7	23715	199.36	44.434	21.622	14.200	10.786	8.885	7.694
8	23924	199.38	44.125	21.352	13.961	10.566	8.678	7.496
9	24091	199.39	43.881	21.138	13.772	10.391	8.514	7.339
10	24222	199.39	43.685	20.967	13.618	10.250	8.380	7.211
12	24427	199.42	43.387	20.705	13.385	10.034	8.176	7.015
15	24632	199.43	43.085	20.438	13.146	9.814	7.968	6.814
20	24837	199.45	42.779	20.167	12.903	9.589	7.754	6.608
25	24959	199.45	42.590	20.003	12.756	9.451	7.623	6.482
30	25041	199.48	42.466	19.892	12.656	9.358	7.534	6.396
40	25146	199.48	42.310	19.751	12.530	9.241	7.422	6.288
60	25254	199.48	42.150	19.611	12.402	9.122	7.309	6.177
120	25358	199.49	41.990	19.469	12.274	9.001	7.193	6.065

Table E.2. Upper 1% points.

$$F_{0.99;\nu 1,\nu 2} = 1 / F_{0.01;\nu 2,\nu 1}$$

	v2									
v1	1	2	3	4	5	6	7	8	9	10
1	4052	98.50	34.116	21.198	16.258	13.745	12.246	11.259	10.562	10.044
2	4999	99.00	30.816	18.000	13.274	10.925	9.547	8.649	8.022	7.559
3	5404	99.16	29.457	16.694	12.060	9.780	8.451	7.591	6.992	6.552
4	5624	99.25	28.710	15.977	11.392	9.148	7.847	7.006	6.422	5.994
5	5764	99.30	28.237	15.522	10.967	8.746	7.460	6.632	6.057	5.636
6	5859	99.33	27.911	15.207	10.672	8.466	7.191	6.371	5.802	5.386
7	5928	99.36	27.671	14.976	10.456	8.260	6.993	6.178	5.613	5.200
8	5981	99.38	27.489	14.799	10.289	8.102	6.840	6.029	5.467	5.057
9	6022	99.39	27.345	14.659	10.158	7.976	6.719	5.911	5.351	4.942
10	6056	99.40	27.228	14.546	10.051	7.874	6.620	5.814	5.257	4.849
12	6107	99.42	27.052	14.374	9.888	7.718	6.469	5.667	5.111	4.706
15	6157	99.43	26.872	14.198	9.722	7.559	6.314	5.515	4.962	4.558
20	6209	99.45	26.690	14.019	9.553	7.396	6.155	5.359	4.808	4.405
25	6240	99.46	26.579	13.911	9.449	7.296	6.058	5.263	4.713	4.311
30	6260	99.47	26.504	13.838	9.379	7.229	5.992	5.198	4.649	4.247
40	6286	99.48	26.411	13.745	9.291	7.143	5.908	5.116	4.567	4.165
60	6313	99.48	26.316	13.652	9.202	7.057	5.824	5.032	4.483	4.082
120	6340	99.49	26.221	13.558	9.112	6.969	5.737	4.946	4.398	3.996

	v2							
v1	12	15	20	15	30	40	60	120
1	4052	98.50	34.116	21.198	16.258	13.745	12.246	11.259
2	4999	99.00	30.816	18.000	13.274	10.925	9.547	8.649
3	5404	99.16	29.457	16.694	12.060	9.780	8.451	7.591
4	5624	99.25	28.710	15.977	11.392	9.148	7.847	7.006
5	5764	99.30	28.237	15.522	10.967	8.746	7.460	6.632
6	5859	99.33	27.911	15.207	10.672	8.466	7.191	6.371
7	5928	99.36	27.671	14.976	10.456	8.260	6.993	6.178
8	5981	99.38	27.489	14.799	10.289	8.102	6.840	6.029
9	6022	99.39	27.345	14.659	10.158	7.976	6.719	5.911
10	6056	99.40	27.228	14.546	10.051	7.874	6.620	5.814
12	6107	99.42	27.052	14.374	9.888	7.718	6.469	5.667
15	6157	99.43	26.872	14.198	9.722	7.559	6.314	5.515
20	6209	99.45	26.690	14.019	9.553	7.396	6.155	5.359
25	6240	99.46	26.579	13.911	9.449	7.296	6.058	5.263
30	6260	99.47	26.504	13.838	9.379	7.229	5.992	5.198
40	6286	99.48	26.411	13.745	9.291	7.143	5.908	5.116
60	6313	99.48	26.316	13.652	9.202	7.057	5.824	5.032
120	6340	99.49	26.221	13.558	9.112	6.969	5.737	4.946

Table E.3. Upper 2.5% points.

$$F_{0.975;\nu1,\nu2} = 1/F_{0.025;\nu2,\nu1}$$

	v2									
v1	1	2	3	4	5	6	7	8	9	10
1	648	38.51	17.443	12.218	10.007	8.813	8.073	7.571	7.209	6.937
2	799	39.00	16.044	10.649	8.434	7.260	6.542	6.059	5.715	5.456
3	864	39.17	15.439	9.979	7.764	6.599	5.890	5.416	5.078	4.826
4	900	39.25	15.101	9.604	7.388	6.227	5.523	5.053	4.718	4.468
5	922	39.30	14.885	9.364	7.146	5.988	5.285	4.817	4.484	4.236
6	937	39.33	14.735	9.197	6.978	5.820	5.119	4.652	4.320	4.072
7	948	39.36	14.624	9.074	6.853	5.695	4.995	4.529	4.197	3.950
8	957	39.37	14.540	8.980	6.757	5.600	4.899	4.433	4.102	3.855
9	963	39.39	14.473	8.905	6.681	5.523	4.823	4.357	4.026	3.779
10	969	39.40	14.419	8.844	6.619	5.461	4.761	4.295	3.964	3.717
12	977	39.41	14.337	8.751	6.525	5.366	4.666	4.200	3.868	3.621
15	985	39.43	14.253	8.657	6.428	5.269	4.568	4.101	3.769	3.522
20	993	39.45	14.167	8.560	6.329	5.168	4.467	3.999	3.667	3.419
25	998	39.46	14.115	8.501	6.268	5.107	4.405	3.937	3.604	3.355
30	1001	39.46	14.081	8.461	6.227	5.065	4.362	3.894	3.560	3.311
40	1006	39.47	14.036	8.411	6.175	5.012	4.309	3.840	3.505	3.255
60	1010	39.48	13.992	8.360	6.123	4.959	4.254	3.784	3.449	3.198
120	1014	39.49	13.947	8.309	6.069	4.904	4.199	3.728	3.392	3.140

	v2							
v1	12	15	20	15	30	40	60	120
1	648	38.51	17.443	12.218	10.007	8.813	8.073	7.571
2	799	39.00	16.044	10.649	8.434	7.260	6.542	6.059
3	864	39.17	15.439	9.979	7.764	6.599	5.890	5.416
4	900	39.25	15.101	9.604	7.388	6.227	5.523	5.053
5	922	39.30	14.885	9.364	7.146	5.988	5.285	4.817
6	937	39.33	14.735	9.197	6.978	5.820	5.119	4.652
7	948	39.36	14.624	9.074	6.853	5.695	4.995	4.529
8	957	39.37	14.540	8.980	6.757	5.600	4.899	4.433
9	963	39.39	14.473	8.905	6.681	5.523	4.823	4.357
10	969	39.40	14.419	8.844	6.619	5.461	4.761	4.295
12	977	39.41	14.337	8.751	6.525	5.366	4.666	4.200
15	985	39.43	14.253	8.657	6.428	5.269	4.568	4.101
20	993	39.45	14.167	8.560	6.329	5.168	4.467	3.999
25	998	39.46	14.115	8.501	6.268	5.107	4.405	3.937
30	1001	39.46	14.081	8.461	6.227	5.065	4.362	3.894
40	1006	39.47	14.036	8.411	6.175	5.012	4.309	3.840
60	1010	39.48	13.992	8.360	6.123	4.959	4.254	3.784
120	1014	39.49	13.947	8.309	6.069	4.904	4.199	3.728

Table E.4. Upper 95% points.

$$F_{0.95;\nu 1,\nu 2} = 1 / F_{0.05;\nu 2,\nu 1}$$

	v2									
v1	1	2	3	4	5	6	7	8	9	10
1	161	18.51	10.128	7.709	6.608	5.987	5.591	5.318	5.117	4.965
2	199	19.00	9.552	6.944	5.786	5.143	4.737	4.459	4.256	4.103
3	216	19.16	9.277	6.591	5.409	4.757	4.347	4.066	3.863	3.708
4	225	19.25	9.117	6.388	5.192	4.534	4.120	3.838	3.633	3.478
5	230	19.30	9.013	6.256	5.050	4.387	3.972	3.688	3.482	3.326
6	234	19.33	8.941	6.163	4.950	4.284	3.866	3.581	3.374	3.217
7	237	19.35	8.887	6.094	4.876	4.207	3.787	3.500	3.293	3.135
8	239	19.37	8.845	6.041	4.818	4.147	3.726	3.438	3.230	3.072
9	241	19.38	8.812	5.999	4.772	4.099	3.677	3.388	3.179	3.020
10	242	19.40	8.785	5.964	4.735	4.060	3.637	3.347	3.137	2.978
12	244	19.41	8.745	5.912	4.678	4.000	3.575	3.284	3.073	2.913
15	246	19.43	8.703	5.858	4.619	3.938	3.511	3.218	3.006	2.845
20	248	19.45	8.660	5.803	4.558	3.874	3.445	3.150	2.936	2.774
25	249	19.46	8.634	5.769	4.521	3.835	3.404	3.108	2.893	2.730
30	250	19.46	8.617	5.746	4.496	3.808	3.376	3.079	2.864	2.700
40	251	19.47	8.594	5.717	4.464	3.774	3.340	3.043	2.826	2.661
60	252	19.48	8.572	5.688	4.431	3.740	3.304	3.005	2.787	2.621
120	253	19.49	8.549	5.658	4.398	3.705	3.267	2.967	2.748	2.580

	v2							
v1	12	15	20	15	30	40	60	120
1	161	18.51	10.128	7.709	6.608	5.987	5.591	5.318
2	199	19.00	9.552	6.944	5.786	5.143	4.737	4.459
3	216	19.16	9.277	6.591	5.409	4.757	4.347	4.066
4	225	19.25	9.117	6.388	5.192	4.534	4.120	3.838
5	230	19.30	9.013	6.256	5.050	4.387	3.972	3.688
6	234	19.33	8.941	6.163	4.950	4.284	3.866	3.581
7	237	19.35	8.887	6.094	4.876	4.207	3.787	3.500
8	239	19.37	8.845	6.041	4.818	4.147	3.726	3.438
9	241	19.38	8.812	5.999	4.772	4.099	3.677	3.388
10	242	19.40	8.785	5.964	4.735	4.060	3.637	3.347
12	244	19.41	8.745	5.912	4.678	4.000	3.575	3.284
15	246	19.43	8.703	5.858	4.619	3.938	3.511	3.218
20	248	19.45	8.660	5.803	4.558	3.874	3.445	3.150
25	249	19.46	8.634	5.769	4.521	3.835	3.404	3.108
30	250	19.46	8.617	5.746	4.496	3.808	3.376	3.079
40	251	19.47	8.594	5.717	4.464	3.774	3.340	3.043
60	252	19.48	8.572	5.688	4.431	3.740	3.304	3.005
120	253	19.49	8.549	5.658	4.398	3.705	3.267	2.967

Table E.5. Upper 10% points.

$$F_{0.9;\nu1,\nu2} = 1/F_{0.1;\nu2,\nu1}$$

v1	v2									
	1	2	3	4	5	6	7	8	9	10
1	40	8.53	5.538	4.545	4.060	3.776	3.589	3.458	3.360	3.285
2	50	9.00	5.462	4.325	3.780	3.463	3.257	3.113	3.006	2.924
3	54	9.16	5.391	4.191	3.619	3.289	3.074	2.924	2.813	2.728
4	56	9.24	5.343	4.107	3.520	3.181	2.961	2.806	2.693	2.605
5	57	9.29	5.309	4.051	3.453	3.108	2.883	2.726	2.611	2.522
6	58	9.33	5.285	4.010	3.405	3.055	2.827	2.668	2.551	2.461
7	59	9.35	5.266	3.979	3.368	3.014	2.785	2.624	2.505	2.414
8	59	9.37	5.252	3.955	3.339	2.983	2.752	2.589	2.469	2.377
9	60	9.38	5.240	3.936	3.316	2.958	2.725	2.561	2.440	2.347
10	60	9.39	5.230	3.920	3.297	2.937	2.703	2.538	2.416	2.323
12	61	9.41	5.216	3.896	3.268	2.905	2.668	2.502	2.379	2.284
15	61	9.42	5.200	3.870	3.238	2.871	2.632	2.464	2.340	2.244
20	62	9.44	5.184	3.844	3.207	2.836	2.595	2.425	2.298	2.201
25	62	9.45	5.175	3.828	3.187	2.815	2.571	2.400	2.272	2.174
30	62	9.46	5.168	3.817	3.174	2.800	2.555	2.383	2.255	2.155
40	63	9.47	5.160	3.804	3.157	2.781	2.535	2.361	2.232	2.132
60	63	9.47	5.151	3.790	3.140	2.762	2.514	2.339	2.208	2.107
120	63	9.48	5.143	3.775	3.123	2.742	2.493	2.316	2.184	2.082

v1	v2							
	12	15	20	15	30	40	60	120
1	40	8.53	5.538	4.545	4.060	3.776	3.589	3.458
2	50	9.00	5.462	4.325	3.780	3.463	3.257	3.113
3	54	9.16	5.391	4.191	3.619	3.289	3.074	2.924
4	56	9.24	5.343	4.107	3.520	3.181	2.961	2.806
5	57	9.29	5.309	4.051	3.453	3.108	2.883	2.726
6	58	9.33	5.285	4.010	3.405	3.055	2.827	2.668
7	59	9.35	5.266	3.979	3.368	3.014	2.785	2.624
8	59	9.37	5.252	3.955	3.339	2.983	2.752	2.589
9	60	9.38	5.240	3.936	3.316	2.958	2.725	2.561
10	60	9.39	5.230	3.920	3.297	2.937	2.703	2.538
12	61	9.41	5.216	3.896	3.268	2.905	2.668	2.502
15	61	9.42	5.200	3.870	3.238	2.871	2.632	2.464
20	62	9.44	5.184	3.844	3.207	2.836	2.595	2.425
25	62	9.45	5.175	3.828	3.187	2.815	2.571	2.400
30	62	9.46	5.168	3.817	3.174	2.800	2.555	2.383
40	63	9.47	5.160	3.804	3.157	2.781	2.535	2.361
60	63	9.47	5.151	3.790	3.140	2.762	2.514	2.339
120	63	9.48	5.143	3.775	3.123	2.742	2.493	2.316

APPENDIX

F

Critical Values of the *t*-Distribution

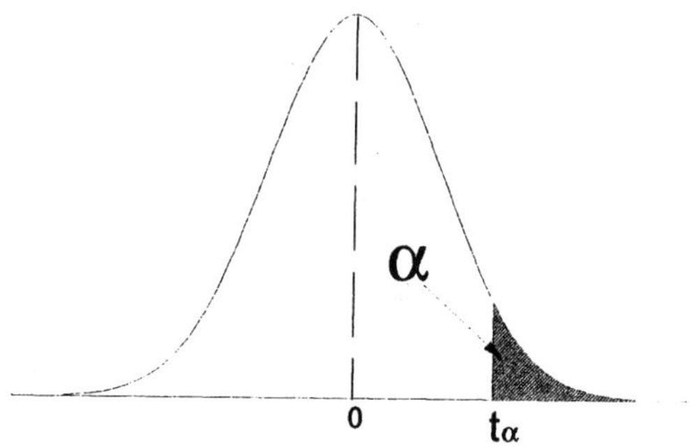

Table F.1. Critical values of the *t*-distribution.

	α				
γ	0.1	0.05	0.025	0.01	0.005
1	3.078	6.314	12.706	31.821	63.657
2	1.886	2.920	4.303	6.965	9.925
3	1.638	2.353	3.182	4.541	5.841
4	1.533	2.132	2.776	3.747	4.604
5	1.476	2.015	2.571	3.365	4.032

Continued on next page...

Table F.1—*continued...*

γ	α 0.1	0.05	0.025	0.01	0.005
6	1.440	1.943	2.447	3.143	3.707
7	1.415	1.895	2.365	2.998	3.499
8	1.397	1.860	2.306	2.896	3.355
9	1.383	1.833	2.262	2.821	3.250
10	1.372	1.812	2.228	2.764	3.169
11	1.363	1.796	2.201	2.718	3.106
12	1.356	1.782	2.179	2.681	3.055
13	1.350	1.771	2.160	2.650	3.012
14	1.345	1.761	2.145	2.624	2.977
15	1.341	1.753	2.131	2.602	2.947
16	1.337	1.746	2.120	2.583	2.921
17	1.333	1.740	2.110	2.567	2.898
18	1.330	1.734	2.101	2.552	2.878
19	1.328	1.729	2.093	2.539	2.861
20	1.325	1.725	2.086	2.528	2.845
21	1.323	1.721	2.080	2.518	2.831
22	1.321	1.717	2.074	2.508	2.819
23	1.319	1.714	2.069	2.500	2.807
24	1.318	1.711	2.064	2.492	2.797
25	1.316	1.708	2.060	2.485	2.787
26	1.315	1.706	2.056	2.479	2.779
27	1.314	1.703	2.052	2.473	2.771
28	1.313	1.701	2.048	2.467	2.763
29	1.311	1.699	2.045	2.462	2.756
∞	1.282	1.645	1.960	2.326	2.576

APPENDIX

G

Poisson Probability Sums

Table G.1. Poisson probability sums.

$$\sum_{x=o}^{r} p(x;\mu)$$

r	μ								
	0.1	0.2	0.3	0.4	0.5	0.6	0.7	0.8	0.9
0	0.9048	0.8187	0.7408	0.6703	0.6065	0.5488	0.4966	0.4493	0.4066
1	0.9953	0.9825	0.9631	0.9384	0.9098	0.8781	0.8442	0.8088	0.7725
2	0.9998	0.9989	0.9964	0.9921	0.9856	0.9769	0.9659	0.9526	0.9371
3	1.0000	0.9999	0.9997	0.9992	0.9982	0.9966	0.9942	0.9909	0.9865
4	1.0000	1.0000	1.0000	0.9999	0.9998	0.9996	0.9992	0.9986	0.9977
5	1.0000	1.0000	1.0000	1.0000	1.0000	1.0000	0.9999	0.9998	0.9997
6	1.0000	1.0000	1.0000	1.0000	1.0000	1.0000	1.0000	1.0000	1.0000

Continued on next page...

Table G.1—*continued...*

r	μ								
	1.0	1.5	2.0	2.5	3.0	3.5	4.0	4.5	5.0
0	0.3679	0.2231	0.1353	0.0821	0.0498	0.0302	0.0183	0.0111	0.0067
1	0.7358	0.5578	0.4060	0.2873	0.1991	0.1359	0.0916	0.0611	0.0404
2	0.9197	0.8088	0.6767	0.5438	0.4232	0.3208	0.2381	0.1736	0.1247
3	0.9810	0.9344	0.8571	0.7576	0.6472	0.5366	0.4335	0.3423	0.2650
4	0.9963	0.9814	0.9473	0.8912	0.8153	0.7254	0.6288	0.5321	0.4405
5	0.9994	0.9955	0.9834	0.9580	0.9161	0.8576	0.7851	0.7029	0.6160
6	0.9999	0.9991	0.9955	0.9858	0.9665	0.9347	0.8893	0.8311	0.7622
7	1.0000	0.9998	0.9989	0.9958	0.9881	0.9733	0.9489	0.9134	0.8666
8	1.0000	1.0000	0.9998	0.9989	0.9962	0.9901	0.9786	0.9597	0.9319
9	1.0000	1.0000	1.0000	0.9997	0.9989	0.9967	0.9919	0.9829	0.9682
10	1.0000	1.0000	1.0000	0.9999	0.9997	0.9990	0.9972	0.9933	0.9863
11	1.0000	1.0000	1.0000	1.0000	0.9999	0.9997	0.9991	0.9976	0.9945
12	1.0000	1.0000	1.0000	1.0000	1.0000	0.9999	0.9997	0.9992	0.9980
13	1.0000	1.0000	1.0000	1.0000	1.0000	1.0000	0.9999	0.9997	0.9993
14	1.0000	1.0000	1.0000	1.0000	1.0000	1.0000	1.0000	0.9999	0.9998
15	1.0000	1.0000	1.0000	1.0000	1.0000	1.0000	1.0000	1.0000	0.9999
16	1.0000	1.0000	1.0000	1.0000	1.0000	1.0000	1.0000	1.0000	1.0000

r	μ								
	5.5	6.0	6.5	7.0	7.5	8.0	8.5	9.0	9.5
0	0.0041	0.0025	0.0015	0.0009	0.0006	0.0003	0.0002	0.0001	0.0001
1	0.0266	0.0174	0.0113	0.0073	0.0047	0.0030	0.0019	0.0012	0.0008
2	0.0884	0.0620	0.0430	0.0296	0.0203	0.0138	0.0093	0.0062	0.0042
3	0.2017	0.1512	0.1118	0.0818	0.0591	0.0424	0.0301	0.0212	0.0149
4	0.3575	0.2851	0.2237	0.1730	0.1321	0.0996	0.0744	0.0550	0.0403
5	0.5289	0.4457	0.3690	0.3007	0.2414	0.1912	0.1496	0.1157	0.0885
6	0.6860	0.6063	0.5265	0.4497	0.3782	0.3134	0.2562	0.2068	0.1649
7	0.8095	0.7440	0.6728	0.5987	0.5246	0.4530	0.3856	0.3239	0.2687
8	0.8944	0.8472	0.7916	0.7291	0.6620	0.5925	0.5231	0.4557	0.3918
9	0.9462	0.9161	0.8774	0.8305	0.7764	0.7166	0.6530	0.5874	0.5218
10	0.9747	0.9574	0.9332	0.9015	0.8622	0.8159	0.7634	0.7060	0.6453
11	0.9890	0.9799	0.9661	0.9467	0.9208	0.8881	0.8487	0.8030	0.7520
12	0.9955	0.9912	0.9840	0.9730	0.9573	0.9362	0.9091	0.8758	0.8364
13	0.9983	0.9964	0.9929	0.9872	0.9784	0.9658	0.9486	0.9261	0.8981
14	0.9994	0.9986	0.9970	0.9943	0.9897	0.9827	0.9726	0.9585	0.9400

Continued on next page...

Table G.1—*continued...*

r	μ								
	5.5	6.0	6.5	7.0	7.5	8.0	8.5	9.0	9.5
15	0.9998	0.9995	0.9988	0.9976	0.9954	0.9918	0.9862	0.9780	0.9665
16	0.9999	0.9998	0.9996	0.9990	0.9980	0.9963	0.9934	0.9889	0.9823
17	1.0000	0.9999	0.9998	0.9996	0.9992	0.9984	0.9970	0.9947	0.9911
18	1.0000	1.0000	0.9999	0.9999	0.9997	0.9993	0.9987	0.9976	0.9957
19	1.0000	1.0000	1.0000	1.0000	0.9999	0.9997	0.9995	0.9989	0.9980
20	1.0000	1.0000	1.0000	1.0000	1.0000	0.9999	0.9998	0.9996	0.9991
21	1.0000	1.0000	1.0000	1.0000	1.0000	1.0000	0.9999	0.9998	0.9996
22	1.0000	1.0000	1.0000	1.0000	1.0000	1.0000	1.0000	0.9999	0.9999
23	1.0000	1.0000	1.0000	1.0000	1.0000	1.0000	1.0000	1.0000	0.9999
24	1.0000	1.0000	1.0000	1.0000	1.0000	1.0000	1.0000	1.0000	1.0000

r	μ								
	10.0	11.0	12.0	13.0	14.0	15.0	16.0	17.0	18.0
0	0.0000	0.0000	0.0000	0.0000	0.0000	0.0000	0.0000	0.0000	0.0000
1	0.0005	0.0002	0.0001	0.0000	0.0000	0.0000	0.0000	0.0000	0.0000
2	0.0028	0.0012	0.0005	0.0002	0.0001	0.0000	0.0000	0.0000	0.0000
3	0.0103	0.0049	0.0023	0.0011	0.0005	0.0002	0.0001	0.0000	0.0000
4	0.0293	0.0151	0.0076	0.0037	0.0018	0.0009	0.0004	0.0002	0.0001
5	0.0671	0.0375	0.0203	0.0107	0.0055	0.0028	0.0014	0.0007	0.0003
6	0.1301	0.0786	0.0458	0.0259	0.0142	0.0076	0.0040	0.0021	0.0010
7	0.2202	0.1432	0.0895	0.0540	0.0316	0.0180	0.0100	0.0054	0.0029
8	0.3328	0.2320	0.1550	0.0998	0.0621	0.0374	0.0220	0.0126	0.0071
9	0.4579	0.3405	0.2424	0.1658	0.1094	0.0699	0.0433	0.0261	0.0154
10	0.5830	0.4599	0.3472	0.2517	0.1757	0.1185	0.0774	0.0491	0.0304
11	0.6968	0.5793	0.4616	0.3532	0.2600	0.1848	0.1270	0.0847	0.0549
12	0.7916	0.6887	0.5760	0.4631	0.3585	0.2676	0.1931	0.1350	0.0917
13	0.8645	0.7813	0.6815	0.5730	0.4644	0.3632	0.2745	0.2009	0.1426
14	0.9165	0.8540	0.7720	0.6751	0.5704	0.4657	0.3675	0.2808	0.2081
15	0.9513	0.9074	0.8444	0.7636	0.6694	0.5681	0.4667	0.3715	0.2867
16	0.9730	0.9441	0.8987	0.8355	0.7559	0.6641	0.5660	0.4677	0.3751
17	0.9857	0.9678	0.9370	0.8905	0.8272	0.7489	0.6593	0.5640	0.4686
18	0.9928	0.9823	0.9626	0.9302	0.8826	0.8195	0.7423	0.6550	0.5622
19	0.9965	0.9907	0.9787	0.9573	0.9235	0.8752	0.8122	0.7363	0.6509
20	0.9984	0.9953	0.9884	0.9750	0.9521	0.9170	0.8682	0.8055	0.7307

Continued on next page...

Table G.1—*continued...*

r	μ								
	10.0	11.0	12.0	13.0	14.0	15.0	16.0	17.0	18.0
21	0.9993	0.9977	0.9939	0.9859	0.9712	0.9469	0.9108	0.8615	0.7991
22	0.9997	0.9990	0.9970	0.9924	0.9833	0.9673	0.9418	0.9047	0.8551
23	0.9999	0.9995	0.9985	0.9960	0.9907	0.9805	0.9633	0.9367	0.8989
24	1.0000	0.9998	0.9993	0.9980	0.9950	0.9888	0.9777	0.9594	0.9317
25	1.0000	0.9999	0.9997	0.9990	0.9974	0.9938	0.9869	0.9748	0.9554
26	1.0000	1.0000	0.9999	0.9995	0.9987	0.9967	0.9925	0.9848	0.9718
27	1.0000	1.0000	0.9999	0.9998	0.9994	0.9983	0.9959	0.9912	0.9827
28	1.0000	1.0000	1.0000	0.9999	0.9997	0.9991	0.9978	0.9950	0.9897
29	1.0000	1.0000	1.0000	1.0000	0.9999	0.9996	0.9989	0.9973	0.9941
30	1.0000	1.0000	1.0000	1.0000	0.9999	0.9998	0.9994	0.9986	0.9967
31	1.0000	1.0000	1.0000	1.0000	1.0000	0.9999	0.9997	0.9993	0.9982
32	1.0000	1.0000	1.0000	1.0000	1.0000	1.0000	0.9999	0.9996	0.9990
33	1.0000	1.0000	1.0000	1.0000	1.0000	1.0000	0.9999	0.9998	0.9995
34	1.0000	1.0000	1.0000	1.0000	1.0000	1.0000	1.0000	0.9999	0.9998
35	1.0000	1.0000	1.0000	1.0000	1.0000	1.0000	1.0000	1.0000	0.9999
36	1.0000	1.0000	1.0000	1.0000	1.0000	1.0000	1.0000	1.0000	0.9999
37	1.0000	1.0000	1.0000	1.0000	1.0000	1.0000	1.0000	1.0000	1.0000

APPENDIX

H

Critical Values of the Chi-Square Distribution

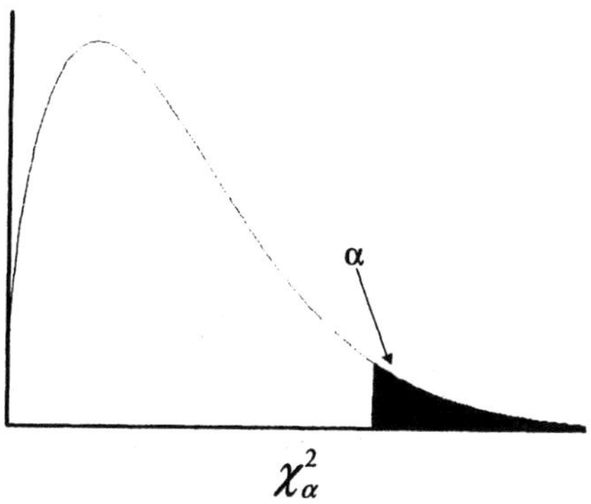

Table H.1. Critical values of the chi-square distribution.

γ	α									
	0.995	0.99	0.98	0.975	0.95	0.90	0.80	0.75	0.70	0.50
	0.00004	0.000	0.001	0.001	0.004	0.016	0.064	0.102	0.148	0.455
2	0.0100	0.020	0.040	0.051	0.103	0.211	0.446	0.575	0.713	1.386
3	0.0717	0.115	0.185	0.216	0.352	0.584	1.005	1.213	1.424	2.366
4	0.207	0.297	0.429	0.484	0.711	1.064	1.649	1.923	2.195	3.357
5	0.412	0.554	0.752	0.831	1.145	1.610	2.343	2.675	3.000	4.351

Continued on next page...

Table H.1—continued…

γ	α									
	0.995	0.99	0.98	0.975	0.95	0.90	0.80	0.75	0.70	0.50
6	0.676	0.872	1.134	1.237	1.635	2.204	3.070	3.455	3.828	5.348
7	0.989	1.239	1.564	1.690	2.167	2.833	3.822	4.255	4.671	6.346
8	1.344	1.646	2.032	2.180	2.733	3.490	4.594	5.071	5.527	7.344
9	1.735	2.088	2.532	2.700	3.325	4.168	5.380	5.899	6.393	8.343
10	2.156	2.558	3.059	3.247	3.940	4.865	6.179	6.737	7.267	9.342
11	2.603	3.053	3.609	3.816	4.575	5.578	6.989	7.584	8.148	10.341
12	3.074	3.571	4.178	4.404	5.226	6.304	7.807	8.438	9.034	11.340
13	3.565	4.107	4.765	5.009	5.892	7.042	8.634	9.299	9.926	12.340
14	4.075	4.660	5.368	5.629	6.571	7.790	9.467	10.165	10.821	13.339
15	4.601	5.229	5.985	6.262	7.261	8.547	10.307	11.037	11.721	14.339
16	5.142	5.812	6.614	6.908	7.962	9.312	11.152	11.912	12.624	15.338
17	5.697	6.408	7.255	7.564	8.672	10.085	12.002	12.792	13.531	16.338
18	6.265	7.015	7.906	8.231	9.390	10.865	12.857	13.675	14.440	17.338
19	6.844	7.633	8.567	8.907	10.117	11.651	13.716	14.562	15.352	18.338
20	7.434	8.260	9.237	9.591	10.851	12.443	14.578	15.452	16.266	19.337
21	8.034	8.897	9.915	10.283	11.591	13.240	15.445	16.344	17.182	20.337
22	8.643	9.542	10.600	10.982	12.338	14.041	16.314	17.240	18.101	21.337
23	9.260	10.196	11.293	11.689	13.091	14.848	17.187	18.137	19.021	22.337
24	9.886	10.856	11.992	12.401	13.848	15.659	18.062	19.037	19.943	23.337
25	10.520	11.524	12.697	13.120	14.611	16.473	18.940	19.939	20.867	24.337
26	11.160	12.198	13.409	13.844	15.379	17.292	19.820	20.843	21.792	25.336
27	11.808	12.879	14.125	14.573	16.151	18.114	20.703	21.749	22.719	26.336
28	12.461	13.565	14.847	15.308	16.928	18.939	21.588	22.657	23.647	27.336
29	13.121	14.256	15.574	16.047	17.708	19.768	22.475	23.567	24.577	28.336
30	13.787	14.953	16.306	16.791	18.493	20.599	23.364	24.478	25.508	29.336

Continued on next page…

Table H.1—*continued...*

	α									
γ	0.30	0.25	0.20	0.10	0.05	0.025	0.02	0.01	0.005	0.001
1	1.074	1.323	1.642	2.706	3.841	5.024	5.412	6.635	7.879	10.828
2	2.408	2.773	3.219	4.605	5.991	7.378	7.824	9.210	10.597	13.816
3	3.665	4.108	4.642	6.251	7.815	9.348	9.837	11.345	12.838	16.266
4	4.878	5.385	5.989	7.779	9.488	11.143	11.668	13.277	14.860	18.467
5	6.064	6.626	7.289	9.236	11.070	12.833	13.388	15.086	16.750	20.515
6	7.231	7.841	8.558	10.645	12.592	14.449	15.033	16.812	18.548	22.458
7	8.383	9.037	9.803	12.017	14.067	16.013	16.622	18.475	20.278	24.322
8	9.524	10.219	11.030	13.362	15.507	17.535	18.168	20.090	21.955	26.124
9	10.656	11.389	12.242	14.684	16.919	19.023	19.679	21.666	23.589	27.877
10	11.781	12.549	13.442	15.987	18.307	20.483	21.161	23.209	25.188	29.588
11	12.899	13.701	14.631	17.275	19.675	21.920	22.618	24.725	26.757	31.264
12	14.011	14.845	15.812	18.549	21.026	23.337	24.054	26.217	28.300	32.909
13	15.119	15.984	16.985	19.812	22.362	24.736	25.472	27.688	29.819	34.528
14	16.222	17.117	18.151	21.064	23.685	26.119	26.873	29.141	31.319	36.123
15	17.322	18.245	19.311	22.307	24.996	27.488	28.259	30.578	32.801	37.697
16	18.418	19.369	20.465	23.542	26.296	28.845	29.633	32.000	34.267	39.252
17	19.511	20.489	21.615	24.769	27.587	30.191	30.995	33.409	35.718	40.790
18	20.601	21.605	22.760	25.989	28.869	31.526	32.346	34.805	37.156	42.312
19	21.689	22.718	23.900	27.204	30.144	32.852	33.687	36.191	38.582	43.820
20	22.775	23.828	25.038	28.412	31.410	34.170	35.020	37.566	39.997	45.315
21	23.858	24.935	26.171	29.615	32.671	35.479	36.343	38.932	41.401	46.797
22	24.939	26.039	27.301	30.813	33.924	36.781	37.659	40.289	42.796	48.268
23	26.018	27.141	28.429	32.007	35.172	38.076	38.968	41.638	44.181	49.728
24	27.096	28.241	29.553	33.196	36.415	39.364	40.270	42.980	45.559	51.179
25	28.172	29.339	30.675	34.382	37.652	40.646	41.566	44.314	46.928	52.620
26	29.246	30.435	31.795	35.563	38.885	41.923	42.856	45.642	48.290	54.052
27	30.319	31.528	32.912	36.741	40.113	43.195	44.140	46.963	49.645	55.476
28	31.391	32.620	34.027	37.916	41.337	44.461	45.419	48.278	50.993	56.892
29	32.461	33.711	35.139	39.087	42.557	45.722	46.693	49.588	52.336	58.301
30	33.530	34.800	36.250	40.256	43.773	46.979	47.962	50.892	53.672	59.703

♦♦♦

APPENDIX

I

Tolerance Interval Factors

Table I.1a. Values of ***k*** for two-sided limits.

n	α = 0.10				α = 0.05				α = 0.01			
	γ=0.10	γ=0.05	γ=0.01	γ=0.001	γ=0.10	γ=0.05	γ=0.01	γ=0.001	γ=0.10	γ=0.05	γ=0.01	γ=0.001
2	15.978	18.800	24.167	30.227	32.019	37.674	48.430	60.573	160.193	188.491	242.300	303.054
3	5.847	6.919	8.974	11.309	8.380	9.916	12.861	16.208	18.930	22.401	29.055	36.616
4	4.166	4.943	6.440	8.149	5.369	6.370	8.299	10.502	9.398	11.150	14.527	18.383
5	3.494	4.152	5.423	6.879	4.275	5.079	6.634	8.415	6.612	7.855	10.260	13.015
6	3.131	3.723	4.870	6.188	3.712	4.414	5.775	7.337	5.337	6.345	8.301	10.548
7	2.902	3.452	4.521	5.750	3.369	4.007	5.248	6.676	4.613	5.488	7.187	9.142
8	2.743	3.264	4.278	5.446	3.316	3.732	4.891	6.226	4.147	4.936	6.468	8.234
9	2.626	3.125	4.098	5.220	2.967	3.532	4.631	5.899	3.822	4.550	5.966	7.600
10	2.535	3.018	3.959	5.046	2.839	3.379	4.433	5.649	3.582	4.265	5.594	7.129
11	2.463	2.933	3.849	4.906	2.737	3.259	4.277	5.452	3.397	4.045	5.308	6.766
12	2.404	2.863	3.758	4.792	2.655	3.162	4.150	5.291	3.250	3.870	5.079	6.477
13	2.355	2.805	3.682	4.697	2.587	3.081	4.044	5.158	3.130	3.727	4.893	6.240
14	2.314	2.756	3.618	4.615	2.529	3.012	3.955	5.045	3.029	3.608	4.737	6.043
15	2.278	2.713	3.562	4.545	2.480	2.954	3.878	4.949	2.945	3.507	4.605	5.876
16	2.246	2.676	3.514	4.484	2.437	2.903	3.812	4.865	2.872	3.421	4.492	5.732
17	2.219	2.643	3.471	4.430	2.400	2.858	3.754	4.791	2.808	3.345	4.393	5.607
18	2.194	2.614	3.433	4.382	2.366	2.819	3.702	4.725	2.753	3.279	4.307	5.497
19	2.172	2.588	3.399	4.339	2.337	2.784	3.656	4.667	2.703	3.221	4.230	5.399
20	2.152	2.564	3.368	4.300	2.310	2.752	3.615	4.614	2.659	3.168	4.161	5.312

Continued on next page...

Table I.1a—*continued...*

n	α = 0.10				α = 0.05				α = 0.01			
	γ=0.10	γ=0.05	γ=0.01	γ=0.001	γ=0.10	γ=0.05	γ=0.01	γ=0.001	γ=0.10	γ=0.05	γ=0.01	γ=0.001
21	2.135	2.543	3.340	4.264	2.286	2.723	3.577	4.567	2.620	3.121	4.100	5.234
22	2.118	2.524	3.315	4.232	2.264	2.697	3.543	4.523	2.584	3.078	4.044	5.163
23	2.103	2.506	3.292	4.203	2.244	2.673	3.512	4.484	2.551	3.040	3.993	5.098
24	2.089	2.480	3.270	4.176	2.225	2.651	3.483	4.447	2.522	3.004	3.947	5.039
25	2.077	2.474	3.251	4.151	2.208	2.631	3.457	4.413	2.494	2.972	3.904	4.985
30	2.025	2.413	3.170	4.049	2.140	2.549	3.350	4.278	2.385	2.841	3.733	4.768
35	1.988	2.368	3.112	3.974	2.090	2.490	3.272	4.179	2.306	2.748	3.611	4.611
40	1.959	2.334	3.066	3.917	2.052	2.445	3.213	4.104	2.247	2.677	3.518	4.493
45	1.935	2.306	3.030	3.871	2.021	2.408	3.165	4.042	2.200	2.621	3.444	4.399
50	1.916	2.284	3.001	3.833	1.996	2.379	3.126	3.993	2.162	2.576	3.385	4.323

Table I.1b. Values of ***k*** for one-sided limits.

n	α = 0.10				α = 0.05				α = 0.01			
	γ=0.10	γ=0.05	γ=0.01	γ=0.001	γ=0.10	γ=0.05	γ=0.01	γ=0.001	γ=0.10	γ=0.05	γ=0.01	γ=0.001
3	4.258	5.310	7.340	9.651	6.158	7.655	10.552	13.857	-	-	-	-
4	3.187	3.957	5.437	7.128	4.163	5.145	7.042	9.215	-	-	-	-
5	2.742	3.400	4.666	6.112	3.407	4.202	5.741	7.501	-	-	-	-
6	2.494	3.091	4.242	5.556	3.006	3.707	50.62	6.612	4.408	5.409	7.334	9.540
7	2.333	2.894	3.972	5.201	2.755	3.399	4.641	6.061	3.856	4.730	6.411	8.348
8	2.219	2.755	3.783	4.955	2.582	3.188	4.353	5.686	3.496	4.287	5.811	7.566
9	2.133	2.649	3.641	4.772	2.454	3.031	4.143	5.414	3.242	3.971	5.389	7.014
10	2.065	2.568	3.532	4.629	2.355	2.911	3.981	5.203	3.048	3.739	5.075	6.603
11	2.012	2.503	3.444	4.515	2.275	2.815	3.852	5.036	2.897	3.557	4.828	6.284
12	1.966	2.448	3.371	4.420	2.210	2.736	3.747	4.900	2.773	3.410	4.633	6.032
13	1.928	2.403	3.310	4.341	2.155	2.670	3.659	4.787	2.677	3.290	4.472	5.826
14	1.895	2.363	3.257	4.274	2.108	2.614	3.585	4.690	2.592	3.189	4.336	5.651
15	1.866	2.329	3.212	4.215	2.068	2.566	3.520	4.607	2.521	3.102	4.224	5.507
16	1.842	2.299	3.172	4.146	2.032	2.523	3.463	4.534	2.458	3.028	4.124	5.374
17	1.820	2.272	3.136	4.118	2.001	2.468	3.415	4.471	2.405	2.962	4.038	5.268
18	1.800	2.249	3.106	4.078	1.974	2.453	3.370	4.415	2.357	2.906	3.961	5.167
19	1.781	2.228	3.078	4.041	1.949	2.423	3.331	4.364	2.315	2.855	3.893	5.078
20	1.765	2.208	3.052	4.009	1.926	2.396	3.295	4.319	2.275	2.807	3.832	5.003

Continued on next page...

Table I.1b—*continued...*

n	α = 0.10				α = 0.05				α = 0.01			
	γ=0.10	γ=0.05	γ=0.01	γ=0.001	γ=0.10	γ=0.05	γ=0.01	γ=0.001	γ=0.10	γ=0.05	γ=0.01	γ=0.001
21	1.750	2.190	3.028	3.979	1.905	2.371	3.262	4.276	2.241	2.768	3.776	4.932
22	1.736	2.174	3.007	3.952	1.887	2.350	3.233	4.238	2.208	2.729	3.727	4.866
23	1.724	2.159	2.987	3.927	1.869	2.329	3.206	4.204	2.179	2.693	3.680	4.806
24	1.712	2.145	2.969	3.904	1.853	2.309	3.181	4.171	2.154	2.663	3.638	4.755
25	1.702	2.132	2.952	3.882	1.838	2.292	3.158	4.143	2.129	2.632	3.601	4.706
30	1.657	2.080	2.884	3.794	1.778	2.220	3.064	4.022	2.029	2.516	3.446	4.508
35	1.623	2.041	2.833	3.730	1.732	2.166	2.994	3.934	1.957	2.431	3.334	4.364
40	1.598	2.010	2.793	3.679	1.697	2.126	2.941	3.866	1.902	2.365	3.250	4.255
45	1.577	1.986	2.762	3.638	1.669	2.092	2.897	3.811	1.857	2.313	3.181	4.168
50	1.560	1.965	2.735	3.604	1.646	2.065	2.963	3.766	1.821	2.296	3.124	4.096

Table I.2. Proportion of population covered with (1–α)% confidence and sample size *n*.

n	α = 0.1	α = 0.05	α = 0.01	α = 0.005
2	0.052	0.026	0.006	0.003
4	0.321	0.249	0.141	0.111
6	0.490	0.419	0.295	0.254
10	0.664	0.606	0.496	0.456
20	0.820	0.784	0.712	0.683
40	0.907	0.887	0.846	0.829
60	0.937	0.924	0.895	0.883
80	0.953	0.943	0.920	0.911
100	0.962	0.954	0.936	0.929
150	0.975	0.969	0.957	0.952
200	0.981	0.977	0.968	0.961
500	0.993	0.991	0.987	0.986
1000	0.997	0.996	0.994	0.993

Table I.3. Sample size required to cover (1–γ)% of the population with (1–α)% confidence.

γ	α = 0.1	α = 0.05	α = 0.01	α = 0.005
0.005	777	947	1325	1483
0.01	388	473	662	740
0.05	77	93	130	146
0.01	38	46	64	72
0.15	25	30	42	47
0.20	18	22	31	34
0.25	15	18	24	27
0.30	12	14	20	22
0.40	6	10	14	16
0.50	7	8	11	12

APPENDIX

J

Useful Integrals

$$\int u\,dv = uv - \int v\,du \tag{J.1}$$

$$\int x^n dx = \frac{x^{n+1}}{n+1} \quad , n \neq -1 \tag{J.2}$$

$$\int \frac{dx}{x} = \ln x \tag{J.3}$$

$$\int e^{ax} dx = \frac{e^{ax}}{a} \tag{J.4}$$

$$\int x e^{ax} dx = \frac{e^{ax}}{a^2}(ax-1) \tag{J.5}$$

$$\int x^m e^{ax} dx = \begin{cases} \dfrac{x^m e^{ax}}{a} - \dfrac{m}{a}\displaystyle\int x^{m-1} e^{ax} dx \\ \qquad or \\ e^{ax} \displaystyle\sum_{r=0}^{m} (-1)^r \frac{m!\,x^{m-r}}{(m-r)!\,a^{r+1}} \end{cases} \tag{J.6}$$

$$\int \ln x\,dx = x \ln x - x \tag{J.7}$$

$$\int x^n \ln(ax)dx = \frac{x^{n+1}}{n+1}\ln ax - \frac{x^{n+1}}{(n+1)^2} \tag{J.8}$$

$$\int \frac{dx}{x\ln x} = \ln(\ln x) \tag{J.9}$$

$$\frac{d}{dc}\int_p^q f(x,c)dx = \int_p^q \frac{d}{dc} f(x,c)dx + f(q,c)\frac{dq}{dc} - f(p,c)\frac{dp}{dc} \tag{J.10}$$

♦♦♦

APPENDIX

K

Government Documents

The United States government used to publish government documents. The following documents can still be obtained from other organizations.

LC-78-2 *Storage Reliability Analysis Report* (Vol. 1, *Electrical and Electronic Devices;* Vol. 2, *Electromechanical Devices;* Vol. 3, *Hydraulic and Pneumatic Devices;* Vol. 4, *Ordnance Devices;* Vol. 5, *Optical Devices and Electro-Optical Devices*)

MIL-HDBK-189 *Reliability Growth Management*

MIL-HDBK-217 *Reliability Prediction of Electronic Equipment*

MIL-HDBK-251 *Reliability/Design Thermal Applications*

MIL-HDBK-263 *Electrostatic Discharge Control Handbook for Protection of Electrical and Electronic Parts, Assemblies and Equipment*

MIL-HDBK-338 *Electronic Reliability Design Handbook*

MIL-HDBK-472 *Maintainability Prediction*

MIL-HDBK-781 *Reliability Test Methods, Plans, and Environments for Engineering Development, Qualification, and Production*

MIL-STD-470 *Maintainability Program Requirements for Systems and Equipment*

MIL-STD-690 *Failure Rate Sampling Plans and Procedures*

MIL-STD-721 *Definition of Terms for Reliability and Maintainability*

MIL-STD-756 *Reliability Modeling and Prediction*

MIL-STD-757 *Reliability Evaluation from Demonstration Data*

MIL-STD-781 *Reliability Testing for Engineering Development, Qualification, and Production*

MIL-STD-785 *Reliability Program for Systems and Equipment, Development and Production*

MIL-STD-790 *Reliability Assurance for Electronic Parts*

MIL-STD-810 *Environmental Test Methods and Engineering Guidelines*

MIL-STD-883 *Test Methods and Procedures for Microelectronics*

MIL-STD-965 *Parts Control Program*

MIL-STD-1472 *Human Engineering Design Criteria for Military Systems, Equipment and Facilities*

MIL-STD-1629 *Procedures for Performing a Failure Mode, Effects and Criticality Analysis*

MIL-STD-1635 *Reliability Growth Testing*

MIL-STD-2068 *Reliability Development Tests*

MIL-STD-2074 *Failure Classification for Reliability Testing*

NPRD-2 *Nonelectric Parts Reliability Data*

RADC-TR-75-22 Nonelectric Reliability Notebook

APPENDIX

L

Financial Formulas

To find the future value (F), given a present value (P), an interest rate (i), and a period (n)

$$F = P(1+i)^n \tag{L.1}$$

To find the present value, given the future value

$$P = F(1+i)^{-n} \tag{L.2}$$

To find the annuity (A), given a future value

$$A = F\left[\frac{i}{(1+i)^n - 1}\right] \tag{L.3}$$

To find the future value given an annuity

$$F = A\left[\frac{(1+i)^n - 1}{i}\right] \tag{L.4}$$

To find the annuity given a present value

$$A = P\left[\frac{i(1+i)^n}{(1+i)^n - 1}\right] \tag{L.5}$$

To find the present value given an annuity

$$P = A\left[\frac{(1+i)^n - 1}{i(1+i)^n}\right] \tag{L.6}$$

$$P = A\left[\frac{(1+i)^n - 1}{i(1+i)^n}\right] \tag{L.6}$$

♦♦♦

APPENDIX

M

For Further Information...

International Quality Federation

The IQF provides a means of information sharing for engineering and quality professionals. The IQF also provides a Six Sigma Black Belt certification. The IQF is a unique society – membership is free!

www.IQFNET.org

American Society for Quality

ASQ helps sets standards for reliability, and administers the Certified Reliability Engineer Examination.

ASQ
600 North Plankinton Avenue
Milwaukee, WI 53203
(800) 248-1946

Reliability journals

- *Annals of Reliability and Maintainability*
- *IEEE Transactions on Reliability*
- *Proceedings of the Annual Reliability and Maintainability Symposium*
- *Quality and Reliability Engineering International*
- *Reliability Review*

- *Technometrics*
- *Applied Statistics*
- *Journal of Quality Technology*

Reliability data sources

- Center for Information and Numerical Data Analysis and Synthesis, (317) 494-6300
- Concrete Technology Information Analysis Center, (601) 634-3269
- Department of Defense Nuclear Information Analysis Center, (805) 963-6400
- Failure Rate Data—Available from *Fleet Missile Systems Analysis and Evaluation Group*, Department of Defense, Corona, CA; (714) 736-4677
- Government and Industry Data Exchange Program, (714) 736-4677
- IEEE Data, (212) 705-7900
- Licensee Event Reports—failure summaries of components from nuclear power plants. Available from the *Nuclear Regulatory Commission*, Washington, DC
- National Electrical Reliability Council—component failure data from power plants. Available from the *National Electrical Reliability Council*, New York, NY
- Non-electric Parts Reliability Data, (315) 330-4151
- Infrared Information Analysis Center, (313) 994-1200
- Metals and Ceramics Information Center, (614) 424-5000
- National Nuclear Data Center, (516) 282-2103
- Plastics Technical Evaluation Center, (201) 724-3189
- Reliability Analysis Center, (315) 330-4151

♦♦♦

APPENDIX

N

Glossary

Accessibility—A measure of the relative ease of admission to the various areas of an item for the purpose of operation or maintenance.

Achieved—Obtained as the result of measurement.

Alignment—Performing the adjustments that are necessary to return an item to specified operation.

Availability—A measure of the degree to which an item is in an operable and committable state at the start of a mission when the mission is called for at an unknown (random) time. (Item state at start of mission includes the combined effects of the readiness-related system R&M parameters, but excludes mission time; see **Dependability**).

Burn-in (Pre-conditioning)—The operation of an item under stress to stabilize its characteristics. Not to be confused with **de-bugging**.

Calibration—A comparison of a measuring device with a known standard. Not to be confused with alignment (see MIL-C-45662).

Chargeable—Within the responsibility of a given organizational entity (applied to terms such as failures, maintenance time etc.).

Checkout—Tests or observations of an item to determine its condition or status.

Corrective Action—A documented design, process, procedure, or materials change implemented and validated to correct the cause of failure or design deficiency.

Criticality—A relative measure of the consequence of a failure mode and its frequency of occurrences.

De-bugging—A process to detect and remedy inadequacies. Not to be confused with terms such as burn-in, fault isolation or screening.

Degradation—A gradual impairment in ability to perform.

Demonstrated—That which has been measured by the use of objective evidence gathered under specified conditions.

Dependability—A measure of the degree to which an item is operable and capable of performing its required function at any (random) time during a specified mission profile, given item availability at the start of the mission. (Item state during a mission includes the combined effects of the mission-related system R & M parameters but excludes non-mission time; see Availability).

Derating—a) Using an item in such a way that applied stresses are below rated values, or b) the lowering of the rating of an item in one stress field.

Direct Maintenance Man-Hours per Maintenance Action (DMMH/MA) —A measure of the maintainability parameter related to item demand for maintenance manpower. The sum of direct maintenance man hours, divided by the total number of maintenance actions (preventative and corrective) during a stated period of time.

Direct Maintenance Man Hours per Maintenance Event (DMMH/ME)—A measure of the maintainability parameter related to item demand for maintenance manpower. The sum of direct maintenance man hours, divided by the total number of maintenance events (preventative and corrective) during a stated period of time.

Disassemble—Opening an item and removing a number of parts or subassemblies to make the item that is to be replaced accessible for removal. This does not include the actual removal of the item to be replaced.

Dormant—The state wherein an item is able to function but is not required to function. Not to be confused with **down-time operable**—the state of being able to perform the intended function. Dormant is the same as "not operating."

Downing event—The event which causes an item to become unavailable to initiate its mission (the transition from up-time to down-time).

Durability—A measure of useful life (a special case of reliability).

Environment—The aggregate of all external and internal conditions (such as temperature, humidity, radiation, magnetic and electric fields, shock vibration , etc.) whether natural or man-made, or self-induced, that influences the form, performance, reliability or survival of an item.

Environmental stress screening (ESS)—A series of tests conducted under environmental stresses to disclose weak parts and workmanship defects for correction.

Failure—The event, or inoperable state, in which any item or part of an item does not , or would not, perform as previously specified.

Failure analysis—Subsequent to a failure, the logical systematic examination of an item, its construction, application, and documentation to identify the failure mode and determine the failure mechanism and its basic course.

Failure catastrophic—A failure that can cause item loss.

Failure critical—A failure, or combination of failures, that prevents an item from performing a specified mission.

Failure dependent—Failure which is caused by the failure of an associated item(s). Not independent.

Failure effect—The consequence(s) a failure mode has on the operation, function, or status of an item. Failure efforts are classified as local effect, next higher level, and end effect.

Failure mode and effects analysis (FMEA)—A procedure by which each potential failure mode in a system is analyzed to determine the results or effects thereof on the system, and to classify each potential failure mode according to its severity.

Failure, independent—Failure which occurs without being caused by the failure of any other item. Not dependent.

Failure, intermittent—Failure for a limited period of time, followed by the item's recovery of its ability to perform within specified limits without any remedial action.

Failure mechanism—The physical, chemical, electrical, thermal or other process which results in failure.

Failure mode—The consequence of the mechanism through which the failure occurs, i.e., short, open, fracture, excessive wear.

Failure, non-chargeable—a) A non-relevant failure, or b) a relevant failure caused by a condition previously specified as not within the responsibility of a given organizational entity (All relevant failures are chargeable to one organizational entity or another).

Failure, non-relevant—a) A failure verified as having been caused by a condition not present in the operational environment, or b) a failure verified as peculiar to an item design that will not enter the operational inventory.

Failure, random—Failure whose occurrence is predictable only in a probabilistic or statistical sense. This applies to all distributions.

Failure rate—The total number of failures within an item population, divided by the total number of life units expended by that population, during a particular measurement interval under stated conditions.

Fault—Immediate cause of failure (e.g., maladjustment, misalignment, defect, etc.).

Fault isolation—the process of determining the location of a fault to the extent necessary to effect repair.

Fault localization—The process of determining the approximate location of a fault.

Inherent R&M value—A measure of reliability or maintainability that includes only the effects of an item design and its application, and assumes an ideal operation and support environment.

Interchange—Removing the item that is to be replaced, and installing the replacement item.

Inventory, active—The group of items assigned to an operational status.

Inventory, inactive—The group of items being held in reserve for possible future assignments to an operational status.

Item—A non-specific term used to denote any product, including systems, materials parts, subassemblies, sets, accessories, etc. (Source: MIL-STD-280).

Life profile—A time-phased description of the events and environments an item experiences from manufacture to final expenditures of removal from the operational inventory, to include one or more mission profiles.

Life units—A measure of use duration applicable to the item (e.g., operating hours, cycles, distance, rounds fired, attempts to operate, etc.).

Maintainability—The measure of the ability of an item to be retained in or restored to specified condition when maintenance is performed by personnel having specified skill levels, using prescribed procedures and resources, at each prescribed level of maintenance and repair.

Maintainability, mission—The measure of the ability of an item to be retained in or restored to specified condition when maintenance is performed during the course of a specified mission profile. (The mission-related system maintainability parameter).

Maintenance—All actions necessary for retaining an item in or restoring it to a specified condition.

Maintenance action—An element of a maintenance event. One or more tasks (i.e. , fault localization, fault isolation, servicing and inspection) necessary to retain an item in or restore it to a specified condition.

Maintenance, corrective—All actions performed as a result of failure, to restore an item to a specified condition. Corrective maintenance can include any or all of the following steps: localization, isolation, disassembly, interchange, reassemble, alignment and checkout.

Maintenance event—One or more maintenance actions required to effect corrective and preventative maintenance due to any type of failure or malfunction, false alarm or scheduled maintenance plan.

Maintenance manning level—The total authorized or assigned personnel, per system at specified levels of maintenance organization.

Maintenance, preventive—All actions performed in an attempt to retain an item in specified condition by providing systematic inspection, detection, and prevention of incipient failures.

Maintenance ratio—A measure of the total maintenance manpower burden required to maintain an item. It is expressed as the cumulative number of man-hours of maintenance expended in direct labor during a given period of the life units divided by the cumulative number of end item life units during the same period.

Maintenance scheduled—Preventive maintenance performed at prescribed points in the item's life.

Maintenance time—An element of down time which excludes modification and delay time.

Maintenance, unscheduled—Corrective maintenance required by item conditions.

Malfunction—The event, or inoperable state, in which any item or part of an item does not, or would not, perform as previously specified. The same as **failure**.

Mean maintenance time—The measure of item maintainability taking into account maintenance policy. The sum of preventive and corrective maintenance times, divided by the sum of scheduled and unscheduled maintenance events, during a stated period of time.

Mean time between demands (MTBD)—A measure of the system reliability parameter related to demand for logistic support. The total number of system life units divided by the total number of item demands on the supply system during a stated period of item, e.g. Shop Replaceable Unit (SRU), Weapon Replaceable Unit (WRU), Line Replacement Unit (LRU), and Shop Replaceable Assembly (SRA).

Mean time between downing events (MTBDE)—A measure of the system reliability parameter related to availability and readiness. The total number of system life units, divided by the total number of events in which the system becomes unavailable to initiate its mission(s), during a stated period of time.

Mean time between failure (MTBF)—A basic measure of reliability for repairable items: The mean number of life units during which all parts of the item perform within their specified limits, during a particular measurement interval under stated conditions.

Mean time between maintenance (MTBM)—A measure of the reliability taking into account maintenance policy. The total number of life units expended by a given time, divided by the total number of maintenance events (scheduled and unscheduled) due to that item.

Mean time between maintenance actions (MTBMA)—A measure of the system reliability parameter related to demand for maintenance manpower. The total number of system life units, divided by the total number of maintenance actions (preventive and corrective) during a stated period of time.

Mean time between removals (MTBR)—A measure of the system reliability parameter related to demand for logistic support. The total number of system life units divided by the total number of items removed from that system during a stated period of time. This term is defined to exclude removals performed to facilitate other maintenance and removals for product improvement.

Mean time to failure (MTTF)—A basic measure of reliability for non-repairable items. The total number of life units of an item divided by the total number of failures within that population, during a particular measurement interval under stated conditions.

Mean time to repair (MTTR)—A basic measure of maintainability. The sum of corrective maintenance times at any specific level of repair, divided by the total number of failures within an item repaired at that level, during a particular interval under stated conditions.

Mean time to restore system (MTTRS)—A measure of the system maintainability parameter related to availability and readiness. The total corrective maintenance time, associated with downing events, divided by the total number of downing events, during a stated period of time (Excludes time for off-system maintenance and repair of detached components).

Mean time to service (MTTS)—A measure of an on-system maintainability characteristic related to servicing that is calculated by dividing the total scheduled crew, operator, driver servicing time by the number of times the item was serviced.

Mission profile—A time-phased description of the events and environments an item experiences from initiation to completion of a specified mission, to include the criteria of mission success or critical failures.

Mission time between critical failures (MTBCF)—A measure of mission reliability. The total amount of mission time, divided by the total number of critical failures during a stated series of missions.

Mission time to restore functions (MTTRF)—A measure of mission maintainability. The total corrective critical failure maintenance time, divided by the total number of critical failures, during the course of a specified mission profile.

Not operating (Dormant)—The state wherein an item is able to function but is not required to function. Not to be confused with **Down time**.

Operable—The state of being able to perform the intended function.

Operational readiness—The ability of a military unit to respond to its operation plan(s) upon receipt of an operations order (A function of assigned strength, item availability, status, or supply, training, etc.).

Operational R&M value—A measure of reliability or maintainability that includes the combined effect of item design, installation, quality, environment, operation, maintenance and repair.

Predicted—That which is expected at some future time, postulated on analysis of past experience and tests.

Reassembly—Assembling the items that were removed during disassembly and closing the reassembled items.

Redundancy—The existence of more than one means for accomplishing a given function. Each means of accomplishing the function need not necessarily be identical.

Redundancy, active—That redundancy wherein all redundant items are operating simultaneously.

Redundancy, standby—That redundancy wherein the alternative means of performing the function is not operating until it is activated upon failure of the primary means of performing the function.

Reliability—a) The duration or probability of failure-free performance under stated conditions, or b) the probability that an item can perform its intended function for a specified interval under stated conditions (For non-redundant items this is equivalent to definition *a*. For redundant items this is equivalent to definition of mission reliability.).

Reliability growth—The improvement in a reliability parameter caused by the successful correction of deficiencies in item design or manufacture.

Reliability mission—The ability of an item to perform its required functions for the duration of a specified "mission profile."

R&M accounting—That set of mathematical tasks which establish and allocate quantitative R&M requirements, and predict and measure quantitative R&M requirements.

R& M engineering—That set of design, development and manufacturing tasks by which reliability and maintainability are achieved.

Repairable item—An item which can be restored to perform all of its required functions by corrective maintenance.

Screening—A process for inspecting items to remove those that are unsatisfactory or those likely to exhibit early failure. Inspection includes visual examination, physical dimension measurement and functional performance measurement under specified environmental conditions.

Servicing—The performance of any act needed to keep an item in operating condition, (i.e. lubricating, fueling, oiling, cleaning, etc.), but not including preventative maintenance of parts or corrective maintenance tasks.

Single point failure—The failure of an item which would result in failure of the system and is not compensated for by redundancy or alternative operational procedure.

Sneak circuit analysis—A procedure conducted to identify latent paths which cause occurrence of unwanted functions or inhibit desired functions assuming all components are functioning properly.

Storage life (Shelf life)—The length of time an item can be stored under specified conditions and still meet specified requirements.

Subsystem—A combination of sets, groups, etc. which performs an operational function within a system and is a major subdivision of the system (Example: Data processing subsystem, guidance subsystem). Source MIL-STD-280.

System, general—A composite of equipment and skills, and techniques capable of performing or supporting an operational role, or both. A complete system includes all equipment, related facilities, material, software, services, and personnel required for its operation and support to the degree that it can be considered self-sufficient in its intended operational environment.

System R&M parameter—A measure of reliability or maintainability in which the units of measurement are directly related to operational readiness, mission success, maintenance manpower cost, or logistic support cost.

Test, acceptance—A test conducted under specified conditions by, or on behalf of, the government, using delivered or deliverable items, in order to determine the item's compliance with specified requirements (Includes acceptance of first production units).

Test measurement and diagnostic equipment (TMDE)—Any system or device used to evaluate the condition of an item to identify or isolate any actual or potential failures.

Test, qualification (Design approval)—A test conducted under specified conditions, by or on behalf of the government, using items representative of the production configuration, in order to determine compliance with item design requirements as a basis for production approval (also known as **Demonstration**).

Testing development (Growth)—A series of tests conducted to disclose deficiencies and to verify that corrective actions will prevent recurrence in the operational inventory. Note that repair of test items does not constitute correction of deficiencies (Also known as **Test-Analyze-And-Fix (TAAF)** testing).

Time—The universal measure of duration. The general word "time" will be modified by an additional term when used in reference to operating time, mission time, test time, etc. In general expressions such as "Mean-Time-Between-Failure (MTBF)," time stands for "life units" which must be more specifically defined whenever the general term refers to a particular item. See Figure G.1 for time relationships.

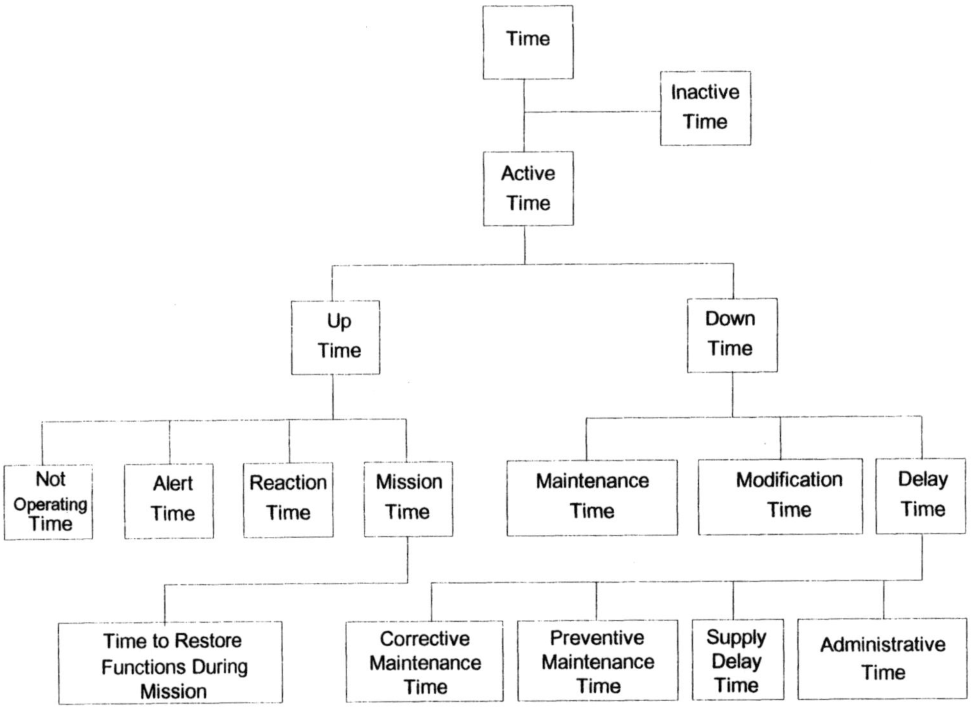

Figure G.1. Time relationships.

Time, active—That time during which an item is in an operational inventory.

Time, administrative—That element of delay time, not included in the supply.

Time, alert—That element of up time during which an item is assumed to be in specified operating condition and is awaiting a command to perform its intended mission.

Time, checkout—That element of maintenance time during which performance of an item is verified to be a specified condition.

Time, delay—That element of down time during which no maintenance is being accomplished on the item because of either supply or administrative delay.

Time, down (Downtime)—That element of active time during which an item is not in condition to perform its required function (Reduces availability and dependability).

Time, inactive—That time during which an item is in reserve.

Time, mission—That element of up time required to perform a stated mission profile.

Time, modification—The time necessary to introduce any specific change(s) to an item to improve its characteristics or to add new ones.

Time, not operating—That element of uptime during which the item is not required to operate.

Time, reaction—That element of uptime needed to initiate a mission, measured from the time command is received.

Time, supply delay—That element of delay time during which a needed replacement item is being obtained.

Time, turn around—That element of maintenance time needed to replenish consumables and check out an item for recommitment.

Time, up (Uptime)—That element of active time during which an item is in condition to perform its required functions.

Uptime ratio—A composite measure of operational availability and dependability that includes the combined effects of item design, installation, quality, environment, operation, maintenance, repair and logistic support. The quotient of uptime divided by uptime plus downtime.

Useful life—The number of life units from manufacture to when the item has an unrepairable failure or unacceptable failure rate.

Utilization rate—The planned or actual number of life units expended, or missions attempted, during a stated interval of calendar time.

Wearout—The process which results in an increase of the failure rate or probability of failure with increasing number of life units.

♦♦♦

Index

-1-
1/2 factorial design....472
1/4 factorial design....473
-2-
2^{k-1} design....472
2-way ANOVA....477
-3-
3-way ANOVA....477
-5-
5% rank....107
-9-
95% rank....107
-A-
Accelerated Testing....224
acceleration factor....224
Acceptance Testing....224
Accessibility....272
Achieved availability....257
AGREE....1, 192, 193
Air Conditioning....354
alpha error....137
alternate hypothesis....474
alternative hypothesis....137
AMSAA Model....206
Analysis of Variance....473, 474
AND Gate....164
AND NOT Gate....164
ANOVA....473, 474
ARINC....192, 272
Arrhenius effect....303
Arrhenius Model....227
ASQC....573
Audio Capabilities....348
Availability....257
availability - exponential failure & repair distributions....260
availability, steady state....260
availability, system....262
average availability....261
average uptime....261
-B-
balanced design....466
Basic event....165
Basic Execution Time Model....198
bathtub curve....266
Bayes' theorem....77
BayesianTesting....248
Bernoulli process....49
beta distribution....64, 68, 134
beta distribution - mean....68
beta distribution - parameter estimation....70
beta distribution - relationship to the gamma distribution....64
beta distribution - relationship to the triangular distribution....68
beta distribution - relationship to the uniform distribution....68
beta distribution - variance....68
beta error....137
beta hazard function....70
Beta hazard function....71
beta probability density function....68
Beta probability density function....69
beta reliability function....70
Beta reliability function....71
binary string bit method....166
binomial distribution....49, 54
binomial distribution - mean....55
binomial distribution - Poisson approximation....54, 56
binomial distribution - variance....55
binomial probability density....54
Block diagrams....290
Boolean algebra....166
built-in test equipment....11
-C-
Catastrophic failure....9
Censored data....96
censoring - left....97
censoring - multiple....97
censoring - right....96
Censoring - type 1....96
Censoring - type II....96
central limit theorem....40
Central tendency....420, 425, 429
characteristic life....59
Chargeable failures....10
Chi-Square Distribution....72
chi-square distribution - mean....73

chi-square distribution - relationship to the beta distribution ... 73
chi-square distribution - relationship to the F distribution ... 73
chi-square distribution - relationship to the gamma distribution ... 72
chi-square distribution - relationship to the normal distribution ... 72, 73
chi-square distribution - variance ... 73
chi-square goodness of fit test ... 125
chi-square probability density function ... 72
closed loop system ... 6
coefficient of determination ... 490
coefficient of determination r^2 ... 488
coefficient of multiple correlation ... 495
co-linear ... 474
collecting data ... 6
Color-coding ... 347
combination testing method ... 166
combinations of random variables ... 83
Complex Systems ... 156
Conditional Event ... 165
confidence interval ... 491
confidence interval - exponential distribution mean ... 100
confidence interval - exponential distribution percentiles ... 102
confidence interval - exponential distribution reliability ... 102
confidence interval - nonparametric for reliability ... 135
confidence interval - variance ... 133
confidence intervals - mean ... 131
confidence intervals - normal distribution parameters ... 111
Confidence intervals - normal distribution percentiles ... 111
Confidence intervals - normal distribution reliability ... 111
Confidence Intervals - proportions ... 133
confidence intervals - Weibull distribution parameters ... 119
Confidence intervals - Weibull distribution percentiles ... 120
Confidence intervals - Weibull distribution reliability ... 120
confidence limit - nonparametric for reliability ... 107
confounding ... 465, 468
Consistency ... 95
continuous data ... 5
Continuous density function ... 33
corrective action ... 7, 16
Corrective action ... 15
corrective-action ... 17
correlation coefficient, r ... 488
covariance of the estimates ... 491
Critical Design Review ... 315
Critical failure ... 9, 11
critical value ... 480
Criticality ... 288
criticality analysis ... 278
Criticality analysis ... 288
criticality number ... 279
cumulative distribution function ... 34, 35
cumulative distribution function - exponential distribution ... 104
cumulative hazard function ... 97
cumulative hazard function - exponential distribution ... 102
cumulative hazard function - normal distribution ... 111
cumulative hazard function - Weibull distribution ... 121
cut set ... 166
cut set, minimum ... 166
-D-
data accuracy ... 14
data collecting ... 5
data collecting form ... 12
data recording form ... 6
degree-of-freedom ... 465
degrees of freedom between treatments ... 476
degrees of freedom for the error ... 476
density function ... 33
density functions ... 33
Department of Defense ... 1
Dependent failure ... 9
dependent variables ... 464
derating ... 297
Derating ... 303
design adequacy ... 273
design of experiments ... 463
design resolution ... 473
Design review ... 314
design review team ... 317

Detection mechanism 288
Detection probabilities 279
Disassembly 271
discrete data 5
discrete density function 33
distributions - reproductive 88
DOD 1
DOE 463
Duane Model 203
Dynamic programming 192

-E-

efficient estimator 95
Elemental Activities 256
Endurance tests 301
Environmental Stress Screening 223, 299
Environmental tests 301
Environments 288
Equipment design failure 9
Equipment manufacturing failure 9
EVD 63
evolutionary operation 502
EVOP 502
Exclusive OR Gate 164
expected value 37
experiment 463
experimental design 463
experimental error 465
Expert Opinion Data 3
exponential - failure rate 50
exponential distribution 49, 50
exponential distribution - maximum likelihood parameter estimation 99
exponential distribution - mean 50
exponential distribution - parameter estimation 99
exponential distribution - variance 52
exponential hazard function 51, 52
exponential probability density function 50
exponential reliability function 50
extreme value distribution 63
Eyring model 232

-F-

F distribution 74
F distribution - mean 75
F distribution - relationship to the normal distribution 75
F distribution - variance 75
F probability density function 74
factorial design 471
factors 463
Fail safe design 367
failure 7
Failure 7
Failure Analysis, Reporting and Corrective Action System 314
Failure cause 288
failure censoring 96
Failure effect 288
failure free testing 102
Failure mode 288
Failure mode and effects analysis 287
failure mode ratio 280
failure modes 277
Failure report form 13
failure review board 6
failure truncation - exponential distribution 101
Failure, catastrophic 292
Failure, critical 292
Failure, marginal 292
Failure, minor 292
failures 7
Failures 7
FARADA 4
Fault isolation 272
Fault tree 162
fault tree analysis, quantitative 169
Fault tree events 165
Fault tree gates 164
Fault-Tree Construction 162
F-distribution 480
F-distribution - relationship to the chi-square distribution 74
Flow diagram 7
FMEA 287
FMECA 277
FMECA planning 289
FMECA-Maintainability information 288
FRACAS 6, 314
fractional factorial designs 472
FRB 6, 7, 16
full factorial design 472
Full Scale Engineering Development Phase 318
Functional block diagrams 290
Functional packaging 272

functions of several variables - mean 82
functions of several variables - mean & variance 81
functions of several variables - variance 82
-G-
gamma distribution 64, 87, 152
gamma distribution - mean 64
gamma distribution - relationship to the chi-square distribution 65
gamma distribution - relationship to the Erlang distribution 65
gamma distribution - relationship to the exponential distribution 64, 65
gamma distribution - relationship to the normal distribution 65
gamma distribution - variance 65
Gamma function 60
gamma hazard function 66
Gamma hazard function 67
gamma probability density function 64
gamma random variables, sum of 87
gamma reliability function 66
Gamma reliability function 66
Gaussian distribution 40
geometric distribution 50, 58
geometric distribution - mean 58
geometric distribution - variance 58
geometric probability density function 58
GIDEP 4, 298
goodness of fit 72
Goodness of fit 125
grouped data 97
-H-
hazard analysis 329
hazard function 37
hazard function - constant 51
Hazard plotting 97
Hazard Plotting 95
Hazard Plotting - exponential distribution 102
Hazard Plotting - normal distribution 111
Hazard Plotting - Weibull distribution 121
hazard probability 331
hazard risk assessment models 334
Hazard severity categories 331
high-stress error 365
Hollander-Proschan test 126
House Event 165
Human factors 343
human factors in field testing 368
human factors in production 363
Human reliability 343
human vibration capability 357
Humidity 354
hypergeometric distribution 49, 57
hypergeometric distribution - mean 57
hypergeometric distribution - variance 57
Hypothesis testing 137
Hypothesis testing - means 137
hypothesis testing - power 137
Hypothesis testing - variance 140
-I-
Identification 272
IEEE Data 4, 574
incomplete beta function 70
Independent failure 9
independent variables 464
Infant mortality failures 265
Inherent Availability 257
instantaneous failure rate 37
integrated testing 300
Interaction 485
Interaction plots 485
interaction tables 468
interactions 464, 465, 472
Interchange 271
Interchangeability 272
Intermittent failure 9
interval data 97
-J-
Jacobian 90
-K-
Kaplan-Meier estimate 98
Kaplan-Meier jump 126
Kaplan-Meier reliability estimate 126
Knobs 351
Kolmogorov-Smirnov test 125
kurtosis 40
-L-
L_{12} design 466
L_4 design 466
L_8 design 466, 469
Label characteristics 348
Labeling 348
lack of memory 51
least squares regression 489, 494

left censoring ... 97
LER Data ... 4
life cycle cost ... 1
Life cycle costs ... 2
likelihood function ... 97
Limited thermal tolerance zones ... 354
Linear graphs ... 468
linear regression ... 105
local information matrix ... 97
local information matrix - normal distribution ... 111
local information matrix - Weibull distribution ... 119
Location-coding ... 346
logarithmic Poisson execution time model 199
lognormal ... 258
lognormal distribution ... 40, 45
lognormal distribution - mean ... 46
Lognormal distribution - parameter estimation ... 118
lognormal distribution - variance ... 46
lognormal hazard function ... 47
lognormal location parameter ... 45
lognormal probability density function ... 46
lognormal reliability function ... 46
lognormal scale parameter ... 45
Lotus® 123™ ... 496

-M-

main effects ... 472
maintainability ... 255
Maintainability ... 258
maintainability, designing for ... 271
Maintainability, designing for ... 271
Major failure ... 11
Malfunction Active Repair Time ... 256
Malfunction annunciation ... 272
Malfunction Repair Time ... 256
Markov analysis ... 187
Matching Moments - Weibull distribution ... 125
mathematical model ... 465
maximize the response variable ... 502
Maximum Corrective Maintenance Time ... 258
Maximum Likelihood ... 95
Maximum Likelihood Estimation ... 97
Maximum likelihood estimation - normal distribution ... 109
Maximum likelihood estimation - Weibull distribution ... 118
MCMT ... 258
mean ... 39
mean rank ... 98
mean time to repair ... 258
Mechanism failure ... 9
median corrective maintenance time ... 258
median rank ... 98, 104
method of least squares ... 489
Microsoft® Excel™ ... 496
MIL-E-5400 ... 305
Military Handbook - 217 ... 4, 146
Military Standard-882 ... 328
MIL-P-11268 ... 305
minimize the response variable ... 502
minimize the variability ... 502
minimum cut set ... 166
Minimum variance ... 95
Minor failure ... 12
mission reliability ... 273
modified order of failure ... 104, 105
Modularization ... 271
Moment estimation ... 98
Moment Estimation ... 95
moment-generating functions ... 39
Monte Carlo Simulation ... 172
MTTR ... 258
Multi-factor ANOVA ... 477
multiple censored ... 97
Multiple failures ... 9
multiple regression ... 493
Multiple regression ... 502

-N-

negative correlation ... 488
NERC ... 4
noise factor ... 464
Nonchargeable failures ... 11
non-linear effects ... 464
non-linear model ... 497
Non-linear regression ... 497
Nonrelevant failures ... 10
normal distribution ... 40
normal hazard function ... 44
normal reliability function ... 41
NPRD ... 4
NPRDS ... 4

null hypothesis....137, 235, 474

-O-

OC curve....238
Occurrence probabilities....278
One-Way ANOVA....474
operating characteristic curve....238
Operational Availability....257
operational readiness....273
optimum schedule for preventive maintenance....267
OR Gate....164
origin....39
orthogonal design....466
over stressing....17
over-stress....302

-P-

parallel system....149
parallel system, shared load....154
parallel, active....149
parallel, passive....149
parallel, pure....149
parallel, shared....149
parameter estimation....95
parameter estimation software....96
Part design failure....9
Part failure rates....281
Part manufacturing failure....9
Part Stress Analysis....146
partial factorial designs....472
parts control....297
Parts Count....146
Parts Count method....147
parts program....298
Parts standardization....298
Parts testing....298
path set....167
Pattern failures....9
Performance tests....301
Phadke....466
Plackett and Burman design....466
Poisson distribution....49, 53, 88
Poisson distribution - mean....53
Poisson distribution - variance....53
Poisson probability density function....53
Poisson process....198
Poisson random variable, sum of....88
population mean....40
population standard deviation....40
positive correlation....488
posterior distribution....79
prediction interval....491
Preliminary Design Review....315
preventive maintenance....265
Primary event....165
Primary failure....9
Primary, Consolidated Data....3
prime number method....166
prior distribution....79
Priority AND Gate....164
probability density function....33, 34
Probability density functions....33
Probability plotting....98
Probability Plotting....95
Probability plotting - exponential distribution....104
Probability Plotting - normal distribution....114
Probability Plotting - Weibull distribution....123
probability plotting - with multiple censored data....107
product safety program....328
Production Phase....318
Production Readiness Review....315
propagation of error....81
Propagation of error....81
Push-buttons....350
PVP....4

-Q-

quasi-symmetrical....68
Quasi-symmetry....69

-R-

r^2 488
Random number generators....173
Random numbers....172
Raw Data....3
Reassembly....271
regression....473
Regression....487
Relevant failures....10
Reliability Allocation....191
Reliability allocation - repairable systems....195
Reliability Analysis Checklist....319
Reliability and Maintenance Analyst....268
Reliability block diagrams....290

reliability data sources, public databases......... 4
Reliability Development/Growth Test....... 300
reliability function 36
Reliability Growth 202
Reliability Growth Testing............................ 223
reliability management.................................. 311
reliability program plan 311, 313
Reliability Qualification Test 300
Reliability Qualification Testing.................. 223
reliability test program.................................. 299
Reliability testing 223
replacement.. 49
RESIDUAL RISK 336
resolution.. 473
response surface methodology..................... 502
responses. 463
right censoring.. 96
risk assessment 331
Risk Priority Number 278
Rotary selector switch 352
r-out-of-n Gate.. 164
r-out-of-n System.. 150
-S-
safety 277
Safety design requirements 330
safety level... 357
sample mean 40
sample standard deviation 40
sample variance 75
screening design 464, 472, 511
Screening designs 465
Secondary failure 9
Sequential sampling - expected trials to reach a decision 240
sequential sampling - OC curve 238
sequential simplex optimization.................. 502
Sequential testing 234
Sequential testing - binomial distribution... 235
Sequential testing - expected time to reach a decision 244
Sequential testing - exponential distribution 240
Sequential testing - OC curve....................... 242
Sequential testing - Type I error 235
Sequential testing - Type II error................ 235
Sequential testing - Weibull distribution..... 245
Sequential testing -expected failures required to reach a decision 244
Series System.. 147
Severity... 288
Severity classification 291
Severity classification categories................. 292
Severity Probabilities.................................... 278
Shape-coding 346
SHARPE... 192
SIGHT Capabilities 344
Simple linear regression 489
simplex 502
Simulation 172
simulation modeling 175
Single censoring 97
Single failure point.. 288
Size-coding 346
skewness... 39
Sneak Analysis 208
Software error failure 9
software hazard risk analysis..................... 337
software reliability.. 196
Special cause 420, 425, 429
standard normal cumulative distribution function 42
standard normal reliability function.............. 42
standby redundant 149, 155
standby redundant system 151
state transition diagram................................ 188
steepest assent 502
step-stress 302
Stress-Strength Systems 159
Student's t Distribution................................. 75
Sufficiency 95
sum of squares between treatments............ 475
sum of squares due to error 475
switching 151
switching, imperfect 154
switching, perfect.. 152
system availability 262, 263
system downtime 257
System downtime.. 257
system effectiveness 272
System effectiveness..................................... 273
system maintainability.................................. 257
System Repair Time 257
-T-
t distribution 75
t distribution - mean.. 76

t distribution - relationship to the F distribution ... 76
t distribution - variance ... 76
t probability density function ... 76, 77
T_{50} ... 46
Taguchi ... 511
Taguchi design ... 466
Test Readiness Review ... 315
test realism ... 301
theory of large numbers ... 40
thermal stress ... 228
Thermal tolerance and comfort zones ... 354
thumbwheel control ... 353
time censoring ... 96
time truncation - exponential distribution .. 100
Toggle switches ... 349
tolerance interval ... 136
touch capabilities ... 345
transferability principle ... 257
transformation ... 499
transformation functions ... 89
Transformation Functions - several variables ... 90
Transformation Functions - single variable . 89
Transformation of random variables ... 88
transition rate matrix ... 188
treatment ... 474
Triangle In ... 165
Triangle Out ... 165
truncation ... 96
t-statistic ... 137
type I asymptotic distribution for maxima and minima ... 63
Type I Censoring ... 96
Type I error ... 137
Type II Censoring ... 96
Type II error ... 137
type III asymptotic distribution for the minima ... 63
-U-
unbiased ... 95
Undetectable failure ... 289
Undeveloped Event ... 165
unit, uniform random number ... 172
Unverified failure ... 10
-V-
variance ... 38
variance of the estimate ... 491
Ventilating ... 353
Verified failure ... 10
vertex ... 502
-W-
Weapons System Effectiveness Industry Advisory Committee ... 272
Wear-out failures ... 265
Weibull distribution ... 59
Weibull distribution - 3 parameter ... 118
Weibull distribution - mean ... 60
Weibull distribution - parameter estimation ... 118
Weibull distribution - scale parameter ... 59
Weibull distribution - shape parameter ... 59
Weibull distribution - variance ... 60
Weibull hazard function ... 61, 62
Weibull probability density function ... 59
Weibull reliability function ... 61, 62